TECHNICAL AND ETHICAL OPTIMIZATION: CHALLENGES AND OPPORTUNITIES FOR THE MODERN ENGINEER AND SCIENTIST

TECHNICAL AND ETHICAL OPTIMIZATION: CHALLENGES AND OPPORTUNITIES FOR THE MODERN ENGINEER AND SCIENTIST

DANIEL A. VALLERO, Ph.D.

Civil and Environmental Engineering, Pratt School of Engineering, Duke University, Durham, NC, United States

ELSEVIER

Elsevier
Radarweg 29, PO Box 211, 1000 AE Amsterdam, Netherlands
125 London Wall, London EC2Y 5AS, United Kingdom
50 Hampshire Street, 5th Floor, Cambridge, MA 02139, United States

ISBN: 978-0-443-44925-3

For Information on all Elsevier publications
visit our website at https://www.elsevier.com/books-and-journals

Publisher: Priscilla Braglia
Senior Acquisitions Editor: Anita Koch
Editorial Project Manager: Sadhna Rajan
Production Project Manager: Rajeswari R.
Cover Designer: Mark Rogers

Typeset by MPS Limited, Chennai, India

Curiosity is a wonderful thing; we can all agree with that.
It leads to innovation.
And improves our nation.
But it also killed the cat.

—DAV

Contents

7. Environmental ethics

8. Land ethics

9. Optimization methods

10. Risk-based optimization

11. Engineering education

12. Planet-scale technologies

13. Human factors engineering

14. Professionalism

15. Technological wisdom

16. Optimizing for success and avoiding failure

17. Research ethics

18. A resilient and sustainable future

Foreword

It is now almost 20 years since Daniel A. Vallero published *Biomedical Ethics for Engineers*. Over this period, the book has remained relevant and is now being updated in view of the enormous changes that have taken place in the world of engineering.

These include changes in technology, infrastructure, products, and, more recently, the development of massive data collection systems and the related artificial intelligence systems. This book provides a link between technological developments, human and planetary health, and ethical considerations.

It has been expanded from the original 10 chapters to 18. These new chapters cover the following areas: science ethics, engineering ethics, land ethics, environmental ethics, research ethics, planet-scale technologies, professionalism conduct, and technological wisdom.

Professionalism and integrity for engineers involve seeking ways to benefit society through engineering projects, all of which must have an ethical foundation. In this respect, Vallero has succeeded. Although this book has been written especially for engineers, its relevance is applicable to all human endeavors, including science, biotechnology, and research of every kind.

It comes at a time when the world is embarking on new and, in some cases, untested projects ranging from geoengineering, genetic engineering, combating climate change, planetary exploration, artificial intelligence systems, and data banks and even to data banks in space.

This book is a necessary addition to libraries of not only engineers but all who actively work in areas that touch on human activities and environmental spaces. It also has much to offer policymakers, researchers, and students who care about the future of humanity.

Trevor M. Letcher
Somerset, United Kingdom
March 2026

Preface

It has been almost 20 years since *Biomedical Ethics for Engineers* [1] was published. A lot has happened since 2007. At that time, I briefly mentioned several of the emerging technologies, but many of them were in their earliest stages, and so many more are now thrust upon us. For example, it would be difficult to find anyone over the age of eight who is not at least casually aware of the promise and threat of generative artificial intelligence (GenAI). Indeed, the stock market and regular newscasts remind us daily. At this writing, Nvidia and SpaceX are the richest companies in the world. There is talk of data centers in space.

There are so many aspects of emerging technologies that a book could be completely dedicated to one or two aspects of a single technology. Instead, I have attempted to cover every aspect of several. As such, I have drawn heavily from my previous research, work, and writings. I give equal attention to the scientific and moral aspects by taking a systems approach. As such, this project has provided me with a chance to update the ethical features of my recent books on the fundamentals of systems science [2], air pollution [3], and water pollution [4].

Indeed, I am unapologetic about the importance of an engineering perspective in understanding the opportunities and vulnerabilities presented by emerging technologies. I intentionally incorporate environmental, social, and technical examples throughout to illustrate how easily mistakes can be made, providing hints at potential problems that are missed. To paraphrase one of my mentors at Duke, the late Aarne Vesilind, everything matters in engineering. And, in ethics.

In my first book, *Engineering the Risks of Hazardous Waste* [5], I quoted Harvard professor William W. Lowrance, who wrote 50 years ago [6]:

> We must hope that society at large will come to appreciate the capabilities and inherent limitations of science and technology; and we must hope that those in the technical world will come to appreciate the non-rational nature and great subtlety of social decisions. The risks are changing. Menaces are upon us. Time is short. Decisions have to be made.... May discussion of these troublesome issues be temperate, imaginative, and effective.

The quote is just as relevant today as it was in 1976. Perhaps even more. Some of the menaces are the same, such as the ever increasing use of emerging technologies in war. Some are different, such as the threats to society posed by GenAI and quantum computing. However, the solutions must still be matched against the threats' nonrational and subtle nature.

I find that the questions are more important than the answers as an effective means of learning complex and ill-posed subject matter, especially those that have both technical and moral import. Especially for large language modeling and machine learning, the queries are arguably more important than the outputs[1]. This mindset blends

[1] This revelation came up in a conversation with my former colleague at the U.S. Environmental Protection Agency, Daniel Chang, a gifted computational chemist.

well with my approach to learning and teaching since my writing and pedagogical methodology are Socratic and eclectic. One thing that I have learned having lived in the American South for over three decades is the value of narrative storytelling. The Socratic approach provides a way to relive events and along the way to learn through the students' own inquisitiveness. Using questioning and uncertainties to elicit the truth is ideally suited to ethics, environmental science, and engineering subject matter.

Most ethical, engineering, and environmental decisions have no unique solutions. Adverse consequences result from a series of highly complex contingencies of decisions and acts. The contingent probabilities of a particular outcome in a specific situation at a particular time can seem to be minuscule. But that specific outcome did in fact occur, so we need to discover why. What are the root causes? What interplay of factors made an unlikely outcome become an actual outcome?

We can turn these questions around. What can be done to achieve a desirable outcome? How can the complicated relationships between drivers and constraints be optimized to give us an ideal and ethical outcome? No single decision tool works in every situation, especially when human beings are involved. We must not try to solve a human problem by optimizing for some measure of central tendency. We may aim to protect the person of the average age, living in the average home, earning the average income, and having the most popularly held set of values. This may make the math easier if we do this, but are we not ignoring the unique needs of the most vulnerable among us? Are we favoring the healthiest at the expense of infirm, the well-off at the expense of the poor, and the least susceptible at the expense of the weakest members of society?

My approach to this book is to offer ways to look at technologies from several ethical perspectives. Sometimes, the best way is to emphasize duty. This duty is, of course, a key aspect of what it means to be a professional, but it also applies to everyone who has obligations to the fellow members of humanity.

Other times, it is preferable to consider the technological challenges from a teleological perspective. Just because we can do something does not mean we should do it. To what end is artificial intelligence increasingly being used, for example? Is it to increase a company's stock value? Is it to have better medicine? Is it to make us all smarter? And what if the end is not so good? What if achieving these "desirable" ends costs society and adds the risk of failures that we are presently not able to forecast? Will it affect the cost and availability of electricity for those who can least afford an increase? What is its effect of the dignity of work? Will it somehow become uncontrollable as we approach singularity?

Sometimes, it is better to consider the morality of what is being valued. How does a technology fit with what we value? What if we gain security but lose a commensurate amount of freedom and privacy? What if something gives us momentary pleasure, but at tremendous long-term costs down the road? What if something increases monetary value at the expense of quality of life or immediate human gratification at environmental costs?

These and other ethical constructs are considered throughout this book. I am a technological optimist. But I am, or hope to be, a realist. There are only a few technologies that I discuss that are inherently immoral, at least as we see them now. However, the abuse and misuse of any technology can lead to adverse outcomes in health, safety, and welfare. That is why we must be clear-eyed about the opportunities and problems they have introduced and those that will soon occur. I hope that this book is a valuable resource for seeing the promise, but engendering the awareness of the potential problems and the need for precaution with emerging technologies. Anything that is new is uncertain, so ample factors of safety are in order.

This book's aim

I have written each chapter as a freestanding resource of one area of expertise, such as engineering education, professional ethics, or bioethics. Thus, these chapters can provide readings for specific topics covered in an engineering ethics course, an ethics of science course, or an environmental engineering capstone course. Several chapters can be used together for an independent study class or for a seminar course. That said, the coverage of this book is comprehensive, so it can serve as the primary textbook for a technology course or a professionalism course in scientific and business curricula.

This book is also written with a more general audience in mind. It is a resource for the informed reader who does not necessarily have to be a practitioner or researcher in an emerging technology. As such, any terminology that is not widely used is defined in context with a more detailed definition in the glossary. Call-out examples and sidebar discussions appear throughout the text to clarify important concepts. This is needed since a systems approach requires the meshing of many disciplines and perspectives. As such, environmental consultants, public interest groups, and other stakeholders should find this book comprehensible and helpful as they address or learn more about the various aspects of emerging technologies.

Daniel A. Vallero

References

[1] D.A. Vallero, Biomedical Ethics for Engineers, The Biomedical Engineering Series, (2007) .

[2] D.A. Vallero, Environmental Systems Science: Theory and Practical Applications, 1st ed, Elsevier, (2020).

[3] D.A. Vallero, Fundamentals of Air Pollution, 6th ed, Academic press, (2025).

[4] D. Vallero, Fundamentals of Water Pollution: Quantifying Pollutant Formation, Transport, Transformation, Fate and Risks, 1st ed, Elsevier, (2024).

[5] D.A. Vallero, J.J. Peirce, Engineering the Risks of Hazardous Wastes, Butterworth-Heinemann, (2003) pp. xxi, 306 p.

[6] W.W. Lowrance, "Of Acceptable Risk: Science and the Determination of Safety," 1976.

Acknowledgments

I have been blessed with so many brilliant and honorable teachers, mentors, and colleagues that it is not possible to thank them all. Of course, since this book is about ethics, I must first mention the unselfish support and encouragement of my late parents, Jim and Berniece Vallero. I am often reminded of lessons instilled from them and from spiritual leaders like Reverends George Orville Ankarlo and Etta Isabelle Ankarlo six or seven decades ago. On the academic front, several quickly come to mind. Charles Hess, Melvin Kazeck, Loran Marlow, Phillip Simon, Bob Mendelson, and the rest of the geosciences faculty at Southern Illinois University introduced me to spatial scientific concepts that I have used throughout my career. Interestingly, I have managed to retain some key concepts from my second major at SIUE, that is, psychology. John McCall and other faculty introduced me to topics that have seasoned my thoughts and writing for decades. It was not until relatively recently that I came to realize that my wife was not the only great thing that I gained from the field of psychology (she was a psych major; of course, she was the best thing to come from that pursuit).

A half century ago, Edward Vest had enough faith in me to hire me as a community planner at the Environmental Protection Agency (EPA) (my first MS was in City and Regional Planning). Ever since, my EPA colleagues have enlightened and challenged me to appreciate the environment in all its manifestations. Notably, Gerald Ankley, Larry Purdue, Gary Foley, and Jay Messer were key mentors in EPA's public health and ecological research programs. While at the EPA, I had the pleasure of working with gifted employees, contractors, and grantees. Most notably, my appreciation of and knowledge about exposure science was enhanced by my partnerships with Paul Lioy and Panos Georgopoulos at the Environmental and Occupational Health Sciences Institute at Rutgers University and Thomas McKone at the Lawrence Berkeley National Laboratory. EPA Exposure Scientist, Peter Egeghy, has been a good friend whose sense of humor and wisdom are a treasure.

Ross McKinney and the civil and environmental engineering faculty at the University of Kansas introduced me to topics that I still take great pleasure in pursuing. Jeff Peirce, Henry Petroski, Henri Gavin, Aarne Vesilind, and other talented faculty at Duke University's Pratt School of Engineering opened me to the richness and rigor of research.

Indeed, Aarne introduced me to the world of engineering and professional ethics when he chose me, a PhD candidate at the time, to lead Duke's Science Technology and Human Values Program while he went on sabbatical. A few years later, he and Seymour Mauskopf, who founded the program, selected me as the permanent program director. Since then, ethics has been woven into most of my writings and systems thinking.

I have greatly benefited from conversations with ethics experts like Michael Liou and Michael Davis at the Illinois Institute of Technology, Caroline Whitbeck at Case

Western Reserve, and Joseph Herkert at North Carolina State University.

Trevor Letcher has been a brilliant collaborator for several books and projects. His mastery of thermodynamics has always been an essential complement to my environmental systems science and engineering ethos. I look forward to other projects with him, including our upcoming third edition of *Waste: A Handbook for Management*.

As has been the case for many book projects, the people at Elsevier are ever creative and helpful.

Most of all, I wish to thank my wife, Janis, for her support and ideas. Many of the topics in this book stemmed from our discussions over our morning coffee. Her expertise in relational ethics is seldom discussed in technical ethics courses. Throughout my career, she has advised me to incorporate trust, mutual respect, and engagement into the requisite codes of professional and technical ethics.

In this book, I have endeavored to cover the most pertinent and teachable topics in optimizing decision-making based on science, engineering, and ethics. I know that there are many other topics that could be covered, so I would appreciate hearing about them at dav1@duke.edu.

Enjoy!

Daniel A. Vallero

CHAPTER

1

Emerging technologies

OUTLINE

The 21st century has seen the emergence of many technologies in every scientific and engineering discipline. All are justified for societal good, such as improved medical procedures, better energy production, enhanced economics, and a cleaner environment. Unfortunately, the advocates for most of these technologies are also the ones who benefit the most and who know most about their inner workings.

As a start to understanding the present and future ethical and societal aspects, let us begin by discussing the most notable emerging and recently changing technologies. All are influencing the engineering and science community, and most are likely to increase their influence on scientific integrity.

1.1 Artificial intelligence

The ability of designed machines and systems to emulate human intelligence is known as artificial intelligence (AI). These emulations include making decisions, recognizing patterns like speech and vision, solving problems, and learning. As such, a major method of

Technical and Ethical Optimization: Challenges and Opportunities for the Modern Engineer and Scientist
DOI: https://doi.org/10.1016/B978-0-443-44925-3.00001-6

achieving these processes is known as machine learning, which is a model that contains instructions—that is, training and testing algorithms—that continuously improve as additional and better data are introduced.

The model is known as a large language model (LLM) because of the vast amount of data needed for the model to sift through and differentiate results that are successful or failures. The training set for an LLM includes text and coding from which the LLM learns to understand information. From this, the model will generate and predict terms and a human-like language that will be used for pattern recognition, context construction, and relationships within the accessed internet data. The testing dataset is used to determine the performance and the ability of the model to generalize the accessed data, especially those not previously accessed by the LLM. That is, the model's training set allows builds, whereas the testing data determine the LLM's ability to learn, and the performance of LLM is determined by its ability to extend the building of scenarios not previously seen by the model [1].

Success is based on the goal of the LLM. For example, if the goal is to differentiate sounds or colors, the algorithms that generate results that approach the actual sounds or colors are performing successfully. The results may be somewhat close to an actual phenomenon at first and, if the LLM is performing successfully, will continue to improve the LLM's algorithms to generate results that are almost the same as the actual phenomenon. This improvement is accomplished through natural language processing that enables summarizing and translating data to generate content, for example, better machine vision and hearing [2,3].

Expert systems were among the first forms of AI that mimic a human's ability to apply a knowledge using an inference engine. The engine consists of logical rules. The rules can be forward chains [4], that is, based on facts in the knowledge base and moving logically to a conclusion, or backward chains [5], that is, beginning with the conclusion and moving backward to the facts that support that conclusion. For example, forward chaining can be used to train a system to paint a fence. The learner must first master the acts needed to remove the paint from the container and to place the correct amount of paint on a brush, then the optimal force needed to move the paint from the brush to the fence board, and then the action required to return the brush to the container and repeat the process until the fence is completely coated. Backward chaining can emulate a physician by starting with a hypothesis that a patient has a particular strain of influenza by determining the patient's symptoms to see if they track to that specific strain of influenza. One rule is that the patient has a fever; another rule is that the patient has a cough. Both rules must be satisfied and verified to be true. However, there can be other strains that have these same symptoms, so additional rules would need to be met to improve the specificity of the diagnosis. Thus, expert systems can become quite complex.

Arguably, the most daunting aspect of AI is singularity, that is, the time when AI is smarter than human intelligence. In some matters, AI has already reached singularity, certainly in terms of computational speed. On the one hand, AI can make current jobs more efficient, since up to 90% of the workforce will soon be augmented by AI [6–8]. On the other hand, this gain in efficiency can be met with unwelcomed effects on the human labor force as AI displaces human jobs [9–11].

Another concern is the need for large amounts of electrical energy needed to support generative AI, which is requiring the siting and building of data centers. These centers are opposed by advocacy groups and nearby neighborhoods who see them as adversely affecting their quality of life, increasing local utility costs, and harming the quality of air and water, including the need for large amounts of water. Indeed, a data center can require up to five million gallons of water per day for evaporative cooling [12].

The Roman Catholic Pope Leo XIV's recent encyclical, *Magnifica Humanitas* [13] (Vatican City: Vatican Press, 2026), invokes the need for data center siting to be predominantly based on the best interests of the smallest and least centralized authority, such as the family or neighborhood, rather than the global, national, or international corporate interests. This is actually one of the core pillars of Catholic Social Teaching, known as subsidiarity. Indeed, this not only applies to AI, but to any technology, such as synthetic biology, corporate farming, and social media. Interestingly, this concept is similar to the political principle of federalism, which distributes power across various levels of government. Both concepts share skepticism about too much central authority, which is a key aspect of checks and balances codified in the U.S. Constitution. Another core pillar, solidarity, the call for each person to care for one another, also applies to any technology that harms a fellow human, like infinite scrolling, social media addiction, and AI hallucinations.

1.2 Blockchain technology

Existing technologies can compartmentalize records into blocks that are linked chronologically, that is, blockchains. Blockchain technology is a shared, decentralized digital ledger that uses cryptography to create a system for tracking data. Copies of the ledger are shared across nodes on the internet. Security is based on the level of verification with each new block.

The main concern with blockchain is its misuse. Blockchain technology has no single, central database but runs on many servers, that is, it has no "on-off switch." Thus, a coding flaw in the blockchain platform can be introduced unintentionally or intentionally, making otherwise confidential data available to unauthorized users, including hackers [14].

1.3 Quantum computing

Quantum mechanics describes natural phenomena at the atomic and subatomic scales. At these scales, matter and energy exist as discrete quanta, that is, packets that behave both as particles and waves, known as the wave-particle duality.

At the subatomic scale, electromagnetic particles move in discrete paths, whereas waves bend around corners (i.e., diffraction) and propagate in ways that differ from those of particles. Particles and waves also differ in how they exchange energy, that is, particles do so by collision at specific points in time and space, whereas waves disperse in space and deposit continuously as the wave front encounters matter. Einstein and Infeld [15] explained this duality:

> But what is light really? Is it a wave or a shower of photons? There seems no likelihood for forming a consistent description of the phenomena of light by a choice of only one of the two languages. It appears we must use sometimes the one theory and sometimes the other, while at times we may use either. We are faced with a new kind of difficulty. We have two contradictory pictures of reality; separately neither of them fully explains the phenomena of light, but together they do.

The electromagnetic radiation (EMR) spectrum varies by wavelength (λ), which indicates a wave. The change in energy between the two states corresponds to a quantum or "photon" of solar radiation, indicating a particle. Thus, a photon is an amount, that is, a bundle of light energy (E). The frequencies (v) of absorption are expressed by Planck's law:

$$E = hv = \frac{hc}{\lambda} \tag{1.1}$$

where h is Planck's constant, c is the speed of light, and v and λ are the frequency and wavelength of the light of the photon (hv), respectively. The symbol for the photon comes from (1.1), since E is the fundamental energy of a single light particle that is directly proportional to its frequency (v), with Planck's constant (h) as the proportionality factor.

Molecules and atoms interact with photons under certain conditions to absorb energy of various wavelengths. Molecules undergo a transition from the ground electronic state to an excited state. For example, the range of λ for visible light is 400–700 nm.[1] Given that $E = \frac{hc}{\lambda}$ and Planck's constant (h) is 6.626×10^{-34} J·s = 4.136×10^{-15} electron volt-seconds (eV·s), then $hc = (4.136 \times 10^{-15}\ \text{eV·s})\ (2.997 \times 10^{8}\ \text{m s}^{-1}) = 1.240 \times 10^{-6}\ \text{eV·m} = 1240\ \text{eV·nm}$. At λ= 400 nm, $E = \frac{1240\ \text{eV} \cdot \text{nm}}{400\ \text{nm}} = 3.10$ eV, and at λ= 700 nm, $E = \frac{1240\ \text{eV} \cdot \text{nm}}{700\ \text{nm}} = 1.77$ eV. Thus, the energy of the photons in visible light ranges from 1.77 to 3.10 eV.

Energy levels of a molecule are described by quantum mechanics as a discrete sequence, depending on the molecule's spectroscopic properties. The discrete sequence, (E_n) [16]$_n$, is the set of all possible energy values (eigenvalues) bound by a quantum system. Consider the possible energy levels of $(E_{n)n}$ specific to a molecule trapped in an energy potential that is metaphorically "in a well" [16–18], if the well corresponds to a one-dimensional interval, [0.1], and x is the spatial variable.

The particle location can be estimated using a probability density function, $p(x)$, derived from the wavefunction $f(x)$ as $p(x) = |f(x)|^2$. The Schrödinger equation governs this function:

$$-\frac{h^2}{2}\frac{d^2f}{m\,dx^2} + V(x)f = Ef \tag{1.2}$$

where m is the particle mass, $V(x)$ is the energy potential that defines the well, E corresponds to the particle energy, and h is Planck's constant, that is, 6.63×10^{-34} J s. The particle can move freely within the well ($V = 0$) but is "trapped," which means that its probability density function is null at the well's boundaries.

The governing equation for f is:

$$\frac{d^2f}{dx^2} = \frac{2mE}{h^2}f, f(0) = f(1) = 0 \tag{1.3}$$

Thus, the solutions are in the form: $f(x) \sim \sin\left(\sqrt{\frac{2mE}{h^2}}\,x\right)$, where $\sqrt{\frac{2mE}{h^2}} = n\pi$ with n being a positive integer. This results in a discrete spectrum of energy levels in which $E_n = n^2 \cdot \frac{h^2}{2m}\pi^2$.

The emission of a photon by a substance with an energy level of E_1 will correspond to transition from E_1 to a lower energy state, for example, E_2. Therefore, $E_2 < E_1$. Conversely,

[1] An eV is a common energy unit in physics, where 1 eV ≈ $1.6e^{-19}$ Joules, An electronvolt-second, that is, eV multiplied by seconds, is a unit of action, which is related to Planck's constant, that is, $h \approx 4{,}136 \times 10^{-15}$eV·s.

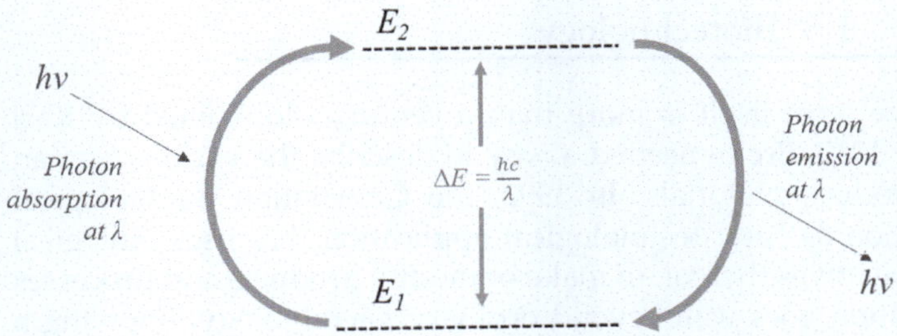

FIGURE 1.1 Planck's law and energy transition. When a photon (hv) is absorbed or emitted, it causes the transition between energy states E_1 and E_2 at a specific wavelength (λ). Note: h is Planck's constant, and c is the speed of light. *Data from: B. Sportisse, Atmospheric pollution. From processes to modelling (2008).*

when a photon is gained, that is, absorbed by the compound, the transition is to a higher energy level, that is, $E_1 < E_2$. Planck's law requires that the photon's wavelength be fixed by the energy transition. Therefore, a photon may only be emitted or absorbed if its wavelength corresponds to a transition that is possible (see Fig. 1.1). The energy gap is inversely proportional to wavelength. So, a smaller λ EMR contains greater amounts of energy.

Superposition is the ability of a qubit, that is, a quantum bit, to exist as both zero or one, or any combination within the range of zero and one. This differs from classical physics, which maintained that a bit must be either zero or one. This means that quantum computing can process immense quantities of possibilities at the same time. The connection between qubits can become entangled when two or more qubits' quantum states, that is, superpositions of zero and one, cannot be distinguished but exist as a shared system. Superposition and entanglement allow for enormous computing power by increasing the potential states that can be explored [19,20]. However, these strengths can present risks that include the diminished reliability of encryption by breaking public keys, compromising verification techniques, weakening communication security, invading the internet of things (IoT) in businesses and residences, harming critical infrastructure, and threatening geopolitical relationships, including insecure military and diplomatic data that must be confidential [21–24].

1.4 Geoengineering

Planetary-scale efforts intended to improve global conditions are known as geoengineering [25]. Most geoengineering proposals have been introduced to address climate change, so they are also known as climate engineering. Geoengineering is not a new idea, but it has taken on new significance because of complications and disappointments with implementing most proposals to mitigate climate change. Proponents of geoengineering usually admit that these measures can be associated with significant risks to the environment and public health, but they believe that they are worth pursuing, given the failure of means of mitigating climate change problems [26]. Opponents, especially some environmental groups, are against any type of geoengineering [27,28]. In the middle are those who argue that geoengineering research should continue, but that specific proposals should not yet be implemented without a better understanding of the risks and costs of geoengineering [29–31].

1.5 Biotechnology

Biotechnology is not new. The term itself is more than a century old, coined by Karl Ereky, a Hungarian engineer, in 1919. Ereky needed a way to describe the various human uses of living organisms to produce materials. In 1992, the Convention on Biological Diversity established the definition as "any technological application that uses biological systems, living organisms, or derivatives thereof, to make or modify products and processes for specific use" [32]. This definition goes well beyond ordinary microbiology, implying a need to consider the chemical processes at work in living systems [33].

Both biomedical and engineering disciplines deal directly and indirectly with living things. Indeed, biomedical engineering and environmental engineering consider biology as a means of understanding and managing the opportunities and risks of applying living organisms to societal needs. The hazards posed directly or indirectly by biotechnologies are discussed (see Tables 1.1 and 1.2). This is especially true for the risks associated with synthetic biology and genetically modified microorganisms [34–42], which extend to both human populations and ecosystems (see the discussion box, "Small Things Matter in Biotechnology").

Discussion box

Small things matter in biotechnology

The Monarch butterfly (*Lepidoptera* spp.) is an exquisitely beautiful creature. What if pollen from genetically manipulated corn (*Zea mays*) poses a threat to its survival?

The agent of concern is the genetically modified soil bacterium *Bacillus thuringiensis* (Bt) [43]. The Bt produces a protein that targets insect pests. Scientists harvest the genetic material that expresses this protein and inject it into plant species to engender pest resistance and other properties deemed beneficial to the plant. However, what risks accompany these benefits?

There is always uncertainty in upscaling from the laboratory. Indeed, no problems were observed in preliminary field studies, merely indicating that Bt-containing pollen fed directly to Monarch larvae is toxic. Later, more intensive studies began to show that the risks were present, but sufficiently low as to be acceptable, given the benefits of insect resistance. The previous assumption was that corn pollen normally travels limited distances and that the pollen does not sufficiently accumulate on milkweed leaves, the almost exclusive diet of Monarch caterpillars. In addition, corn pollen production ordinarily does not occur simultaneously with the active feeding by these Monarch larvae. As such, the US Environmental Protection Agency concluded that the planting of Bt corn was not a threat to this species. However, complex factors in this and similar ecosystem risk cases must include seasonality, sensitivity, and biology of the specific species, which requires ample field studies, keeping in mind the strong possibility of combinations of events that can lead to unforeseen adverse outcomes.

Subtle changes can make the difference between a successful remediation effort and a failure. Fittingly, the butterfly is the metaphor for chaos, that is, the so-called "butterfly effect," which postulates "sensitive

(cont'd)

dependence upon initial conditions," as a postulate of chaos theory [44]. A small change for good or bad can reap exponential rewards and costs. Edward Lorenz, at a 1963 New York Academy of Sciences meeting, related the comments of a "meteorologist who had remarked that if the theory were correct, one flap of a seagull's wings would be enough to alter the course of the weather forever." Lorenz later revised the seagull example to be that of a butterfly [45]. He argued that outcomes are affected by seemingly small events cascading through time.

The importance of seemingly minor items within a systematic approach occurred at the Iron Gate Dam in Europe. The Black Sea is an enclosed catchment basin, receiving freshwater and sediment from rivers that drain much of Europe and parts of Asia, so it is highly sensitive to nutrient enrichment, that is, eutrophication [46,47]. The Danube River receives effluents from eight European countries, flows into the Black Sea, and is the largest source of stream-borne nutrients. In less than a decade, the system changed from a biologically diverse system to one dominated by jellyfish (*Aurelia* and the comb jelly *Mnemiopsis*). Unintentionally introduced in the mid-1980s, this invasion resulted in the destruction of fisheries by the early 1990s due to human perturbations in the coastal ecosystems and in the drainage basins of the rivers, including changes to the hydrologic character of out-flowing rivers. The biggest of these was the damming of the Danube in 1972 by the Iron Gates, approximately 1000 km upstream from the Black Sea [48]. In addition, urban and industrial development, heavy use of commercial fertilizers, overfishing, and the introduction of exotic, invasive organisms (e.g., *Mnemiopsis*) contributed to the problem. After 1970, this change in nutrient concentrations induced phytoplankton blooms during the warm months and changed the dominance to non-siliceous species that were not a first choice as food for meso-zooplankton. The decreased fish stocks further increased the dominance of the jellyfish, since they competed better than the game fish for the same food. Ironically, since the mid-1990s, the ecosystems have begun to improve, mainly due to increased nutrient (phosphorus and nitrogen) loading. In most situations, it is desirable to decrease this loading to prevent eutrophication. But in this instance, the added nutrients allowed certain plankton and bottom-dwelling organisms, that is, benthos, to re-colonize. The abundance of jellyfish has also stabilized, with a concomitant increase in anchovy eggs and larvae [49].

Nutrient limitation occurs when the presence of a chemical, such as phosphorus or nitrogen, is insufficient to sustain the growth of a community or species. Usually, marine systems are nitrogen limited whereas freshwater plankton systems are phosphorus limited. Numerous freshwater organisms can fix atmospheric nitrogen but, with minor exceptions, the nitrogen is impeded in marine water. The nutrient requirements differ by species. A disturbance in the ratio of nitrogen, phosphorus, silica, and even iron changes the biotic composition of a particular plankton community.

Often, all four nutrients can be considered limiting. For instance, the lack of silica limits diatoms. This was first observed in natural blooms off Cape Mendocino in the United States and subsequently observed in

(cont'd)

the northwestern part of the Black Sea, after closing the Iron Gates dam. The case also demonstrates that economics is crucial, since the marine ecosystem improvement directly corresponds to the decline of the economies of Central and Eastern European nations in the 1990s [49].

The lesson from the butterfly and jellyfish is that small changes, even those intended to benefit society, may contribute to unexpectedly large effects. This is a lesson for any emergent technology.

Genetic research embodies practicality. Of course, researchers are interested in advancing knowledge. For engineers, this research also has an eye toward supporting the practice. Society demands that the state of the science be advanced as rapidly as possible *and* that all possible negative aspects are identified and acknowledged and, of these, that no dangerous side effects ensue in time. The engineering practitioner and researcher are also adept at optimizing among numerous variables for the best design outcomes. Emergent areas are associated with some degree of peril. An international query of top scientists asked which technologies are needed to help developing countries (see Table 1.3) [53]. Each expert was asked the following questions about the specific technologies:

- Impact. How much difference will the technology make in improving health?

TABLE 1.1 European Federation of Biotechnology's classes of risks posed by genetically modified microorganisms.

Hazard level	Description of microbial hazards
Least	Never identified as causative agents of disease in humans and that offer any threat to the environment.
Hazardous when contained, low human population risk	May cause disease in humans and which might, therefore, offer a hazard to laboratory workers. They are unlikely to spread in the environment. Prophylactics are available, and treatment is effective.
Severe when contained, moderate human population risk	Severe threat to the health of laboratory workers, but a comparatively small risk to the population at large. Prophylactics are available and treatment is effective.
High human population risk	Severe illness in humans and serious hazard to laboratory workers and to people at large. In general, effective prophylactics are not available and no effective treatment is known.
Greatest ecological and human population risk	Most severe threat to the environment, beyond humans. May be responsible for heavy economic losses. Includes several classes, Epl, Ep2, Ep3 (see Table 1.2 for descriptions) to accommodate plant pathogens.

Data from: B. Jank, A. Berthold, S. Alber, O. Doblhoff-Dier, Assessing the impacts of genetically modified microorganisms, The International Journal of Life Cycle Assessment vol. 4, (no. 5) (1999) 251–252.

TABLE 1.2 European Federation of Biotechnology classes of microorganisms causing diseases in plants [50–52].

European Federation of Biotechnology class	Description of microbes in class
Ep 1.	May cause diseases in plants but have only local significance. They may be mentioned in a list of pathogens for the individual countries concerned. Very often, they are endemic plant pathogens and do not require any special physical containment. However, it may be advisable to employ good microbiological techniques
Ep 2.	Known to cause outbreaks of disease in crops as well as in ornamental plants. These pathogens are subject to regulations for species listed by authorities in the country concerned
Ep 3.	Mentioned in quarantine lists. Importation and handling are generally forbidden. The regulatory authorities must be consulted by prospective users

Data from: H. Lelieveld et al., Safe biotechnology. 7. Classification of microorganisms on the basis of hazard, Applied Microbiology and Biotechnology vol. 45, (no. 6) (1996) 723–729; W. Frommer, and the Working Party on Safety in Biotechnology of the European Federation of Biotechnology (1994) Safe biotechnology. 3. Safety precautions for handling microorganisms of different risk classes, Applied Microbiology and Biotechnology vol. 30, (1989) 541–552; P. Doblhoff-Dier et al., Safe Biotechnology 10: DNA content of biotechnological process waste. The Safety in Biotechnology Working Party on the European Federation of Biotechnology, Trends in Biotechnology 2000).

TABLE 1.3 Ranking by global health experts of top 10 biotechnologies needed to improve health in developing countries.

Final ranking	Biotechnology
1	Modified molecular technologies for affordable, simple diagnosis of infectious diseases
2	Recombinant technologies to develop vaccines against infectious diseases
3	Technologies for more efficient drug and vaccine delivery systems
4	Technologies for environmental improvement (sanitation, clean water, bioremediation)
5	Sequencing pathogen genomes to understand their biology and to identify new antimicrobials
6	Female-controlled protection against sexually transmitted diseases, both with and without contraceptive effect
7	Bioinformatics to identify drug targets and to examine pathogen–host interactions
8	Genetically modified crops with increased nutrients to counter specific deficiencies
9	Recombinant technology to make therapeutic products (e.g., insulin, interferons) more affordable
10	Combinatorial chemistry for drug discovery

Data from: A.S. Daar, H. Thorsteinsdóttir, D.K. Martin, A.C. Smith, S. Nast, P.A. Singer, Top ten biotechnologies for improving health in developing countries, Nature Genetics vol. 32, (no. 2) (2002) 229–232.

- Appropriateness. Will it be affordable, robust, and adjustable to health care settings in developing countries, and will it be socially, culturally, and politically acceptable?
- Burden. Will it address the most pressing health needs?
- Feasibility. Can it realistically be developed and deployed within a time frame of 5–10 years?

- Knowledge gap. Does the technology advance health by creating new knowledge?
- Indirect benefits. Does it address issues such as environmental improvement and income generation that have indirect, positive effects on health?

All the areas of need involve biotechnologies, either directly (as in numbers 5 and 8, the need for improved sequencing and genetically modified organisms) or indirectly (e.g., improved environmental tools). Thus, bioengineers are at the forefront of technological progress and will continue to play an increasingly important role in the future. The concomitant societal challenges require that every engineer fully understands the implications and possible drawbacks of these technological breakthroughs. Key among them will be biotechnical advances at smaller scales, well below the cell and approaching the molecular level.

Technological processes at these scales require engineers to improve their grasp of the potential environmental implications. As biotechnologies advance, so will the concomitant societal challenges. Many engineering and science disciplines will be involved, requiring a better appreciation for and ability to predict the implications and possible drawbacks of technological developments. Key among them will be biotechnical advances at smaller scales that approach the molecular level. Biotechnological processes at these scales, for example, require engineers to improve their grasp of the potential environmental implications.

1.6 Genetic engineering and synthetic biology

Viruses, bacteria, and other microbes are remarkable biochemical factories. For much of human history, they have existed as difficult-to-explain black boxes of complex processes. These processes allow the microbes alive and provide the ability to adapt to various hostile environments. Some produce spores; many have durability and protracted latency periods; all can reproduce rapidly in large numbers, awaiting more favorable environmental conditions. The various systems that allow for this efficient survival have become increasingly better understood in recent decades, to the point that cellular and subcellular processes of uptake and absorption, nutrient distribution, metabolism, and product elimination have been characterized, at least empirically. More recently, the genetic materials of deoxyribonucleic acid (DNA) and the various forms of ribonucleic acids (RNA) have been mapped. As genes have become better understood, so has the likelihood of their being manipulated. Such manipulation is the stuff of biotechnology.

Biotechnologies at first were merely passive [35,54,55] and adaptive. As evidence, even before they were called environmental engineers, sanitary engineers observed that natural systems like surface waters and soil broke down organic materials into simpler, inorganic compounds, which are now known as biodegradation [33,56]. Based on the lessons learned from the biochemistry of bacteria and fungi active in natural biodegradation, engineers developed bioremediation technologies, that is, techniques to cleanup soil, surface water, or groundwater contaminated with organic compounds, especially volatile organic compounds and halogenated forms like dioxins, pesticides, pharmaceuticals, and most recently, the per- and polyfluoroalkyl substances (PFAS) [57–69]. In other words, biological processes are providing a remedy for some very important societal problems, that is, polluted resources.

Evolving research indicated that additional microbial genera can use organic matter detritus on forest floors, suspended organic material in water, and adsorbed onto soil particles as sources of energy needed for growth, metabolism, and reproduction. The engineering community correctly hypothesized that concentrated systems could do the same with society's organic wastes, giving rise to the design of trickling filters, oxidation ponds, and other wastewater treatment systems that are just amplified versions of the biochemical processes in natural systems.

Passive systems use the microbes already found in nature but are induced to acclimate to the organic compounds that need to be broken down. The microbial population's preference results from withholding all energy sources except the contaminant to be treated. In the presence of oxygen, the stoichiometry is:

$$\text{Organic matter} + O_2 \xrightarrow{\text{Microbes}} CO_2 + H_2O + \text{Cell matter biomass} + \text{Other end products} \tag{1.4}$$

Ultimately, the microbes degrade complex organic wastes to CO_2 and water in the presence of molecular oxygen, or to CH_4 and water in the absence of O_2, known as aerobic and anaerobic digestion, respectively.

Passive biotechnology is widely used. For example, passive biotechnologies allow for large-scale agriculture, including hybrid crops, nutrient cycling in agriculture, and vaccines in medicine [56]. More recently, biotechnologists have favored active systems to achieve such societal objectives in a timelier manner than the slowly evolved passive approaches. Whereas passive systems evolved over millennia, applications of active biotechnologies have grown exponentially in just a few decades. In addition to the speed at which active systems are being introduced, many new biotechnologies require techniques that allow genetic changes not possible with mere acclimation of natural organisms but require manipulations at the cellular scale, that is, synthetic biology [70]. The downside of active genetic manipulation was captured by astrophysicist, Eva Novotny [71]:

> Science has reached a point where the imagination and technical capabilities of scientists are running ahead faster than society can evaluate and control the outcomes of their achievements. The perception of many scientists is that all that can be done in science should be done – and if we do not do it, a competitor will. But their theoretical models of the natural world do not encompass the complexities of the real natural world. Nature works in profoundly subtle, intricately balanced and interconnected ways that the human race does not yet fully appreciate. It is for this reason that independent scientists urge caution before we release into the environment and into our own bodies, crops and foods that have been developed by crossing not only dissimilar species but even kingdoms. The long-term consequences cannot be predicted.

Society will judge the success of the various biotechnologies based on their results and faults. Synthetic biosystems that improve and protect life will be deemed successes. Otherwise, failures occur when adverse outcomes outweigh the benefits. Genetic manipulation of organisms will be gauged on the balance between risks and reliability related to all possible societal results. The uncertainty of outcomes from emerging biotechnologies presents a unique challenge. Could the application of the technology lead to a risk that is distributed disproportionately throughout society? Will the effects be irreversible and widespread? A researcher may become so invested in the possible beneficial aspects of the research that negative outcomes are given less attention (see Fig. 1.2).

Most emerging technological uncertainties begin by observing a lack or paucity of case studies, templates, and precedents from which to draw when making decisions about whether to continue or pause in their more widespread applications. Genetic engineering essentially constricted the changes in the microbe to "knocking in" or "knocking out" a desired specific trait, for example, tumor suppression [72], immune responses [73], degradation of a chemical contaminant [74], or making a substance more sustainable and biodegradable [75]. For most genetic engineering efforts to date, arguments and concerns centered around that single change. However, synthetic biology goes beyond knocking in or out a single trait. Instead, a whole new creature is built. This means that the uncertainty propagates with each change from the unedited version, that is, the chassis organism, to the formation of the synthetic organism [76].

FIGURE 1.2 Manipulating genetic material at the cellular level often has a heavy dose of uncertainty. Usually, genetically manipulated organisms vary in their competitiveness in natural systems. Always, a healthy respect for possible negative implications is needed, given possibility that the manipulated species being introduced to the human population or an ecosystem can lead to unexpected consequences [77]. *From: D. Vallero, Environmental biotechnology: a biosystems approach, Academic Press, 2010 used with permission.*

Ethical decisions as to whether it is right or wrong to synthesize an organism to provide a good or to perform a service to humankind do not have precedents. The decision-making processes are data-poor and knowledge-poor. The simpler the organism at the outset, the more unusual the modified organism becomes, and the less likely present human body and ecosystem regulatory processes will recognize them [78]. Thus, uncertainty increases moving from the natural progenitor to the adapted species to the genetically engineered microbe and, finally, to the synthetic organism. Thus, the gauging the potential downstream harm becomes less certain moving from editing the original "chassis" organism to the synthesized one. This was observed by Munthe [79]:

> ... gene editing technology may allow for precise and effective artificial mutation production (by effecting exact deletions of very small parts of a genetic sequence, thereby reliably and precisely boosting or shutting off very specific cellular functions) that is more effective and accurate than traditional genetic modification technologies while leaving no trace behind of such interference, thereby being indistinguishable from the way in which genetic changes in organisms and cells are being produced artificially using traditional techniques or spontaneously in nature. It also can boost the precision and effectiveness of adding entirely new components into a genome, regardless of where these come from – other organisms of the same species, other species, or synthetically produced genetic material.

The precision and accuracy here refer to the target organism, but not to the ethical outcomes. Researchers are often challenged with intellectual "tunnel vision" regarding the value of their work. This myopia is supported by strong, societal rationales. Researchers correctly observe the need for more efficacious medicine and treatments, better waste treatment technologies, and safer food. Thus, synthesis research is supported by an anticipated value to society that rests in what is "best" for society. This is what ethicists refer to as a "teleological" argument. The work is good because the expected outcome is good. However, the precision and accuracy of predicting the onset of other outcomes, for example, side effects and unintended consequences, are distressingly lacking the necessary data, information, knowledge, and wisdom to predict this greatest good for society.

The bioengineering advances vary from relatively well-studied adaptations of microbial populations to the most complex and uncertain gene-editing biotechnologies. Ethical awareness is arguably easier to enhance for the more accepted and established technologies, for example, making their applications efficient and fair, than for gene editing and other synthetic processes.

1.7 Nanotechnology

Technologies that work at the nanoscale, that is, 100 nm or less in at least one dimension, are known as nanotechnologies. These efforts aim to take advantage of the special physical and chemical properties of substances as they approach the molecular scale, including electromagnetics, materials science, and physiology. However, given the large uncertainties of risks of operating at these minuscule scales, there is concern that the benefits could be outweighed by the societal and individual costs and risks [80–85].

Martin Rees presented a worst-case, doomsday scenario related to the ultimate, existential risk posed by nanotechnology, known as the grey goo scenario [86]. A combination of AI, synthetic biology, and nanobots has presented the remote possibility of leading to an "extinction technology" in which an omnivorous cell gains an unchecked ability to exponentially replicate itself if part of their design is to be, thereby using all organic matter as food. No other "life" on earth would exist if this "doomsday" scenario were to occur [86].

Although, thankfully, the grey goo scenario is far-fetched, the elements of the risks are there. For example, programmed cell death is the normal process for living organisms, known as apoptosis [87]. However, this normal process can be and is overridden by common physiological processes, notably during tumorigenesis, the process of producing uncontrolled and rapidly dividing cells [88]. Cancer is a mutation of genetic material of normal cells, altering the normal balance between cell proliferation and apoptosis in favor of tumorigenesis. These cells may mutate into malignant tumors, which can metastasize into other organs. During tumorigenesis, billions of cells that theoretically originate from an initial cell that is cloned avoid apoptosis and become genetically altered into a different cell that may be malignant. Thus, a similar evasion of apoptosis by a nanobot could occur.

The synergism of nanobot technology, synthetic biology, and AI is, at least, something to be concerned about. The creature formed from this confluence of technologies, known as "biote-bots," raises the potential for [89]:

> ... extremely small, nano-/pico-aerial vehicles controlled locally or remotely by artificial intelligence mindsets to deliver, on specific hosts and tissues, either diverse bioagents produced by conventional and synthetic (micro)biology, including xenobiota or bionic microbiota or existing microbiota selected from natural reservoirs. The accuracy in delivery would leverage minute quantities of pathogens to cause mass-scale bioevents. Such hybrids (biote-bots) would increase the effectiveness of unfit but virulent pathogens, preserve the carried biota for the trip and contain bioagents' weaponization footprint to levels below the detection threshold of current regimes, while complicating immune response and denying preinfection detection and identification. To respond, we suggest that novel diagnostics and surveillance amenities are needed, prompting cooperation of experts from Medicine, medical instruments/diagnostics, artificial Intelligence and from disciplines tackling cybernetics, remote sensing, surveying and tracking.

1.8 Human enhancement and eugenics

Editable humans and selective genetics can be seen as necessary means of survival, such as a way to prevent a future case of cancer, or as horrible eugenics, such as the genocide of "unfit" humans. We have seen the former in promising research to prevent certain cancers and sickle cell disease using clustered regularly interspaced short palindromic repeats (CRISPR) gene editing. We have also seen the latter in Iceland's aggressive screening that has led to nearly 100% abortion rates targeting the birth of Down syndrome babies [90–94].

Recent newspaper articles have reported the attempts by rich and powerful people to replicate themselves in future generations by using advanced reproductive and genetic technologies. Their arguments range from vanity to creating a cadre of elites with superior intelligence to "improve" the human gene pool [95,96].

1.9 Robotics

The ability of designed machines and systems to replace humans in various tasks can improve productivity, as well as safety and quality assurance in many settings [97]. For example, construction is a promising area for robotics [98–102], especially in prefabrication and clash detection and prediction using building information modeling (BIM) to predict encounters in a virtual, three-dimensional space before the onset of construction. It often combines big data and historical project data with AI and machine learning design constraint algorithms to predict conflicts based on the spatial relationships of the many building components. Thus, robotics allows for automatically identifying and prioritizing critical clashes to enhance efficiency and safety [103]. Robotics, combined with diagnostic tools, is now common surgical standard of care [104]. Robotics also is being increasing used in environmental practices, such as sorting to improve recycling rates [105].

These benefits, however, are accompanied by risks and costs. These include human job replacement and security concerns [106]. Combining AI with robotics in healthcare creates massive amounts of sensitive data, where a breach can lead to the loss of patient trust and data integrity [107,108].

Perhaps the best guide for robotics comes from science fiction literature, as encapsulated by Asimov's rules [109]:

- Do not harm humans or allow them to come to harm.
- Obey human orders unless they conflict with the First Law.
- Protect your own existence unless it conflicts with the First or Second Law; and
- Do not harm humanity, or by inaction, allow humanity to come to harm, superseding all others.

1.10 Drones

Drones have become ubiquitous in just a few years. They are found everywhere, from children's toys in residential backyards to sports arenas, to battlefields. They are basically flying devices that are controlled remotely. They often carry cameras. Larger drones can carry geographic positioning systems (GPS), surveillance equipment, lasers, and weaponry.

Drones can be very useful in survey damage following disasters, monitoring environmental quality, and as early warning systems. Because of the many drones in use, aviation regulations have been promulgated to protect aircraft. In the United States, for example, these include [110]:

- Flying below 400 feet
- Remaining clear of surrounding obstacles
- Always staying within visual line of sight
- Remaining clear of and not interfering with manned aircraft operations
- Staying more than five miles from an airport without prior permission from the airport and control tower
- Not flying near people or stadiums

- Not exceeding 55 pounds
- Flying only during daylight without prior approval from the Federal Aviation Administration to fly in the dark
- Not flying around Washington, DC
- Not flying in national parks.

The loss of privacy and the removal of human control in lethal decisions are arguably the two major ethical issues associated with drone use.

1.11 Digital surveillance

Society can benefit from enhanced law enforcement that applies digital techniques, like facial recognition and monitoring a person's or group's online history, data usage, and communications, for example, phone records [111,112]. However, as is true for drone use, digital surveillance methods increase the likelihood of loss of privacy. The stored data and "digital footprints" are also vulnerable to hacking and misuse.

At the individual scale, digital surveillance increases the risk of compulsive behaviors, such as impulse buying, as the purchasing data targets products likely to be attractive to a particular person [113]. It can also lead to discrimination against certain people in terms of access to financial opportunities, healthcare, and insurance, as well as blacklisting and "redlining" [114–116]. Surveillance may also be misused by law enforcement and other authorities, including constitutional violations [117,118].

1.12 Conclusion

This chapter introduces and summarizes several emerging technologies. This sets the stage for the next chapter's introduction to ethical theories as they relate to the current technological environment.

References

[1] C. Paduraru, A. Staicu, A. Stefanescu, LLM-based methods for the creation of unit tests in game development, Procedia Computer Science vol. 246, (2024) 2459–2468.

[2] M. Laad, R. Maurya, N. Saiyed, Unveiling the vision: A comprehensive review of computer vision in AI and ML, in: 2024 International conference on advances in data engineering and intelligent computing systems (ADICS), IEEE, 2024, pp. 1–6.

[3] Y. Matsuzaka, R. Yashiro, AI-based computer vision techniques and expert systems, AI vol. 4, (no. 1) (2023) 289–302.

[4] F. Sabry, Forward chaining: fundamentals and applications, One Billion Knowledgeable (2023).

[5] M. Kazemi, N. Kim, D. Bhatia, X. Xu, D. Ramachandran, Lambada: backward chaining for automated reasoning in natural language, in: Proceedings of the 61st Annual Meeting of the Association for Computational Linguistics (Volume 1: Long Papers), 2023, pp. 6547–6568.

[6] S. Paul, L. Yuan, H.K. Jain, L.P. Robert, Jr, J. Spohrer, et al., Intelligence augmentation: human factors in ai and future of work, AIS Transactions on Human-Computer Interaction vol. 14, (no. 3) (2022) 426–445.

[7] D. Harborth, K. Kümpers, Intelligence augmentation: rethinking the future of work by leveraging human performance and abilities, Virtual Reality vol. 26, (no. 3) (2022) 849–870.

[8] S. Dégallier-Rochat, M. Kurpicz-Briki, N. Endrissat, O. Yatsenko, Human augmentation, not replacement: a research agenda for AI and robotics in the industry, Frontiers in Robotics and AI vol. 9, (2022) 997386.

[9] A. Rawashdeh, The consequences of artificial intelligence: an investigation into the impact of AI on job displacement in accounting, Journal of Science and Technology Policy Management vol. 16, (no. 3) (2025) 506–535.

[10] R. Tiwari, The impact of AI and machine learning on job displacement and employment opportunities, International Journal of Engineering Technologies and Management Research vol. 7, (no. 1) (2023) 1–9.

[11] P. Adhikari, Exploring the nexus between artificial intelligence and job displacement: a literature review, Journal of National Development vol. 37, (no. 1) (2024) 1–13.

[12] M. Yañez-Barnuevo, "Data centers and water consumption," *Environmental and Energy Study Institute*, vol. 55, 2025.

[13] P. Leo XIV, *Magnifica Humanitas*. Vatican City: Vatican Press, 2026.

[14] M. Somers, The risks and unintended consequences of blockchain, *Ideas Made to Matter*. [Online]. Available from: https://mitsloan.mit.edu/ideas-made-to-matter/risks-and-unintended-consequences-blockchain#:~:text=go%20with%20them.-,Distributed%20control,the%20key%20holder%2C%20Madnick%20said.

[15] L. Infeld, The Evolution of Physics, CUP Archive, (1971).

[16] S.D. Ilčev, Global Satellite Meteorological Observation (GSMO) Theory, Springer, (2018).

[17] B. Sportisse, Atmospheric Pollution. From processes to modelling (2008).

[18] B. Sportisse, Fundamentals in Air Pollution: From Processes to Modelling, Springer Science & Business Media, (2009).

[19] A. Khrennikov, Roots of quantum computing supremacy: superposition, entanglement, or complementarity? The European Physical Journal Special Topics vol. 230, (no. 4) (2021) 1053–1057.

[20] P. Malik, The Emperor's Quantum Computer Illustrated: An Alternative Light-Centered Interpretation of Quanta, Superposition, Entanglement and the Computing that Arises from it, Illustrated with Art, Possibilities Publishing, (2024).

[21] A. Vance, Cybersecurity risks and opportunities in the quantum computing age: a study, Quantum Journal of Engineering, Science and Technology vol. 5, (no. 3) (2024) 31–39.

[22] F. Raheman, The future of cybersecurity in the age of quantum computers, Future Internet vol. 14, (no. 11) (2022) 335.

[23] M.J. H. Faruk, S. Tahora, M. Tasnim, H. Shahriar, N. Sakib, A review of quantum cybersecurity: threats, risks and opportunities, in: 2022 1st International Conference on AI in Cybersecurity (ICAIC), IEEE, 2022, pp. 1–8.

[24] Y. Baseri, V. Chouhan, A. Ghorbani, Cybersecurity in the Quantum Era: Assessing the Impact of Quantum Computing on Infrastructure (2024) *arXiv preprint arXiv:2404.10659*.

[25] J. Shepherd, Geoengineering the Climate: Science, Governance and Uncertainty, Royal Society, (2009).

[26] A. Robock, Benefits and risks of stratospheric solar radiation management for climate intervention (geoengineering), The Bridge vol. 50, (no. 1) (2020) 59–67.

[27] A. Corner, N. Pidgeon, Geoengineering the climate: the social and ethical implications, Environment: Science and Policy for Sustainable Development vol. 52, (no. 1) (2010) 24–37.

[28] Vasquez, Solar geoengineering could wreak havoc on the planet vasquez, Sierra, the Magazine of the Sierra Club (2023) [Online]. Available https://www.sierraclub.org/sierra/2023-3-fall/stress-test/solar-geoengineering-could-wreak-havoc-planet.

[29] D.A. Vallero, D.B. Resnik, Geoengineering: enhancing cloud albedo, Access Science (2013).

[30] A.A. Tsonis, A. Kirwan Jr, Geoengineering from the standpoint of uncertainty and related risks: science or science fiction? Stochastic Environmental Research and Risk Assessment vol. 37, (no. 9) (2023) 3325–3327.

[31] D.B. Resnik, D.A. Vallero, Geoengineering: an idea whose time has come? Journal of Earth Science & Climatic Change (2011).
[32] A. Sasson, Medical Biotechnology: Achievements, Prospects and Perceptions, United Nations University Press, (2005).
[33] D.A. Vallero, C.K. Gunsch, Applications and implications of emerging biotechnologies in environmental engineering, Journal of Environmental Engineering vol. 146, (no. 6) (2020) 03120005.
[34] B. Jank, A. Berthold, S. Alber, O. Doblhoff-Dier, Assessing the impacts of genetically modified microorganisms, The International Journal of Life Cycle Assessment vol. 4, (no. 5) (1999) 251–252.
[35] S.R. Kim, et al., Improvement of the approval system for the development and experimentation of living modified organisms in synthetic biology, Public Health Weekly Report vol. 16, (no. 32) (2023) 1141.
[36] T. Sun, J. Song, M. Wang, C. Zhao, W. Zhang, Challenges and recent progress in the governance of biosecurity risks in the era of synthetic biology, Journal of Biosafety and Biosecurity vol. 4, (no. 1) (2022) 59–67.
[37] J. Li, H. Zhao, L. Zheng, W. An, Advances in synthetic biology and biosafety governance, Frontiers in Bioengineering and Biotechnology vol. 9, (2021). Available from https://doi.org/10.3389/fbioe.2021.598087 Review.
[38] B.D. Trump, et al., Building biosecurity for synthetic biology, Molecular Systems Biology vol. 16, (no. 7) (2020) e9723.
[39] C.L. Cummings, J. Kuzma, Societal risk evaluation scheme (SRES): scenario-based multi-criteria evaluation of synthetic biology applications, PLoS One vol. 12, (no. 1) (2017) e0168564.
[40] H. König, D. Frank, R. Heil, C. Coenen, Synthetic biology's multiple dimensions of benefits and risks: implications for governance and policies, Synthetic Biology, Springer, (2016), pp. 217–232.
[41] M.M. Epstein, T. Vermeire, Scientific opinion on risk assessment of synthetic biology, Trends in Biotechnology vol. 34, (no. 8) (2016) 601–603.
[42] G.K. Gronvall, Mitigating the Risks of Synthetic Biology (2015).
[43] J. Losey, L. Rayor, M. Carter, Transgenic Pollen Harms Monarch Larvae Nature (1999) ed: London.
[44] R.C. Hilborn, Chaos and Nonlinear Dynamics: An Introduction for Scientists and Engineers, Oxford university press, (2000).
[45] E. Lorenz, The butterfly effect, The Chaos Avant-garde: Memories of the Early Days of Chaos Theory, World Scientific, (2000), pp. 91–94.
[46] F. Popescu, et al., Analysis of surface water quality and sediments content on danube basin in djerdap-iron gate protected areas, Water vol. 14, (no. 19) (2022) 2991.
[47] C. Nistor, I. Săvulescu, B.-A. Mihai, L. Zaharia, M. Vîrghileanu, S. Carablaisă, The impact of large dams on fluvial sedimentation: the iron gates reservoir on the Danube river, Acta Geographica Slovenica vol. 61, (no. 1) (2021) 41–55.
[48] C. Lancelot, J. Staneva, D. Van Eeckhout, J.-M. Beckers, E. Stanev, Modelling the Danube-influenced north-western continental shelf of the Black Sea. II: ecosystem response to changes in nutrient delivery by the Danube River after its damming in 1972, Estuarine, Coastal and Shelf Science vol. 54, (no. 3) (2002) 473–499.
[49] D. Vallero, Paradigms Lost: Learning From Environmental Mistakes, Mishaps and Misdeeds, Butterworth-Heinemann, (2005).
[50] H. Lelieveld, et al., Safe biotechnology. 7. Classification of microorganisms on the basis of hazard, Applied Microbiology and Biotechnology vol. 45, (no. 6) (1996) 723–729.
[51] W. Frommer, and the working party on safety in biotechnology of the European Federation of Biotechnology (1994) safe biotechnology. 3. Safety precautions for handling microorganisms of different risk classes, Applied Microbiology and Biotechnology vol. 30, (1989) 541–552.
[52] P. Doblhoff-Dier, et al., Safe biotechnology 10: DNA content of biotechnological process waste. The Safety in Biotechnology Working Party on the European Federation of Biotechnology, Trends in Biotechnology (2000).
[53] A.S. Daar, H. Thorsteinsdóttir, D.K. Martin, A.C. Smith, S. Nast, P.A. Singer, Top ten biotechnologies for improving health in developing countries, Nature Genetics vol. 32, (no. 2) (2002) 229–232.
[54] S. Hirschi, T.R. Ward, W.P. Meier, D.J. Müller, D. Fotiadis, Synthetic biology: bottom-up assembly of molecular systems, Chemical Reviews vol. 122, (no. 21) (2022) 16294–16328.

[55] S. Jaiswal, P. Shukla, Alternative strategies for microbial remediation of pollutants via synthetic biology, Frontiers in Microbiology vol. 11, (2020) 808.
[56] D. Vallero, Environmental Biotechnology: A Biosystems Approach, Elsevier Science, (2015).
[57] J. Zhang, S. Zhang, B. Liu, Degradation technologies and mechanisms of dioxins in municipal solid waste incineration fly ash: a review, Journal of Cleaner Production vol. 250, (2020) 119507.
[58] S.K. Kirkok, J.K. Kibet, T.K. Kinyanjui, F.I. Okanga, A review of persistent organic pollutants: dioxins, furans, and their associated nitrogenated analogues, SN Applied Sciences vol. 2, (2020) 1–20.
[59] S. Kanan, F. Samara, Dioxins and furans: a review from chemical and environmental perspectives, Trends in Environmental Analytical Chemistry vol. 17, (2018) 1–13. Available from https://doi.org/10.1016/j.teac.2017.12.001.
[60] A. Schecter, L. Birnbaum, J.J. Ryan, J.D. Constable, Dioxins: an overview, Environmental Research vol. 101, (no. 3) (2006) 419–428.
[61] National Center for Health Statistics, National Health and Nutrition Examination Survey - 2013-14 - Data Documentation, Codebook, and Frequencies - Perfluoroalkyl and Polyfluoroalkyl Substances (PFAS_J), NCHS, (2023). Available from https://wwwn.cdc.gov/Nchs/Nhanes/2013-2014/SSPFSU_H.htm(accessed February 23.
[62] E. P. A. U.S, Per- and Polyfluoroalkyl Substances (PFAS): Proposed PFAS National Primary Drinking Water Regulation, U.S. EPA, (2023). Available from https://www.epa.gov/sdwa/and-polyfluoroalkyl-substances-pfasaccessed October 6.
[63] A. Berhanu, I. Mutanda, J. Taolin, M.A. Qaria, B. Yang, D. Zhu, A review of microbial degradation of per-and polyfluoroalkyl substances (PFAS): biotransformation routes and enzymes, Science of the Total Environment (2022) 160010.
[64] L.P. Burkhard, Evaluation of published bioconcentration factor (BCF) and bioaccumulation factor (BAF) data for per-and polyfluoroalkyl substances across aquatic species, Environmental Toxicology and Chemistry (2021).
[65] J. Glüge, et al., An overview of the uses of per- and polyfluoroalkyl substances (PFAS) (in eng), Environmental Science: Processes & Impacts vol. 22, (no. 12) (2020) 2345–2373. Available from https://doi.org/10.1039/d0em00291g.
[66] A.M. Calafat, L.-Y. Wong, Z. Kuklenyik, J.A. Reidy, L.L. Needham, Polyfluoroalkyl chemicals in the US population: data from the National Health and Nutrition Examination Survey (NHANES) 2003–2004 and comparisons with NHANES 1999–2000, Environmental Health Perspectives vol. 115, (no. 11) (2007) 1596–1602.
[67] U.S. Environmental Protection Agency, Technical Overview of Volatile Organic Compounds (2024) Available from https://www.epa.gov/indoor-air-quality-iaq/technical-overview-volatile-organic-compounds#8 (accessed June 1).
[68] A. Mukhopadhyay, S. Duttagupta, A. Mukherjee, Emerging organic contaminants in global community drinking water sources and supply: a review of occurrence, processes and remediation, Journal of Environmental Chemical Engineering vol. 10, (no. 3) (2022) 107560.
[69] D. Marathe, S. Balbudhe, K. Kumari, Persistent organic pollutants: a global issue, a global response," in, Persistent Organic Pollutants, CRC Press, (2021), pp. 1–32.
[70] B.D. Trump, C.L. Cummings, S. Galaitsi, J. Kuzma, I. Linkov, Synthetic biology: perspectives on risk analysis, governance, communication, and ELSI, Synthetic Biology 2020: Frontiers in Risk Analysis and Governance, Springer, (2020), pp. 1–18.
[71] E. Novotny, Letter to Prime Minister re: 114 Biotechnology Scientists (2003) [Online]. Available https://www.sgr.org.uk/resources/letter-prime-minister-re-114-biotechnology-scientists.
[72] Y. Lu, et al., Genome-scale CRISPR-Cas9 knockout screening in hepatocellular carcinoma with lenvatinib resistance, Cell Death Discovery vol. 7, (no. 1) (2021) 359.
[73] V.D. Castro, J. Galaine, R. Loyon, Y. Godet, CRISPR-Cas gene knockouts to optimize engineered T cells for cancer immunotherapy, Cancer Gene Therapy vol. 31, (no. 8) (2024) 1124–1134.
[74] S. Sharma, et al., Microbial remediation of polluted environment by using recombinant *E. coli*: a review, Biotechnology for the Environment vol. 1, (no. 1) (2024) 8.
[75] S. Garg, M. Kim, D. Romero-Suarez, Current advancements in fungal engineering technologies for sustainable development goals, Trends in Microbiology vol. 33, (no. 3) (2025) 285–301.

[76] D.T.C. Chan, J. Bjerg, H.C. Bernstein, Broad-host-range synthetic biology: rethinking microbial chassis as a design variable, ACS Synthetic Biology vol. 14, (no. 10) (2025) 3815–3821.
[77] D. Vallero, Environmental Biotechnology: A Biosystems Approach, Academic Press, (2010).
[78] D.J. Berry, Synthesis and the Organism: Biology, Chemistry, and Engineering (2018).
[79] C. Munthe, Precaution and Ethics: Handling Risks, Uncertainties and Knowledge Gaps in the Regulation of New Biotechnologies, Federal Office for Buildings and Publications and Logistics (FOBL), (2017).
[80] D.A. Vallero, Mapping exposure onto nanoscale toxicity measures, in: J.R. Lead, S.H. Doak, M.J.D. Clift (Eds.), Nanotoxicology in Humans and the Environment, Springer International Publishing, Cham, (2021), pp. 141–191.
[81] S. Arnaldi, A. Muratorio, Nanotechnology, Uncertainty and Regulation. A Guest Editorial, Springer, (2013).
[82] National Nanotechnology Initiative, National Nanotechnology Initiative Strategic Plan, National Science and Technology Council, (2011).
[83] J.R. Lead, E. Smith, Environmental and Human Health Impacts of Nanotechnology, John Wiley & Sons, (2009).
[84] K.A. Dunphy Guzmán, M.R. Taylor, J.F. Banfield, Environmental Risks of Nanotechnology: National Nanotechnology Initiative Funding, 2000– 2004, ACS Publications, (2006).
[85] M. Nasrollahzadeh, S.M. Sajadi, Risks of nanotechnology to human life, Interface science and technology vol. 28, Elsevier, (2019), pp. 323–336.
[86] M.J. Rees, Our Final Hour: A Scientist's Warning: How Terror, Error, and Environmental Disaster Threaten Humankind's Future in This Century–on Earth and Beyond, Basic Books (AZ), (2003).
[87] M. Mustafa, et al., Apoptosis: a comprehensive overview of signaling pathways, morphological changes, and physiological significance and therapeutic implications, Cells vol. 13, (no. 22) (2024) 1838.
[88] O. Morana, W. Wood, C.D. Gregory, The apoptosis paradox in cancer, International Journal of Molecular Sciences vol. 23, (no. 3) (2022) 1328.
[89] M.E. Kambouris, Y. Manoussopoulos, A. Velegraki, G.P. Patrinos, The biote-bot hybrid. The ultimate biothreat merging nanobots, AI-enabled cybernetics and synthetic biology, Future Medicine AI vol. 1, (no. 1) (2023).
[90] R. Thorneycroft, Prenatal testing, down syndrome, and selective termination: a (critical) criminology of genocide? Diversity in Criminology and Criminal Justice Studies, Emerald Publishing Limited, (2022), pp. 167–181.
[91] M. Gemma, The CRISPR crossroad: a cure for genetic disease or beginning of "neo-eugenics"? Stevens Institute of Technology (2021).
[92] N.I. Kara, Screening syndromes out: updating the international" Genocide" vernacular for a changing technological age, North Carolina Journal of International Law vol. 45, (2019) 163.
[93] A. Knight, J. Miller, Prenatal Genetic Testing, Abortion, and Disability Justice, Oxford University Press, (2023).
[94] N.I. Kara, Screening syndromes out: updating the international "genocide" vernacular for a changing technological age, North Carolina Journal of International Law vol. 45, (no. 1) (2020).
[95] G.-F. Umeh, A.C. Umeh, Francis galton's eugenics as a means to a new world: an ethical approach, Nnadiebube Journal of Education in Africa vol. 9, (no. 1) (2024).
[96] R. Slorach, From eugenics to scientific racism, International Socialism Journal vol. 165, (2020) 133–154.
[97] C.-J. Liang, T.-H. Le, Y. Ham, B.R.K. Mantha, M.H. Cheng, J.J. Lin, Ethics of artificial intelligence and robotics in the architecture, engineering, and construction industry, Automation in Construction vol. 162, (2024) 105369. Available from https://doi.org/10.1016/j.autcon.2024.105369.
[98] F. Bademosi, R.R. Issa, Factors influencing adoption and integration of construction robotics and automation technology in the US, Journal of Construction Engineering and Management vol. 147, (no. 8) (2021) 04021075.
[99] M. Pan, Y. Yang, Z. Zheng, W. Pan, Artificial intelligence and robotics for prefabricated and modular construction: a systematic literature review, Journal of Construction Engineering and Management vol. 148, (no. 9) (2022) 03122004.
[100] S. Parascho, Construction robotics: from automation to collaboration, Annual Review of Control, Robotics, and Autonomous Systems vol. 6, (no. 1) (2023) 183–204.

[101] P. Pradhananga, M. ElZomor, G. Santi Kasabdji, Identifying the challenges to adopting robotics in the US construction industry, Journal of Construction Engineering and Management vol. 147, (no. 5) (2021) 05021003.
[102] S. Zhao, et al., Application and development of autonomous robots in concrete construction: Challenges and opportunities, Drones vol. 6, (no. 12) (2022) 424.
[103] P.P. Borase, Revolutionizing construction: the synergy of VDC/BIM, laser scanning and dusty robotics, 2024 Innovation in the Construction Industry (2024).
[104] S. Ness, T.R. Xuan, O.O. Oguntibeju, Influence of AI: robotics in healthcare, Asian Journal of Research in Computer Science vol. 17, (no. 5) (2024) 222–237.
[105] L.Q. Thao, An automated waste management system using artificial intelligence and robotics, Journal of Material Cycles and Waste Management 25 (6) (2023) 3791–3800. Available from https://doi. org/10.1007/s10163-023-01796-4.
[106] Q. Chen, B. García de Soto, B.T. Adey, Construction automation: research areas, industry concerns and suggestions for advancement, Automation in Construction vol. 94, (2018) 22–38. Available from https://doi. org/10.1016/j.autcon.2018.05.028.
[107] K.E. Emam, L. Arbuckle, Anonymizing Health Data: Case Studies and Methods to Get You Started, O'Reilly Media, Inc, (2013).
[108] C. Elendu, et al., Ethical implications of AI and robotics in healthcare: a review, Medicine vol. 102, (no. 50) (2023) e36671.
[109] I. Asimov, Forward the Foundation, Spectra, (1994).
[110] A. Hopkins, The Ethical Debate on Drones (2017).
[111] F. Bacalu, Digital policing tools as social control technologies: data-driven predictive algorithms, automated facial recognition surveillance, and law enforcement biometrics, Analysis and Metaphysics no. 20 (2021) 74–88.
[112] I. Karina, M. Gultom, H.S. Flora, J. Sidabalok, E.N. Butarbutar, Implications of digital technology for criminal law enforcement: challenges and opportunities in the age of digitalization, UNES Law Review vol. 6, (no. 4) (2024) 10039–10047.
[113] S.E. Beatty, M.E. Ferrell, Impulse buying: modeling its precursors, Journal of Retailing vol. 74, (no. 2) (1998) 169–191.
[114] R. Clarke, Risks inherent in the digital surveillance economy: a research agenda, Journal of Information Technology vol. 34, (no. 1) (2019) 59–80.
[115] M.L. Wang, C.M. Gago, K. Rodriguez, Digital redlining—the invisible structural determinant of health, JAMA: The Journal of the American Medical Association vol. 331, (no. 15) (2024) 1267–1268.
[116] L.S. Bessette, Digital redlining, minimal computing, and equity, Critical Digital Pedagogy in Higher Education (2023) 69–79.
[117] M.C. Wheatley, Ethics of surveillance technologies: balancing privacy and security in a digital age, Premier Journal of Data Science vol. 1, (2024) 100001.
[118] C. Jiaojiao, Law enforcement challenges and perspectives within the system of constitutional oversight, Экономика и социум no. 3-1 (130) (2025) 1017–1022.

CHAPTER

2

Ethical constructs

OUTLINE

The Brain—is wider than the Sky—For—put them side by side—The one the other will contain With ease—and You—beside—The Brain is deeper than the sea—For—hold them—Blue to Blue—The one the other will absorb—As Sponges—Buckets—do—The Brain is just the weight of God—For—Heft them—Pound for Pound—And they will differ—if they do—As Syllable from Sound— ***Emily Dickinson (1830–86)[1]***

2.1 Introduction

Technologists do not have to be reminded of the importance of imagination in their endeavors. Dickinson poetically and persuasively elucidates the fact that the mind is almost limitless in its ability to envision the opportunities from the smallest to the largest realms of reason. Given enough reliable information, humans can see beyond the constraints of what is to what can be. Imagination is arguably the greatest asset of the technologists who possess

Technical and Ethical Optimization: Challenges and Opportunities for the Modern Engineer and Scientist
DOI: https://doi.org/10.1016/B978-0-443-44925-3.00004-1

the mindset to design because what they can see does not yet exist. The purpose of design is to take that vision and build a pathway to a real outcome.

In addition to discovery, imagination can also be a useful device to help to understand the nature of ethical challenges to be encountered by the technologist and to provide a means of identifying ways to avert ethical problems before they arise. This is a key aspect of failure engineering, that is, the interdisciplinary field that investigates the root and ancillary causes of failures. Such anticipations and expectations pave the way for safer and more reliable systems. For example, engineers had to change their perspectives on possible failures following the World Trade Center attacks. Before, engineers primarily designed buildings to withstand natural disasters and unintentional events like a plane lost in a fog. After, facilities had to be hardened against intentional events, like terrorism [2]. However, more than physical hardening is needed. There needs to be thoughtful imagination of what can happen, even if it is a very low probability. Robert Prieto summed this up quite well [3].

> Critical infrastructure must be designed to resist attack and catastrophic failure. Immediately following the attacks and the subsequent collapse of the World Trade Center towers, there were those who called for high-profile buildings and other critical infrastructure to be designed to stop airplanes. This, simply put, is utter nonsense. Unless we are prepared to live in a heavily engineered environment more closely resembling the complex of caves in Afghanistan, we will not design buildings to stop planes. The challenge is to keep airplanes away from buildings and to root out those who challenge our way of life at the source. ...Each engineering disaster, whether natural or manmade, has taught us lessons. Over time, these lessons are disseminated within a subset of our profession or within an industry segment. Some lessons are only understood with the fullness of time and often offer a deeper understanding of the real challenges we as engineers face. Short-term code modifications often satisfy the tendency to overreact in the short term but, as understanding develops, may fail to sufficiently react in the long term.

Some technological failures cause little harm, such as an automobile coating that lasts only a few years. It is certainly a disappointment but not catastrophic. Others, can indeed be disastrous, such as a medical device that fails to operate as designed, resulting in injury or death. Furthermore, there is a big ethical difference between a mistake and misconduct [4]. It lies in intent. An honest mistake happens often, usually with little damage. Misconduct occurs when the inventor or technologist deliberately fabricates, falsifies, or otherwise creates a dangerous product or hides a design flaw [5]. Misconduct also includes flaws that should have been detected and shared but were not. For professionals, this includes gross negligence, incompetence, or unsafe practices and products [6–8].

Most scientists, engineers, and technologists are at least somewhat more familiar with the works of Newton, Boyle, Einstein, and Bohr, certainly more than they are with the teachings of philosophers like Hammurabi, Hippocrates, Aristotle, Hobbes, Kant, Mill, and Rawls. However, four centuries ago, philosophy was central to all studies, including the physical and biological "sciences." In fact, the physical and natural sciences were yet to be distinguished as such and merely fell within the realm of "natural philosophy" [9–18]. The Ancient and Renaissance scientists did not distinguish ethics and other aspects of philosophy from physics and the other foundational sciences that underpin technologies. As such, it may be useful to consider our careers as technologists a bit more comprehensively as a starting point for understanding morality as it applies to emerging technologies.

A comprehensive view that incorporates ethics into a science-based technology development process allows the consideration of certain topics not often covered in science and engineering texts. Good actions versus unethical decision-making, moral versus immoral acts, and obligations versus prohibitions are understood by most professionals at some intuitive level, but unlike our colleagues in the humanities and social sciences, we are more likely to avoid considering them theoretically and, worse, when we are deciding on the best way to move forward with a technology's development and applications. However, the classical works of Aristotle, Aquinas, Kant, et al. make a strong case that life is a mix of available virtues and vices. Virtue is the power to do good or a habit of doing good. One of Aristotle's most memorable lines is that "Excellence is habit" [19,20]. If we do good, we are more likely, according to Aristotle, to keep doing good. Conversely, vice is the power and habit of doing evil. The road to perdition is often a series of small vices that have ruined careers and livelihoods and even led to jail time [21].

2.2 The syllogism

The scientific method depends on objectivity. This can contrast with the ethical decision making given the subjectivity or relational nature of good and evil. As a result, there is some discomfort in scientific and engineering circles, where meaning on certainty and consistency of definition is crucial to problem solving. This problem has been noticed for quite some time. As evidence, in 1923, Mead put it succinctly [22]:

> But if we have succeeded in applying science to our mechanical task, and in this have accomplished prodigies, we do not seem to have succeeded in applying scientific method to the formulation of our ends and purposes.

Mead and others, including the author, find that the scientific facts are a necessary but a wholly insufficient component of any ethical case analysis. To give this thinking some structure, philosophers tell us that we can determine whether a moral argument is valid, although not necessarily correct, by parsing the argument into a so-called syllogism consisting of four parts [23–29]:

1. the factual premise
2. the connecting premise (i.e., factual to evaluative)
3. the evaluative premise
4. the moral conclusion.

For example, the facts may show that exposing people to a chemical compound at a certain dosage (e.g., one part per million) leads to a specific form of cancer in one in every 10,000 people. We also know that, from a public health perspective, intentionally allowing people to contract cancer because of some human activity is morally wrong. Thus the syllogism would be:

1. Factual premise: Exposure to chemical X at 1 ppm leads to cancer.
2. Connecting premise: Release of 10 kg per day of chemical X leads to 1 ppm exposure for people living near an industrial plant.

3. Evaluative premise: Executive decisions that allow industrial releases that lead to cancer are morally wrong.
4. Moral conclusion: Therefore, corporate executives who decide to release 10 or more kilograms of chemical X from their plants, and government officials who allow this to happen, are acting immorally.

Science provides the factual premise and part of the connecting premise, but social mores and norms provide the evaluative premise and lead to the moral conclusion. However, factual uncertainties will disrupt the validity of the argument. Scientific uncertainties are brought about both by variability and error.[1] Spatial and temporal variability are ever present. Each technology's development and use consist of a unique set of factors, dependent variables, situations, and scenarios, so that what has occurred will never be completely repeated. Every volume of material and every use is unique at the nanoscale. These keep changing with time and usage. Environmental, meteorological, financial, and demographic conditions in 2021 are different from conditions in 2026. Of course, there are errors. Some are random in that the conditions that lead to the outcome are partially explained by chance and events that were neither predictable nor correctable, although we can try to explain them statistically, for example, by assuming normal distributions or using Markov Chain Monte Carlo approaches [31–38].

The other type of error is systematic, also known as bias. Bias can be technical, such as a thermometer that always reads 0.001 degrees higher than the real temperature. Unlike random error, where repeated tests lower the error, no matter how many times this meter is used, the bias remains. There is also observational bias. Each person sees through a mental and moral prism different from anyone else's. This prism is the result of the person's own experiences and expertise. This prism is the perception of what is real and what is important. For a scientist or engineer, the personal bias is heavily weighted in sound science, or at least what that person believes to be sound science, as opposed to "junk science" or "advocacy science."[2]

Sound science requires sufficient precision and accuracy in presenting the facts. Precision describes how refined and repeatable an operation can be performed, such as the exactness in the instruments and methods used to obtain a result. It is an indication of the uniformity or reproducibility of a result. This can be likened to shooting arrows,[3] with

[1] Another way to look at uncertainty is that it is a function of variability and ignorance. This was articulated in the review of ecological case studies in US Environmental Protection Agency [30]. L. Ginzburg, one of the peer reviewers, pointed out that "variability includes stochasticity arising from temporal and spatial heterogeneity in environmental factors and among exposed individuals. Ignorance includes measurement error, indecision about the form of the mathematical model, or appropriate level of abstraction." Thus variability can be lessened by increased attention, for example, empirical evidence, and "translated into risk (i.e., probability) by the application of a probabilistic model," but ignorance cannot. Ignorance simply translates into confidence intervals, or "error bounds" on any statement of risk.

[2] See, for example: The Physics Gallery: The Physical Principles of Unworkable Devices: https://dsimanek.vialattea.net/museum/physgal.htm. Donald E. Simanek's humorous but informative site on why perpetual motion machines cannot work, that is, their inventors assumed erroneous "principles." This site is instructive to environmental decision makers to beware of "junk science." Sometimes, a good way to learn why something works the way it does is to consider all the reasons that it fails to work.

[3] My apologies to the originator of this analogy, who deserves much credit for this teaching device. The target is a widely used method to describe precision and accuracy.

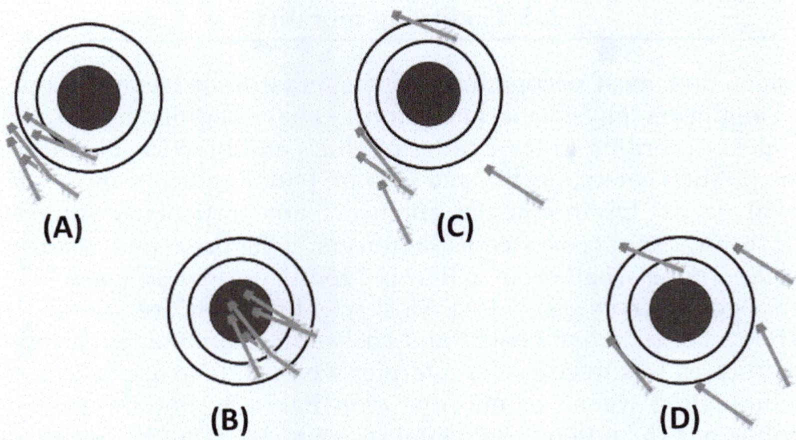

FIGURE 2.1 Precision and accuracy. The bull's eye represents the true value. Targets (A) and (B) demonstrate data sets that are precise; Targets (B) and (D) data sets that are accurate; and Targets (C) and (D) data sets that are imprecise. Target (B) is the ideal data set, which is precise and accurate.

each arrow representing a data point. Targets (A) and (B) in Fig. 2.1 show equal precision. If the center of the target, that is, the bullseye, is the "true value," data set (B) is more accurate than (A). If we are consistently missing the bull's eye in the same direction at the same distance, this is an example of bias or systematic error. The good news is that if we are aware that we are missing the bullseye (e.g., by comparing our results to those of known standards when using our analytical equipment), we can calibrate and adjust the equipment. To stay with our archery analogy, the archer would move her sight up and to the right.

Thus accuracy is an expression of the extent to which a study conforms to some defined standard, that is, the true value. So, accuracy expresses the quality of what is found, whereas precision expresses the quality of the operation by which the finding was obtained. The two other scenarios of data quality are shown in Targets (C) and (D). Thus the four possibilities are that our data is precise but inaccurate (Target A), precise and accurate (Target B), imprecise and inaccurate (Target C), and imprecise and accurate (Target D).

At first, Target (D) may seem unlikely, but it is not uncommon. The difference between Targets (B) and (D) is simply that (D) has more "spread" in the data. For example, the variance and standard deviation of (D) are much larger than those of (B). However, their measures of central tendency, that is, the means, are nearly the same. So, both data sets are giving us the right answer, but almost all the data points in (B) are near the true value. None of the data points in (D) are near the true value, but the mean (average location on the target) is near the center of the bullseye, so it has the same accuracy as Target (B) but with much less precision. To have a credible enterprise, the precision and accuracy of the facts used as a basis for a decision must be known [39].

2.3 Codifying morality

All professions and most occupations have standards of behavior, but not all are codified. Most engineers, physicians, and attorneys have codified norms, that is, ethical canons and codes. According to their codes of ethics and practice, engineers must hold paramount the public's safety, health, and welfare [40]. Engineers apply the sciences to address societal needs. Environmental engineers are particularly interested in protecting public health and ecosystems. Engineers, scientists, and managers employ techniques to determine whether an action is "good" or "not so good." Or, it may be better stated as "good versus good" [41]. Most scientific decisions are not "good versus bad", but a choice between one benefit and cost ratio to another such ratio. There are also tradeoffs, such as less freedom for a farmer who wants to apply a pesticide that is effective in eliminating weeds to improve crop harvests, but the pesticide may be banned for public health and environmental reasons. As such, the health and environmental decision tools used to support a regulatory action have inherent societal conflicts and weaknesses.

Certainly, decisions and actions that unnecessarily expose people and ecosystems to toxic substances, that is, bad environmental quality decisions, must be avoided. Mindless and misapplied science and technology are immoral. It is most likely that ethical breaches related to air pollution are predominantly unintentional. However, willful dishonest malpractice is rare; yet, unfortunately, it does occur. Often, the breach is not related to science or engineering per se but related to an administrative, financial, or other ancillary aspect of environmental quality management, for example, inappropriate billable hours (e.g., double billing), violations of contract specifications, and inappropriate disclosures. These often fall under the fundamental canon that requires engineers to act as faithful agents and trustees [40].

Engineers and scientists may believe that if the work is sufficiently important, for example, removing carcinogens from the environment, that they will be excused if they cut corners or otherwise violate an ethical principle. After all, they are results-oriented and may succumb to the fallacy that ends justify the means, even unethical means. Indeed, the author has observed the woes of very competent scientists who believe that the primacy of science will inoculate them from ethical misconduct. It certainly does not.

Even the choice of tool or model is laden with ethics. For example, an engineer may have used a model so frequently in the past that little or no thought goes into whether the model is appropriate for a given situation. The engineer has lost objectivity in the model selection process due to confirmation bias, that is, only choosing methods or concepts that agree with one's own perspective, avoiding those that may be better or more appropriate but do not follow one's previously conceived paradigm. Engineers belong to a helping profession, trying to achieve outcomes that serve humankind by altering some aspect of the natural or built environment. However, the way this is done can lead to problems. Sometimes, engineers, scientists, and managers become practically myopic, solving an air pollution problem but, in the process, causing other, potentially worse problems. Examples include removing fly ash and sulfur dioxide from flue gases but myopically storing the sludge and ash in pits that contaminate surface and groundwater; requiring additives to fuels to reduce the emissions of carbon monoxide, but which themselves become air and water pollutants; and introducing cutting-edge treatment technologies that rely on and diminish the availability of finitely available elements that are needed elsewhere by society, for example, in medical devices. As the philosopher Georg Wilhelm Friedrich Hegel put it, "The essential tragic fact is not so much the war of good with evil as the war of good with good" [42].

In addition to "good versus good," there are also the examples of "good versus good enough." Environmental scientists are challenged with the question, "How clean is clean?" Perhaps the colloquial question, "Are we done yet?" is more appropriate.

Everything in this book aims to improve awareness in technological decision making, but some approaches are better than others in terms of credible science and feasibility. Indeed, the famous engineer, Norman Augustine, has stated that "…Engineers who make bad decisions don't know that they are confronting ethical issues" [43].

Of course, technical decisions must be underpinned by sound science and engineering. The public must be able to that trust practitioners are using the best available information and appropriate methods. Unfortunately, this trust is put into jeopardy when practitioners cut corners and otherwise do not adhere to the ethical standards. The recent, and arguably most notorious example of this breach of trust is "Dieselgate," in which Volkswagen engineers and managers were accused in 2015 of intentionally programming turbocharged direct injection diesel engines to activate their emissions controls only during laboratory emissions testing, which caused the vehicles' oxides of nitrogen (NO_x) output to meet US standards during regulatory testing but emit up to 40 times more NO_x in everyday driving [44]. Volkswagen personnel are said to have deployed this programming software in about eleven million cars worldwide in model years 2009 through 2015.

2.4 Responsible conduct of research

Researchers and practitioners must balance the advancement of the technologies with shortcomings before and after an emerging technology is released [45]. This means that while the technology is operational, all negative aspects must be prevented and when a problem occurs, the technologist and owner are transparent in notifying the users and the public of the problem and the means of rectifying it. This is difficult for many emerging technologies, given that predictions are often determined from extrapolations of limited datasets using models with varying amounts of uncertainty [46].

Indeed, this ethical dilemma is more common for journalists and anthropologists than for biologists, engineers, and biomedical researchers. Journalism and anthropology expect researchers to be thrust into the lives of those they are studying. Commonly, however, they are asked to drop the notepad or camera and to help the objects of their research. Researchers in the physical and natural sciences are trained to maintain objectivity. Conversely, engineers and physicians are trained to hold paramount the safety, health, and welfare of the public and patients, respectively. This leads to an ethical problem when maintaining objectivity in some way detracts from the paramount calling of these professions.

An example of this dilemma occurred in the 1990's when the U.S. Environmental Protection Agency's Environmental Monitoring and Assessment Program (EMAP) was building a network of monitoring sites throughout the contiguous US. The sites were designed to be sited objectively and confidentially. However, the question arose as to whether the scientist responsible for choosing and setting up the monitoring site should report any pollution, especially illegal activities like pesticide misuse at the site. Several scientists thought it better to ignore these activities because it would bias the siting and confidentiality of the data. However, this point of view was overruled because of the primacy of law and environmental protection. This is a case where strict adherence to the scientific method is subordinate and inferior to other societal needs.

2.5 Truth

> Tragically, the AI impersonation fraud is part of a much larger and much longer lasting undermining of truth. At one extreme are those individuals who speak loftily of "my truth", as if it were private property, exempt from challenge of facts or logic. But a privately owned truth is irrelevant to communication between people. More important are whole institutions – including education and the news media – whose basic reason for existing is to convey truth, but who cannot resist the temptation to seek power instead. ***Thomas Sowell [47]***

At a workshop of the National Academy of Engineering, John Ahearne, who served as Executive Director of Sigma Xi, the Scientific Research Society, warned that scientists should never be arguing about the facts. Facts simply exist. Their meanings can certainly be debated, and attention given to their various interpretations, but facts must not be ignored or changed simply to fit an ideology or paradigm. The facts may make some or many uncomfortable, but objectivity is an inviolable aspect of sound science and reason. During a professional ethics class, a debate arose among a few environmental science students at Duke regarding the ethics of the findings of a journal article regarding sea turtles. Some students sympathized with Stephen Schneider, a Stanford University climatologist and former government advisor, who argued some years ago that scientists ought to "offer up scary scenarios, make simplified dramatic statements, and make little mention of any doubts we may have. Each of us must decide what the right balance is between being effective and being honest" [48]. Ahearne argues that a distorted version of "utilitarian science" is dangerous because it violates a first canon of science, that is, honest inquiry and reporting. Philosopher of science, C.P. Snow articulated this principle some 50 years ago [49]:

> The only ethical principle which has made science possible is that the truth shall be told all the time. If we do not penalise false statements made in error, we open up the way, don't you see, for false statements by intention. And of course a false statement of fact made deliberately, is the most serious crime a scientist can commit.

Not every scientist and technical professional agrees with Snow. The debate that precipitated the Duke students' quandary was actually over the seemingly innocuous field of biological taxonomy. There was a surprising acceptance by some of the justification for using morally unacceptable means to achieve the greater good.[4] The journal *Conservation Biology* published several scientific, philosophical, and ethical perspectives on whether to misuse science to promote the larger goal (ecological conservation) to protect the black sea turtle. Even though the taxonomy is scientifically incorrect, that is, the black sea turtle is not a unique species, some writers called for a "geopolitical taxonomy" [51–55]. The analogy of war was invoked as a justification, with one writer declaring that "it is acceptable to tell lies to deceive the enemy." The debate moderators asked a telling question: "Should legitimate scientific results then be withheld, modified, or 'spun' to serve conservation goals?" Continuing with the war analogy, some scientists likened the deceptive taxonomy to propaganda needed to prevent advances by the enemy. The problem is that, as Snow would put it, once you stop telling the truth, you have lost credibility as scientists, even if the deception is for a noble cause [51]. Two writers, Kristin Shrader-Frechette and Earl D. McCoy, emphasized that credible science requires that "... in virtually all cases in professional ethics, the public has the right to know the truth when human or environmental welfare is at issue" [54].

Aristotle tried to clarify the dichotomy of good and evil by devising lists of virtues and vices, which amount to a taxonomy of good and evil. One of the many achievements of Aristotle was his keen insight into the similarities among various kinds of living things. He categorized organisms into two kingdoms: plants and animals. Others no doubt made such observations, but Aristotle documented them. He formalized and systematized this taxonomy. Such a taxonomic perspective also found its way into Aristotle's moral philosophy.

We all will not agree on which of the virtues and vices are best or even whether something is a virtue or a vice (e.g., loyalty), but one concept does seem to come to the fore in most major religions and moral philosophies: empathy. Putting oneself in another's situation is a good metric for virtuous acts. The "Golden Rule" is at the heart of Immanuel Kant's "categorical imperative." With apologies to Kant, here is a simplified way to describe the categorical imperative: When deciding whether to act in a certain way, ask if your action (or inaction) will make for a better world if all others in your situation acted in the same way. An individual action's virtue or vice is seen in a comprehensive manner. It is not whether one should cut corners and use deficient materials when designing a device, it is whether everyone in the same situation should do likewise. A corollary to this concept is what our colleague Elizabeth Kiss of Duke's Kenan Center for Ethics calls the "Six O'clock News" imperative. That is, when deciding whether an action is ethical or not, consider how my friends and family would feel if they heard about all of its details on tonight's TV news. That may cause me to consider more fully the possible externalities and consequences of one's decision (and maybe even tempt me to "over-design").

[4] The principal source for this discussion is: [50].

2.6 Psychological aspects of ethics

Technologists are familiar with conditions and constraints. For example, we know from chaos theory the importance of initial conditions on outcomes. However, outcomes can be changed by boundary conditions and constraints imposed by the technologist's interventions. Chaos concepts can also apply to moral decision-making. At a most basic level, a person's path toward moral decision-making is a continuum. A person's conscience influences behavior.[5] A conscience develops rather rapidly in human development and is enhanced constantly throughout life. As a child learns to communicate, especially with the first caregivers, the conscience is controlled and informed by external sources. Humans learn to do what they are told by an authority figure or learn by watching and imitating others. As they mature, however, the conscience, to a greater extent, is controlled internally. A mature conscience results from internalization. At first, the human baby is completely dependent on outside sources, then grows to become independent, but eventually behaves as interdependent members of a community.

> Another doctrine repugnant to Civil Society is, that whatsoever a man does against his Conscience is Sin; For a man's Conscience, and his Judgment is the same thing; and as the Judgment, so also the Conscience may be erroneous; once a man lives in a commonwealth, the Law is the public Conscience, by which he hath already undertaken to be guided. Otherwise in such diversity, as there is of private Consciences, which are but private opinions, the Commonwealth must needs be distracted, and no man dare to obey the Sovereign Power, farther than it shall seem good in his own eyes.... ***Thomas Hobbes (1651) [58]***

Individual, personal conscience is the building block for what the 17th-century natural philosopher Thomas Hobbes considered a requirement for "social contract." Hobbes argued that humans are egoistic and self-serving by nature, so an efficient society first instructs its individual members to behave in ways that support the community. Hobbes thought that society had to overcome the individual person's "state of nature" with the selfish, brutish individual desires that would lead a community to anarchy. This argument sees the role of conscience as twofold: to benefit the individual member (private conscience) and to benefit society as a whole (the commonwealth).

2.7 Thought experiments

Throughout this book, ethical questions relevant to the technologies are considered, especially those that are new and emergent. Few, if any, of these questions will be answered to even a modicum of consensus. That is the nature of complex problems with many unknowns. However, the very process of inquiry may help to give context and structure to some otherwise amorphous issues. In this chapter, we will take the Socratic approach and

[5] Arguably, most engineers lack a substantial amount of formal training in psychology and the behavioral sciences, so no previous background in these areas is needed for this text. However, I have observed many technical people who have substantial "people skills." Also, most have gained knowledge and have read extensively in these areas after their baccalaureate education. An interesting and valuable change in engineering education has been greater emphasis on the student's grasp of the social sciences and the humanities. See: [56,57].

ask important questions. Indeed, I believe Socrates thought the questions were usually more important than the answers to such questions [59]. So, let us begin with an essential query.

2.7.1 When is an action morally permissible, and what kinds of behavior are morally obligatory?

It is a balmy Saturday morning in the North Carolina Piedmont, where a hundred or so new graduate students were gathered in the auditorium of Duke University's Old Chemistry Building. It was orientation week in late August, when Duke requires that every Ph.D.-seeking graduate student participate in a full day of research ethics training.

The Responsible Conduct of Research (RCR) training program developed by the Graduate School had inauspicious beginnings. A few years earlier, Duke, like other prominent research universities, risked losing funding from the National Institutes of Health (NIH) because their research programs could not demonstrate that their researchers were receiving adequate training in research ethics. So, as is often the case, Duke did not simply respond to the criticism and threat by instituting a *pro forma* RCR program to satisfy the core research ethics elements to be reviewed by the NIH auditors but decided to use RCR as a "teachable moment" and galvanizing event. Among the creative approaches to be adopted was to engage the students as soon as they arrived in Durham. Part of this commitment was to invite provocative speakers to evoke responses and to stir the imagination of the students. This first gathering was hoped to instill in each Ph.D. student a self-directive, proactive, life-long ethos of responsible research. To begin this journey on that Saturday, the students were asked to participate in two thought experiments.

In this vein, rather than try the familiar ethical analytical approach of tediously (and often boringly) delineating, point by point, each possible ethical aspect of science and engineering, a tool often applied by philosophers was introduced, that is, the thought experiment.

Thought experiments are imagined scenarios that provide structure to explore ideas, usually in the absence of sufficient experiential information [60]. After the students had conducted a couple of projects and were given some advice on how to avoid ethical dilemmas, Duke University philosophy professor, Alex Rosenberg,[6] presented the following thought experiment that was introduced by J.J. Thomson [61]:

Thought Experiment 1:

You are riding in a trolley car in San Francisco. You are the only passenger. Being the inquisitive type that you are, you watch the trolley driver's methods of controlling speed, changing tracks, and other aspects of handling the machinery. You ascend a hill, arrive at the top, and begin your descent. To your surprise and horror, during the downhill acceleration, the trolley driver jumps from the car, leaving you alone.

Being a take-charge person, you move to the driver's seat and quickly find out why the driver jettisoned. The brakes have completely failed. The driver must have had a strong understanding of the laws of motion, since he jumped just before the car accelerated to the velocity at which anyone jumping from the car would suffer mortal injuries. In other words, the trolley passed that point of nonfatal injury to the jumper. You note that the failure was a complete electrical malfunction, so that the horn and warning devices have also become inoperative. The only remaining control available to you is the mechanical lever that allows the driver to change tracks at a junction. The car continues to accelerate until you see on the track in front of you four workers (all the same size, sex, age, and demographics) who are standing on the tracks with their backs to you. There is no way that they can hear you

[6] I have modified Rosenberg's and Thomson's thought experiments a bit to make them more specifically relevant to technological ethics, but the results reported are those observed in the RCR workshop, August 2004.

due to the noisy equipment around them and the fact that your horn does not work. You also happen to notice that there is a switch track in front of you, just behind the workers, that allows you to change tracks. However, a single worker (of the same size, sex, age, and demographic as the other four workers) is standing on the track in front of you to which you can switch.
The students are asked, "What do you do?"

Your ethical decision consists of only two choices. As an ethicist might ask, what is *morally permissible?* What is *morally obligatory?* At least for most professions like medicine and engineering, the terms differ in meaning [62,63]. A morally permissible action or decision is one that is acceptable since it does not violate an ethical standard, within the boundaries of what is allowed by ethical principles, duties, or codes of conduct, but not necessarily required. A morally permissible action is the lowest level for moral behavior, defined by external standards. In contrast, moral obligation goes beyond the minimal permissibility and is driven by conscience and empathy [63].

The contrast between moral obligation is humorously demonstrated in the movie, *Office Space* [64]. Joanna, played by Jennifer Aniston, is passive-aggressively reprimanded by Stan (played by Mike Judge), the manager of Chotchie's restaurant, for not wearing enough "flair." According to Stan, Joanna meets the permissible standard of 15 pieces of flair but should go beyond it, like her obnoxious coworker, Brian (played by Todd Duffey), who wears 37 pieces. In Stan's mind, Joanna meets the morally permissible standard but not the morally obligatory standard, because a dedicated waitstaff member "should" wear more than the minimum amount of flair. Joanna disagrees and quits.

Back to the thought experiment: The group of students unanimously declared that switching the tracks was morally permissible since you would kill one person to save four. Most also agreed that you were morally obliged (not just permitted) to take the step.

The original version of this thought experiment was conceived by philosopher Philippa Ruth Foot [65–67] (born 1926). She is one of the leaders of the contemporary movement in virtue ethics. Virtue ethics centers around what a person should become, that is, the goal of ethics is *eudaimonia*, the Aristotelian concept of "success." The 20th-century movement is known as the *aretaic* turn, from the Greek term *arête*, meaning "excellence." This contrasts with utilitarian and other "consequentialist" ethical models, where ethics is predominantly determined by outcome. Rosenberg then offered a second thought experiment:

Thought Experiment 2:
You are a respected biomedical engineering major who has gone on to become a world-renowned surgeon at St. Bob the Scientist Hospital in San Francisco. You have kept up your skills in engineering and blended them with your surgical expertise to design a transplant device, based on nanotechnologies and artificial intelligence, that allows you to transplant any organ from one person to another in ten minutes with absolutely no rejection (Hey, it's a thought experiment, and as we have learned from philosophers and theoretical physicists, we can assume this). The transplant device is up and running at St. Bob's.
You are also a charitable physician. In fact, the reason you chose St. Bob's is that it has two distinct wings. Wing 1 is a world-class hospital that treats the rich and famous. Wing 2 is a public clinic that allows anyone to walk in and receive treatment. You practice in both wings, including dedicating *pro bono* services in Wing 2 every Tuesday from 9:00 p.m. to 1:00 a.m. Wednesday. Many of the patients you see in Wing 2 are completely destitute and have no family or friends. They are loners. Recently, a Wing 2 patient was brought in to see you from a local homeless shelter. You find in your interview with him that he has no family, no job, and no interest in bettering himself. You gave him a checkup and found him to be in excellent physical health, including a check of his lungs, liver, and pancreas.

Earlier in the evening, in Wing 1, you saw several patients. Four of these patients were in dire need of transplants. One needed a pancreas, one a liver, and two needed lungs. In fact, if they do not receive them by midnight, all four will die. Your transplant machine could easily save them if the organs were available from a healthy donor, such as your Wing 2 patient.
What should you do? What is morally permissible in this case?

The Ph.D. students were flummoxed. Almost all said that it is morally *impermissible* to take the organs from the one person (and in the process kill him) to save the four. But most could not say why or put their finger on the difference between the two scenarios.

Both situations involve an interface of technology with ethics. The ethical decision includes a crude technology in the first (the level and switch track) and an advanced technology in the second (the nanotechnology-AI-based transplant machine). In both, we *can* change the situation, but the bioethical question is, *"Should we?"*

2.8 Trust

The students' dilemma invites another moral question.

2.8.1 What must a technologist do to be trusted?

On their face, the previous cases are identical questions about what is moral. That is, is one permitted to kill one to save four? So, what is the big difference in the decision-making? If we are talking about a decision about anything but life, the question is almost ridiculous. When the decision involves a human being, arithmetic is an inadequate metric. Even when other organisms, like pets, there must be more than numbers to trigger a decision. Same goes for ecosystems. So, when it comes to optimized decisions for living things, as in bioethics, such questions are often complicated and debatable. Arguably, the difference basically comes down to *trust*.

Exercise

Teachable moment: trust

Consider the people whom you can trust. The reason for trust can be based on *intrinsic* qualities, such as to your knowledge, they have never lied to you or are there every time you need them. The reason for your trust can also be based on *extrinsic* constraints, such as they are duty-bound by law or must adhere to a professional standard (e.g., a physician, architect, or engineer). Often, they are combinations of the two (e.g., a trusted family doctor or a structural engineer you have hired in the past).

Explain how such trust applies to the practice of engineering and the applications of emerging technologies, like robotics in surgery. Discuss the types and degrees of controls that the engineering profession has over its members (extrinsic controls), and how much the social contract with engineering depends on characteristics of the individual engineer (intrinsic controls).

2.8.2 Credat emptor

The thought experiments demonstrate, at least at some intuitive level, that trust is one of the distinguishing attributes of the technologist. Arguably, the engineer's principal client is the public. As with any professional, clients need know little about the practice of engineering because, owing to the expertise, they have delegated the authority to the professional. For example, society allows a patient to undergo brain surgery even if that person has no understanding of the fundamentals of brain surgery. Likewise, society cedes authority to professional engineers (PE) for design decisions. With that authority comes a commensurate amount of responsibility, and when things go wrong, culpability. The first canon of most engineering ethical codes requires that we "hold paramount" the health and welfare of the public. The public is an aggregate, not an "average." So, leaving out any segment violates this credo. Thus, even though most people favor an approach, it is still up to the engineer to ensure that any recommended approach is scientifically sound. In other words, no design, even a very popular one, is to be recommended unless it meets acceptable, scientifically sound standards. Conversely, the most scientifically sound approach may have unacceptable social side effects. Once again, the engineer is put in the position of balancing tradeoffs.

Often, texts, manuals, and handbooks are valuable, but only when experience and good listening skills are added to the mix can wise (and ethical) decisions be made. First-century thinking linked maturity to "self-control" or "temperance" (Greek *kratos* for "strength"). St. Peter, for example, considered knowledge a prerequisite for temperance. Thus from a professional point of view, we can extend this reasoning to say that one can really only understand and appropriately apply scientific theory and principles after practicing them for a sufficient number of trials and with time.[7] This is the structure required by most professions. For example, engineers who intend to practice must first submit to a rigorous curriculum approved and accredited by the Accreditation Board for Engineering and Technology, then must sit for the Future Engineers (FE) examination. After some years of professional tutelage by and intellectual osmosis with seasoned, licensed professionals, the engineer has demonstrated the *kratos* (strength) to sit for the PE exam. Only after the future engineer successfully passes the PE exam does the State license the engineer and the National Society for Professional Engineering certify that the engineer is a "professional engineer" and eligible to use the initials PE after one's name. The professional is now schooled beyond textbook knowledge and knows more about why many problems cannot be generalized. In professionalism as in ethics, one size does not fit all. In fact, like the engineer, any mentored technical professional must have humility [68–70], beginning every new technology with the knowledge and self-awareness that one does not know everything needed to be known. To paraphrase Dirty Harry Callahan (played by Clint Eastwood), "a technologist has to know his limitations" [71].

[7] I realize that St. Peter was referring to one's spirituality, but anyone who has even casually studied his life would see that he fully integrated the physical and spiritual person.

2.8.3 The principle of double effect

It is tempting to begin an immediate analysis of the students' responses for right and wrong elements. That is the typical case-based approach. Philosophers and ethicists commonly apply casuistry [72,73], an approach to determine right and wrong decisions or behavior by analyzing cases that illustrate general ethical rules. Among the problems with this approach is that it forces one to judge. Engineers and technologists, on the other hand, are problem solvers and designers, not judges. Thus they are more comfortable building a solution than to sit in judgment. However, to prime the pump, let us begin to think about why the students responded as they did.

The first thought experiment introduces us to an important consideration in ethics, the so-called "principle of double effect." Think of this as an equation with two constraints. The first constraint stems from Socrates' moral challenge to "first do no harm." The second constraint was first fully articulated by the 13th-century Italian theologian, Saint Thomas Aquinas [74]:

> "Good is to be done and promoted and evil is to be avoided."

In its simplest form, the doctrine says that an act that leads to negative side effects is permitted, but deliberate harm, even for good causes, is wrong. This seems similar in both scenarios, but the second thought experiment involves actively harming one person, whereas the first thought experiment involves harm as a side effect. In the trolley case, you are just redirecting the harm. However, in the second case, you have to do something to the homeless man to save the four. In the first case, no person has any more rights than anyone else not to be crushed, but in the second case, the homeless man has a right not to be killed, even though his organs would be put to good use.

These distinctions among professional decisions point to the fact that engineers, to some degree during their careers, will face risk tradeoffs and the double effects. For example, in designing a device and introducing a drug, some persons (e.g., immuno-compromised) may be harmed by its use. If every member of a certain group is harmed at the expense of another group's benefit, this could arguably be conceived as intentional harm. The way to mitigate this harm is to fully disclose the shortcomings of the device and to work on improvements that will decrease the likelihood of harm. The device must do inherent good, for example, provide insulin to diabetics or deliver drugs to ill patients. In other words, the designer must not actually intend to accomplish the bad effect (harming immuno-compromised people). The ill effect is simply unavoidable in the effort to do good. If another approach would avoid the bad and still accomplish the good, then such an option is the preferred and obligatory act.

Another provision of the double effect is that the good effect must be at least as directly an effect of the action as the negative effect. In particular, the bad must not *cause* the good effect. The benefit of the bad must not outweigh the benefit of the good. Finally, another ethically acceptable approach without the side effects must not be available.

A bioethical example of the double effect is that of the vaccine. A government or manufacturer normally tries to know the population risk of administering a vaccine to the public prior to the vaccine's release. Most are expected to benefit, with a small number experiencing adverse side effects. From this small group, a subset of vaccine recipients will die. The lives are saved because of the administration of the vaccine, not as a result of the deaths of those who die from side effects. The side effects must not advance any goals of the drug manufacturer. Thus the side effects are not intended as a means to any other outcome. Finally, the proportion of lives saved must be much larger compared to the harm the vaccine incurs, for example, lives lost. If so, this satisfies the requirement that benefits outweigh the negative outcomes. It would not be permissible to produce and administer the vaccine with these side effects (deaths) if another better means of preventing the disease is available.

Ideally, any decision to implement a large-scale action should be based on a thorough understanding of the benefits and risks. Projects should only be initiated when there is sufficient evidence that the potential benefits outweigh the risks, and that developing the technology will not present serious harm [75]. Unfortunately, there is substantial temporal and spatial uncertainty about risks of most emerging technologies, due to their novelty and complexity. Being emergent, there are no direct measures of a technology's safety and utility. Large uncertainties about their potential success and little-known impacts on human health, the environment, and the economy make the risks difficult to predict or the ways to manage any changes at the outset. Even if there is confidence in high reward and low risk, ethics goes beyond risk assessment to include a wide array of societal needs and expectations [76–78].

Because the benefits and risks of most emerging technologies are uncertain at this point, a precautionary approach is warranted. That is, the onus should be that for any action with the potential for severe and irreversible consequences, the technologist must provide adequate evidence of its safety. This evidence may begin with smaller-scale, lower-risk projects preceding increasingly larger ones.

Traditional cost-benefit analyses have not been conducted at the global scale. Even for relatively small-scale projects, these analyses often cannot quantify social scientific factors, such as political and economic variables.

2.9 Values

A person's conscience drives that person's values. Similarly, the collective conscience of a group codifies their values, for example, a professional association. At the community level, values are codified into laws and regulations. However, society seldom has complete consensus on many moral issues. For example, much debate has occurred within the legal community as to whether the use of lethal injection to execute convicted criminals is "cruel and unusual punishment." The debate includes positions across a wide spectrum, ranging from lethal injection being morally reprehensible to its being morally obligatory. The US Constitution prohibits cruel and unusual punishment, so it is not surprising that the debate reached the United States Supreme Court. In the spring of 2006, the Court refused to examine this issue, but a large controversy was brewing. A central, ethical issue is concerned with the two-step system of injection, which may cause extreme pain if the first chemical, an anesthetic, is not properly injected. Thus when the actual lethal dose of the second chemical is injected, the resulting pain puts the state in the position of a cruel executioner.

Perhaps another thought experiment will help us consider the complexity of some values debates [79]:

Thought Experiment:

Dante is a chemical engineer employed by InfernoChem, Inc. (ICI). Dante's expertise is in small-scale reactors. Recently, he developed a manifold system that allows two different chemicals to be injected at specific preset times. His device is particularly useful for injecting catalysts into reactors. For example, a polymerization step in one of the reactors requires that the first catalyst be injected when the reactor temperature is 50°C and the second catalyst be injected 10 minutes later when the reactor temperature reaches 80°C. For the past 18 months, ICI has required that Dante continuously miniaturize the two-step injection system. In fact, the company awarded Dante two monetary awards for his progress. Recently, the system was reduced in fluid volume to 25 cubic centimeters.

Dante's perspective is that of a devout Roman Catholic. He opposes capital punishment for religious and personal reasons. He briefly mentioned these convictions to Tom, his immediate supervisor, on two occasions after executions were covered in the local news. Dante discussed this in a very matter of fact and dispassionate way.

Tom is aware that the reason for the short deadline to miniaturize Dante's injection system is that ICI has a contract with the State Bureau of Prisons to produce a single injection system that would ensure prisoners receiving lethal injections receive the entire dose of the anesthetic drug before the toxic dose is administered. However, based on Dante's opposition to the death penalty, Tom has chosen not to disclose the actual application of the new technology.

Was Tom morally permitted to withhold the company's plans from Dante? What is the appropriate thing for Dante to do?

Exercise

Capital punishment, abortion, and the definition of human life

The American Medical Association (AMA) policy on abortion is:

The Principles of Medical Ethics of the AMA do not prohibit a physician from performing an abortion in accordance with good medical practice and under circumstances that do not violate the law. (III, IV) [80].

And

... (1) abortion is a medical procedure and should be performed only by a duly licensed physician and surgeon in conformance with standards of good medical practice and the Medical Practice Act of his state; and (2) no physician or other professional personnel shall be required to perform an act violative of good medical judgment. Neither physician, hospital, nor hospital personnel shall be required to perform any act violative of personally held moral principles. In these circumstances, good medical practice requires only that the physician or other professional withdraw from the case, so long as the withdrawal is consistent with good medical practice. ("H-5.995 Abortion | AMA")

Compare and contrast this policy with the AMA policy on capital punishment:

An individual's opinion on capital punishment is the personal moral decision of the individual. A physician, as a member of a profession dedicated to preserving life when there is hope of doing so, should not be a participant in a legally authorized execution. Physician participation in execution is defined generally as actions which would fall into one or more of the following categories: (1) an action which would directly cause the death of the condemned; (2) an action which would assist, supervise, or contribute to the ability of another individual to directly cause the death of the condemned; (3) an action which could automatically cause an execution to be carried out on a condemned prisoner.

Compare this policy to an engineering code of ethics. For example, the civil engineering codes of ethics include language similar to

(cont'd)

holding paramount the safety, health, and welfare of the public. Identify commonalities and any inconsistencies between the two policies, especially the differences between the engineer's paramount focus on the public good and the physician's focus on the patient.

In particular, consider that for lethal injection executions, the AMA prohibits a physician from aiding the execution by:

- selecting injection sites.
- starting intravenous lines as a port for a lethal injection device.
- prescribing, preparing, administering, or supervising injection drugs or their doses or types; inspecting, testing, or maintaining lethal injection devices; and,
- consulting with or supervising lethal injection personnel.

Some have argued that, in an abortion, the doctor does not "kill" the unborn child intentionally but makes the "choice" to evict the child from the womb. This position argues that the unborn child's right to life does not "entail that the child *in utero* is morally entitled to the use of the mother's body for life support." Of course, such strained logic rests on the conclusion that the unborn child is neither living nor human and, as such, is undeserving of protection and life preservation.

Questions:

1. Compare the technical and moral differences and similarities between the prohibitions against participating in an execution and the applications of the technologies of abortion.
2. Is the child *in utero* living? Is the child *human*? If so, explain the bioethical distinction between calling abortion an act of killing versus a choice to evict the child.

One of the means of determining whether an ethical decision is sound is to see if we are being morally consistent in our values. For example, some have pointed out inconsistencies in those who support the death penalty, yet ideologically oppose abortion, embryonic stem cell research, euthanasia, and assisted suicide, as well as other acts against humans at the beginning and end of life. The counterargument is that in these other situations, innocents are being killed, while in the case of capital punishment, if due process is followed, the person being executed is treated fairly. Again, it comes down to what is most cherished, what has the most *value*. If human life is sacred, as most believe, however they define it, can it ever become unsacred, as in the case of a convicted murderer?

2.10 Quantifying ethics

Engineers and other technologists often communicate with mathematics, so they may be tempted to quantify concepts that are largely subjective, such as goodness and morality. This book offers some rough ways to do that but with the caveat that ethics does not lend itself well to quantification. Usually, the best that can be done is to use equations as guidance and screening tools for ethical decision-making.

The concept of the felicific calculus was developed by one of the founders of utilitarianism, Jeremy Bentham, to quantify the moral rightness of a decision or action [81–83]. It calculates the total pleasure or utility versus the pain induced by that pleasure or utility. The quantification includes intensity, duration, certainty, propinquity, fecundity, purity, and extent to estimate the greatest good for the greatest number. Intensity is the strength of the pleasure and pain. Duration is the time that they will last. Certainty is the likelihood that the pleasure and pain will last. Propinquity or remoteness is how soon they will occur. Purity is the likelihood that the pain can be avoided. Extent is the number of people who are or may be affected.

Similar to the felicific calculus, the moral importance equation was formulated by Francis Hutcheson as an attempt to quantify morality [84]:

$$M = B \times A$$

where,

M is the moral importance of an action or decision; B is the benevolence of the agent; and A is the ability of the agent

Thus the importance of the decision is based on the goodness and likelihood of the outcome. For example, a public works department chair must decide on a way to handle solid waste in their town.

Action A (lower value): He spends the money on a contract with an out-of-state waste handler to remove the waste out of town.

Action B (higher value): He spends the same amount of money to build a landfill that meets state standards, and implements a sustainable recycling and composting system.

According to the moral importance equation, Action B is more moral because it does not shift the problem to other people and provides a sustainable solution that is completely within the span of control for the town. As such, it provides greater good for a larger number of people.

Other attempts have been made to quantify ethics, including the so-called human equation, introduced by Muhammad Al-Khwarizmi [85], with an algorithm that includes ethics, beauty, wealth, and authority. Another is the "Five P's of Ethical Power", consisting of purpose, pride, patience, persistence, and perspective [86].

Any of these tools can be useful in ethical optimization, but with the caution that the results are highly subjective and qualitative.

2.11 Conclusions

This chapter introduces several ethical constructs from among the myriad systems proposed by philosophers, ethicists, and scientists over the millennia. This sets the stage for engineering ethics, which is more directly related to moral decisions as they relate to emerging technologies.

References

[1] Emily Dickinson, The Brain – is wider than the sky (1862).
[2] S.K. Pfatteicher, "A New Era: The Limits of Engineering Expertise in a Post-9/11 World," in: Proceedings of the 2007 IEEE International Symposium on Technology and Society, Institute of Electrical and Electronics Engineers, 2007, pp. 1–4.
[3] R. Prieto, "The Three R's: Lessons Learned from September 11, 2001," in: Terrorism: Reducing Vulnerabilities and Improving Responses – US-Russian Workshop Proceedings, National Research Council, 2004.
[4] National Science Foundation, Responsible conduct of research, Federal Register vol. 74, (no. 160) (2009) 42126–42128 20 August 2009.
[5] L. Myllyaho, M. Raatikainen, T. Männistö, J.K. Nurminen, T. Mikkonen, On misbehaviour and fault tolerance in machine learning systems, Journal of Systems and Software vol. 183, (2022) 111096.
[6] L. Bouter, Research misconduct and questionable research practices form a continuum, Accountability in Research vol. 31, (no. 8) (2024) 1255–1259.
[7] D.B. Resnik, L.M. Rasmussen, G.E. Kissling, An international study of research misconduct policies, Accountability in Research vol. 22, (no. 5) (2015) 249–266.
[8] R. Wolniak, Engineering ethics-main principles, Zeszyty Naukowe. Organizacja i Zarządzanie/Politechnika Śląska (2022).
[9] A. Gare, Natural philosophy and the sciences: challenging science's tunnel vision, Philosophies vol. 3, (no. 4) (2018) 33.
[10] T.G. Chondros, Natural philosophy and the development of mechanics and engineering from the 5th century bc to middle-ages, FME Transactions vol. 45, (no. 4) (2017).
[11] R. French, A. Cunningham, Before Science: The Invention of the Friars' Natural Philosophy, Routledge, (2016).
[12] D. Neves, I.L. Salazar, R.D. Almeida, R.M. Silva, Molecular mechanisms of ischemia and glutamate excitotoxicity, Life Sciences (2023) 121814.
[13] M. Silva, Know how and cultural context: From ideas to facts and from facts to ideas, Modernity, Frontiers and Revolutions, CRC Press, (2018), pp. 3–10.
[14] P.D. Omodeo, The engineer and the philosopher, Rethinking Stevin, Stevin Rethinking (2021) 25.
[15] D.F. Channell, The emergence of the engineering sciences: an historical analysis, Philosophy of Technology and Engineering Sciences, Elsevier, (2009), pp. 117–154.
[16] P.D. Omodeo, Practices and theories of contingency in Renaissance approaches to nature, Contingency and Natural Order in Early Modern Science, Springer, (2019), pp. 93–113.
[17] W.R. Laird, The scope of renaissance mechanics, Osiris vol. 2, (1986) 43–68.
[18] A. Blair, Natural Philosophy and the'New Science' (1999).
[19] A. Aristotle, The Ethics of Aristotle, BoD–Books on Demand, (2019).
[20] Aristotle, Aristotle. Collector's Library (2009).
[21] S.N. Grovogui, Time, technology, and the imperial eye: Perdition on the road to redemption in international relations theory, Time, Temporality and Violence in International Relations, Routledge, (2016), pp. 45–60.
[22] G.H. Mead, Scientific method and the moral sciences, The International Journal of Ethics vol. 33, (no. 3) (1923) 229–247.
[23] S. Khemlani, P.N. Johnson-Laird, Theories of the syllogism: a meta-analysis, Psychological Bulletin vol. 138, (no. 3) (2012) 427.
[24] M. Bucciarelli, P.N. Johnson-Laird, Strategies in syllogistic reasoning, Cognitive Science vol. 23, (no. 3) (1999) 247–303.
[25] B.G. Bara, M. Bucciarelli, P.N. Johnson-Laird, Development of syllogistic reasoning, The American Journal of Psychology (1995) 157–193.
[26] J.S.B. Evans, J.L. Barston, P. Pollard, On the conflict between logic and belief in syllogistic reasoning, Memory & Cognition vol. 11, (no. 3) (1983) 295–306.
[27] T.J. Smiley, What is a syllogism? Journal of Philosophical Logic (1973) 136–154.
[28] W.D. Ross, The discovery of the syllogism, The Philosophical Review vol. 48, (no. 3) (1939) 251–272.
[29] P. Shorey, The origin of the syllogism, Classical Philology vol. 19, (no. 1) (1924) 1–19.
[30] U. S. Environmental Protection Agency, Guidelines for Ecological Risk Assessment, Washington, DC, EPA/630/R-92/001. Available from: https://www.epa.gov/sites/default/files/2014-11/documents/framework_eco_assessment.pdf, 1998 (accessed 14.05.24).

[31] H. Sharifi Moghadam, S. Alimohammadi, Calibration and uncertainty analysis of freundlich and langmuir isotherms using the markov chain monte carlo (MCMC) approach, Journal of Water and Wastewater; Ab va Fazilab (in persian) vol. 35, (no. 4) (2024) 23–48.

[32] O. Lhotský, et al., The effects of hydraulic/pneumatic fracturing-enhanced remediation (FRAC-IN) at a site contaminated by chlorinated ethenes: a case study, Journal of Hazardous Materials vol. 417, (2021) 125883. Available from https://doi.org/10.1016/j.jhazmat.2021.125883, (2021/09/05/).

[33] U. Sadana, A. Chenreddy, E. Delage, A. Forel, E. Frejinger, T. Vidal, A survey of contextual optimization methods for decision-making under uncertainty, European Journal of Operational Research vol. 320, (no. 2) (2025) 271–289.

[34] G.H. Nielsen, et al., Application of probabilistic methods to address variability and uncertainty in estimating risks for non-cancer health effects, Environmental Health vol. 21, (no. Suppl 1) (2023) 129.

[35] S. Jilke, Measuring technological uncertainty and technological complexity: scale development and an assessment of reliability and validity, International Journal of Innovation Science vol. 13, (no. 3) (2021) 381–400.

[36] I. Scoones, What is Uncertainty and Why Does It Matter? (2019).

[37] P. Abbaszadeh, H. Moradkhani, D.N. Daescu, The quest for model uncertainty quantification: a hybrid ensemble and variational data assimilation framework, Water Resources Research vol. 55, (no. 3) (2019) 2407–2431.

[38] J.B. Roca, P. Vaishnav, M.G. Morgan, J. Mendonça, E. Fuchs, When risks cannot be seen: regulating uncertainty in emerging technologies, Research Policy vol. 46, (no. 7) (2017) 1215–1233.

[39] Elements of a Quality Assurance Project Plan (QAPP) For Collecting, Identifying and Evaluating Existing Scientific Data/Information. Available from: https://www.epa.gov/osa/elements-quality-assurance-project-plan-qapp-collecting-identifying-and-evaluating-existing, 2017..

[40] National Society of Professional Engineers. NSPE Code of Ethics for Engineers. Available from: https://www.nspe.org/resources/ethics/code-ethics, 2016 (accessed 31.08.18).

[41] D. Vallero, Environmental Biotechnology: A Biosystems Approach, Elsevier Science, (2015).

[42] G.W.F. Hegel, A. Paolucci, H. Paolucci, Hegel on Tragedy, Harper & Row, New York and London, (1975).

[43] N.R. Augustine, Ethics and the second law of thermodynamics, Bridge vol. 32, (no. 3) (2002) 4–7.

[44] B. Chappell, It was installed for this purpose, VW's US CEO Tells Congress about Defeat Device, *National Public Radio*, (2015).

[45] D.A. Vallero, This section is excerpted and adapted from: "ethics and air pollution in the 21st century: biomedical and environmental perspectives", Ethics in Biology, Engineering and Medicine: An International Journal vol. 4, (no. 3) (2013).

[46] D.A. Vallero, Ethics and air pollution in the 21 st century: biomedical and environmental perspectives, Ethics in Biology, Engineering and Medicine: An International Journal vol. 4, (no. 3) (2013).

[47] T. Sowell, My experience with artificial intelligence, ed, Wall Street Journal, Dow Jones and Company, New York, NY, (2026), p. A15.

[48] J. Schell, Our fragile earth, October 1989, Discover (1987) 44–50.

[49] C.P. Snow, The Search, Victor Gollancz Ltd, London, UK, (1934).

[50] M.B.R.E.C. Chamberlain, M.L. Gibeau, Human perspectives and conservation of grizzly bears in Banff National Park, Canada, Conservation Biology vol. 26, (no. 3) (2012) 420–431. Available from https://doi.org/10.1111/j.1523-1739.2012.01856.xDigital.

[51] S.A.A.B. Karl, Evolutionary significant units versus geopolitical taxonomy: molecular systematics of an endangered sea turtle (genus Chelonia), B.W, Conservation Biology vol. 13, (no. 5) (1999) 990–999. Available from https://doi.org/10.1046/j.1523-1739.1999.97352.x.

[52] P.C.H. Pritchard, Comments on evolutionary significant units versus geopolitical taxonomy, Conservation Biology vol. 13, (1999) 1000 03.

[53] Q.J. Grady, Using character concordance to define taxonomic and conservation units, JM, Conservation Biology vol. 13, (no. 5) (1999) 1004–1007.

[54] K.A.M. Shrader-Frechette, Molecular systematics, ethics, and biological decision making under uncertainty, E.D. Conservation Biology vol. 13, (no. 5) (1999) 1008–1012.

[55] B.W.B.A.S.A. Karl, In war, truth is the first casualty, Conservation Biology vol. 13, (no. 5) (1999) 1013–1016.

[56] A. Labib, M.R. Abdi, S. Hadleigh-Dunn, M. Yazdani, Evidence-based models to support humanitarian operations and crisis management, Decision Making: Applications in Management and Engineering vol. 5, (no. 1) (2022) 113–134.

[57] J. Donald, S. Lachapelle, T. Sasso, G. Gonzales-Morales, K. Augusto, J. McIsaac, On the place of the humanities and social sciences in the engineering curriculum: a Canadian perspective, Global Journal of Engineering Education vol. 19, (no. 1) (2017) 6–18.
[58] T. Hobbes, Leviathan: Or, the Matter, Form, and Power of a Commonwealth Ecclesiastical and Civil, G. Routledge and sons, (1894).
[59] F. Leal, Every Socratic dialogue is about ethics, Socratic Dialogue and ethics vol. 9, (2005) 42.
[60] R. Cooper, Thought experiments, Metaphilosophy vol. 36, (no. 3) (2005) 328–347.
[61] J.J. Thomson, Killing, Letting Die and the Trolley Problem" and" the Trolley Problem, Harvard University Press, Cambridge, Mass, (1986), pp. 78–116.
[62] M.F. Blanc, Moral permissibility and legitimacy of the use of coercive interrogations: implications for the intelligence professional, participating health-care professionals, and society, International Journal of Intelligence Ethics vol. 1, (no. 1) (2010) 122–135.
[63] J.K. Alexander, Promising, professional obligations, and the refusal to provide service, HEC F vol. 17, (2005) 178.
[64] M. Judge, et al., Office Space. Twentieth Century Fox Home Entertainment (2005).
[65] P. Foot, Virtues and Vices and Other Essays in Moral Philosophy, Oxford University Press, (2002).
[66] P. Foot, Rationality and virtue, Norms, Values, and Society, Springer, (1994), pp. 205–216.
[67] P. Foot, Utilitarianism and the virtues, Mind; A Quarterly Review of Psychology and Philosophy vol. 94, (no. 374) (1985) 196–209.
[68] B. Michalec, The Humility Paradigm and the Introduction of the Professional Humility Concept (2020) ed.
[69] M. Alfano, M.P. Lynch, A. Tanesini, The Routledge Handbook of Philosophy of Humility, Routledge, New York, (2021).
[70] B. Michalec, A. Xyrichis, C. Arenson, "Professional Humility": Introducing a New Framework to Advance Interprofessionalism, Taylor & Francis, (2024), pp. 587–592 vol. 38.
[71] H. Callahan, Magnum Force," *Movie), Hollywood: Warner Brothers Production*, (1973).
[72] C. Ginzburg, L. Biasiori, A historical approach to casuistry, Norms and Exceptions in Comparative Perspective, *Bloomsbury Academic*, Londres-New York, (2019).
[73] J. Boyle, Casuistry, Handbook of Bioethics: Taking Stock of the Field From a Philosophical Perspective, Springer, (2004), pp. 75–88.
[74] T. Aquinas, *Summa theologica*. 1702.
[75] D.A. Vallero, Societal adaptation to climate change, Climate Change, Elsevier, (2021), pp. 799–815.
[76] D.A. Vallero, Chapter 8 - Decision support tools, in: D.A. Vallero (Ed.), Environmental Systems Science, Elsevier, (2021), pp. 293–357.
[77] K. Shrader-Frechette, Risk Analysis and Scientific Method: Methodological and Ethical Problems with Evaluating Societal Hazards, Springer Science & Business Media, (2012).
[78] V.P. Paglioni, Towards ethical risk assessment, Advances in Reliability, *Safety and Security*, (2024).
[79] D.A. Vallero, Biomedical ethics for engineers, The Biomedical Engineering Series (2007).
[80] American Medical Association, "Policy E-2.01 Abortion," ed, 1997.
[81] J. Koh, Bentham and effective Altruism, Revue d'études benthamiennes no. 25 (2024).
[82] A. Baujard, A return to Bentham's felicific calculus: from moral welfarism to technical non-welfarism, The European Journal of the History of Economic Thought vol. 16, (no. 3) (2009) 431–453.
[83] V. Bianchini, The felicific calculus and the art of life according to James and John Stuart Mill, James Mill, John Stuart Mill, and the History of Economic Thought, Routledge, (2023), pp. 45–57.
[84] G. Bullough, The Seventeenth Century: Studies in the History of English Thought and Literature from Bacon to Pope; Milton and Others; Stratford-upon-Avon Studies 6: Restoration Theatre; Dryden's Heroic Drama, Oxford University Press, (1966) ed.
[85] A. Masharrafa, M. Ahmad, Al-Khwārizmī, Muhammad ibn Mūsa. Kitāb al-mukhtasar fī hisāb al-jabr wa'lmuqābala (The compendious book on calculation by completion and balancing), ed: Cairo: al-Qahirah, Reprinted, (1968).
[86] N.A. Obodo, S.C. Ndema, The Five P's Of Ethical Power (2021).

CHAPTER

3

Science ethics

OUTLINE

3.1 Introduction: discovery by design and design by discovery

For the most part, the only facts of interest to science are those I call general, that is, repeatable and reproducible: these form the touchstones of theories and laws. ***R.G. Newton (1997) [1].***

The importance of scientific integrity is key to any ethical decision. This should come as no surprise to a scientist or to anyone engaging in the scientific method. However, it is relatively easy to become so committed to a cause, idea, or opinion that we scientists are tempted and, too often, give in to the temptation to override scientific facts to support an ideology. Like many ethical breaches, this often comes not overnight, but over a long, perhaps decades-long slide. Indeed, in the beginning, the arguments were, in fact, based on scientific evidence, but incrementally took on greater attention to ancillary values, like available funds, popular opinion, and political activism. The warning to those deciding on the ethics of an emerging technology is best articulated by the words of the physicist, Richard Feynman [2]:

The first principle is that you must not fool yourself – and you are the easiest person to fool. So, you have to be very careful about that. After you've not fooled yourself, it's easy not to fool other scientists. You just

Technical and Ethical Optimization: Challenges and Opportunities for the Modern Engineer and Scientist
DOI: https://doi.org/10.1016/B978-0-443-44925-3.00003-X

> have to be honest in a conventional way after that. I would like to add something that's not essential to the science, but something I kind of believe, which is that you should not fool the layman when you're talking as a scientist… I'm talking about a specific, extra type of integrity that is not lying, but bending over backwards to show how you're maybe wrong, that you ought to have when acting as a scientist. And this is our responsibility as scientists, certainly to other scientists, and I think to laymen.

Well said, Professor!

3.1.1 Scientific principles and truth

At its most basic level, every environmental insult and technological disaster can be described by the laws of motion and thermodynamics. The laws of motion describe the mechanics of how things move from one place to another. Thus, the movement of pollutants from a landfill or hazardous waste site can be explained by fluid mechanics. Possibly less obvious, even energy-related phenomena, like the destruction of the ozone layer and global climate change, must account for motion [3,4]. For example, the ozone-destroying chlorine (Cl) atoms find their way to the stratosphere owing to differences in density within the atmosphere and general molecular motion. Greenhouse gases, like carbon dioxide (CO_2) and methane (CH_4), leave their terrestrial sources and mix in the troposphere. Similarly, the loss of habitat due to changes in coastal development, for example, more impervious surfaces, changes the fluid dynamics, diminishing immediate infiltration into the soil that results in greater runoff [5], as well as scouring of sand and other barriers as the ocean reclaims the coastline [6].

The laws of thermodynamics explain that energy and matter must be conserved, that is, every molecule and erg of energy must be accounted for in every system, including the human body and an ecosystem. Thus, the various wavelengths of energy and the mass of greenhouse gases enter arrangements that warm or cool various parts of the atmosphere and ocean. On a smaller scale, the solar energy of summer and fall is absorbed by Earth, air, and water and is converted to mechanical energy manifested in a tornado or hurricane. The breach of a tank results in changes in temperature, pressure, and density relationships, which can lead to leaking of hazardous materials or explosions.

Physical, chemical, and biological laws and principles are necessary components of an environmental root cause analysis but cannot fully explain any environmental problem. Indeed, the physical sciences must be combined with the principles of social sciences. Thus, consideration of the interplay among science, engineering, and technology will begin our discussion of science ethics.

Success and failure in science often begin with the efficiency and effectiveness of a system. Pollution results from combinations of inefficiency and ineffectiveness. The first is a thermodynamics term. The second is a management term. Efficiency is a measure of how well a thermodynamic system is working in terms of the amount of mass and energy taken in by the system versus the amount of mass or energy coming out. In contrast, effectiveness has to do with a particular set of values placed on that system and the extent to which the system performs with respect to providing those values.

A physicist and psychologist may look at the facts associated with an environmental problem and arrive at differing root causes and factors. Indeed, a physicist and a chemist, both physical scientists, may also look at the same problem and disagree on how to describe and give weight to the importance of the events that led to the problem. Some of this difference among scientists has to do with the experts' academic and professional preparation and experiences.

Scientists first generate rules based on observations of the world around them; the way things work. From these rules, the reasons for an event or outcome can be deduced. Unlike most mathematicians, engineers and technologists are not restricted to deductive reasoning but can also apply inductive and intuitive reasoning [7–9]. Inductive reasoning or "abstraction" approaches a problem from the opposite viewpoint, starting with something concrete to form a more abstract ideal. Philosophers have argued for centuries regarding the value of inductive reasoning [10]. Since induction is the process that takes specific facts, findings, or cases and then generally applies them to construct new concepts and ideas, abstraction leaves out specific details, unifying them into a whole based on a defined principle. For example, a brown-feathered goose, a white-feathered goose, and a polka dot-feathered goose can all be integrated because each is a goose, albeit with differences. The feather color, then, can be eliminated under the principle or criterion of being a goose (i.e., "gooseness"). Thus, the color is not relevant in this reasoning. A brown goose, brown bear, and brown paper bag can be integrated under the criterion of having brown color. The other aspects besides "brownness" of each item's characteristics are not relevant in this case, so they are omitted. In the 18th century, the Scottish philosopher, David Hume, postulated the so-called "problem of induction" [9]. To paraphrase, Hume was asking, "Why should things that we may be observing on a regular basis continue to hold in the future?" In other words, there is no justification in using induction because there is no reason that the conclusion of any inductive argument is valid. Like the scientific revolutionaries a couple of centuries earlier, Hume rejected *a priori* reason, since humans are incapable of fully and directly comprehending the laws of nature. This can only be accomplished *a posteriori*, through experience, as articulated by Robert Boyle in the 17th century [11,12].

Arguably, Hume would have considered this inductive syllogism problematic [3]:

> Every time I add the metal nickel (Ni) to my activated sludge, the bacteria grow more rapidly. Therefore, the next time that I add Ni to the sludge, my bacteria's growth rate will increase.

There are many reasons why the Ni addition may not lead to increased growth, such as different strains not adapting to an enzymatic need for Ni, temperature changes that may induce changed behaviors which render the Ni addition ineffective, and incomplete mixing that does not allow the microbes sufficient access to the Ni. Under the regular (expected?) conditions in the treatment plant, the cause-and-effect relationship is repeated without fail. Mathematicians may have a harder time with this expectation, but is it really any different from pressing your brake pedal and expecting the car to stop? The answer is "yes" because there is always a probability (hopefully very low) that a leak in the master cylinder or brake line could cause the hydraulics to fail and the car would not stop when the brake pedal is depressed, but such probabilities do not render, in the author's opinion, inductive reasoning unusable in scientific reasoning.

Intuitive reasoning incorporates deductive and inductive reasoning to build a knowledge base about how things work and under what conditions they fail. This is a very common type of reasoning for engineers and physicians, since the specifics of the individual cases that were successful and unsuccessful, which built the professional's repertoire, are long forgotten and homogenized into an experiential dataset. Intuition is a direct perception of meaning without conscious reasoning [13–15]. A problem with intuition can arise if there is no update over time and new experiences, that is, the failure to recognize new findings and scientific paradigm shifts [16–20].

Technical types are not limited in determining whether an environmental or technological problem is the result of a particular inefficiency or ineffectiveness. For example, engineers commonly base environmental and other risk assessments on previous experience and observation, that is, as a baseline for acceptability of actions and decisions that have led up to the present set of conditions. Once they have this understanding, they may apply it by using the rule to create something, that is, a technology designed to reach some end, from manufacturing insulin to making a crop more resistant to pests to degrading toxic waste. Thus, out of scientific discovery, the engineer and designer apply knowledge, and the application requires at least one technology. In turn, the applications and technology enhance scientific knowledge by more precise, accurate, and representative observations:

> Technology is the outcome of engineering; it is rare that science translates directly to technology, just as it is not true that engineering is just applied science [21].

Therefore, discovery can lead to the design, and the design can lead to discovery.

3.2 Research ethics

> Research is not an organized profession in the same way as law or medicine. Researchers learn best practices in a number of ways and in different settings. The norms for responsible conduct can vary from field to field. Add to this the growing body of local, state, and Federal regulations and you have a situation that can test the professional savvy of any researcher. ***Nicholas H. Steneck [22] Emeritus Professor, University of Michigan***

Researchers are motivated by explaining the previously unexplained, so they are never satisfied with the *status quo*. Scientific inquiry extends the understanding of poorly explained phenomena. For example, in environmental science, there are relatively few "data-rich" pollutants, but myriad "data-poor" pollutants. So, computational and informatic tools have been developed to use information from a data-rich chemical compound to draw knowledge about other data-poor chemicals that share a few attributes with the data-rich chemical. For example, a quantitative structure-activity relationship (QSAR) is established between the better known and the lesser known [23–26].

Research is conducted with integrity whenever the research process is carried out with honesty, robustness, and transparency at every step [27]. These steps include a continuum of proposing, conducting, evaluating, and reporting findings. Although research integrity should be integral to any scientific endeavor and is inherently an inviolable aspect of good science, the scientific and regulatory community has developed rules that must be followed and monitored for compliance, including codes required by scientific disciplines, professional associations, and government agencies. Sadly, the guidance, rules, regulations, and standard operating procedures are often minimal standards that fall short of what it means to be a researcher of integrity [27,28].

The 21st century has seen a surge in fraudulent research, reproducibility fiascos, dubious practices, and other breaches of research integrity that have threatened the credibility of even reliable scientific results. The public has critical questions about the trustworthiness of scientific outputs in all fields, but in particular biomedical science, potentially leading to a "domino effect" of the loss of trust in medical technologies and care [27].

Similar losses of credibility have occurred in the climate and atmospheric sciences [29]. The loss of trust is exacerbated when advocacy has trumped sound science in constructing models, presenting results, and predicting future climate scenarios. The author recently conducted a search of articles to see if the climate change advocates have gained the self-awareness to admit, at least in part, their strident view of an unquestioned consensus about the causes and blame in recent decades regarding global warming. The search found no articles that admitted any fault in the scientific literature for what the scientists had done, only blame for those who do not concur completely with the climate science community. The introductions to these articles often acknowledged skepticism and mistrust by large segments of society but were dismissive of the possibility that the skeptics were correct in at least some facets of the debate. This is more akin to dogma than scientific objectivity. The following hypothetic conversation appeared in a blog to demonstrate the so-called Kuperberg's Parable, which describes at least some of the source of the groupthink [30] and the lack of appreciation for viewing a complex problem through more than one lens [31]:

Doctor (D): You ought to stop smoking, among other reasons because smoking causes lung cancer.

Patient (P): Are you sure? I like to smoke. It also creates jobs.

D: Yes, the science is settled.

P: All right, if the science is settled, can you tell me when I will get lung cancer if I continue to smoke?

D: No, of course not, it's not that precise.

P: Okay, how many cigarettes can I safely smoke?

D: I can't tell you that, although I wouldn't recommend smoking at all.

P: Do you know that I will get lung cancer at all no matter how much I smoke?

D: No, it's a statistical risk. But smoking also causes heart disease.

P: I certainly know smokers with heart disease, but I also know non-smokers with heart disease. Even if I do get heart disease, would you really know that it's because I smoke?

D: No, not necessarily; it's a statistical effect.

P: If it's statistical, then you do know that correlation is not causation, right?

D: Yes, but you can also see the direct effect of smoking on lungs of smokers in autopsies.

P: Some of whom lived a long time, you already admitted.

D: Yes, but there is a lot of research to back this up.

P: Look, I'm not a research scientist, I'm interested in my case. You have an extended medical record for me with X-rays, computerized axial tomography (CAT) scans, blood tests, you name it. You can gather more data about me if you like. Yet you're hedging everything you have to say.

D: Of course, there's always more to learn about the human body. But it's a settled recommendation that smoking is bad for you.

P: It sounds like the science is anything but settled. I'm not interested in hypothetical recommendations. Why don't you get back to me when you actually know what you're talking about. In the meantime, I will continue to smoke, because as I said, I enjoy it. And by the way, since you're so concerned about my health, I believe in healthy skepticism.

An argument can be made that both the doctor and the patient have correct facts and logic. The same set of facts along with some differing perspectives and educational training, somehow led to widely variant conclusions. One can hope that the doctor, who in the parable represents the so-called "scientific consensus," can find a way to appreciate the discourse. Indeed, the science linking cigarette smoke to cancer is much more settled than that of the link between fossil fuels as an existential threat to the human population. This is not to say that there is no link, but that the climate link is more complex and less direct than that of tobacco smoking and health. For example, there are many variables that can be offsetting of the predicted outcome, such as the relationship between increased carbon dioxide (CO_2) concentrations in the atmosphere and plant growth that absorbs the CO_2 and respires greater amounts of molecular oxygen via photosynthesis, or the improvements in geoengineering and other technologies that address increases in global greenhouse gases [32–39].

The point here is not to deny or accept all global climate hypotheses presented by the consensus scientists, but to remind every scientist, engineer, and technologist that we can be wrong about at least some aspects of our position and to avoid all confirmation biases that blind us to our own lack of objectivity [40]. Too often, for example, those opposed to advances in artificial intelligence and other emerging technologies are labeled as modern-day Luddites. Sometimes, this label is applied to even those who are not opposed to the consensus but are simply questioning the outcomes that are being predicted, especially when many of those making the predictions are far from honest brokers and often have large financial stakes and ego investment in the outcomes.

One of the problems with emerging technologies is that the advancement of the technology often precedes sufficient ethical considerations and vetting, that is, the so-called "ethics gap" [41]. For example, during the COVID-19 epidemic, many public health officials lacked transparency and clear explanations of interventions and other decisions that affected privacy and healthcare requirements like mask mandates, mandatory testing, and work life, stating they were simply "following the science," yet the data models upon which they were relying lacked sufficient accuracy and reliability [41]. At the time, sensible people were publicly shamed when they varied in their recommendations on how best to manage the epidemic. Even worse, some policymakers suggested that for their own good, the public needed to be told "noble lies" that intentionally misrepresented facts to support their decisions [42].

Of course, every biomedical, environmental, or other failure has an anthropogenic causal component in the event tree that led to the failure. Certainly, the components must include physical and biological factors but ignoring or underweighting the human factors results in an incomplete root cause analysis and limited risk assessment. It was not that long ago, for example, that climatologists and meteorologists had little hope of predicting patterns in time and space in a manner much better than using primitive folklore and other methods. In fact, some of the folklore methods that worked well have been explained through scientific investigation. Thus, scientific investigation is fueled by curiosity, which leads to innovation. A particular challenge is why something works well in tightly controlled environments, like a laboratory, but not in the field or the marketplace.

Undeniably, comfort with the *status quo* works against innovation. There are usually strong motivations, both financial and technological, to become risk-averse. The *status quo* is an easier route than pushing envelopes. The resistance may be rooted in the sheer difference between the concept of what is considered "normal" versus "abnormal." The adage, "if it isn't broke, don't fix it" runs counter to the technologist's and inventor's *raison d'etre*.

3.3 Scientific advancement

Envisaging the scientific mindset of the 18th century, it is apparent that environmental perceptions have markedly changed, evoking the need and appreciation for scientific breakthroughs. Treating waste before releasing it into a river was not normal. Such waste may not have even been all that harmful since the natural microbial population and the rest of the biological system could degrade the relatively dilute waste efficiently. Correspondingly, most of the chemical compounds released into the river were not synthetic, so the microbes had already adapted capacities to biodegrade them. Thus, there was not a driving need to design a wastewater treatment plant like there was in the 20th century. However, the petrochemical revolution and industrial expansion took hold, and with it the observation that the natural processes were no longer able to withstand the onslaught of chemical loading. Thus, the "normal" benchmark changed, so that the science and engineering of water quality had to advance.

Physics has had to advance to explain the movement of contaminants within and between environmental compartments. Biology has had to change to explain agents of plagues and epidemics. Hydrology became a discipline to describe water flow and characteristics associated with storms and floods.

Scientific advancement ultimately leads to a paradigm shift, the term coined in the late 20th century by Thomas S. Kuhn [19]. He changed the meaning of the word "paradigm," extending it to an accepted specific set of scientific practices [19,20]. The scientific paradigm consists of what is to be observed and analyzed, the questions that arise pertaining to this scientific subject matter, to whom such questions are to be asked, and how the results of the investigations into this subject matter will be interpreted. The paradigm can be harmful if incorrect theories and information become accepted by the scientific and engineering communities. Excessive comfort with the *status quo*, that is, xenophobia, can sometimes be attributed to a community's well-organized protections against differences, that is, the previously mentioned "groupthink" [30].

Innovations occur when a need or opportunity arises. "Necessity is the mother of invention." Environmental and medical science advancements have followed a progression like other research advances since the mid-20th century. In 1944, representing US policy at the time, Vannevar Bush[1] observed that "basic research is performed without thought of practical ends." According to Bush, basic research contributes to "general knowledge and an understanding of nature and its laws." Seeing an inevitable conflict between research to increase understanding and research geared toward use, he held that "applied research invariably drives out pure" [43]. Today, Bush's "rugged individual approach" has been

[1] Bush was Franklin D. Roosevelt's director of the wartime Office of Scientific Research and Development. This section includes his response when asked to consider the role of science in peacetime in "Science, the Endless Frontier."

largely supplanted by teamwork [44–47]. Certainly, any emerging technology thrives on the synergies of teams. Here, the emphasis has evolved toward a cooperative approach. The view recognizes that, to be effective, groups of people must be both technically competent and able to collaborate to realize a common objective [48].

Basic research seeks to broaden the understanding of the phenomena of a scientific field—it is guided by the quest to further knowledge. Applied science has benefited, for example, from advances in remote sensing, which were advanced by a better understanding of particle-wave phenomena in basic physics research. Louis Pasteur sought to understand germ theory, that is, basic research, which led to his application of this knowledge to lower the spoilage rate of vinegar, beer, wine, and milk; all examples of applied research [43].

The disparity between basic and applied research is captured in the linear model of the dynamic form of the postwar paradigm. It is important to keep in mind, though, that in the knowledge flow model, each of the successive stages depends upon the stage before it (Fig. 3.1).

The simple linear model of scientific advances in Fig. 3.1 has come to be called "technology transfer" as it describes the movement from basic science to technology. The first step in this process is basic research, which charts the course for practical application, eliminates dead ends, and enables the applied scientist and engineer to reach their goals quickly, efficiently, and economically. For example, most of the work of Niels Bohr would fall within the far-left box. Then, applied research involves the elaboration and the application of the known. Here, scientists convert the possible into the actual. The final stage in the technological sequence, development, is the stage where scientists systematically adapt research findings into useful materials, devices, systems, methods, processes, etc. [43].

The characterization of transition from basic to applied science has been criticized as being limited to one-way flow from scientific discovery to technological innovation, which does not seem to fit with 21st-century science. Science does not always lead to engineering, which subsequently drives the need for technology. There are many exceptions. For example, engineering has driven basic science, for example, biomedical science's "black boxes" that progressively, but never completely, become understood by bioscience researchers. One can argue that black boxes are more common in applied biology than in applied physics and applied chemistry. This has been apparent with the application of basic biological principles to the COVID-19 pandemic. Simply understanding the structure of DNA and RNA has not been sufficient to arrive at safe and effective mRNA vaccines and viral therapies [49]. Biomedical scientists and engineers who work with real problems, such as virus detection and treatment, tissue engineering, and drug delivery, have helped to shrink these knowledge gaps [50–52]. Technology has driven both science and engineering.

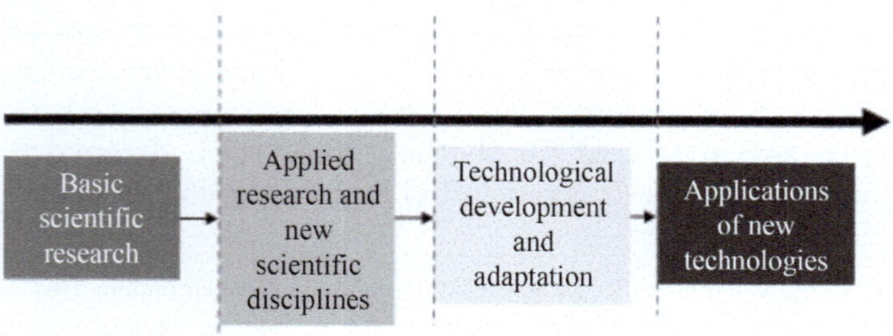

FIGURE 3.1 Progression from basic research to applied research to development of technologies. Source: *Data from: D.E. Stokes, Pasteur's quadrant: Basic science and technological innovation, Brookings Institution Press, 2011.*

A new device allows scientists to embark on whole new areas of research, for example, the polymerase chain reaction (PCR) in rDNA research, and engineers to conceive new designs, for example, the DNA markers that allow for enhanced bioremediation projects. This is reminiscent of the advances in astronomy following Galileo's invention of the telescope [53–55].

The supposition that science exists entirely outside of technology is rather absurd in today's way of thinking. In fact, throughout history, there is seen a reverse flow, a flow from technology to the advancement of science. The innovation of the calculus and the inventions of the microscope and telescope, and later examples of fractal dimensions and rDNA, illustrate that science has progressively become more technology-derived [43]. Biotechnology is a prime example of the technology advancing the science and vice versa. Nanotechnology is another example, which was made possible by advances in electron microscopy and experimental physics [56–58].

The relationship between basic and applied sciences is not universally held within all scientific communities. Some agree that:

> The terms basic and applied are, in another sense, not opposites. Work directed toward applied goals can be highly fundamental in character in that it has an important impact on the conceptual structure or outlook of a field. Moreover, the fact that research of such a nature that it can be applied does not mean that it is not also basic [43].

Environmental protection and biomedical advancement must be based on sound science and require a synthesis of the goals of understanding and use. This combines theory and practice. The one-dimensional model of Fig. 3.1 consists of a line with basic research on one end and applied research on the other, as though the two were opposites. Environmental science, then, calls for a balanced and strong commitment to understanding the theory, for example, abiotic and microbiological processes, and to practice, for example, controlling the for desired effects using these processes. Root cause analysis and improvements would cover the entire line segment in Fig. 3.1. Arguably, two points within a spectrum better represent Pasteur: one at the basic research end of the spectrum and another at the applied research end of the spectrum.

Fig. 3.2 addresses this weakness by compartmentalizing scientific endeavors. For example, within a university follows a flow like that in Fig. 3.3. The science departments are concerned with knowledge-building, the engineering departments with applied knowledge to understand how to solve society's problems, and the university designer is interested in finding innovative ways to use this knowledge. For example, a civil engineer working for the US Army Corps of Engineers is applying general scientific principles to upgrade levee designs but may want to apply historical knowledge and personal experience to arrive at better designs that improve water retention and soil strength. The engineer is behaving much like Thomas Edison, who was most interested in utility and less interested in knowledge for knowledge's sake. In addition, the engineer must work closely with administrators and managers who purchase the systems. This is not to say that innovations do not come from the southwest box in Fig. 3.3, because they clearly do. It simply means that their measures of success at many resource agencies call for major investments in operation and maintenance. In fact, the quadrants must all have feedback loops to one another.

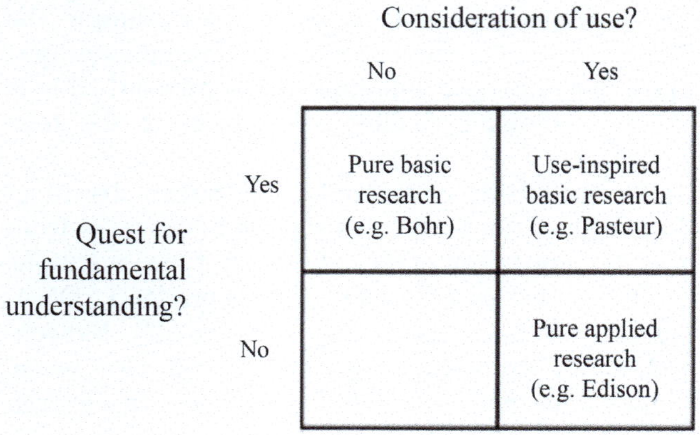

FIGURE 3.2 Research based on knowledge and utility considerations. Source: *Data from: D.E. Stokes, Pasteur's quadrant: Basic science and technological innovation, Brookings Institution Press, 2011.*

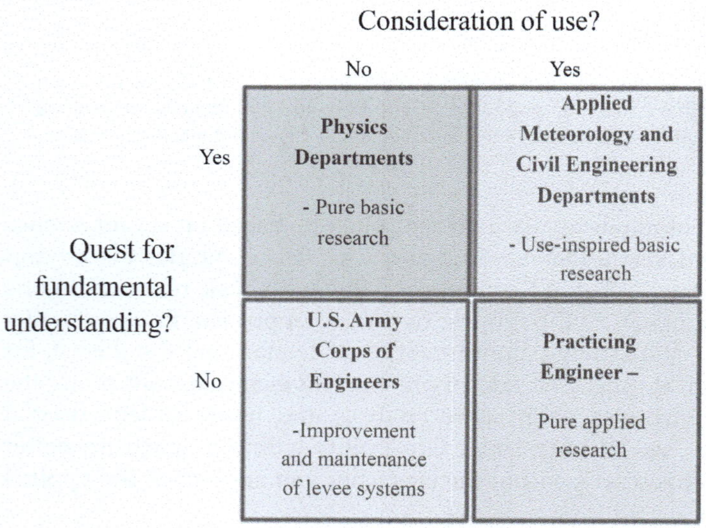

FIGURE 3.3 University research categorized according to knowledge and utility objectives. Source: *Data from: D.E. Stokes, Pasteur's quadrant: Basic science and technological innovation, Brookings Institution Press, 2011.*

This view can also apply to symbiotic relationships among institutions. Duke University is located at one of the points of Research Triangle in North Carolina. The other two points are the University of North Carolina—Chapel Hill and North Carolina State University. All three schools have engineering programs, but their emphasis differs somewhat. Duke is recognized as a world leader in basic research, but its engineering school tends to place a greater emphasis on application of these sciences, for example, adaptive structures during an earthquake. For example, there is much collaboration between Duke's schools of engineering and environment. The School of Public Health houses the engineering programs at the University of North Carolina. As expected, this engineering research tends to focus on addressing and preventing environmental problems that would result from such issues, such as damage to confined animal feeding operations and nutrient loading of rivers. North Carolina State is the first place that the State of North Carolina looks for designers, so the engineers graduating from NC State are ready to design as soon as they receive their diplomas. However, NC State also has an excellent engineering research program that applies the basic sciences to solve societal problems. All of this occurs within the scientific community of the Research Triangle, exemplified by Research Triangle Park (RTP), which includes centers supported by private and public entities that have a particular interest in mind. In this way, the RTP researchers are looking for new products and better processes. The RTP is arguably in the "Edison" quadrant in Pasteur's quadrant, although research in the other two quadrants is ongoing in the RTP laboratories.

The degree to which a given body of research seeks to expand understanding is represented on the vertical axes of Figs. 3.2 and 3.3, and the degree to which the research is driven by considerations of use is represented on the horizontal axes. A body of research that is equally committed to potential utility *and* advancing fundamental understanding is represented as "use-inspired" research [43].

During the last quarter of the 20th century, advances and new environmental applications of science, engineering, along with their associated technologies, began to coalesce into a whole new way to see the world, at least new to most of Western civilization. Ancient cultures on all continents, including the Judeo-Christian belief systems, had warned that humans could destroy the resources bestowed upon us unless the view of stewards and caretakers of the environment was taken seriously. Scientifically based progress was one of the major factors behind the exponential growth of threats to the environment. Environmental controls grew out of the same science, which is now part of a widely accepted environmental ethos.

Most practitioners and researchers see their work as beneficial and self-justified. In other words, their specific work is justified by the general societal need for better medicine, food, environmental quality, better products, and more efficient manufacturing. This can be problematic if even important societal needs, such as advancing science, are sought through unethical means, as recent cases in cooking, trimming, and forging data can attest. Some who have engaged in these unethical acts saw, or at least claimed to have seen, their deceptions as being necessary to advance science for an overall good. This is in opposition to sound science and public trust. A quote attributed to both Snow [59] and Sayers [60] succinctly captures this argument:

> If we do not penalize false statements made in error, we open up the way, don't you see, for false statements by intention. And of course, a false statement of fact, made deliberately, is the most serious crime a scientist can commit.

The societal demands can be very compelling. That is why Snow likely saw the need to warn most scientists of a problem about which most scientists would reflexively agree. After all, scientific ethics seem quite straightforward. Science is the explanation of the physical world, while engineering encompasses applications of science to achieve results. Thus, what we have learned by trial and error has incrementally grown into what is now standard practice of environmental science and engineering. It would seem that following this scientific method honestly is an obvious tenet for any researcher or practitioner of science.

Trial and error learning certainly has advanced our understanding of societal problems, but often at a great cost in terms of the loss of lives and diseases associated with mistakes, poorly informed decisions, and the lack of appreciation for potential engineering, environmental, and biomedical effects. That is why scientists must be careful to guard the public trust. Science and engineering are not popularity contests. Unfortunately, that seems to be one worrisome aspect of many public debates.

Scientists can be recruited, sometimes unknowingly, as advocates for one cause or another. Many of these causes are worthwhile. Many studies are quite good and specifically directed toward a tightly defined research objective. However, research is seldom purely driven by objective, scientific need. Research priorities can be directed by both science and policy. For example, research proposals may be written to fit with what policymakers want; even if that means that the better and more relevant research would be in another area. In fact, recent discussions about scientific consensus related to climate change can be troubling. Seldom does the scientific community have complete consensus on anything except the basic principles (and even these are suspect in quantum mechanics and mathematics). Indeed, there is arguably consensus about many aspects of climate, such as the greenhouse effect and solar radiation. Not every aspect of climate change is at a consensus, that is, unanimity in any scientific discipline is seldom reached [61].

In disaster assessments, sometimes the only sure statement is that a factor either makes a disaster better or worse. For example, increased development along shorelines will likely worsen the impact of hurricanes. The exact amount of worsening is impossible to predict with complete precision. However, the fact that a greater number of vulnerable structures of economic value exist is unquestionable. Also, the decrease in buffering and changes in hydrology due to development will likely increase damage. The march toward credible science can be painstakingly slow.

The foregoing discussion calls for systematic view of benefits and risks. The good news is that popular culture has come to appreciate the systematic relationship between the sciences, engineering, and technologies. As evidence, the concept of "spaceship earth," that is, our planet consists of a finite life support system and that our air, water, food, soil, and ecosystems were not infinitely elastic in their ability to absorb humanity's willful disregard. It is also good news that scientific curiosity and the need to provide stronger evidence about our hypotheses push researchers to revisit some previously neglected details that, in the long run, may well prevent future disasters.[2]

3.4 Turtles all the way down

As mentioned in Chapter 2, several years ago, scientists debated whether to keep the existing taxonomic classifications of the black and green sea turtles. For decades, biologists had believed there were two distinct species, the green (*Chelonia mydas*) and the black turtle

[2] This discussion is excerpted and paraphrased from the author's unpublished collaboration with the late prominent exposure scientist, Paul J. Lioy, of the Environmental and Occupational Health Sciences Institute at Rutgers University. I miss him dearly.

(*Chelonia agassizii*). However, with advances in DNA and other techniques to characterize genetic relationships, it was becoming increasingly obvious that the turtles were so close genetically that they are, in fact, the same species. Such debates often go beyond academics and raise important policy questions. In this case, the question is whether protection under the Endangered Species Act (ESA) should continue to be extended to the not-so-abundant black form of the more abundant green sea turtle. The truly scientifically derived information based on DNA strongly indicates that the black sea turtle should not be considered an evolutionarily distinguishable species from the green sea turtle, and the only distinction is that of a "geopolitical species," one made without regard to "morphological, genetic and reproductive criteria" [62]. The scientific method would seem to have ended this debate among scientists, but it did not. The debate raged on, with scientists asking some troubling questions:

1. Should the black sea turtle be re-classified as a color morph of the green sea turtle, possibly jeopardizing its protections under the ESA?
2. Should the ESA guidelines for what to protect be modified to extend protection to identifiable subpopulations in cases such as this, where there is no geographic or genetic evidence that they are evolutionarily distinct from the main population?
3. Where should the line be drawn between species and subspecies on the scale of genetic distinctiveness is open to some debate?
4. Should scientists interested in conserving the black form exploit that uncertainty to press for continued protection of the black turtles?

The journal *Conservation Biology* decided to put these questions before its readership by publishing a series of articles containing scientific, philosophical, and ethical perspectives on how science should be used in relation to the goal of turtle conservation.[3] Many scientists agreed that the new genetic information was trustworthy, but they worried that it would lead to some unintended and negative consequences for turtle conservation. Some writers called for a "geopolitical taxonomy," where the black sea turtle would continue to be considered a distinct species, even though the genetic information made that view difficult to justify. The analogy of war was invoked as a justification, with one writer declaring that "it is acceptable to tell lies to deceive the enemy." The debate moderators asked a telling question: "Should legitimate scientific results then be withheld, modified, or 'spun' to serve conservation goals?" Continuing with the war analogy, some scientists likened the deceptive taxonomy to propaganda needed to prevent advances by the enemy [63,64]. That is, the erroneous conclusion that violation of scientific principles using disinformation would be justified by some noble end, that is, protecting the turtle

[3] The articles were written by various authors with differing perspectives on the matter, including: S.A. Karl and B.W. Bowen (1999); P.C.H. Pritchard, 1999, "Comments on Evolutionary Significant Units versus Geopolitical Taxonomy," *Conservation Biology*, 13: 1000–03; J.M. Grady and J.M. Quattro,1999, "Using Character Concordance to Define Taxonomic and Conservation Units," *Conservation Biology*, 13: 1004–07; K. Shrader-Frechette and E.D. McCoy, 1999, "Molecular Systematics, Ethics, and Biological Decision Making under Uncertainty," *Conservation Biology* 13, 5:1008–10; and B.W. Bowen and S.A. Karl, 1999, "In War, Truth is the First Casualty." *Conservation Biology* 13, 5: 1013–16.

under existing law, with its arguably arbitrary endangerment thresholds. This should lead to ethical queries, including:

1. Is there a greater good at stake here that supersedes the usual standards of sound science?
2. Is it scientifically acceptable for the proponents of a social goal, such as turtle conservation, to exploit scientific uncertainty in service of that goal? To withhold or distort scientific information?
3. Conservation biology has been described as a "crisis discipline" because it is focused on an urgent and practical end, conservation of biodiversity. Does this create an inherent conflict for conservation biologists?
4. Can scientists play both technical and advocacy roles in an ethical way?
5. How could this sort of controversy be handled?
6. What general lessons about the responsibilities of scientists in society might be derived from this case?

Some scientists have sympathized with Stephen Schneider, a Stanford University climatologist and former government advisor, who has argued that [65]:

> On the one hand, as scientists we are ethically bound to the scientific method, in effect promising to tell the truth, the whole truth, and nothing but - which means that we must include all the doubts, the caveats, the ifs, ands, and buts. On the other hand, we are not just scientists but human beings as well. And like most people we'd like to see the world a better place, which in this context translates into our working to reduce the risk of potentially disastrous climatic change. To do that we need to get some broadbased support, to capture the public's imagination. That, of course, entails getting loads of media coverage. So we have to offer up scary scenarios, make simplified dramatic statements, and make little mention of any doubts we may have. This "double ethical bind" we frequently find ourselves in cannot be solved by any formula. Each of us must decide what the right balance is between being effective and being honest. I hope that means being both.

Indeed, such a distorted version of "utilitarian science" is dangerous. It violates a first canon of science, honest inquiry and reporting. According to Snow [66], it was "crime." Thankfully, not every environmental scientist and practitioner concurs. The utilitarian argument is used to justify a morally unacceptable means to achieve the greater good [67]. The journal *Conservation Biology* published several scientific, philosophical, and ethical perspectives on whether to misuse science to promote the larger goal (conservation) to protect the black sea turtle. Again, even though the taxonomy is scientifically incorrect, that is, the black sea turtle is not a unique species, some writers called for a "geopolitical taxonomy" [62], and the analogy of war was invoked as a justification, with one writer declaring that "it is acceptable to tell lies to deceive the enemy." Those moderating the debate asked a telling question: "Should legitimate scientific results then be withheld, modified, or 'spun' to serve conservation goals?" Continuing with the war analogy, some scientists likened the deceptive taxonomy to propaganda needed to prevent advances by the enemy. Misinformation is a weapon, they seem to argue, and weapons are morally neutral. Only the application and intent of the weapon determine the morality. The problem is that, as Snow would put it, once you stop telling the truth, you have lost credibility as scientists, even if the deception is for a noble cause [68]. Two writers, Kristin Shrader-Frechette and

Earl D. McCoy, emphasized that credible science requires that "... in virtually all cases in professional ethics, the public has the right to know the truth when human or environmental welfare is at issue" [69].

Ends justifying means is common in many venues. For example, politicians recently proclaimed that they do not want to "waste a good crisis." As evidence, a financial crisis offers a rather unique opportunity to take public actions that would not normally be acceptable to the majority. But that's politics and not science. Starting to sound like politicians is a red flag. Scientists should worry that we are losing objectivity and, as a result, credibility. The public often paints with a broad brush. So, even the meticulous and carefully implemented science may be difficult to distinguish from "junk science."

Perhaps the most troubling part of this debate is that it is occurring at all. Once we start treating the truth as a commodity, who can blame the public for losing confidence in science and its practitioners? Certainly, not C.P. Snow. There are indeed legitimate disagreements about scientific methodologies. For example, advocacy is surely not every scientist's basis for their disagreement with the single species designation of green and black turtles, but rather on the credibility of the science underpinning the arguments. Disagreement is a vital part of scientific research. Scientific skepticism must be encouraged. In fact, the Federal Research Misconduct Policy explicitly states that research misconduct "does not include difference of opinion" [70].

3.5 Replication of research

It can be amusing, however, that investigators who obtain resources from the federal government and other agencies say that they have no conflicts of interest or biases. This is not always correct since an investigator can be responding to specific request for data or results by a government agency. In addition, when negative results are found for a hypothesis, it is doubtful that the grantee or contractor will continue to receive more resources or have their results easily accepted for publication in the scientific literature.

It is remarkable that a well-designed and conducted negative study in the environmental health sciences is often not treated with the same level of enthusiasm as a positive finding. In addition, sometimes the authors of submitted manuscripts are so reluctant to state that the results were truly negative. Disappointing, but true. However, efforts by those who find negative results and persist in publishing those results are to be commended. They are providing a great service to the scientific community and the public at large in advancing efforts to ensure scientific integrity. It would be gratifying to have as a lead story someday that states the following: "study shows that X chemical does not cause disease Y". That sort of honesty does resonate with the public, but probably not with much of the media or funding organizations. In contrast, too often headlines are written about a cure or some major breakthrough in science when only a small problem exists in the totality of the aspects associated with a problem.

3.6 A *posteriori* research

Robert Boyle emphasized experimentation, publication, and peer review as requirements of sound modern science. These remain intact today. Lacking *a posteriori* evidence to support it, a hypothesis is merely conjecture. Without peer review, findings may go public with erroneous data and recommendations. Without publication, the scientific community cannot evaluate the scientific method being applied. Thus, when journals do not encourage replication of results, they invite problems in all three requirements. When Boyle's admonitions are not heeded, we do not do science right, and we do not do the right science.[4]

3.7 Research misconduct

Misconduct comes in many forms, but only three have been codified in research regulations. These are fabrication, falsification, or plagiarism in proposing, performing, or reviewing research, or in reporting research results [28]:

1. Fabrication is making up data or results and recording or reporting them in the literature. ("What is Research Misconduct? Part 3: Fabrication")
2. Falsification is the manipulation of research materials, equipment, or processes, or changing or omitting of data or results such that the research is not accurately represented in the research record. ("Definition of Research Misconduct | ORI - The Office of Research Integrity")
3. Plagiarism is the appropriation of another person's ideas, processes, results, or words without giving appropriate credit.

As mentioned, unintentional, honest errors or differences of opinion are not considered acts of research misconduct. The determination of misconduct is based on a preponderance of evidence, often in an oversight review by regulators (see case summary: Zhang, Liping).

Misconduct case study[5]

Based on the evidence from an investigation conducted by Baylor College of Medicine (BCM), the US Office of Research Integrity's (ORI's) oversight review of BCM's investigation, and additional evidence obtained and analysis conducted by ORI during its oversight review, ORI found that Liping Zhang, PhD (Respondent), former Assistant Professor in the Department of Medicine, Section of Nephrology, BCM, engaged in research misconduct under 42 C.F.R. Part 93 in research supported by US Public Health Service (PHS) funds, specifically National Institute of Diabetes and Digestive and Kidney Diseases (NIDDK), National Institutes of

[4] Please pardon my grammar.

[5] From: [71].

(cont'd)

Health (NIH), grants R37 DK037175 and P30 DK079638, and National Cancer Institute (NCI), NIH, grant P30 CA016672.

ORI found by a preponderance of the evidence that Respondent intentionally and knowingly falsified and/or fabricated western blot images and microscopy images by manipulating the images, using unrelated images, or reusing and relabeling the same images to falsely represent different experimental results in three (3) PHS-supported unpublished manuscripts submitted for publication and four (4) grant applications submitted for PHS funds. ORI found that these acts constitute a significant departure from accepted practices of the relevant research community. The affected manuscripts and grant applications are: ("Case Summary: Zhang, Liping | ORI - The Office of Research Integrity")

- Nucleolar protein 66 functions as a novel negative regulator of satellite cells. Original manuscript submitted to the **European Molecular Biology Organization (***EMBO) Reports* in October 2017 (hereafter referred to as "*EMBO Rep*. First Submission").
- Nucleolar protein 66 functions as a novel negative regulator of satellite cells. Second submission to *EMBO Reports* in April 2018 (hereafter referred to as "*EMBO Rep*. Second Submission").
- NO66, an endogenous mediator of muscle growth, suppresses myogenic genes by forming repressive complex with RBBP4 and HDAC2. Original manuscript submitted to *Cell Reports* in July 2015 (hereafter referred to as "*Cell Rep*. Submission").
- R01 DK037175-32, "Protein nutrition in experimental uremia," submitted to NIDDK, NIH, on March 3, 2017, awarded from May 2, 2018 to March 31, 2022.
- R01 AR070845-01A1, "A new pathway to muscle atrophy in chronic kidney disease," submitted to National Institute of Arthritis and Musculoskeletal and Skin Diseases (NIAMS), NIH, on November 3, 2016, administratively withdrawn by NIAMS on March 1, 2019.
- R01 AR069533-01A1, "Epigenetic control of skeletal muscle development and aging-related muscle regeneration," submitted to NIAMS, NIH, on March 1, 2016, administratively withdrawn by NIAMS on November 1, 2018.
- R01 DK116886-01, "A demethylase-dependent mechanism regulates whitening of brown adipocytes and obesity," submitted to NIDDK, NIH, on June 2, 2017, administratively withdrawn by NIAMS on November 2, 2019.

Specifically, ORI found by a preponderance of the evidence that Respondent engaged in research misconduct by intentionally and knowingly falsifying and/or fabricating:

- a western blot image of *EMBO Rep*. Second submission by manipulating a part of the source image to falsely represent an increase in myosin heavy chain expression in differentiation medium
- immunofluorescent microscopy images of *EMBO Rep*. in the NIH grant application R01 AR070845-01A1 by manipulating

(cont'd)

parts of the source images to falsely represent the absence of NO66 and Pax7 coexpression in Pax7+ cells

- western blot images of *EMBO Rep*. First submission, Figure Expanded View (EV) 5 C of *EMBO Rep*. Second submission manipulated parts of the source images to falsely represent: (1) a lack of expression of retinoblastoma binding protein 4 in the immunoglobulin G (IgG) sample, and (2) a positive expression of histone deacetylase 2 in the sample treated with anti-HA antibody
- First submission and second submission reused and relabeled western blot images to falsely represent both myogenin and glyceraldehyde-3-phosphate dehydrogenase (GAPDH) expression as follows:
  - An image of bands representing GAPDH expression was reused and relabeled to represent myogenin expression
  - An image of bands representing IGF1R expression was reused and relabeled to represent GAPDH expression
- Figures reused and relabeled one western blot image to falsely represent the results of two replicated experiments
- Figures in the NIH grant applications R01 DK116886-0, R01 DK037175-32, R01 AR070845-01A1, and R01 AR069533-01A1 reused and relabeled three sets of bands from the same source western blot images to falsely represent the expression of different proteins as follows:
  - An image of a band was reused and relabeled in *EMBO Rep* First submission of R01 DK116886-01, R01 DK037175-32, and R01 AR070845-01A1 to represent GAL4
  - An image of another band was reused and relabeled in R01 DK116886-01, R01 DK037175-32, and R01 AR070845-01A1 to represent NO66
  - An image of two other bands was reused and relabeled to represent H3
    - First and second bands in bottom row of the *EMBO Journal Rep*. were used to represent GAPDH

On December 6, 2024, based on the information in the administrative record, the U.S. Department of Health and Human Services (HHS) suspension and debarment official (SDO) proposed a two-year period of debarment under 2 C.F.R. § 180.800(d)—"Any other cause that is so serious or compelling in nature that it affects your present responsibility" to protect the Federal Government's interest. HHS provided Respondent the opportunity to contest the proposed debarment under 42 C.F.R. Part 93 by requesting a hearing before an administrative law judge with the HHS Departmental Appeals Board or, alternatively, in lieu of requesting a hearing, to contest under 2 C.F.R. Part 180. Respondent did not contest within the prescribed 30-day notice period. Accordingly, the following administrative actions have been implemented:

1. For a period of two (2) years, beginning on January 16, 2025, Respondent is debarred from participating in "covered transactions" as defined in 42 C.F.R. § 180.200, and procurement transactions

(cont'd)

covered under the Federal Acquisition Regulation (48 C.F.R. chapter 1).

2. For a period of two (2) years, beginning on January 7, 2025, Respondent is prohibited from serving in any advisory capacity to PHS, including, but not limited to, service on any PHS advisory committee, board, and/or peer review committee, or as a consultant.

3.8 Unsound science

Misconduct is a severe breach of research ethics, but other unethical forms exist. Arguably, many are violations of the tenets of the scientific method. Such issues can affect the value of adherence to the scientific method. For example, much of the public now relies on the internet for almost all their information. It is inevitable that with the availability of the internet to students and researchers, good and bad habits and outcomes are possible. Access to volumes of information is of great value in finding information on innumerable topics. The problem arises when the information is not necessarily vetted prior to use, which requires an assessment by the potential user after each click to a website. The other problem is that even correct information may be aggregated in ways that detract from knowledge. Data simply exist. They do not become information until they are manipulated by users. Such information does not become knowledge without being integrated meaningfully by those who understand their context. Unfortunately, anyone can "google" a keyword and gather mounds of data, even those who have no idea what such data mean. There was an excellent commercial where people were asked a simple question but answered in highly technical yet irrelevant ways (e.g., in response to a question of what's for dinner, they may respond that the Irish hen eats only green potato caterpillars during an Aurora Borealis…). It can be worse than information overload when it becomes information nonsense.

Obviously, the lack of an objective assessment can lead to the use of misinformation and fraudulent or doctored information in an analysis or examination of an issue or use of information. Again, our educational system has not caught up with how to handle the volumes of information and how effectively to discriminate real versus false concepts and theories. Information has always been available, but never to the degree that it is currently accessible to young minds. In the past, I made it mandatory that students do not use the internet for information gathering. It is a cause not worth pursuing since most reputable scientific journals now publish online. However, the need for an accurate reference and the need to discriminate between open access journals with or without sufficient peer review is essential if science is to grow and yield new discoveries. Students can spend their lives surfing but miss the perfect wave.

Some students and others have been seduced into thinking that if it is not on the internet, it is not worth reading. This leads to a big problem for science; the notion that worthy science began in the 2000s, An appreciation for fundamental scientific achievements needs to be inculcated in the next generation. Since most people now use the internet extensively and many get their information almost exclusively online, scientific journals must accelerate scanning of historical volumes to online portals, as is being done increasingly with books. This will reduce the probability that individuals will try to re-invent well-known successes and failures in science and engineering. Clearly, a documented history of individual scientific disciplines must not be lost. The body of work should also be placed into the context of the major achievements, that is, the value of information and principles that established the science, the practice of each science discipline over time, and the uncertainties and questions that remain. With such a foundation, the public will have a better appreciation of the past and a firm foundation so that new discoveries will have a better opportunity to evolve [72].

With the continuing addition of popular information and academic courses within many disciplines, there is a risk of degrading in-depth inquiry. Universities and other levels of education are forgoing the basics, and for nonmajors, they are offering courses on a multidisciplinary topic and present snippets from a variety of disciplines to explain specific points and major concepts. It is very disconcerting that individuals are being steered in this direction if the basics are being ignored, diminishing the student's ability to discern the truth and to use critical thinking, especially when only one side of the issue is taught, or an alternate explanation is not given serious consideration.

It is important to keep in mind that there is always the potential for attendant, undesired outcomes in the development and application of emerging technologies. Furthermore, even if a technology is acceptable, there can still be an unacceptable cumulative outcome. In other words, there is a need to recognize a technology within concept of systematic success. A multistep process can be used to approach the various types of unintended outcomes that may result from any engineering application [73], but particularly for emergent technologies and discoveries:

- Properly accounting for uncertainty [74];
- Basing risks and benefits on various types of information and understanding the limitations of using secondary sources of data;
- Comparing risks and benefits that may result consistently among the alternative actions [75,76];
- Employing a tiered, intelligent, adaptive, and flexible approach to risk and benefit assessments; often beginning with a higher-tier scoping effort and adding needed details as information is continuously re-evaluated [77,78];
- Evaluating the value of information on which decisions will be based, including assignment of quality and relevance thresholds below which the data and information should be avoided or, at a minimum, given appropriate caveats.
- Determining the approach to validate and to evaluate the models and other tools used to assess and to predict the use of the new technology [79];
- Considering the potential cumulative risks from exposures to all types of stressors, that is, physical, chemical, and biological [80,81], especially regarding possible effects stemming from the release and use of a technology on human health and the integrity of ecosystems over time and space [82].

As seen with the issue of global climate change, there are biases on both sides of many scientific issues. However, the revelations from communication fiascos like "Climategate" suggest that society is reassessing scientific claims that appear to be driven not exclusively by science, but also by political or other agendas [72]. As stated earlier, this type of struggle has existed in the past, but the speed at which society and science are transferring all types of information and personal opinion makes the future of allowing good science to prevail difficult to achieve and sustain without thoroughly evaluating all sides of the science objectively. A major mistake can lead to unnecessary and even counterproductive expenditures of resources [72]. In the case of biomedical technologies, errors can lead to bodily harm and death.

One ongoing recommendation is that all members of a modern society need a modicum of mathematics, digital literacy, and critical thinking skills [83]. Too much data are being generated to allow only the elites to explain complex decisions being made that will affect everyone's lives and livelihoods. A simple explanation based on rudimentary statistical analyses is insufficient for future discoveries and the potential impacts on daily life.

3.9 Conclusions

Scientific advancement has led to amazing technologies and knowledge of how the physical world works. Changes in knowledge have been amazing over such a few centuries and exponentially increasing in recent decades. However, the advances in physical, medical, and social sciences have met with ethical breaches and misconduct along the way. This chapter has presented an array of observations of how these have occurred, along with recommendations on addressing them.

References

[1] R.G. Newton, The Truth of Science: Physical Theories and Reality, Harvard University Press, Cambridge, Massachusetts, (1997).

[2] R. Feynman, Cargo cult science, Engineering & Science vol. 37, (no. 7) (1974) 10–13.

[3] D.A. Vallero, T.M. Letcher, Unraveling Environmental Disasters, Second ed, Elsevier Academic Press, Amsterdam, The Netherlands, (2023).

[4] D.A. Vallero, T.M. Letcher, Engineering risks and failures: lessons learned from environmental disasters, Leadership and Management in Engineering vol. 12, (no. 4) (2012) 199–209.

[5] D.M.V.H. Nair, Amarnath, Impacts of Impervious Surfaces on the Environment (2015).

[6] B.M. Sumer, R.J.S. Whitehouse, A. Tørum, Scour around coastal structures: a summary of recent research, Coastal Engineering vol. 44, (no. 2) (2001) 153–190. Available from https://doi.org/10.1016/S0378-3839(01)00024-2.

[7] D. NeiderORCiD, Combining deductive and inductive reasoning to build reliable systems, Computer Sciences, Rheinland-Pfälzische Technische Universität Kaiserslautern-Landau, (2022).

[8] M. Lange, Hume and the problem of induction, Handbook of the History of Logic vol. 10, Elsevier, (2011), pp. 43–91.

[9] D. Hume, The problem of induction, Philosophy of Science: An Historical Anthology vol. 15, (2009) 219.

[10] D.A. Vallero, Environmental Systems Science: Theory and Practical Applications, 1st ed., Elsevier Academic Press, (2021).

[11] M.L. Martínez, Leviathan and the air pump: a historical study of the emergence of the laboratory style and the creation of phenomena, Transversal: International Journal for the Historiography of Science vol. 18, (2025). Available from https://doi.org/10.24117/2526-2270.2025.i18.07. no. June.
[12] S. Shapin, S. Schaffer, Leviathan and the Air-Pump: Hobbes, Boyle, and the Experimental Life (New in Paper), Princeton University Press, (2011).
[13] M.R. Taleyarkhan, A.M. Lucietto, N.L. Hobson, T.M. Azevedo, Approach to problem solving and use of intuition by engineering technology students, Journal of Global Education and Research vol. 7, (no. 1) (2023) 81–98.
[14] E. Miskioğlu, K.M. Martin, A.R. Carberry, Work in progress: experts' perceptions of engineering intuition,", in 2020 ASEE Virtual Annual Conference (2020).
[15] R.M. Hogarth, Educating Intuition, University of Chicago Press, (2001).
[16] B. Zheng et al., Agent4S: the transformation of research paradigms from the perspective of large language models. arXiv preprint arXiv:2506.23692, 2025.
[17] J.A. Moshiri, Emerging paradigms: exploring the latest trends in science, technology, and business, International Multidisciplinary Journal of Science, Technology & Business vol. 2, (no. 1) (2023) 29–34.
[18] D. Vallero, Paradigms Lost: Learning From Environmental Mistakes, Mishaps and Misdeeds, Butterworth-Heinemann, (2005).
[19] T. Kuhn, What are Scientific Revolutions, MIT, Boston, (1981).
[20] T.S. Kuhn, D. Hawkins, The structure of scientific revolutions, American Journal of Physics vol. 31, (no. 7) (1963) 554–555.
[21] National Academy of Engineering, The Engineer of 2020: Visions of Engineering in the New Century, National Academies Press, Washington, DC, (2004).
[22] N.H. Steneck, ORI Introduction to the Responsible Conduct of Research. Department of Health and Human Services, Office of the Secretary, Office of ..., 2003.
[23] I.A. Balabin, R.S. Judson, Exploring non-linear distance metrics in the structure–activity space: QSAR models for human estrogen receptor, Journal of Cheminformatics vol. 10, (no. 1) (2018) 47. Available from https://doi.org/10.1186/s13321-018-0300-0.
[24] U.S. Environmental Protection Agency, Quantitative Structure Activity Relationship (2015). Available from http://www.epa.gov/nrmrl/std/qsar/qsar.html. (accessed March 19.
[25] D. Wang, et al., A hybrid framework for improving uncertainty quantification in deep learning-based QSAR regression modeling, Journal of Cheminformatics vol. 13, (no. 1) (2021) 69. Available from https://doi.org/10.1186/s13321-021-00551-x.
[26] H. Zhu, T.M. Martin, L. Ye, A. Sedykh, D.M. Young, A. Tropsha, Quantitative structure– activity relationship modeling of rat acute toxicity by oral exposure, Chemical Research in Toxicology vol. 22, (no. 12) (2009) 1913–1921.
[27] A.C.V. Armond, K.D. Cobey, D. Moher, Research Integrity definitions and challenges, Journal of Clinical Epidemiology vol. 171, (2024) 111367. Available from https://doi.org/10.1016/j.jclinepi.2024.111367.
[28] E. National Academies of Sciences and Medicine, Fostering Integrity in Research, National Academies Press, (2017).
[29] V. Cologna, J. Kotcher, N.G. Mede, J. Besley, E.W. Maibach, N. Oreskes, Trust in climate science and climate scientists: a narrative review, PLOS Climate vol. 3, (no. 5) (2024) e0000400. Available from https://doi.org/10.1371/journal.pclm.0000400.
[30] I.L. Janis, Groupthink and group dynamics: a social psychological analysis of defective policy decisions, Policy Studies Journal vol. 2, (no. 1) (1973) 19.
[31] G. Kuperberg, S. Aaronson (Ed.), Facebook post: Kuperberg Parable, vol. 2026, Shtetl-Optimized, (2026)ed:.
[32] K. De Pryck, M. Boettcher, The rise, fall and rebirth of ocean carbon sequestration as a climate 'solution, Global Environmental Change vol. 85, (2024) 102820. Available from https://doi.org/10.1016/j.gloenvcha.2024.102820.
[33] A.A. Tsonis, A. Kirwan Jr, , Geoengineering from the standpoint of uncertainty and related risks: science or science fiction? Stochastic Environmental Research and Risk Assessment vol. 37, (no. 9) (2023) 3325–3327.
[34] D. Visioni, D.G. MacMartin, B. Kravitz, Is turning down the sun a good proxy for stratospheric sulfate geoengineering? Journal of Geophysical Research: Atmospheres vol. 126, (no. 5) (2021) p. e2020JD033952.

[35] A. Robock, Benefits and risks of stratospheric solar radiation management for climate intervention (geoengineering), The Bridge vol. 50, (no. 1) (2020) 59–67.
[36] D.A. Vallero, D.B. Resnik, Geoengineering: enhancing cloud albedo, Access Science (2013).
[37] D.B. Resnik, D.A. Vallero, Geoengineering: an idea whose time has come? Journal of Earth Science & Climatic Change (2011).
[38] D.W. Keith, Photophoretic levitation of engineered aerosols for geoengineering, Proceedings of the National Academy of Sciences vol. 107, (no. 38) (2010) 16428–16431.
[39] J. Shepherd, Geoengineering the Climate: Science, Governance and Uncertainty, Royal Society, (2009).
[40] S.E. Gorman, J.M. Gorman, Confirmation Bias, in: S.E. Gorman, J.M. Gorman (Eds.), Denying to the Grave: Why We Ignore the Science That Will Save Us, Revised and Updated Edition, Oxford University Press, (2021) https://doi.org/10.1093/oso/9780197547458.003.0004.
[41] F.A. Parasidis, Closing the public health ethics gap, E, New England Journal of Medicine vol. 387, (no. 11) (2022) 961–963. Available from https://doi.org/10.1056/NEJMp220754.
[42] K. Powell, V. Prasad, The noble lies of COVID-19, in Future Tense (2021) ed.
[43] D.E. Stokes, Pasteur's Quadrant: Basic Science and Technological Innovation, Brookings Institution Press, (2011).
[44] D.D. Beaver, The many faces of collaboration and teamwork in scientific research: updated reflections on scientific collaboration, COLLNET Journal of Scientometrics and Information Management vol. 7, (no. 1) (2013) 45–54.
[45] G.P. Bush, L.H. Hattery, Teamwork and creativity in research, Administrative Science Quarterly (1956) 361–372.
[46] W.O. Hagstrom, Traditional and modern forms of scientific teamwork, Administrative Science Quarterly (1964) 241–263.
[47] J.J. Winnink, R.J. Tijssen, A.F. van Raan, Theory-changing breakthroughs in science: the impact of research teamwork on scientific discoveries, Journal of the Association for Information Science and Technology vol. 67, (no. 5) (2016) 1210–1223.
[48] J. Fernandez, Understanding group dynamics, Business Line vol. 2, (2002) 2002.
[49] J.K. Lee, H.S. Han, Coronavirus disease 2019 and mRNA vaccines: what's next - miRNA? (in eng), Clin Exp Pediatr vol. 65, (no. 6) (2022) 302–303. Available from https://doi.org/10.3345/cep.2022.00283.
[50] S. Olofsson, M. Mehrian, R. Calandra, L. Geris, M.P. Deisenroth, R. Misener, Bayesian multiobjective optimisation with mixed analytical and black-box functions: application to tissue engineering, IEEE Transactions on Biomedical Engineering vol. 66, (no. 3) (2018) 727–739.
[51] R. Langer, N.A. Peppas, Advances in biomaterials, drug delivery, and bionanotechnology, AIChE Journal vol. 49, (no. 12) (2003) 2990–3006.
[52] N. Ogata, S.W. Kim, J. Feijen, T. Okano, Advanced Biomaterials in Biomedical Engineering and Drug Delivery Systems, Springer Science & Business Media, (2012).
[53] Y. Zik, Galileo and the telescope, Nuncius / Istituto e Museo di Storia Della Scienza vol. 14, (no. 1) (1999) 31–67.
[54] T.S. McKechnie, History of the telescope and its remarkable contribution to scientific discovery (and the 400-year journey from galileo to a rigorous general theory of imaging through earth's turbulent atmosphere), General Theory of Light Propagation and Imaging Through the Atmosphere, Springer, (2022), pp. 1–22.
[55] O. Gingerich, Galileo, the impact of the telescope, and the birth of modern astronomy, Proceedings of the American Philosophical Society vol. 155, (no. 2) (2011) 134–141.
[56] S. Mukherjee, S. Kaushik, P. Singh, S. Shweta, I.P. Sahu, Transmission electron microscopy for biomedical nanotechnology, Analytical Techniques for Biomedical Nanotechnology (2023) 12-1–12-19.
[57] A. Marais, Nanotechnology: a perspective from quantum biology, Global Health Care Equivalency in the Age of Nanotechnology, Nanomedicine and Artificial Intelligence, CRC Press, (2026), pp. 67–92.
[58] M.E. Rodriguez, The impact of physics on advancements in nanotechnology, Austra & Lian Journal of Basic Sciences vol. 5, (no. 6) (2024) 11–20.
[59] C.P. Snow, The Search. House of Stratus (2000).
[60] D.L. Sayers, M. Cumming, Gaudy night, Royal Society for the Blind of South Australia (1937).
[61] P.J. Lioy, D.A. Vallero, The scientific method in an era of advocacy, Ethics in Biology, Engineering and Medicine: An International Journal vol. 1, (no. 4) (2010).

[62] S.A.A.B. Karl, Evolutionary significant units versus geopolitical taxonomy: molecular systematics of an endangered sea turtle (genus Chelonia), B.W. Conservation Biology vol. 13, (no. 5) (1999) 990–999. Available from https://doi.org/10.1046/j.1523-1739.1999.97352.x.
[63] K. Shrader-Frechette, Risk Analysis and Scientific Method: Methodological and Ethical Problems with Evaluating Societal Hazards, Springer Science & Business Media, (2012).
[64] K. Shrader-Frechette, Ethics and environmental advocacy, Environmental Ethics and Forestry: A Reader (2000) 209.
[65] J. Schell, Our fragile earth, October 1989, Discover 44 (50) (1987).
[66] C.P. Snow, The Search, Victor Gollancz Ltd, London, UK, (1934).
[67] M.B.R.E.C. Chamberlain, M.L. Gibeau, Human perspectives and conservation of grizzly bears in Banff National Park, Canada, Conservation Biology vol. 26, (no. 3) (2012) 420–431 https://doi.org/10.1111/j.1523-1739.2012.01856.xDigital.
[68] B.W.B.A.S.A. Karl, In war, truth is the first casualty, Conservation Biology vol. 13, (no. 5) (1999) 1013–1016.
[69] K.A.M. Shrader-Frechette, Molecular systematics, ethics, and biological decision making under uncertainty, E.D. Conservation Biology vol. 13, (no. 5) (1999) 1008–1012.
[70] Federal Research Misconduct Policy. Available from https://ori.hhs.gov/federal-research-misconduct-policy, 2000.
[71] Office of Research Integrity, Case Summary: Zhang, Liping, H. a. H. Services, Ed., ed. Washington, DC: Government Printing Office, 2025.
[72] D.A. Vallero, Biomedical ethics for engineers, The Biomedical Engineering Series (2007).
[73] R.A. Hill, C. Sendashonga, General principles for risk assessment of living modified organisms: lessons from chemical risk assessment, Environmental Biosafety Research vol. 2, (no. 02) (2003) 81–88.
[74] ER-TG-03-1 7/00, Preparing information on risks, costs and benefits for applications under the Hazardous Substances and New Organisms Act 1996, 2000.
[75] B. Hong, et al., Model-based tolerance intervals derived from cumulative historical composition data: application for substantial equivalence assessment of a genetically modified crop, Journal of Agricultural and Food Chemistry vol. 62, (no. 40) (2014) 9916–9926.
[76] P. Macdonald, Developing workable regulatory frameworks for the environmental release of transgenic plants, Collection of biosafety reviews (2012).
[77] ER-TG-02-1 1/100, Assessment of effects of hazardous substances and new organisms on human health, 2000.
[78] R.A. Hill, From Science to Decision-making: The Applicability of Bayesian Methods to Risk Assessment, Human and Ecological Risk Assessment: An International Journal 2 (1996).
[79] B. Breckling, et al., Risk indication of genetically modified organisms (GMO): modelling environmental exposure and dispersal across different scales: oilseed rape in Northern Germany as an integrated case study, Ecological Indicators vol. 11, (no. 4) (2011) 936–941.
[80] J.G. Lundgren, J.J. Duan, RNAi-based insecticidal crops: potential effects on nontarget species, Bioscience vol. 63, (no. 8) (2013) 657–665.
[81] C.V. Rider, et al., Cumulative risk: toxicity and interactions of physical and chemical stressors, Toxicological Sciences 137 (1) (2014) 3–11.
[82] K. Sexton, Cumulative risk assessment: an overview of methodological approaches for evaluating combined health effects from exposure to multiple environmental stressors, International Journal of Environmental Research and Public Health vol. 9, (no. 2) (2012) 370–390.
[83] K. Tang, Eugene, T. Song, Z. Zhu, J. Li, Y.-C. Lee, AI literacy education for older adults: motivations, challenges and preferences, in Proceedings of the Extended Abstracts of the CHI Conference on Human Factors in Computing Systems, 2025, pp. 1–15.

CHAPTER

4

Engineering ethics☆

OUTLINE

Engineering is an active process of solving problems and building new things using computers and other technologies. ***Duke University's K-PhD Program [2]***

If we could remember those early efforts of ours to raise ourselves up among the towers of legs of our parents and their friends, then we can begin to appreciate the task and the achievements of engineers, whether they be called builders in Babylon or scientists in Los Alamos. For all of their efforts are to one end: to make something stand that has not stood before, to reassemble Nature into something new, and above all to obviate failure in the effort. ***Henry Petroski [3].***

4.1 Introduction

The Ancient Greeks considered excellence to be a dichotomous phenomenon. It requires both skill and character. An engineer's academic and professional preparation is designed to

☆ This chapter is an update to and advancement of Chapter 1 in: [1] D. A. Vallero, "Biomedical Ethics for Engineers," *The Biomedical Engineering Series*, 2007.

Technical and Ethical Optimization: Challenges and Opportunities for the Modern Engineer and Scientist
DOI: https://doi.org/10.1016/B978-0-443-44925-3.00013-2

address the former and some of the latter. For example, the engineering curriculum continues to place high demands on the student's grasp of mathematical and scientific concepts. This emphasis has never subsided. However, the engineering curriculum of today increasingly requires an understanding of the social sciences and humanities and demands greater core competencies in interpersonal skills.

Engineering is an active endeavor. It is a process that takes the raw materials and energy that exist at a given time and creates new things and improves things that already exist. Engineers solve problems and build. Engineers do things.[1] They are seldom satisfied merely with possessing information or mastering skills. They are client-focused practitioners, but also innovators. Indeed, the band Smash Mouth described this drive in their song, *All Star*:

> … Get the show on, get paid (And all that glitters is gold) Only shootin' stars break the mold.

The real test of the engineer is how and when to put this knowledge and aptitude into practice. This has been apparent with the Duke University engineering students who recently returned from developing nations after participating in Engineers Without Borders projects. They were joyous in their opportunities to apply the theoretical information in real projects. This eagerness to apply what engineers know is a gift that has truly made the world a better place. Some argue, for example, that many of the improvements in public health, safety, and quality of life are largely attributed to engineers [4]. To wit, engineers have prevented countless diseases and deaths, not to mention vast improvements and extended lifetime expectancies for billions of people, with the design of innovative systems that clean water, air, and soil, treat wastewater, and improve the food supply. They have designed safer transportation systems and improved the resilience of communities against wildfires, storms, and other disasters. Indeed, many of the healthcare systems on which physicians and other medical staff rely daily have been designed by clinical and biomedical engineers. In North America and Europe, for example, public works projects designed by engineers have given us clean water, which has prevented many of the diseases responsible for most premature deaths

[1] This profundity was stated by the late P. Aarne Vesilind, RL Rooke Professor of Engineering, Bucknell University, at the conference, "Engineers Working for Peace", Lewisburg, Pennsylvania, November 15, 2003. Vesilind made the comment as a reminder of the practicality of the profession and the need to respect the ethos of engineers when addressing societal issues, like peace and justice. He was the person who introduced me to professional ethics while I was a Ph.D. student at Duke, having enough trust in me to take over Duke's Program in Science, Technology, and Human Values while he was to be on a half-sabbatical in 1997. I must admit that I was merely a week ahead of my students in learning the basics of Duke's Ethics in Professions course, but it led to some of the most engaging and challenging events in my academic and professional life. I will always be grateful to Aarne.

and disabilities so common a hundred years ago.[2] Pollution control equipment has allowed for cleaner air. Vehicle and transportation designs continue to improve the safety of travel. Chemical engineering advances have improved product manufacturing, leading to higher quality and safer consumer products and pharmaceuticals. Indeed, biomedical devices and systems have lengthened and improved the quality of life for billions.

The engineering call will continue to strengthen. The knowledge and creativity of the engineer will have to grow in proportion to the increased societal expectations. Kristina Johnson, Dean of Duke University's Pratt School of Engineering, characterized the engineer's future obligation to society [2]:

> … as an engineering dean, I'd argue that it our responsibility as good citizens of the planet to solve many of our global problems, such as developing renewable energy sources, purifying water, sustaining the environment, providing low-cost health care and vaccines for infectious diseases, to mention a few. Coupled with global climate issues, transportation and urbanization, we need all the technical horsepower we can educate.

4.1.1 Factors of safety

As in all aspects of engineering, of course, challenges and obstacles lie ahead. The design challenges are familiar to the engineer, such as the unique conditions of situations that make them "think outside of the box." An equation or model seems to work well in most cases, but those few where it fails can be the difference between good engineering and poor design, success and failure, and even life and death. This is not hyperbole. An insufficient factor of safety (γ) in a design not only risks an unhappy clientele but also poses a threat to health and safety. And recall that the principal client of the engineer is the public [5].

Factors of safety (FOS) are a requirement for any engineering design. For example, in structural engineering, the factor is typically expressed simply as [6]:

$$\gamma = \frac{F_{\text{failure}}}{F_{\text{load}}} \tag{4.1}$$

Where, F_{fail}=failure load (N, lb_f) and F_{allow}=allowable load (N, lb_f).

[2] The concept of "clean" is subject to debate within the engineering community. It parallels the questions about safety. When is a device or drug sufficiently "safe" to move to the production stage? Environmental engineers ask a similar question, "How clean is clean?" We wonder when we have done a sufficient job of cleaning up a spill or hazardous waste site. It is often not possible to have nondetectable concentrations of a pollutant, especially since analytical chemistry and other scientific disciplines continue to improve. Commonly, a threshold for cancer risk to a population is one in a million excess cancers. In cleanup situations, the tolerable risk may be much higher (e.g., 1 in 10,000). However, one may find that the contaminant is so difficult to remove that we almost give up on dealing with the contamination and put in measures to prevent exposures, that is, fencing the area in and prohibiting access. This is often done as a first step in site remediation but is unsatisfying and controversial (and usually politically and legally unacceptable). Thus even if costs are high and technology unreliable, the engineer must find suitable and creative ways to clean up the mess and meet risk-based standards.

Note that the units in the numerator and denominator are the same, so the factor is dimensionless and unitless. For example, the buckling failure load of a steel column is estimated to be 10,000 N. If the safety factor is 5, then the equation can be rearranged to calculate the allowable load [6]:

$$F_{\text{allow}} = \frac{F_{\text{failure}}}{\gamma} \tag{4.2}$$

$$F_{\text{allow}} = \frac{10,000N}{5} = 2000N$$

Some equipment FOS are shown in Table 4.1.

"Design Factors of Safety are often published in technical standards but there is no dedicated standard to the subject." ("Factors of Safety - FOS - The Engineering ToolBox") Note that for statutory items such as cranes and pressure vessels, FOS are specified in the design codes [6].

A sufficient engineering factor of safety can mean the difference between life and death in critical infrastructure. The safety factor acts as a buffer between calculated design failure points, like load for a building or lift for an aircraft, with values that depend on material reliability and environmental severity.

Selecting an appropriate factor involves a rigorous and interdisciplinary method in which technical safety margins, human factors, and organizational accountability are balanced.

TABLE 4.1 Factors of safety (*FOS*) ranges for equipment.

Equipment	FOS
Aircraft components	1.5–2.5
Boilers	3.5–6
Bolts	8.5
Cast-iron wheels	20
Engine components	6–8
Heavy-duty shafting	10–12
Lifting equipment, for example, hooks	8–9
Pressure vessels	3.5–6
Turbine components—static	6–8
Turbine components—rotating	2–3
Spring, large, heavy-duty	4.5
Structural steelwork in buildings	4–6
Structural steelwork in bridges	5–7
Wire ropes	8–9

Data from: The Engineering Toolbox, Factors of Safety. https://www.engineeringtoolbox.com/factors-safety-fos-d_1624.html (accessed 20.01.26.)

Given that every design condition is unique, simply applying historical data may be insufficient, so the engineer must go beyond calculations and models to provide an acceptable level of safety. The keys are that the design provides enough redundancy and reliability to perform consistently over the design's lifetime, including operation and maintenance, and is available to prevent undesired, adverse outcomes. This is often accomplished by proactively assessing risks and identifying potential failures, even if they are very unlikely.

In the engineering college, the introductory aspects of FOS are often taught using academic exercises in which the safety factor is issued in a problem statement or is the result of a calculation, for example, a stress analysis. As the student's grasp of the concepts from these exercises evolves to become engineering practice, the student begins to appreciate that the safety factor is a design parameter established by the design engineer using engineering analysis, However, understanding the more subtle aspects of FOS requires mentorship. This goes beyond theoretical and academic knowlege. From the time observing and learning from seasoned professionals, the student begins to acquire engineering judgment, which is based on an appreciation for [7]:

- the uncertainties associated with the physical components, like materials, that are used in the design;
- the uncertainties associated with the operational and manufacturing techniques used in the creation of the product;
- the uncertainties in the analysis that established the part dimensions;
- the operation and maintenance of the product during its lifetime;
- the risks and costs associated with potential failure during its lifetime, including decommissioning and other end-of-product-life actions;
- the relevant standards and codes; and
- the professional and moral obligations to, and the trust of, all clients, especially the public.

Engineers commonly experience the challenge of the lessons to be learned when moving from a prototype to an actual application. Slight changes in scale or complexity (i.e., effects of some unknown variable) limit the application of a design. One specific challenge, the subject of this book, is the ethical challenge; specifically, the challenge of how engineering decisions and actions affect human life, for good or ill.

4.2 The good engineer

Given the previous discussion of the factor of safety where engineering judgment is required, it is important to ensure the engineer is qualified to make such judgments. In other words, does the engineer possess the skills and character to make such a judgment? A query in an artificial intelligence (AI) search asking, "What makes a good engineer?" yielded the following:

> A good engineer combines strong technical skills with essential soft skills, excelling at creative problem-solving, logical thinking, and detailed analysis, while also possessing excellent communication, teamwork, and adaptability to learn continuously in a changing world, staying pragmatic and humble to effectively build and improve systems.

The famous physicist Freeman Dyson (born 1923) said something similar about a good scientist, as well as a good engineer and technologist [8]:

> A good scientist is a person with original ideas. A good engineer is a person who makes a design that works with as few original ideas as possible. There are no *prima donnas* in engineering.

Both address the competence and creativity of the engineer but omit the aspect of professional ethics. Dyson's continuance comes a bit closer to the needed synergy of competence and ethics when he added [8]:

> If we had a reliable way to label our toys good and bad, it would be easy to regulate technology wisely. But we can rarely see far enough ahead to know which road leads to damnation. Whoever concerns himself with big technology, either to push it forward or to stop it, is gambling in human lives.

A slight change to the AI query replacing "good" with "ethical" yielded:

> An ethical engineer prioritizes public safety, health, and welfare above all else, acting with honesty, integrity, and competence by adhering to professional codes, being truthful in public statements, managing conflicts of interest, and taking responsibility for their work, including learning from mistakes to ensure public trust and sustainable, responsible innovation.

This is a much more satisfying definition. However, it is a bit concerning that the "good" definition did not also include everything in the "ethical" definition.

Exercise

1. Give two examples in bioengineering that demonstrate Dyson's first point and two emerging technologies in which we currently need to heed the warning of his second quote.
2. What does this tell us about risk-taking, especially the optimum between necessary versus irresponsible risks?

Engineering is certainly based in the physical, chemical, and biological sciences. However, it also requires art and aesthetics. I once had a conversation with the very competent engineer, Robert Koke, of the US Environmental Protection Agency's Kansas City regional office when I was preparing to build a house. I asked if he would be willing to draw up some plans and use his professional engineering stamp on the design. He humorously told me something to the effect that the house would be fine in terms of function but would have the form of a locomotive. I decided to use architectural plans instead.

Another interesting vignette occurred in a Duke auditorium where senior engineering projects were presented. Henry Petroski's student was providing insights on a bridge project when a question arose about the role of architecture in bridge design. The student did his best to answer, but Petroski had heard enough. He said something akin to the importance of paint and lighting. The room filled with laughter, but he was quite serious and assured everyone that aesthetics is an important consideration in design. In this case, colors and light are key safety aspects of a bridge.

This carries over to emerging technologies in the sense that the most efficient product may not receive the acceptance it deserves if the product does not have at least a modicum of attractiveness. Unfortunately, there have been rather inefficient products that found success only because of their aesthetic appeal.

Thus in addition to the aesthetic aspects of a technology, it must be underpinned by sound science. Herbert Hoover, one of only two US presidents trained as an engineer (the other is Jimmy Carter), eloquently captured this balance in his memoirs (see "The Profession of Engineering").

The profession of engineering [9]

Herbert Hoover

It is a great profession. There is satisfaction in watching a figment of the imagination emerge through the aid of science into a plan on paper. Then it moves to realization in stone, metal, or energy. Then it brings jobs and homes to men. Then it elevates the standards of living and adds to the comforts of life. That is the engineer's high privilege.

The great liability of the engineer compared to men of other professions is that his works are out in the open where all can see them. His acts, step by step, are in hard substance. He cannot bury his mistakes in the grave like the doctors. He cannot argue them into thin air or blame the judge like the lawyers. He cannot, like the architects, cover his failures with trees and vines. He cannot, like the politicians, screen his shortcomings by blaming his opponents and hope that the people will forget. The engineer simply cannot deny that he did it. If his works do not work, he is damned. That is the phantasmagoria that haunts his nights and dogs his days. He comes from the job at the end of the day resolved to calculate it again. He wakes in the night in a cold sweat and puts something on paper that looks silly in the morning. All day he shivers at the thought of the bugs that will inevitably appear to jolt his smooth consummation.

On the other hand, unlike the doctor, his is not a life among the weak. Unlike the soldier, destruction is not his purpose. Unlike the lawyer, quarrels are not his daily bread. To the engineer falls the job of clothing the bare bones of science with life, comfort, and hope.

No doubt, as years go by, people forget which engineer did it, even if they ever knew. Or some politician puts his name on it. Or they credit it to some promoter who used other people's money to finance it. But the engineer himself looks back at the unending stream of goodness that flows from his successes with satisfaction that few professions may know. And the verdict of his fellow professionals is all the accolade he wants.

Enough said.

4.3 Discovery through design and design through discovery

Engineering, like poetry, is an attempt to approach perfection. And engineers, like poets, are seldom completely satisfied with their creations. ***Henry Petroski (1985) [3]***

Engineering is not only active but, as Petroski reminds us, a system filled with feedback loops and continuous improvement. Engineers are reminded by clients, the press, and the public when they fall short, but engineers are sufficiently optimistic to recognize their progress. This drive for continuous improvement was formalized by another engineer, W. Edwards Deming, who established the total quality management (TQM) process. TQM is defined as "a management approach of an organization, centered on quality, based on the participation of all its members and aiming at long-term success through customer satisfaction, and benefits to all members of the organization and to society" [10]. This is a strong mandate. However, it should not imply that engineering advances are linear. In fact, growth is incremental and is often accompanied by setbacks, hypothetically shown in Fig. 4.1.

It is interesting that TQM is not only about efficiency but also about ethics. In fact, a common problem is that TQM, when carried out correctly, must balance engineering and economic success with morally sound practices. The latter has generally received less focus than the former. In many cases, business has been antagonistic to and mutually exclusive to ethical behavior. Conversely, stakeholder theory has risen to offer the

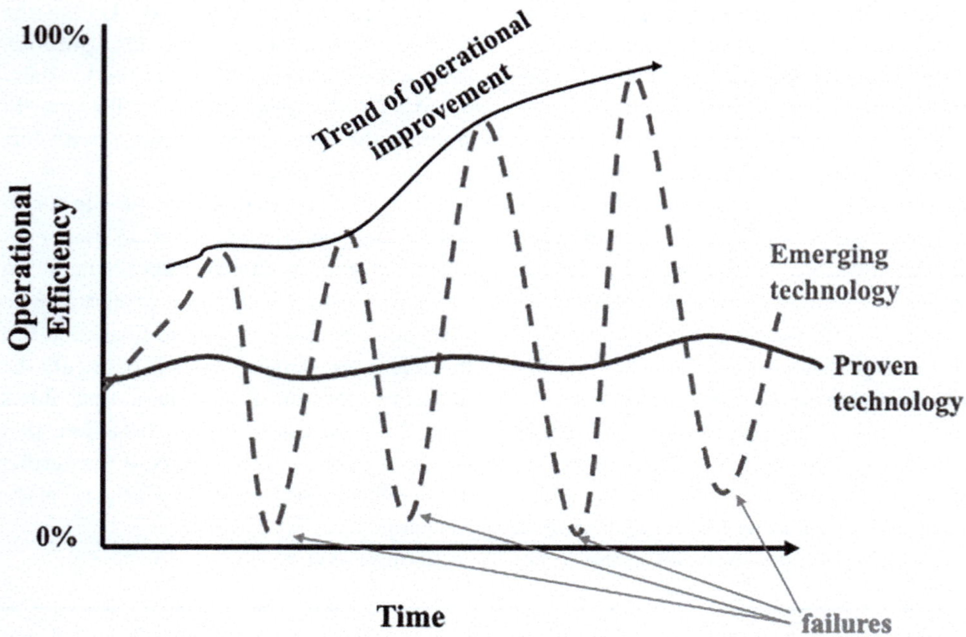

FIGURE 4.1 Hypothetical failure rate of new versus proven technologies. *Data from D.A. Vallero, Biomedical Ethics for Engineers, The Biomedical Engineering Series (2007).*

premise that business and ethics are inclusive of each other, moral concepts must drive the organization if it wants to be successful in whatever enterprise is its core business. Any TQM enterprise must define what business it is in and must embody continuous improvement. In addition, it must instill trust, empowerment, cooperation, and client satisfaction. All are elements of an ethical and efficient enterprise. Deming's 14 points of an effective enterprise reinforce the idea that success is a combination of business efficiency and ethics [11,12]:

1. Create constancy of purpose: Focus on long-term improvement for competitiveness and jobs, not just short-term profits.
2. Adopt the new philosophy: Acknowledge and embrace a new era of quality, demanding leadership for change.
3. Cease dependence on inspection: Build quality into the process to prevent defects, rather than just catching them later.
4. End price tag decisions: Minimize total cost by working with a single, trusted supplier, not just the cheapest.
5. Improve constantly and forever: Continuously improve all systems of planning, production, and service.
6. Institute training on the job: Develop skills for everyone, including modern supervision methods.
7. Institute leadership: Focus supervisors on helping people and machines improve, not just meeting numbers.
8. Drive out fear: Create an environment in which employees can work effectively and share ideas.
9. Break down barriers: Foster teamwork and communication among departments.
10. Eliminate slogans and targets: Remove exhortations and numerical targets that hinder quality, focusing on process improvement instead.
11. Eliminate numerical quotas: Remove quotas and numerical goals that focus on sheer numbers, not on quality.
12. Remove barriers to pride: Eliminate annual ratings/merit systems, allowing workers pride in their work.
13. Institute education and self-improvement: Provide vigorous programs for learning and growth for everyone.
14. Put everyone in the company to work: Involve everyone in the transformation toward quality.

In retrospect, most technological disasters could have been prevented if a company optimized based on physical, financial, *and* ethical drivers and constraints. Disasters like the Ford Pinto and Takata airbags would not have occurred. Research has shown that the principal causes of such failures were inherent flaws in their ethical systems [13]. This was due to the failure of management to design and apply strong enforcement means to inculcate ethical and moral principles when making quality decisions. Implementation of systems like TQM, Six Sigma, and ISO 9000 on their own is unable to prevent companies from experiencing these if sufficient attention is not given to ethical outcomes, that is, safer products, less harm to the environment, better quality of life during the lifetime of a product, and human health.

This is not an argument for a moratorium on innovation, but a call for a thorough evaluation and honest forecasting of all possible outcomes prior to the release of a technology. Especially in biomedical and computational applications, engineers push the envelope. They must. The growth in medical science is directly attributable to new technology, pharmacology, and systems. The same goes for computational sciences, environmental sciences, and all disciplines of engineering and design. The unforeseen setbacks depicted in Fig. 4.1 go with the territory, but the public does not take kindly to those that could have been prevented if reasonable steps had been taken. It hurts the professional and the profession because a preventable failure delays the overall advance in engineering solutions to problems, as shown in Fig. 4.2. It is even worse if the setbacks are the result of mishaps and mistakes. And, since engineering requires intensive and extensive training and experience, most situations ripe for mistakes and mishaps are only detectable to fellow engineers. This was well put by Petroski [3]:

> Engineers ... are not superhuman. They make mistakes in their assumptions, in their calculations, in their conclusions. That they make mistakes is forgivable; that they catch them is imperative. Thus, it is the essence of modern engineering not only to be able to check one's own work but also to have one's work checked and to be able to check the work of others.

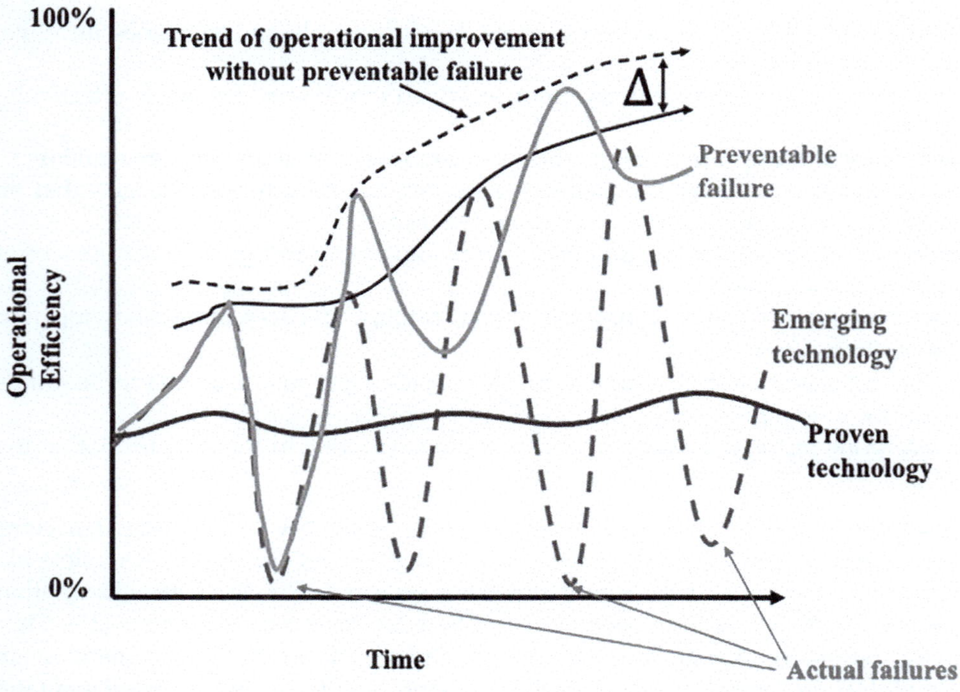

FIGURE 4.2 Hypothetical effect of preventable failure on overall advance of a new technology. The solid line represents the technological advance had the preventable failure not occurred. The delta is the difference in technology development attributed to the preventable mistake. Without the failure, the technological progress would be improved by Δ. *Data from D.A. Vallero, Biomedical Ethics for Engineers, The Biomedical Engineering Series (2007).*

4.4 Professionalism

Engineering has existed for millennia, such as the design of irrigation systems in Mesopotamia 6000 years ago and pyramids in Egypt more than 2000 years ago. In the 14th century, warfare engineering emerged in a form similar to modern military engineering. Civil and mechanical engineering became formal disciplines during the Industrial Revolution in the 18th and 19th centuries. Electrical, aerospace, and automotive engineering emerged early in the 20th century. Later in the century, the invention of the transistor and then computers changed every engineering discipline.

Thus engineering has been around for a long time, but a profession requires more than simple competence. A profession includes a core of specialized knowledge and training as well as work that is and is perceived to be of great value. The professional also has the trust of the client, which is delineated in a code of professional ethics. It also includes collegiality and trust among members, with each member maintaining a large amount of autonomy and self-regulation [14].

4.4.1 Due diligence

Although they are related and often occur in the same enterprise, engineering due diligence differs from business and financial due diligence [15]. Engineering due diligence focuses on the efficiency and risks of a technology's viability, processes, coding, infrastructure, hardware, and physical, chemical, and biological properties. Business due diligence focuses on the validation of business health, for example, the market, financial, legal, operational, and strategic elements of a new technology. Interestingly, these textbook definitions omit any moral or ethical content.

Engineering due diligence includes an exhaustive investigation and analysis of a property, project, or company's structural, environmental, operational, and other aspects of a design or project before any transaction or major decision, using the best available data, information, and professional knowledge. Due diligence identifies potential and actual risks, verifies that a plan is workable and feasible, and ensures safety during the earliest stages of a project and throughout the project or product's lifetime. The engineer must know all relevant codes, ordinances, and rules and confirm compliance.

Due diligence also requires transparency regarding the data and information considered and used in selecting all actions, including site assessments, reviews of financial viability, and code checks, that is, verifying that a design meets standards, ensures safety, and functions properly. For example, the code check would include a structural analysis consisting of a manual review and software to calculate and predict stress and stability for a building, which assures that clients understand potential liabilities and costs, protects them from unexpected problems, and ensures that all practices are ethical and safe. For example, for the cleanup of a polluted land site, due diligence regarding the land's suitability for various uses must be evaluated, for example, a project to return an abandoned industrial hazardous waste site to social use, that is, a brownfield site [16]. This consists of knowledge and adherence to zoning requirements, site conditions, environmental hazards, and estimates of remediation and waste removal costs. The environmental site assessment is a due diligence report of the condition of real estate that is provided to clients and government officials. The due diligence information can be found in records like maps and aerial photographs, site visits of above and below ground conditions, interviews, and professional judgment [17–19].

Phase I assessments are comprised of a review of the property's historical uses, interviews with neighbors and employees, requests under the Freedom of Information Act, government environmental databases, as well as a site inspection by a trained environmental professional. A Phase II assessment requires more data, for example, collecting site samples to perform a quantitative assessment of the quality of the air, soil, surface water, and groundwater [17,18]. Due diligence also requires that existing buildings are inspected and assessed for safety, necessary repairs, code compliance, structural conditions, and the presence of hazardous materials like asbestos and polychlorinated biphenyls (PCBs) [20,21].

Some aspects of due diligence apply to many occupations. However, for a professional engineer or physician, due diligence is inviolable. Perhaps the lack of due diligence in nonengineering professions can be instructive. Duke University is blessed with some of the world's best physicians and medical personnel. As a research institute, it often receives some of the most challenging medical cases, as was the case for Jesica Santillan, a teenager in need of a heart transplant. The surgeon in charge, Dr. James Jagger, had an impeccable record, and the hospital is world renowned for such a surgery. Unfortunately, something went terribly wrong. The heart that was transplanted was of a different blood type than that of the patient. The heart was rejected, and even after another heart was located and transplanted, Jesica died on February 22, 2003, due to the complications brought on by the initial rejection [22].

The logical and ethical question is how something so vital, crucial, and easy to know, like blood type, could be overlooked? It appears to be a systematic error. The system of checks and balances failed. And the professional, that is, the surgeon, was ultimately responsible and accountable for this or any other failure on his watch.

The mistakes emphasize the broader need for due diligence by the professional. The key lesson for health providers and others, including engineers, is that systems can fail and need to be continuously monitored and updated to prevent a seemingly small error that can lead to terrible outcomes. There is always a human element to any system. In this case, Duke's standard operating procedures depended on surgeons and organ-procuring agencies never making mistakes. The best way to improve safety measures is to focus on how a system can cause even an unlikely error and to prevent that error [23].

What else can we learn from the Santillan case? One lesson is that a system is only as good as the rigor and vigilance applied to it. There is really no such thing as "auto pilot" when it comes to systems. Aristotle helps us here. He contended that the whole is greater than the sum of its parts. This is painfully true in many public health disasters. Each person or group may be doing an adequate or even superlative job, but there is no guarantee that simply adding each of the parts will lead to success. The adage that things "fall through the cracks" is a vivid metaphor. The first mate may be doing a great job in open waters but may not be sufficiently trained in dire straits when the captain is away from the bridge. A first response team may be adequately trained for forest fires (where water is a very good substance for firefighting) but may not be properly suited for a spill of an oxidizing agent (where applying water can make matters considerably more dangerous). Without someone with a "global view" to oversee the whole response, the perfectly adequate and even exemplary personnel may contribute to the failure.

Systems are often needed in the daily lives of professionals. These systems must be tested and inspected continuously. Any step in the critical path that leads to failure is important. In fact, the more seemingly "mundane" the task, the less likely people are to think a lot about it. So, these small details may be the largest areas of vulnerability. Like the butterfly effect in chaos theory, the chain of events or critical path of one's decision will ultimately determine whether a desirable or an undesirable outcome will occur. One must wonder how many meetings before the Santillian case had significant discussions on how to ensure that the blood type is always properly labeled, knowing that even one breach would lead to dire consequences. One can venture that such a discussion began to occur much more frequently in preop meetings, as well as in hospital board meetings, at this site and arguably throughout the world. The public and professional peers will judge whether a fellow professional is applying due diligence or whether the designs and projects are impaired by something that *should* have been known and properly considered.

A defining aspect of professionalism is trust. In this case, trust in the institution was gravely damaged. Special care is needed to restore trust after it is lost or damaged. In a sense, this case reminds one of philosopher Immanuel Kant's categorical imperative, that is, an ethical action is one that can and should become a universal law, and one that everyone would follow for the betterment of society. In this instance, the imperative is to be honest and diligent to ensure that all systems in use are monitored and improved to the point that an error does not lead to harm [24].

Professionalism requires responsible investigation and ethical behavior as absolute, nonnegotiable obligations, and not merely situational or strategic choices. For example, AI has many promising attributes, but the potential disadvantages can be direct and indirect. Directly, defective, ineffectively trained, or poorly understood algorithms may produce erroneous results, which may have wide-scale impacts if they are used to set policies and serve as the basis for decisions about health, safety, and welfare. Indirectly, the increasing use of AI can exacerbate the weaknesses of environmental professionals who have overreliance and incomplete understanding, overconfidence, and lack of due diligence regarding AI model results. Many of these negative phenomena have already been witnessed in other industries, such as commercial aviation and the automotive industry [25].

AI can be properly deployed to support professional decision-making and used to assimilate and interpret data as part of the public good, but like any technology, it can be misused and provide erroneous results [26]. Unfortunately, as is often the case, AI technological development has already preceded the ethical considerations needed to assure the public that dire outcomes will not occur [27].

4.5 Technology, engineering, and economics

The National Academy of Engineering recently declared that "engineers and their inventions and innovations have helped shape the changes that have made our lives more productive and fruitful" [28]. But what does it mean to become more fruitful? To begin, we can consider the economic implications.

In economics, technology can be considered a tool of empowerment, as it empowers producers to generate more output from given levels of the two inputs, labor and capital. In this sense, just as a catalyst in chemistry is a substance that increases the rate of reaction, technology is a catalyst for production of output—as it increases the amount of output we get from given inputs. Incidentally, this analogy does not hold completely in the economics of technology, since in chemistry the catalyst is a chemical substance that increases the rate of a reaction without being consumed. However, technologies do indeed become consumed, that is, they become antiquated and in need of replacement and upgrades.

Technology allows for the use of more advanced capital, which results in better and faster ways to create output. Producers are rendered more efficient as they can produce more output given the same amount of input. This results in greater profit, which leads to economic growth. For example, if we consider the basic supply-demand model, we see that improvements in technology cause the supply curve to shift out to the right, meaning that producers will create more output for any given price (see Fig. 4.3). This raises equilibrium output and lowers equilibrium price. Equilibrium, in this case, is used to refer to the price level at which the aggregate supply curve (an upward-sloping line that illustrates how much producers would be willing to supply at any given price) crosses the aggregate demand curve (the downward-sloping line that shows how much of a good consumers would demand at any given price). This price level is called the "equilibrium" price because here, the amount supplied equals the amount demanded. It is a "balance," as any further increase in supply would disrupt the equilibrium. The same connotation of "balance" is seen in the usage of "equilibrium" in chemistry and thermodynamics, which is the study of work, heat, and energy in a system.

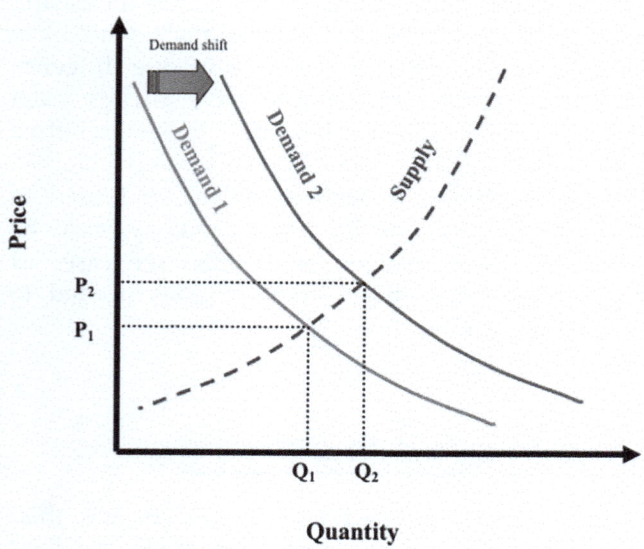

FIGURE 4.3 Supply-demand economic model. The increase in demand results in a relative increase in price and quantity needed to reach a new equilibrium point on the supply curve. *Data from D.A. Vallero, Biomedical Ethics for Engineers, The Biomedical Engineering Series, 2007.*

An object is said to be in thermodynamic equilibrium when it is in thermal, chemical, and mechanical equilibrium. It is observed that some properties of an object can change when the object is heated or cooled. Should two objects be brought into physical contact, an initial change in the properties of both objects results. For instance, heat, a form of energy, is transferred between these two objects when they are brought into contact with one another. Eventually, this transfer of energy stops. At this point, the objects are said to be in thermodynamic equilibrium [29]. Thus in both the economic and thermodynamic definitions of equilibrium, the system is said to be in balance; nothing changes unless acted upon by an exogenous force. In the case of market equilibrium, these endogenous forces act to shift the supply curve or the demand curve, subsequently changing where they cross to form the equilibrium price. Some examples of exogenous forces are increased costs for producers, changes in the regulatory environment, or changes in the wealth of consumers.

This supply-demand relationship is analogous to safety, as shown in Fig. 4.4. Like price in the economic model, the safety of an engineered device or system translates into costs. However, this is not a linear relationship. An unsafe product is not inexpensive if all the factors are included in the cost calculations. Primary costs of production are only part of the equation. Premature obsolescence and failures can lead to significant costs of lawsuits, lack of public trust, and recalls and warranty costs [30].

Technological advances often improve use of scarce resources, which are then allocated through the economic system, facilitating the achievement of Pareto optimality, the state in which resources are allocated so efficiently that no one entity can be more well-off without making at least one other entity worse off. It is a valuable concept in economics and multiobjective optimization, striving for outcomes where any change to benefit one entity imposes no cost on another. However, there are times when Pareto efficiency is unfair, such as in the protection of children and the infirm, who need to be treated differently than the

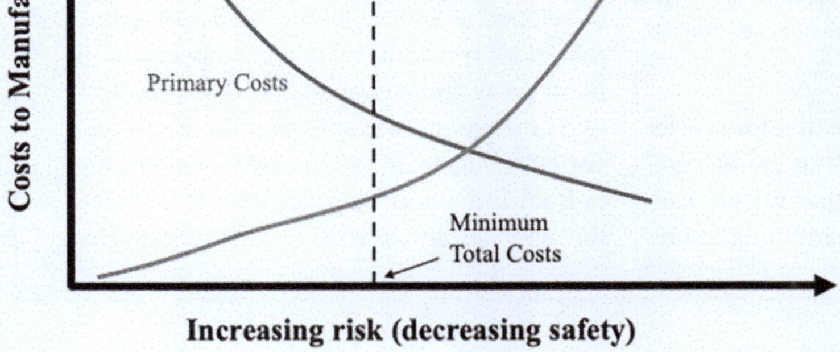

FIGURE 4.4 Safety and risks associated with primary and secondary costs. Increased safety can be gained by considering secondary costs in product and system design. *Data from M. Martin, R. Schinzinger, Engineering Ethics, McGraw Hill, Boston, 1996.*

rest of a population [31]. In this case, treating everyone as though they have no weaknesses lacks justice [32,33].

Technology broadens the horizon through which economics operates. When firms invest, they increase capital; and increasing input means more output and more economic growth. Depreciation of capital stock yields less output. The term sustainability in economics is akin to the concept of environmental sustainability. In economics, it is used to describe capital withstanding time and continuing to function to facilitate the production of input. In environmental engineering, sustainability relates to designing a system that maintains its desired efficiency and effectiveness over time and space. The positive side of emerging technologies is that they allow higher levels of economic sustainability for capital. The downside is that any problems can persist and even propagate through time and space.

Teachable moment: the dismal science versus the technological optimist

Economists and engineers differ in their worldviews. British political economist Robert Malthus[3] (1766–1834) predicted that starvation would result as projected population growth exceeded the rate of increase in the food supply. His forecast was based on the population growth at an exponential rate, whereas the food supply growth rate would be linear. These predictions greatly underestimated the role of technology in increasing the global food supply, causing many to doubt such economic models. In a later incarnation of Malthusian thinking, Paul Ehrlich's *Population Bomb* gives an exceedingly grim prognosis for the future:

Each year food production in underdeveloped countries falls a bit further behind burgeoning population growth, and people go to bed a little hungrier. While there are temporary or local reversals of this trend, it now seems inevitable that it will continue to its logical conclusion: mass starvation *[34].*

Not only does Ehrlich state that the world is headed toward calamity, but he is convinced that there is nothing anyone can really do that will provide anything more than temporary abatement. If one were to focus on Ehrlich's attitude toward technology as part of the solution to the impending problem, he would see Ehrlich's *technological pessimism*, so to speak. Ehrlich's lack of confidence in technology to deal with the problems plaguing the future is perhaps seen most explicitly on page 20: "But, you say, surely Science (with a capital 'S') will find a way for us to occupy the other planets of our solar system and eventually of other stars before we get all that crowded" [34]. Ehrlich was sure that "the battle to feed humanity is over." He insisted that India would be unable to provide sustenance for the 200-million-person influx in its population by 1980. He was wrong, thanks to the green revolution and Norman Borlaug. Borlaug and his team engaged in a program that developed a special breed of dwarf wheat that was resistant to a wide spectrum of plant pests and diseases, and that produced two or three more grains than the traditional varieties. His team then taught local farmers in both India and Pakistan how to cultivate the new strain of wheat. This astonishing

[3] His actual given name was Thomas Robert Malthus.

(cont'd)

increase in the production of wheat within a few years has come to be called the green revolution, its inception credited to Norman Borlaug [35]. Since 1968, when Ehrlich published his frightful predictions, "India's population has more than doubled, its wheat production has more than tripled, and its economy has grown nine-fold" [35]. Pakistan has progressed from harvesting 3.4 million tons of wheat each year to around 18 million, and India has made similar impressive movements from 11 million tons to 60 million [35].

As this predictive framework earned economics the title of "dismal science," engineers look upon the same problems with more technical optimism. Engineers "mess up" the Malthusian curve by finding ways to accomplish this (e.g., Borlaug "messed up" Ehrlich's predictions). Ehrlich criticizes the Pope for saying, "You must strive to multiply bread so that it suffices for the tables of mankind and not, rather, favor an artificial control of birth, which would be irrational, in order to diminish the number of guests at the banquet of life" [34]. Ehrlich feels that this representative *bread* cannot indeed be multiplied: Can we expect great increases in food production to occur through the placing of more land under cultivation? The answer is most definitely *"no"* [34]. The truth is, however, that Borlaug did in fact multiply the bread, and that engineers feel they *can* continue to design technologies that help support the growing population and the expanding needs of mankind. In *The Engineer of 2020*, one finds descriptions of the various ways engineers in the future will help to solve the very same problems Ehrlich (and the Malthusian model in general) are concerned with. Where Ehrlich felt technology's role in solving the problem would only be seen through how "Improved technology has greatly increased the potential of war as a population control device" [34], engineers look toward technology not as a "means for self-extermination" [34] but rather they opt to use it to *support and improve* life in the future.

Having examined Ehrlich's prognosis for the 1970s and 1980s, consider now a prognosis made by engineers for 2020. According to *The Engineer of 2020*, the world's population will approach eight billion people; much of this increase will be seen in groups that are today considered underdeveloped countries, mainly in Asia and Africa. The report goes on to say that "for the first time in history, the majority of people, mostly poor will reside in urban centers, mostly in countries that lack the economic, social, and physical infrastructures to support a burgeoning population" [28]. Engineers, however, see in the challenges posed by a highly crowded and densely populated world of 2020 an opportunity for "the application of thoughtfully constructed solutions through the work of engineers" [28]. Likewise, engineers look upon the necessity for improved healthcare delivery in the world of the future with confidence. They feel that they will be able to make advanced medical technologies accessible to this ever-growing global population base. In the developed world of 20 years from now, they see *positive* implications on human health due to improved air quality, the control and cleanup of hazardous waste sites, and focused efforts to treat diseases like malaria and AIDS [28].

(cont'd)

Engineers believe that they can solve the problems posed by the future [28], as opposed to views like the one posed by Ehrlich, who sees a future where "Small pockets of Homo sapiens hold on for a while in the Southern Hemisphere, but slowly die out as social systems break down, radiation poisoning takes effect, climactic changes kill crops, livestock dies off, and various man-made plagues spread. The most intelligent creatures ultimately surviving this period are cockroaches" [34]. Indeed, this dramatic example serves to illustrate the differences between engineers' technical optimism and doomsayers' technical *pessimism*, so to speak. But what does the future call for—the proverbial idealist or the modern-day skeptic? Mankind can either resign itself to failure and deem the problems it will come to face unsolvable, or it can press forward, attempting to solve the problems it faces and overcome tomorrow's challenges. While the question is subjective, it becomes clear that to have progress, most engineers choose the latter option.

Questions:

1. Is the moniker "dismal scientist" truly applicable to economists? Explain.
2. What are some downsides to technological optimism?
3. What is the difference between high density and crowding (or "overcrowding")?
4. Answer the following three questions:

a. What concerns are there about several nations' so-called "die with dignity" initiatives and laws, especially regarding the likelihood of coercion, particularly of the poor and desperate who may choose this given inadequate palliative care? ... or of the potential "slippery slopes," disability rights, mental health vulnerabilities, and inadequate social support systems? [36]

b. How can preimplantation genetic diagnosis (PGD), or more commonly preimplantation genetic testing (PGT), create ethical problems? In Canada, for example, PGT operates under the *Assisted Human Reproduction Act* (AHRA) to prohibit certain uses of this technology while leaving others, particularly clinical applications, to be managed through professional guidelines rather than comprehensive federal regulations. Explain some of the potential downsides regarding the use of PGT to encourage embryo destruction, eugenics, and so-called "designer babies" and "savior siblings" [37–42].

c. How do these two policies stand up against the "do no harm" expectation of the Hippocratic Oath, especially Article 7: "Into whatsoever houses I enter, I will enter to help the sick and I will abstain from all intentional wrong-doing and harm, especially from abusing the bodies of man or woman, bond or free"?

Defend your agreement or disagreement with the following statement:

Unfortunately, governmental policies can establish legally binding mandates as to

(cont'd)

when, how, and why innumerable innocent human beings shall be consigned or abandoned to death, reminiscent of the early slippery slope actions that led to the eugenic[4] movement embraced by many scientists in the United Kingdom, the United States, and Germany in the 1930s, which ultimately led to the heinous actions of the Nazis to produce a "master race." Thus, physicians and healthcare professionals who participate in euthanasia or in prenatal diagnosis may not see their efforts as "weeding out" the unfit, thus denying them their inalienable right to life. Stemming overpopulation is simply the environmental rationalization for eugenics. It is truly unfortunate that some nations, for example, are proud of their efforts to eradicate Down syndrome births. For example, in Iceland, nearly 100% of pregnant women with a positive prenatal diagnosis have chosen to abort the baby. Have they never met a Down syndrome human being? [43–45]

4.6 Optimizing ethics

Like practically everything else in engineering, best serving the public is a matter of optimization. The variables that we choose to give large weights will often drive the design. Treating cancer, providing devices to aid cardiovascular functioning, and delivering effective drug therapies are noble and necessary ends. The engineer must continue to advance the state of the science in these high-priority areas. But the public is a complicated entity, and each human body is uniquely exquisite. Thus, any possible adverse effects must be recognized. These should be incorporated and properly weighted when we optimize benefits. We must weigh these benefits against possible hazards and societal costs.

Everything engineers do is to protect and to enhance life, no matter the specific discipline. Thus all engineers must be mindful of their work's impact on human life. While their keenest interest is usually in human life, they must also be aware of how their work affects other species. In fact, some of the greatest resistance against scientific and engineering research is the result of the perceived indifference to pain and suffering in nonhuman animals.

Even microbial life must be respected for its intrinsic and instrumental value in our laboratories and in the environment, as well as the potential harm that can occur with emerging technologies like genetic engineering and synthetic biology [46–50]. One of the scariest scenarios is that of unchecked self-replication of unicellular species, as many engineering researchers are currently pushing the envelopes of nanotechnology, for example. This includes using microbes to manufacture complex pharmaceuticals and

[4] In 1883, Francis Galton, Charles Darwin's cousin, coined the term "eugenics." He reportedly objected to charity because it encouraged the poor to have more children. Such elitism is an example of social engineering gone amok.

chemical products that either cannot be manufactured or much less efficient in abiotic chemical reactors. However, many credible researchers fear the attendant risks of such nano-machinery, including the creation of unintended toxic byproducts and dangerous new strains of microbes [46,51].

The life aspects of engineering are, for the most part, the major success stories of engineering. For example, more than any other profession, engineers have prevented (in some cases, eliminated) devastating diseases with their public works projects, such as wastewater treatment, sanitary landfills, hazardous waste facilities, air quality controls, and drinking water supplies. Engineers sometimes need to be reminded that their work serves life.

The "medical" part of biomedical engineering implies a strong link between the medical and engineering professions. Like engineers, medical practitioners apply the basic sciences to achieve results. Thus the design of devices and structures is part of the larger healthcare provision. It is quite interesting to watch the growth and evolution of professions. They seem to oscillate between stages of specialization and contraction. Presently, both seem to be occurring in biomedical engineering. Medical doctors have become highly specialized, but their responsibilities have increasingly called for broader accountability. While the individual practitioner may lead one area of healthcare for the patient, they must build systems to ensure that all the other specialties effectively work together to provide the best care for each patient. Some of the most dramatic ethical and legal failures have resulted not from the practitioner's incompetence in their area of specialization, but from their lack of oversight and quality assurance of others who are part of the comprehensive care. Engineers are part of this system of care. In fact, some of the major advances in devices have come about through the close relationships with medical practitioners, such as the collaborations between teaching hospitals and biomedical engineering programs (like those at Duke, Johns Hopkins, and Stanford, to name a few). The engineer brings several assets to the team, including practicality, creativity, adaptability, and a long-term view.

The next term, ethics, can be summed up by Socrates as the way we ought to live. For engineers, this can be modified slightly. Engineering ethics is the way we ought to practice. The fundamental canons of the National Society of Professional Engineers (NSPE) code of ethics [52] capture what engineers "ought" to do. It states that engineers, in the fulfillment of their professional duties, shall:

1. Hold paramount the safety, health, and welfare of the public.
2. Perform services only in areas of their competence.
3. Issue public statements only in an objective and truthful manner.
4. Act for each employer or client as faithful agents or trustees.
5. Avoid deceptive acts.
6. Conduct themselves honorably, responsibly, ethically, and lawfully to enhance the honor, reputation, and usefulness of the profession.

The canons are the professional equivalents of "morality," which refers to societal norms about conduct that is acceptable (virtuous/good) and unacceptable (evil/bad). These norms are shared by members of society to provide stability as determined by consensus [53]. Philosophers consider professional codes of ethics and their respective canons to be normative ethics, which is the philosophical study of ethics concerned with classifying actions as right and wrong without bias. Normative ethics is contrasted with descriptive ethics, which is the study of what a group believes to be right and wrong and how it enforces conduct. Normative ethics regards ethics as a set of norms related to actions. Descriptive ethics deals with what "is" and normative ethics addresses "what should be."

Philosopher Bernard Gert categorizes behaviors into what he calls a "common morality," which is a system that thoughtful people use implicitly to make moral judgments [54]. According to Gert, humans strive to avoid five basic harms: death; pain; disability; loss of freedom; and loss of pleasure. Arguably, the job of the engineer is to design devices, structures, and systems that mitigate such harms in society. Similarly, Gert identifies ten moral rules of common morality:

1. Do not kill.
2. Do not cause pain.
3. Do not disable.
4. Do not deprive of freedom.
5. Do not deprive of pleasure.
6. Do not deceive.
7. Keep your promises.
8. Do not cheat.
9. Obey the law.
10. Do your duty.

Most of these rules are proscriptive. Only rules 7, 9, and 10 are prescriptive, telling us what to do rather than what not to do. The first five directly prohibit the infliction of harm on others. The next five indirectly lead to prevention of harm. Interestingly, these rules track quite closely with the tenets and canons of the engineering profession (see Table 4.2).

The Gert model is good news for engineering ethics. Numerous ethical theories can form the basis for engineering ethics and moral judgment. Immanuel Kant is known for defining ethics as a sense of duty. Thomas Hobbes presented ethics within the framework of a social contract, with elements reminiscent of Gert's common morality. John Stuart Mill considered ethics to be about the goodness of action or decision as the basis for utilitarianism. Philosophers and ethicists spend much effort and intellectual energy deciphering these and other theories as paradigms for ethical decision-making. Engineers can learn much from these points of view, but, in large measure, engineering ethics is an amalgam of various elements of many theories. As evidence, the American Society of Mechanical Engineers (ASME) has succinctly bracketed ethical behavior into three models [55]:

<u>Malpractice, or minimalist, model</u>: In some ways, this is really not an ethical model in that the engineer is only acting in ways that are required to keep his or her license or professional membership. As such, it is more of a *legalistic* model. The engineer operating within this framework is concerned exclusively with adhering to standards and meeting requirements of the profession and any other applicable rules, laws, or codes. This is often a retroactive or backward-looking model, finding fault after failures, problems, or accidents happen. Any ethical breach is assigned based upon design, building, operation, or other engineering steps that have failed to meet recognized professional standards. This is a common approach in failure engineering and in ethical review board considerations. It is also the basis of numerous engineering case studies.

TABLE 4.2 Canons of the National Society of Professional Engineers compared to Gert's rules of morality.

Engineers shall:	Most closely linked to rules of morality identified by Gert
1. Hold paramount the safety, health, and welfare of the public.	• Do not kill. • Do not cause pain. • Do not disable. • Do not deprive of pleasure. • Do not deprive of freedom.
2. Perform services only in areas of their competence.	• Do not deceive. • Keep your promises. • Do not cheat. • Obey the law. • Do your duty.
3. Issue public statements only in an objective and truthful manner.	• Do not deceive.
4. Act for each employer or client as faithful agents or trustees.	• Do not deprive of pleasure. • Keep your promises. • Do not cheat. • Do your duty.
5. Avoid deceptive acts.	• Do not deceive. • Keep your promises. • Do not cheat.
6. Conduct themselves honorably, responsibly, ethically, and lawfully to enhance the honor, reputation, and usefulness of the profession.	• Do your duty. • Obey the law. • Keep your promises.

Data from National Society of Professional Engineers, NSPE Code of Ethics for Engineers. https://www.nspe.org/resources/ethics/code-ethics, 2016 (accessed 24,01.23); B. Gert, Common Morality: Deciding What to Do, Oxford University Press, 2004.

Reasonable-care, or due-care, model: This model goes a step further than the minimalist model, calling upon the engineer to take reasonable precautions and to provide care in the practice of the profession. Interestingly, every major philosophical theory of ethics includes such a provision, such as the harm principle in utilitarianism, the veil [56–58] of ignorance in social contract ethics, and the categorical imperative in duty ethics. It also applies a legal mechanism, known as the *reasonable person standard*. Right or wrong is determined by whether the engineer's action would be seen as ethical or unethical according to a "standard of reasonableness as seen by a normal, prudent nonprofessional." Note that this is not the "reasonable engineer standard," Thus the reasonable person standard adds an onus to the profession, that is, not only should an action be acceptable to one's peers in the profession, but to those outside of engineering. An action could very well be legal, and even professionally permissible, but may still fall below the ethical threshold if reasonable people consider it to be wrong.

Good Works Model: This is the highest level of professional behavior. An ethical technologist, of course, must follow the standards and codes and ensure that a product or process not only provides for performance but also meets the assurances of optimal safety, protection of public health, or providing for social welfare.

Rawls [59] introduced an insightful ethical decision-making tool that relates well to emerging technology development, that is, the so-called "veil of ignorance." The metaphorical veil in front of the technologist prevents them from knowing who risks the greatest harm from a data-poor, new technology. Thus the decision to continue with the technology is based on its potential impact on the most vulnerable current and future members of society, for example, pregnant women, the unborn, neonates, children, the infirm, the immunocompromised, the poor, and the elderly [5]. Like the Christian concept based on Matthew Gospel 25:40, which calls for the protection of the "least of these" [60,61], the technologist is obligated to consider more than the present and target users but also the technology's potential impacts on future and distant peoples [62]. This consideration must include both the ends and the means of developing a new technology, especially during research and expected uses and misuses. The responsibility is daunting and unique since the developer can only prevent an ethical breach in its early stages, such as responsible research conduct and good laboratory practice, with personal control dropping after the introduction and beta testing [63]. Like most ethical frameworks, the easiest and most straightforward step is awareness. Decision-making is more complex, and the behavior of users is increasingly difficult to predict [64–66].

Various ethical models are available to support technological decisions and actions. They differ among professions but can be generalized. As mentioned, the good works model requires that a truly ethical engineer or technologist goes beyond abiding by the law or preventing harm. An ethical technologist, of course, must follow the standards and codes and ensure that a product or process not only provides for performance but also meets the assurances of optimal safety, protection of public health, or providing for social welfare [67]. An analytical tool related to the good works model is the *net goodness* determination, which estimates the goodness or wrongness of an action by weighing its morality, likelihood, and importance [68].

This good works model is rooted in the moral development theories such as those expounded by educational ethicists [69–77], who noted that moral action is a complex process entailing four components: moral awareness (or sensitivity), moral judgment, moral motivation, and moral character. The technologist must first be aware that the situation is moral in nature; that is, at least that the actions considered would have consequences for others. Second, the technologist must have the ability to judge which of the potential actions would yield the best outcome, considering those likely to be affected. Third, the technologist must be motivated to prioritize moral values above other sorts of values, such as wealth or power. Fourth, the technologist must have the strength of character to follow through on the decision to act morally.

Arguably, age and education are the two most important factors in determining a person's likelihood of behaving morally, that is, of being morally aware, making moral judgments, prioritizing moral values, and following through on moral decisions [73]. These are strong indicators of experience and seem to be particularly critical in making moral judgments. A person's ability to make moral decisions tends to grow with maturity as they pursue further education, generally reaching its final and highest stage of development in early adulthood. This is one of the reasons that professions like engineering and medicine require a progression from book learning and intellectual awareness in undergraduate coursework to professional practice in the workplace under the tutelage of a seasoned professional before receiving professional credentialing. This theory of moral development is illustrated in Table 4.3

The normative model of technological development can be developed along the same lines as Kohlberg's model. The moral need to consider the impact one's actions will have on others forms the basis for the normative model we are proposing. Pursuing an activity with the goal of obeying the law has as its driving force the avoidance of punishment, and pursuing an activity with the goal of improving profitability is a goal clearly in line with stockholders' desires; presumably, customers', suppliers', and employees' desires must also be met at some level. And finally, pursuing an activity with the goal of "doing the right thing," behaving in a way that is morally right and just, can be the highest level of professional behavior. This normative model of ethical engineering can be illustrated as Fig. 4.5.

Kohlberg's model of moral development (Table 4.3) and the model for engineering professional growth are dramatically similar. Avoiding punishment in the moral development model aligns with the major focus of avoiding problems early in one's career. The preconventional level and early career experiences have similar driving forces.

At the second level in the moral development model, there is a focus on pleasing one's peers and adherence to community values, while in the professionalism model, the engineer must balance the needs of clients and fellow professionals with those of society at large. Engineering services and products must be of high quality and be profitable, but the focus is shifting away from self-centeredness and personal well-being toward external goals.

Finally, at the highest level of moral development, a concern with universal moral principles begins to govern actions, while for the corporate model, fundamental moral principles have to do more with professionalism than with corporate decisions. The driving

TABLE 4.3 Kohlberg's stages of moral development.

Preconventional level	1. Punishment-obedience orientation 2. Personal reward orientation
Conventional level	3. "good boy"-"nice girl" orientation 4. Law and order orientation
Postconventional level	5. Social contract orientation 6. Universal ethical principle orientation

Note that these steps are progressive. During the two earliest stages of moral development, which are combined under the heading "pre-conventional level," a person is primarily motivated by the desire to seek pleasure and avoid pain. The "conventional level" consists of stages three and four. In stage three, the actor is motivated by the consequences that actions have for peers and their feelings about these actions. During stage four, the actor considers how the wider community will view the actions and be affected by them. Few people reach the "post-conventional" stage, requiring an even broader perspective, where their moral decision-making is guided by universal moral principles [24,59,78].

Data from R. Duska, M. Whelan, Moral Development: A Guide to Piaget and Kohlberg, 1975; L. Kohlberg, Stages of moral development, Moral Education 1, (51) (1971) 23–92.

FIGURE 4.5 Comparison of Kohlberg's moral development stages to professional development in engineering. *Data from D.A. Vallero, Biomedical Ethics for Engineers, The Biomedical Engineering Series, 2007; D.A. Vallero, Beyond responsible conduct in research: new pedagogies to address macroethics of nanobiotechnologies, Journal of Long-Term Effects of Medical Implants 17, (1) (2007). [1,79].*

force or motivation is trying to do the right thing for an internal moral need, versus merely a concern with legality and financial considerations. The collective behaviors of each engineer set the example for the profession, that is, a bottom-up, "microethical" set of values that drive the public's trust in the engineering profession. Of course, any profession also has "macroethical" responsibilities to ensure that the behavior of each member is moral [80–84]. These shared values are codified in guidelines and rules of professional conduct, such as those of the NSPE and the American Medical Association.

Professional growth is enhanced when engineers and technologists base their decisions on sound business and engineering principles. Ethical content should not be an afterthought but integrated within the technology design decision-making process. The exemplars of technology recognize the broad impacts their decisions may have, and act so that their actions will be in the best interest of not only the technologist's own needs and the interests of the organization where the technologist works, but also the broader society and even future generations of technologists influenced by exemplary behavior.

Much of the ethics training emphasizes preconventional thinking, with a focus on adherence to codes, laws, and regulations within the organizational climate, especially profitability and timeliness of product releases. This benefits the technologist and organization, but is only a step toward full professionalism, and sometimes works against exemplary behavior. As mentioned, the engineering codes of ethics require that the principal client of the professional engineer is "the public" [5]. This ethical canon has primacy over all others. So, anything the professional engineer does cannot violate this canon by choosing a behavior that supplants the public interest with personal gain.

No matter how competent, objective, honest, and faithful, the engineer must not jeopardize public safety, health, or welfare. Here, the primary focus of the physician and that of the engineer begins to diverge. The medical practitioner's principal client is the patient, whereas the primary client of the engineer is the public. Nothing is more important to the engineer than the health and safety of the public. The engineer, especially the biomedical engineer, must navigate both professional codes. As evidence, the Biomedical Engineering Society recently recognized this with its approval of a new code of ethics in 2004 (see Appendix 2). The code recognizes that the biomedical engineer practices at the confluence of "expertise and responsibilities in engineering, science, technology, and medicine" [85]. Mirroring the NSPE code, the biomedical engineering community reminds its members that "public health and welfare are paramount considerations."[5]

Public safety and health considerations affect the design process directly. Almost every design now requires at least some attention to sustainability and environmental impacts. Biomedical designs are not excluded. For example, recent changes in drug delivery have been required, such as moving away from the use of greenhouse gas propellants like chlorofluorocarbons (CFCs) and instead using pressure differential systems (such as physical pumps) to deliver medicines. This may seem like a small thing or even a nuisance to those who must use them, but it reflects an appreciation for the importance of incremental effects.

One inhaler does little to affect the ozone layer or threaten the global climate, but millions of inhalers can produce enough halogenated and other compounds that the threat must be considered in designing medical devices. To the best of our abilities, we must ensure that what we design is sustainable over its useful lifetime. This requires that the engineer think about the life cycle, not only during use but also when the use is complete. Programs such as "design for recycling" (DFR) and "design for disassembly" (DFD) allow the engineer to consider the consequences of various design options in space and time. They also help designers to pilot new systems and to consider scale effects when ramping up to full production of devices.

Revisiting the code sheds light on how an emerging technological decision can be optimized. The code is a collection of drivers that are inputs to the optimization [86]. The code must be authoritative in governing the conduct of every member of the profession regarding what is morally permissible. This governance applies to the entire group of members who are acting as moral agents, not judges. Each canon in the code establishes standards of conduct beyond the individual engineer's common sense, traditions, and

[5] This wording is curious. It omits "public safety." However, safety is included under professional obligations that biomedical engineers "use their knowledge, skills, and abilities to enhance the safety, health, and welfare of the public." The other interesting word choice is "consideration." Some of us would prefer "obligations" instead. These compromises may indicate the realities of straddling the design and medical professions. For example, there may be times when the individual patient needs supersede those of the general public, and vice versa.

customs. The code must be a gateway and guide to new members and a reminder to established members of their moral responsibilities, that is, the code includes mnemonic prompts.

Ethical optimization extends the meaning of optimization beyond operational efficiency and reliability. An emerging technology, of course, must be doable based on scientific principles. Ethical optimization, however, also requires that an ethical critical path be built into the design process, parallel with and embodied in the overall technology design concept-to-completion and discovery-to-market flow path [87–89].

If at any decision point in the design it becomes apparent that an ethical breach may occur, the process must be halted. Characterizing as many eventualities as possible and potential outcomes in the critical path is an essential part of the design of a technology long before its introduction and continuing throughout the entire technology life cycle [90]. For example, engineers working with AI, nanotechnologies, bioreactors, genetically modified organisms and materials, and robotics must consider all possible avenues of failure, both operational and moral. Some obviously fall very much within the realm of "unknown knowns" or even the "unknown unknowns" [91–96]. For example, at what point in the root cause analysis of the ethical breach in the "Dieselgate" scandal does the blame begin, where unlawful software was used in Volkswagen automobile emission testing systems to give fraudulent readings of oxides of nitrogen (NO_x)? Was the software intentionally coded to defraud, or was the code misused to give incorrect results? If the former, the original coder was not to blame. But should the coder's superiors have considered the possibility of misuse? Is this a case where someone in the decision tree should have assessed the possibility of misuse even though they were not the one who misused the code? If so, that is the point in the critical path where a correction should have been made based on the potential ethical breach. Somewhere in the technological critical path, the decision was made to misuse the testing system, resulting in a 40-fold emission increase of NO_x [97–100].

All failsafe mechanisms must be in place *and* are operational. Quality assurance officers note that testing for an unlikely but potentially devastating event is difficult. Just because no incidents have yet occurred does not mean everything is okay, and it could mean that the prospective risk assessment does not include every factor that may fail. This is at the heart of engineering reliability.

Because engineering is technical, it depends on scientific breakthroughs. Science and technology are changing drastically and irrevocably. The engineer must stay abreast of new developments. This is particularly challenging in emerging fields like biomedical and computational technologies, where the scale of interest continues to contract with specialization. It was not that long ago when organs were the most refined scale of interest, giving way to organelles and cells. Now, the "nano" scale is receiving the most attention, where structures and systems have design units that are a few angstroms.

4.7 Engineering: integrated and specialized

Professional specialization has both advantages and disadvantages. The principal advantage is that the practicing engineer can focus more sharply on a specific discipline compared to a generalist. The principal disadvantage is that integrating the different parts can be challenging. For example, in a very complex design, only a few people can see the overall goals. Thus those working in specific areas may not readily see duplication or gaps that they assume are being addressed by others.

A classic example of the shortcomings is a case in the video, *Professional Ethics and Engineering*, produced by Duke's Center for Applied Ethics [101]. In it, a seasoned engineer is interviewed by a professional review panel and asked whether he knew that the foundation of a building being constructed was mismatched to the soil type. He said that he did, but it was none of his business, since it was the job of the soil engineers. The panel reminded him that people died because of this failure. As the Santillian case reminds us, medical scenarios can also suffer, since the whole is greater than the sum of its parts.

The work of technical professions is both the effect and cause of modern life. When undergoing medical treatment and procedures, people expect physicians, nurses, emergency personnel, and other health care providers to be current and capable. Society's infrastructures, buildings, roads, electronic communications, and other modern necessities and conveniences are expected to perform as designed by competent engineers and planners. But how does society ensure that these expectations are met? Much of the answer to this question is that society cedes a substantial amount of trust to a relatively small group of experts, the professionals in increasingly complex and complicated disciplines that have grown out of the technological advances that began in the middle of the 20th century and grew exponentially in their waning decades.

Professions, including engineering, are not neatly subdivided as they once were. A visit to the hospital shows that not only do many physicians specialize in particular areas of medicine (e.g., neuromedicine, oncology, and geriatrics), but all these physicians must rely on chemists, radiologists, and tomographic experts to obtain data about their patients (e.g., from serum analysis, magnetic resonance and CT scans, and sonography). In fact, many of the solutions (cures?) to health problems require an intricate cacophony among doctors, biomedical engineers, and technicians, as well as public and community health professionals, epidemiologists, and environmental engineers to prevent and control many of the diseases, making treatment unnecessary. For example, a drug delivery system requires the understanding of the biochemical needs of the patient, the fluid mechanics of the pharmacology, and the actual design of the apparatus. This is a continuum among science, engineering, and technology.

Within this highly complex, contemporary environment, practitioners must ensure that they are doing what is best for the profession and what is best for the patient and client. This *best practice* varies by profession and even within a single professional discipline, so the actual codified *rules* (codes of ethics, either explicit or implicit) must be tailored to the needs of each group. However, many of the ethical standards are quite similar for most technical professions. For example, people want to know that the professional is trustworthy. Trustworthiness is a function of how skilled the professional is in the chosen field and how ethical the person is in practice. Thus the professional possesses two basic attributes: subject matter knowledge and character. Maximizing these two attributes enhances professionalism.

4.8 Who is a professional?

There is debate on just who is a professional. The author often asks the students enrolled in his Professional Ethics course to provide examples of professionals. The list always includes physicians, engineers, airline pilots, and lawyers, and usually includes accountants. A few students consider clergy and military officers (not usually enlisted personnel) to be professionals. Some include businesspersons, teachers, and scientists. Only a small minority include professional athletes, although even most of those admit this is because the group includes the term "professionals." Several other disciplines are included, but support diminishes after the first few. The query is quite unscientific, simply asking for their opinions, but this is interesting since the students are not given any criteria from which to label something as a professional, yet they are readily equipped to answer. They are simply asked, and a list is generated.

The intuitive powers of students (actually, of people in general) are quite amazing. They usually can differentiate very complicated subject matter (e.g., pollutant types, risk, values, and obligatory moral behavior), but they often cannot tell you why. In other words, they cannot explain their methodology, but they clearly use one. So, the author delves a bit more deeply by asking the students to tell him why one group is professional, and another is not. They usually note readily that it is not that one is necessarily more "valuable" than the other, or that ease or difficulty is a determinant factor. Certain highly technical, critical, difficult, and respected "jobs" like an aircraft mechanic are not generally considered "professional." Why not?

Most of us will admit that a certain threshold of expertise is needed to ascribe the label "professional" to someone. However, as our aircraft mechanic example demonstrates, expertise is clearly a necessary but insufficient quality of professionalism. All professionals must be experts in the field, but not all experts are professionals. One distinguishing characteristic of a professional is the level of accountability and degree of responsibility. In our aircraft example, the mechanic is a highly trained expert in a particular area, but within a tightly defined span of control and realm of responsibility. However, the airline pilot is uniquely responsible for all aspects of the plane. That is, everything the mechanic does, everything the copilot does, and everything about the plane, the weather, the flying conditions, and whatever it takes to transport the plane, passengers, and cargo safely to the destination is the responsibility of the pilot. The captain is also accountable for everything that transpires on his or her "watch." Any organization, like any system, has a set of norms and mores that are distributed throughout its membership. Indeed, the mechanic who does not follow protocol will be reprimanded, usually severely, but the captain shares the blame, if for no other reason than the "system" being led by the captain does not adequately ensure high-quality performance by the mechanic. Nor can the pilot defer and deflect blame to the airline company. No company policy or business decision should detract from the professional responsibilities of the pilot. In a word, the pilot remains responsible. The pilot is accountable for the whole flight experience.

4.9 What is technical?

Although this book is all about emerging technologies, we have gotten this far in our discussion without defining what is meant by "technical"; better late than never. With a better idea of what it means to be a professional, we can endeavor to characterize certain professionals as technical professionals. The technical professional must have a mastery of technical subject matter. But what does this mean? Is the ability to play a video game or listen to an electronic device a technical skill? Most of us would not think so. Is the ability to run sonographic software and hardware a technical skill? Most would agree. However, is this ability enough to be considered a technical professional? Many would say, "no, it simply means the person is a skillful technician."

4.9.1 Systematics: incorporating ethics in the design process

The key to engineering success is ensuring that all the right factors are considered in the design phase and that these factors are properly implemented and monitored throughout the project. Integrated engineering approaches require that the engineer's responsibilities extend well beyond the construction, operation, and maintenance stages. Such an approach has been articulated by the ASME. One way to visualize the so-called design for the environment (DFE) recommended by the ASME is to use an integrated matrix [102] (see Table 4.4). This allows the engineer to see the technical and ethical considerations associated with each component of the design, as well as the relationships among these components. For example, health risks, social expectations, environmental impacts, and other societal risks and benefits associated with a device, structure, product, or activity can be visualized at various stages of the manufacturing, marketing, and application processes. This yields a few two-dimensional matrices (see Fig. 4.6) for each relevant design component. Each respective cell indicates both the importance of that component and the confidence (expressed as scientific certainty) that the engineer can have about the underlying information used to assess the importance (see legend to Fig. 4.6).

The matrix approach is qualitative, but it allows comparisons of alternatives that would otherwise be incomparable, which is often the case in bioethics. To some extent, even numerical values can be assigned to each cell to compare them quantitatively, but the results are at the discretion of the analyst, who determines how different areas are weighted. The matrix approach can also focus on design for a more specific measure, such as energy efficiency or product safety, and can be extended to corporate activities as a system.

4.9.2 Warfare

One of the quandaries of engineering and technology is that innovations can get away from us. The technologist may assume that her invention will be used for healthcare or environmental protection but later finds that it has military applications. Being antiwar or opposed to the military-industrial complex, she may wish that her technology would not be used in this way but is powerless to prevent it.

TABLE 4.4 Functions that must be integrated into an engineering design.

1. Baseline studies of natural and built environments
2. Analysis of project alternatives
3. Feasibility studies
4. Environmental impact studies
5. Assistance in project planning, approval, and financing
6. Design and development of systems, processes, and products
7. Design and development of construction plans
8. Project management
9. Construction supervision and testing
10. Process design
11. Startup operations and training
12. Assistance in operations
13. Management consulting
14. Environmental monitoring
15. Decommissioning of facilities
16. Restoration of sites for other uses
17. Resource management
18. Measuring progress for sustainable development

Data from American Society of Mechanical Engineers, "PDHonline Course R139 (2 PDH): Ethics in ME Professional Practice" ASME, Fairfax, Virginia, 2020. Accessed: January 21, 2026. [Online]. Available: https://pdhonline.com/courses/r139/ASME%20Professional%20Practice%20Curriculum%20Eng%20Ethics.pdf.

This is not to say that any technology used for warfare is unethical. Indeed, even Roman Catholic teaching allows for a just war. For example, Paragraph 2309 of the Catechism of the Catholic Church stipulates the just war doctrine [103]:

- "the damage inflicted by the aggressor on the nation or community of nations must be lasting, grave, and certain"; ("A Short Primer on Just War Theory - Catholic Insight")
- "all other means of putting an end to it must have been shown to be impractical or ineffective"; ("Catechism of the Catholic Church") ("Catechism of the Catholic Church")
- there must be serious prospects of success; and,
- the use of arms must not produce evils and disorders graver than the evil to be eliminated. "The power of modern means of destruction weighs very heavily in evaluating this condition." ("United States Catholic Catechism for Adults") ("United States Catholic Catechism for Adults")

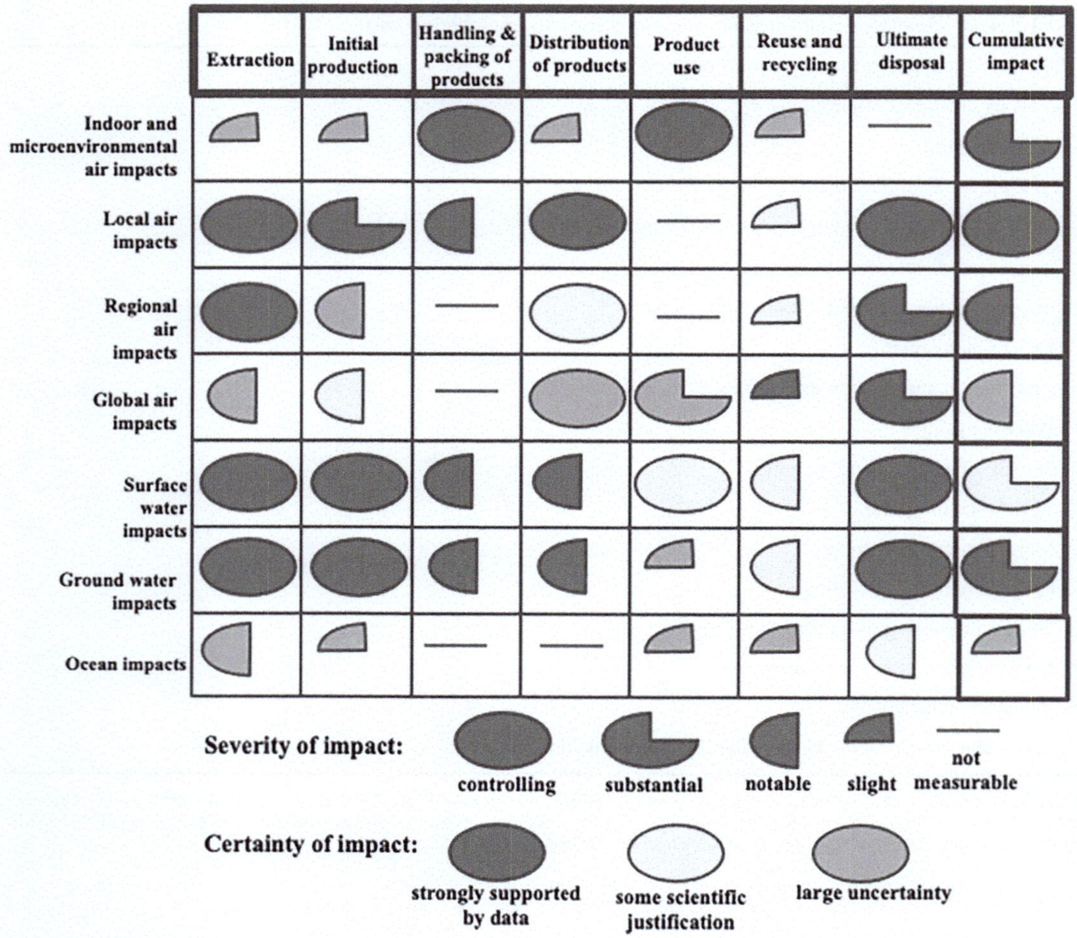

FIGURE 4.6 An example of an integrated engineering matrix; in this instance, applied to sustainable designs. From: [104]. *Data from American Society of Mechanical Engineers, "PDHonline Course R139 (2 PDH): Ethics in ME Professional Practice" ASME, Fairfax, Virginia, 2020. Accessed: January 21, 2026. [Online]. Available: https://pdhonline.com/courses/r139/ASME%20Professional%20Practice%20Curriculum%20Eng%20Ethics.pdf.*

The doctrine continues to state that the evaluation of these conditions for moral legitimacy belongs to the prudential judgment of those who have responsibility for the common good. Note the similarity to the previously discussed double effect principle.

There are also similarities to dual use. Every engineer and technologist needs to recognize that much of their responsibility is affected by geopolitical realities and a global economy. Thus there must be respect and appreciation for how cultures differ, but they should be mindful of the possibility of misuse of innovations and emerging technologies.

Dual use comes in two types: a good use and a bad use. The good type is any science, engineering, and technology that is designed to provide a governmental benefit, for example, improved materials to support a military aircraft or spacecraft, but as a result, provides useful materials and products for the public, for example, improved coatings for vehicles. The bad type is any research and technology that simultaneously benefits and places society at risk, like biotechnological advances that improve vaccines but also increase the risks of bioterrorism.

Dual-use fears can be seen as a form of violation of due diligence. For example, extra care is needed in using similar strains and species in genetically modified microbes, such as the substantial differences and similarities among various strains of *Bacillus* spp. The technologist must give due diligence to ensure that even a slight change in conditions may elicit unexpected and unwanted outcomes, such as weapon-grade or resistant strains of bacteria and viruses [26].

4.10 Conclusions

This chapter has discussed several aspects of what it means to be an ethical engineer. Contemporary society demands many things from its professionals. For those in technical fields, the two prominent expectations are trust and competence. Both are required and built into the codes of practice and ethics of each technical discipline. They are career-long and, indeed, lifelong.

References

[1] D.A. Vallero, Biomedical ethics for engineers, The Biomedical Engineering Series (2007).
[2] K. Johnson, Editorial "we need to keep leading students into science, math", *The Durham (NC) Herald-Sun*, ed, McClatchy Co, Durham, North Carolina, (2006), p. 1.
[3] H. Petroski, To Engineer is Human: The Role of Failure in Successful Design, St Martins Press, (1985).
[4] G. Constable, B. Somerville, A Century of Innovation: Twenty Engineering Achievements That Transformed Our Lives, Joseph Henry Press, (2003).
[5] P.A. Vesilind, A.S. Gunn, Hold Paramount: The Engineer's Responsibility to Society, Cengage Learning, (2010).
[6] The Engineering Toolbox, Factors of Safety. https://www.engineeringtoolbox.com/factors-safety-fos-d_1624.html (accessed 20.01.26).
[7] J.C. Musto, The safety factor: case studies in engineering judgment, International Journal of Mechanical Engineering Education 38 (4) (2010) 286–296. Available from https://doi.org/10.7227/IJMEE.38.4.22010/10/01.
[8] F. Dyson, H. Bethe, Disturbing the Universe, American Institute of Physics, (1979).
[9] H.C. Hoover, The Memoirs of Herbert Hoover: 1874-1920. Years of Adventure, Macmillan, (1951).
[10] T.J. Albrecht, ISO 9002 implementation: lessons learned, Quality Digest 14 (1994) 55–61.
[11] W.E. Deming, W. Edwards Deming, Madonna University, (1991).
[12] W.E. Deming's, Points for Total Quality Management, ed.
[13] E.C. Etienne, Takata's exploding airbags: lessons from a quality debacle, Journal of Business Law 8 (2) (2020) 1–18.
[14] M. Losavio, K.C. Seigfried-Spellar, J.J. Sloan Iii, Why digital forensics is not a profession and how it can become one, Criminal Justice Studies 29 (2) (2016) 143–162. Available from https://doi.org/10.1080/1478601X.2016.1170281. 2016/04/02.
[15] E.G. Krasnow, J.M. Pelkey, J.W. King, Considerations unique to the purchase and sale of broadcast stations, Media Law & Policy 19 (2009) 149.

[16] J.H. Knox, N. Tronolone, Environmental justice as environmental human rights, Vanderbilt Journal of Transnational Law 57 (2024) 153.

[17] W.J. Pierro, When environmental due diligence enters' phase II, American Bankers Association. ABA Banking Journal 84 (9) (1992) 26.

[18] D.B. Lawrence, M.A. Torf, Environmental due diligence beyond the phase I, Advocate (05154987) 58 (5) (2015).

[19] D.F. Rohrman, M.J. Hoffman, Environmental assessments as components of pre-acquisition due diligence, Proceedings of the 45th Industrial Waste Conference May 1990, Purdue University, CRC Press, (2018), pp. 141–162.

[20] A. Sivo, Comparative analysis of phase I environmental due diligence findings across industrial, biogas, and brownfield sectors, Politecnico di Torino (2025).

[21] B.J. Berry Jr, M.B. Carmeli, T.N. McManus, Environmental due diligence and lender liability, Commercial Law Bulletin 6 (1991) 15.

[22] D. Resnick, The Jesica Santillan tragedy: lessons learned, The Hastings Center Report 33 (4) (2003) 15–20.

[23] D.H. Havens, L. Boroughs, To err is human": a report from the Institute of Medicine, Journal of Pediatric Health Care 14 (2) (2000) 77–80.

[24] I. Kant, Groundwork of the Metaphysics of Morals (1785), New Haven, 1785.

[25] S.S. Panesar, M. Kliot, R. Parrish, J. Fernandez-Miranda, Y. Cagle, G.W. Britz, Promises and perils of artificial intelligence in neurosurgery, Neurosurgery 87 (1) (2020) 33–44.

[26] D.A. Vallero, T.M. Letcher, Society, Unraveling Environmental Disasters, second ed, Elsevier Academic Press, Amsterdam, The Netherlands, (2023), p. 720 ch.

[27] J. Steinhoff, AI ethics as subordinated innovation network, AI & Society 39 (4) (2024) 1995–2007.

[28] National Academy of Engineering, The Engineer of 2020: Visions of Engineering in the New Century, National Academies Press, Washington, DC, (2004).

[29] M.J. De Oliveira, Equilibrium Thermodynamics, Springer, (2013).

[30] M. Martin, R. Schinzinger, Engineering Ethics, McGraw Hill, Boston, (1996).

[31] A. Sen, On ethics & economics, Blackwell Publishing, Maiden, MA, (1987), pp. 349–378. American Economic Review, vol. 89.

[32] D.A. Vallero, Chapter 11 - Economics and feasibility assessments, in: D.A. Vallero (Ed.), Environmental Systems Science, Elsevier, (2021), pp. 509–542.

[33] D.A. Farber, The problematics of the pareto principle, 2003.

[34] P. Ehrlich, The Population Bomb, Ballantine, New York, (1968).

[35] S. Singh, Norman Borlaug: A Billion Lives Saved, A World Connected, AgBioWorld, (2006). Available from http://www.agbioworld.org/biotech.

[36] C.R.B. Maboloc, Dying with dignity and the argument from conscience, Prajñā Vihāra 25 (1) (2024) 62.

[37] A. Robertson, Biotechnology and the governance of life: the case of pre-implantation genetic diagnosis, Critical Interventions in the Ethics of Healthcare, Routledge, (2016), pp. 61–80.

[38] C. Obi, S.-T. Ozara, N.E. Chukwukelu, An ethical evaluation of eugenics vis a vis the practice of pre-implantation genetic diagnosis.

[39] N.E. Putra, Preimplantation genetics diagnosis: ethical and legal aspects, International Conference on Law, Economics and Health (ICLEH 2020), Atlantis Press, (2020), pp. 525–528.

[40] B. Keogh, Preimplantation genetic testing: the conflict between reproductive autonomy and disability rights.

[41] G.-F. Umeh, A.C. Umeh, Francis Galton's Eugenics as a means to a new world: an ethical approach, Nnadiebube Journal of Education in Africa 9 (1) (2024).

[42] S. NS, Biotechnology and public policy, Law Relating to Biotechnology, *2nd Ed.*, Springer, Perspectives from USA, UK, EU and India, (2026), pp. 329–374.

[43] R. Thorneycroft, Prenatal testing, down syndrome, and selective termination: a (critical) criminology of genocide? Diversity in Criminology and Criminal Justice Studies, Emerald Publishing Limited, (2022), pp. 167–181.

[44] Á. Stefánsdóttir, 'A world without Down's syndrome'–an evaluation of foetal diagnosis in light of the ethos of medicine, Disability & Society 37 (5) (2022) 809–826.

[45] N.I. Kara, Screening syndromes out: updating the international "genocide" vernacular for a changing technological age, North Carolina Journal of International Law 45 (1) (2020).

[46] Y. Suchikova, Risks and realities of speculative ethics: lessons from nanotechnology for the artificial intelligence discourse, NanoEthics 19 (3) (2025) 15.
[47] M.E. Kambouris, Y. Manoussopoulos, A. Velegraki, G.P. Patrinos, The biote-bot hybrid. The ultimate biothreat merging nanobots, AI-enabled cybernetics and synthetic biology, Future Medicine AI 1 (1) (2023).
[48] S. Anitha, V. Trisha, S. Varsha, B.R.M.L. Manuhya, An overview on nanobots ands its applications, European Chemical Bulletin 12 (8) (2023) 364–376.
[49] A. Ali, M. Qasim, M.U. Dilawar, Z.F. Khan, Y.K. Jadoon, T. Faiz, Nanorobotics: next level of military technology, 2022 International Conference on Business Analytics for Technology and Security (ICBATS, IEEE, (2022), pp. 1–7.
[50] S.J. Evans, P.M. Vecchiarelli, M.J. Clift, S.H. Doak, J.R. Lead, Overview of nanotoxicology in humans and the environment; developments, challenges and impacts, Nanotoxicology in Humans and the Environment, Springer, (2021), pp. 1–40.
[51] T.A. Hemphill, Augmented humanity, Regulation 48 (2025) 57.
[52] National Society of Professional Engineers, NSPE Code of ethics for engineers. https://www.nspe.org/resources/ethics/code-ethics, 2016 (accessed 24.01.23).
[53] T.L. Beauchamp, J.F. Childress, Principles of Biomedical Ethics, Oxford University Press, USA, (2001).
[54] B. Gert, Common Morality: Deciding What to Do, Oxford University Press, (2004).
[55] American Society of Mechanical Engineers, PDHonline Course R139 (2 PDH): Ethics in ME Professional Practice. ASME, Fairfax, Virginia. Available from https://pdhonline.com/courses/r139/ASME%20Professional%20Practice%20Curriculum%20Eng%20Ethics.pdf, 2020 (Accessed 21.01.2026).
[56] J.V. Platz, The Veil of Ignorance in Rawlsian Theory (2017).
[57] H. Runcheva, John Rawls: justice as fairness behind the veil of ignorance, Iustinianus Primus Law Review 4 (2013) 1.
[58] J. Rawls, Justice as rational choice behind a veil of ignorance, Alternative Political Perspectives, second ed., Wadsworth Publishing Company, (1996).
[59] J. Rawls, A Theory of Justice, Belknap Press of Harvard University Press, Cambridge, MA, (1971).
[60] J. McTavish, Why the church says "yes" to life and "no" to IVF, The Linacre Quarterly 89 (4) (2022) 450–454.
[61] M.M. Hogan, E.M. Hogan, A.K. Hogan, M.J. Hogan, Medical Ethics in the Catholic Tradition: Perspectives from Philosophy, Theology, Medicine, and the Law, Taylor & Francis, (2025).
[62] P.A. Vesilind, A.S. Gunn, Engineering, Ethics, and the Environment, Cambridge University Press, (1998).
[63] D. Vallero, Ethical decisions in emergent science, engineering and technologies, in: S.J. Miag (Ed.), in Emerging Informatics-Innovative Concepts and Applications, INTECH Open Access Publisher, (2012).
[64] D.B. Resnik, Does RCR education make students more ethical, and is this the right question to ask? Accountability in Research 21 (4) (2014) 211–217.
[65] N. Pickus, J.A. Reuben, Debating Moral Education: Rethinking the Role of the Modern University, Duke University Press, (2010).
[66] N.M. Pickus, T. Dostert, Rethinking the role of ethics in public affairs education, Journal of Public Affairs Education 8 (1) (2002) 9–23.
[67] J.M. Stansbury, S. Sonenshein, Positive business ethics: Grounding and elaborating a theory of good works, in: Handbook of Positive Organizational Scholarship, 2012, pp. 340–352.
[68] W. Jordan, A virtue ethics approach to engineering ethics, in: 2006 Annual Conference & Exposition, 2006, pp. 11.142. 1–11.142. 11.
[69] L. Kohlberg, Resolving moral conflicts within the just, Moral Dilemmas and Ethical Reasoning (2017) 71.
[70] A. Colby, L. Kohlberg, A. Abrahami, J. Gibbs, A. Higgins, The Measurement of Moral Judgment, Cambridge University Press, (2011).
[71] A. Colby, et al., The Measurement of Moral Judgement: Volume 2, Standard Issue Scoring Manual, Cambridge University Press, (1987).
[72] J. Reimer, D.P. Paolitto, R.H. Hersh, Promoting Moral Growth: From Piaget to Kohlberg, Longman, New York, (1983).
[73] R. Duska, M. Whelan, Moral Development: A Guide to Piaget and Kohlberg (1975).
[74] L. Kohlberg, Stages of moral development, Moral Education 1 (51) (1971) 23–92.
[75] J.R. Rest, Development in Judging Moral Issues, U of Minnesota Press, (1989).
[76] J.R. Rest, Moral Development: Advances in Research and Theory (1986).

[77] J. Piaget, The Moral Judgment of the Child, Routledge, (2013).
[78] I. Kant, The Metaphysics of Morals (1797) (1996) na.
[79] D.A. Vallero, Beyond responsible conduct in research: new pedagogies to address macroethics of nanobiotechnologies, Journal of Long-Term Effects of Medical Implants 17 (1) (2007).
[80] D.A. Vallero, Macroethics and engineering leadership, Leadership and Management in Engineering 8 (4) (2008) 287–296.
[81] J.R. Herkert, Ways of thinking about and teaching ethical problem solving: microethics and macroethics in engineering, Science and Engineering Ethics 11 (3) (2005) 373–385.
[82] J.R. Herkert, Future directions in engineering ethics research: microethics, macroethics and the role of professional societies, Science and Engineering Ethics 7 (3) (2001) 403–414.
[83] J.R. Herkert, Professional societies, microethics, and macroethics: product liability as an ethical issue in engineering design, International Journal of Engineering Education 19 (1) (2003) 163–167.
[84] B. Allenby, Macroethical systems and sustainability science, Sustainability Science 1 (1) (2006) 7–13.
[85] Biomedical Engineering Society, BMES Code of Ethics, ed, 2021.
[86] M. Davis, Codes of engineering ethics: recent trends, in: K. Laas, M. Davis, E. Hildt (Eds.), Codes of Ethics and Ethical Guidelines: Emerging Technologies, Changing Fields, Springer International Publishing, Cham, (2022), pp. 35–54.
[87] M. Stremlau, B.S. Slusher, The Potential of Academic Drug Discovery: Successes and Challenges, ed. Taylor & Francis, 2025.
[88] P.B. Panchal, Use of integrated intelligence scheduling system (IISS) for heavy civil construction projects, Journal of Emerging Technologies and Innovative Research 12 (5) (2025) a800–a815.
[89] J.T. Mitchel, V. Hays, M. Shatzoff, G. Park, The critical path initiative meets medical devices, Applied Clinical Trials 16 (3) (2007) 48.
[90] D. De Silva, D. Alahakoon, An artificial intelligence life cycle: from conception to production, Patterns 3 (6) (2022).
[91] S. Hajikazemi, A. Ekambaram, B. Andersen, Y.J. Zidane, The Black Swan–knowing the unknown in projects, Procedia-Social and Behavioral Sciences 226 (2016) 184–192.
[92] K.D. Grieger, S.F. Hansen, A. Baun, The known unknowns of nanomaterials: describing and characterizing uncertainty within environmental, health and safety risks, Nanotoxicology 3 (3) (2009) 222–233.
[93] B. Sinclair-Desgagné, Chasing unknown unknowns with the available data, International Journal of Data Science and Analytics (2025) 1–17.
[94] J. Li, H. Zhao, L. Zheng, W. An, Advances in synthetic biology and biosafety governance, (in English), Frontiers in Bioengineering and Biotechnology (2021). Available from https://doi.org/10.3389/fbioe.2021.598087Review 9, 2021–April–30.
[95] D. Rumsfeld, Known and Unknown: A Memoir, Penguin, (2011).
[96] D. Rumsfeld, Knowns & Unknowns, ed: JSTOR, 2002.
[97] L. Moratis, Dieselgate, Encyclopedia of Sustainable Management, Springer, (2023), pp. 1146–1151.
[98] C. Florio, A.F. Sproviero, The will to accountability in the Volkswagen dieselgate, 13th Annual Conference of the EuroMed Academy of Business-Business Theory and Practice Across Industries and Markets, EuroMed Press, (2020), pp. 456–468.
[99] E. Mujkic, D. Klingner, Dieselgate: how hubris and bad leadership caused the biggest scandal in automotive history, Public Integrity 21 (4) (2019) 365–377.
[100] L. Bovens, The ethics of dieselgate, Midwest Studies in Philosophy (2016).
[101] K. Dill, S. Wells, P.A. Vesilind, Professional Ethics and Engineering, Program in Science, Technology and Human Values, Duke University, (1999).
[102] American Society of Mechanical Engineers, Design for sustainability. ASME. Available from http://www.professionalpractice.asme.org/communications/sustainability/2.htm (accessed 23.05.06).
[103] C. Answers, Just war doctrine, Catholic. com (2005).
[104] Daniel A. Vallero, Fundamental of Air Pollution, 6th, Academic Press, Amsterdam, NV, 202 ISBN: 9780443314469.

CHAPTER

5

Bioethics

OUTLINE

5.1 Introduction

Bioethics is the set of moral principles and values (the *ethics* part) needed to respect, to protect, and to enhance life (the *bio* part). Engineers, medical practitioners, and all technical professionals must be clear about meanings and must avoid any trace of ambiguity. A stray mark on a blueprint or a misreading of a prescription can lead to harmful outcomes, even death. That said, there may be a few parts of the bioethics definition that are missing, such as words like medicine, health, and biotechnologies. They are certainly embedded in the biological aspects of the definition, but bioethics is much more, considering the risks and rewards.

The term was coined by Van Rensselaer Potter II (1911–2001). Although Potter was a biochemist, his thinking was much like that of an engineer, that is, his thinking was rational

Technical and Ethical Optimization: Challenges and Opportunities for the Modern Engineer and Scientist
DOI: https://doi.org/10.1016/B978-0-443-44925-3.00008-9

and fact-based. His original 1971 definition was one rooted in integration. Potter considered bioethics to bridge science and the humanities to serve the best interests of human health and to protect the environment [1]:

> From the outset it has been clear that bioethics must be built on an interdisciplinary or multidisciplinary base. I have proposed two major areas with interests that appear to be separate but which need each other: medical bioethics and ecological bioethics. Medical bioethics and ecological bioethics are non-overlapping in the sense that medical bioethics is chiefly concerned with short-term views: the options open to individuals and their physicians in their attempts to prolong life.... Ecological bioethics clearly has a long-term view that is concerned with what we must do to preserve the ecosystem in a form that is compatible with the continued existence of the human species.

Biomedicine, engineering, and the development and application of emerging biotechnologies share a common feature; they call for balance. Meeting the needs and demands of the patient, the client, and the public must be in balance with the actual and potential risks. Ethics and regulation must be balanced [2].

Society demands that the state of the science be advanced as rapidly as possible *and* that no dangerous side effects ensue. Most engineers and biotechnologists appreciate the value of pushing the biotechnological envelopes. Both are adept at optimizing among numerous variables for the best design outcomes. However, emergent technologies are always associated with some degree of peril. A query of top scientists addressed this issue [3], focusing on those biotechnologies intended to help developing countries. The queries in the survey addressed both the need and the risk of emerging biotechnologies. As such, the study included both the societal and technological areas of greatest potential value (see Table 5.1).

Each international expert was asked the following questions about the specific technologies:

- Impact. How much difference will the technology make in improving health?
- Appropriateness. Will it be affordable, robust, and adjustable to health care settings in developing countries, and will it be socially, culturally, and politically acceptable?
- Burden. Will it address the most pressing health needs?
- Feasibility. Can it realistically be developed and deployed in a time frame of 5–10 years?
- Knowledge gap. Does the technology advance health by creating new knowledge?
- Indirect benefits. Does it address issues such as environmental improvement and income generation that have indirect, positive effects on health?

The three highest priority areas require major advances in biomedical engineering. The fourth area is within the domain of environmental and civil engineering. The fifth is a challenge for genetic and tissue engineers. The sixth technology falls within biomedical engineering research and clinical engineering. The seventh combines the work of computer engineers and biomedical engineers, including artificial intelligence and machine learning. The eighth is a blend of agricultural and biomedical engineering with food sciences. The ninth area calls for advances in biomedical, clinical, and tissue engineering, and the tenth expresses the need for innovations in computational pharmacological modeling, material sciences, biomedical engineering, and chemical

TABLE 5.1 2002 ranking by global health experts of top 10 biotechnologies needed to improve health in developing countries.

Final ranking	Biotechnology
1	Modified molecular technologies for affordable, simple diagnosis of infectious diseases
2	Recombinant technologies to develop vaccines against infectious diseases
3	Technologies for more efficient drug and vaccine delivery systems
4	Technologies for environmental improvement (sanitation, clean water, and bioremediation)
5	Sequencing pathogen genomes to understand their biology and to identify new antimicrobials
6	Female-controlled protection against sexually transmitted diseases, both with and without contraceptive effect
7	Bioinformatics to identify drug targets and to examine pathogen–host interactions
8	Genetically modified crops with increased nutrients to counter specific deficiencies
9	Recombinant technology to make therapeutic products (e.g., insulin, interferons) more affordable
10	Combinatorial chemistry for drug discovery

Data from A.S. Daar, H. Thorsteinsdóttir, D.K. Martin, A.C. Smith, S. Nast, P.A. Singer, Top ten biotechnologies for improving health in developing countries, Nature genetics 32, (2) (2002) 229–232.

engineering. This is evidence that bioethics is a growing concern for all engineering disciplines. Notably, each of these technological areas is associated with bioethical issues, but in unique ways.

Both parts of the term "biomedical" are important. As mentioned, "bio" connotes life. The dictionary definition of this amalgamation denotes "life or living organisms, or systems derived from them" [4]. This is an engineering-friendly definition. In fact, the discipline of "bioengineering" relates to the "operation on industrial scale of biochemical processes… and is usually now termed biochemical engineering" [4]. Interestingly, this appears to be a distinction between what molecular biologists and biochemists do versus what engineers do with the same information. Bioengineering is "application of the physical sciences and engineering to the study of the functioning of the human body and to the treatment and correction of medical conditions" [4]. This tracks closely with the definition of "biomedical engineering."

Thus, as agents of technological progress, engineers are in a pivotal position [5]. Technology will continue to play an important role that will increase exponentially. The associated societal challenges require that every engineer and technologist must fully understand the implications, possible disadvantages, and side effects of these technological breakthroughs. Key among them will be biotechnical advances at smaller scales, well below the cell and approaching the molecular level.

The combination of nanotechnologies and biotechnologies will propagate even greater uncertainties than each will separately. Technological processes at these scales require that engineers and technologists be ever improving their grasp of the potential ethical implications. The essence of life processes is at stake.

Professional perspectives of bioethics will vary. For example, the physician and the engineer differ in whom they view as their principal client. For the engineer, it is the public and the need to ensure its members are provided with safety, health, and welfare. The American Medical Association, however, has established principles that prioritize the individual patient's welfare, autonomy, and justice [6]. From a bottom-up perspective, the engineer and the physician do not differ, since if each engineer and each physician upholds their professional trust, the outcome would be the same, that is, a safer and healthier public. Biomedical engineers must operate and work in the union of both professions. Educating future biomedical engineers and technologists must be based on several philosophical schools, including consequentialism, deontology, and virtue ethics. The best learning approach is to consider modern case studies that emphasize the role of ethical and scientific foresight as it applies to the engineer's role in human genome editing and brain-computer interfaces, for example, with an appreciation of the need for emerging medical technologies that will occur during an engineer's professional life, for example, machine learning and artificial intelligence in health care decision-making and stem cell technologies [7].

5.2 Major bioethical issues

Technical work, especially engineering practice and research, is deeply committed to and involved in the advancing technologies that will benefit humankind. However, this commitment and involvement call for deliberate and serious consideration of actual and potential ethical issues. The President's Council on Bioethics has summarized the dichotomy between the promise and ethical challenges [8]:

> For roughly half a century, and at an ever-accelerating pace, biomedical science has been gaining wondrous new knowledge of the workings of living beings, from small to great. Increasingly, it is providing precise and sophisticated knowledge of the workings also of the human body and mind. Such knowledge of how things work often leads to new technological powers to control or alter these workings, powers generally sought in order to treat human disease and relieve suffering. But, once available, powers sought for one purpose are frequently usable for others. The same technological capacity to influence and control bodily processes for medical ends may lead (wittingly or unwittingly) to non-therapeutic uses, including "enhancements" of normal life processes or even alterations in "human nature." Moreover, as a result of anticipated knowledge of genetics and developmental biology, these transforming powers may soon be able to transmit such alterations to future generations.

Many of the ethical breaches in biomedicine have been the result of disrespect. As discussed in Chapter 17, the Nazi trials and other notorious unethical investigations are evidence of disrespecting people. Respect for persons necessitates intellectual honesty and truthfulness with human subjects. A key way to ensure respect for patients and subjects is the requirement that they give their informed consent to participate in a research study or to undergo a procedure that involves some risk. Indeed, informed consent is a balance between respect for persons and advancing knowledge. For example, certain aspects of full and complete disclosure of *a priori* consent may be waived in a study if a particular disclosure is likely to render the research invalid. In numerous instances, the subjects can honestly be informed that they are being invited to participate in research that embodies certain facets that will not be made known until the study has been completed. The waiver is granted by the Institutional Review Board (IRB) if the study poses minimal risk, will not adversely affect the welfare and rights of the participant, is not doable without the exemption, and, if appropriate, the participant will receive relevant information later [9]. Thus, informed consent consists of both the relevant information about the research and the actual and cognizant consent.

Informed consent goes beyond research. Prior to most surgeries, for example, patients are asked to sign a form that states the patient understands the risks associated with that procedure. The probability of each adverse outcome is delineated with a percentage or some expression of odds. For example, an operation may have a 1%–5% chance of fever and extended hospital stay, a 0.1% chance of some hearing loss, and a 0.0001% chance of death. These data are based on epidemiological studies of complications in this type of surgery, which the patient is about to receive, resulting in death of one in 10,000 cases. This is not the same as that patient's risk, which is a function of the patient's unique vulnerabilities and strengths. It is likely a general reflection of *all* cases that have been investigated. For example, if the patient is 25 years of age and in good health, the individual risk of death is much lower than that of an 89-year-old cancer patient. So, the patient's stratum of the population may have a one in a million chance of death, while the elderly, ill cohort's stratum has a one in 500 chance.

The need for such consent became apparent in the aftermath of Nazi Germany's unethical treatment of prisoners of war, especially the "medical" procedures to which they were subjected. Other unethical medical treatments took place in the United States under the banner of "research," notably the withholding of treatment and dishonesty of researchers in treating syphilis in African Americans in Tuskegee, Mississippi, the mistreatment of mentally handicapped patients, and sterilization of poor women of childbearing age. Such inhumane and immoral practices called for increased scrutiny and accountability of medical and other scientific research involving human subjects. One of the mandates to become well established is that a patient, client, or subject be thoroughly informed about any risks.

Certainly, people need to be informed about the risks of important decisions, and medical decisions are among the highest priority decisions one can make.

One challenge is asking the subject or patient enough, but not too much, to be truly informed to give consent. If 10-year-old child is to be included in a clinical trial, prudence and practical experience tell us that she will not be sufficiently prepared intellectually and morally for such a decision. We delegate such decisions to her guardian, as defined by regulation and the courts. The unsettling question on her behalf is whether the guardian is sufficiently informed and otherwise inclined to make the decision for her.

Overloading the patient or subject with technical information can lead to problems with enrollment in a study or, worse, having a patient forgo necessary treatment. Good practice avoids needlessly frightening patients while being completely honest about the risks.

Full disclosure is only valuable if the information is properly and precisely interpreted by the person making the decision.

Informed consent is an example of the balancing the art and science of biotechnological applications. There have been many instances where the balance has been successful, but there are still a few that are suspect. So, let us consider some of these technological areas that illustrate bioethical concerns.

5.2.1 Neuroethics

Ethicists, philosophers, and scientists continuously struggle with defining just what constitutes personhood, but virtually every definition includes the human mind. So, the mind-brain connection includes elements of both ethics and neuroscience. Nanotechnology and other emergent applications of neurotechnology can affect who a person is. The medical definition of neuroethics is a bit pedestrian; that is, ethical aspects of neuromedicine.

Arguably, neuroethics is more than a subset of biomedical ethics. Manipulations of neural tissue affect who we are, including our free will. This has intrigued ethicists for millennia. Some neuroscientists broaden the definition of neuroethics to include aspects of responsibility, personhood, and human agency. These issues are already upon us [10]:

> "How will the rapid growth of human brain/machine interfacing – a combination of neuroscience and informatics (cyborgery) – change how we live and think? These are not esoteric or science1fiction; we aren't talking about some science- fiction prospects of human cloning, but prospects and problems that will become increasingly sharply present for us and our children in the next ten to twenty years."

Every neuroethical issue is entangled with the concept of "self." Indeed, self is more than an aggregation of neurons, or even their synaptic functions. The challenge is how to characterize the mind and self, while merging the scientific method with ethics. But this is nothing new. Rene Descartes tried but died before completing work on this conundrum [11]. The effort continued with the rationalists, who held a view closest to Descartes' [12,13]. Next, David Hume and the empiricists tried and the search continues with the modern utilitarian and deontological viewpoints [14]. All these views fall short in dealing with the self, so they also fail in providing an ethical framework for neuroethics. For example, Hume's logical positivism contended that all knowledge is based on logical inference from simple "protocol sentences" grounded in observable facts [15]. However, this argument is self-defeating since even this postulation cannot be observed. Perhaps, this argues that mind and brain are separate but interrelated concepts.

Thus, neuroethics must reconcile two perspectives [16]: (1) the doctrine of psychophysical parallelism, which holds that a mental state is always accompanied by a neural activity, and (2) the concept of self, which holds that the self or person is more than the mental functions, but also includes temperament or motivation, which are affected by physiological systems other than neural (e.g., endocrine) and, some would argue, nonphysiological factors (e.g., the soul). The first perspective is the most widely held by the neuroscience community, but the second is the one that is embraced by most people, and which is at the heart of many faith traditions. This difference is fodder for ethical conflict between the professional (e.g., the treating physician or hospital staff) and the patient. It also explains in part the intense conflict in coma and other cases, such as those where a person is in a permanent vegetative state (PVS), but whose loved ones "know" that person is still living. For example, Roman Catholic tradition holds that "the administration of water and food, even when provided by artificial means, always represents a natural means of preserving life, not a medical act" [17]. However, the nutrition and hydration may be discontinued if all medical interventions are deemed futile or if these actions impose disproportionate and excessive burden on the patient [18].

5.2.2 Organ transplantation

> The debate over organ transplantation touches on many of the deepest issues in bioethics: the obligation of healing the sick and its limits; the blessing and the burden of medical progress; the dignity and integrity of bodily life; the dangers of turning the body, dead or alive, into just another commodity; the importance of individual consent and the limits of human autonomy; and the difficult ethical and prudential judgments required when making public policy in areas that are both morally complex and deeply important. It is no exaggeration to say that our attitudes about organ transplantation say much about the kind of society that we are, both for better and for worse. ***President's Council on Bioethics (2003) [19]***

> Most of the naturally occurring systems, such as human beings and other living creatures, and "unnatural" systems, such as refrigerators are open systems… ***Mehrdad Massoudi, Carnegie Mellon University (2016) [20]***

Many biologists consider most characteristics of the human body to be no different from those of any other mammalian system. Many engineers consider the human body to be just another thermodynamic system, albeit a very complex one. Thus, it may be easy for technologists to perceive the moral irrelevance of organ transplants to many people. That can be a problem. The controversy arises when scientists and others ignore or downplay the importance of moral attitudes and strongly held convictions of people about the sacredness of the human body down to each cell.

The author recalls a meeting of government scientists and officials shortly after the attack on the World Trade Center towers in 2001. There was some discussion about the dust and residues at Ground Zero. The discussion was quite matter-of-fact and centered around the seemingly mundane work of how to collect samples for chemical and physical analysis. At some point in the meeting, one of the young scientists stormed out of the room. After discussing the matter with him later, he shared that we were discussing the remains of human beings with disregard for the sanctity and sacredness of every molecule in the dust. The author had to admit that it had not crossed his mind, but in retrospect, it was truly an attitude of disrespect on the part of the participants, not willful, but still disrespectful. Such indifference is in opposition to many people and all faith traditions that see the human body as something much more than a bunch of cells.

This attitude is not so different from the debates about psychophysical parallelism between the body and the mind [21], especially as it conveys a sense of self. This elicits the bioethical question: Are we devaluing humanity if we have a free exchange of organs? The moral question asked by the President's Council on Bioethics whether it is possible that in ignoring societal taboos, we have diminished humanity, has it "lessened us, dehumanized us, and corrupted us?" This leads to the Council's more practical questions [19]:

1. What is the most ethically responsible and prudent public policy for procuring cadaveric organs?
2. Should the current law be changed, modified, or preserved?

The engineer and technologist should be aware that such questions are being asked and should not assume that all, or even the majority, of people share the scientific community's prevailing perspective on organ transplantation. In fact, this may be the first time that many readers have heard about there even being an ethical issue associated with transplants. Suffice to say that there is diversity of opinion and in the likelihood that many are uneasy with the current system of transplants and may even be in outright opposition to the procedure. The reasons for opposition include lack of knowledge of how the organ donation process works, religious beliefs, fear of damage to body integrity, distrust of professionals, fear of organ trafficking, discomfort with the irreversibility of brain death, and fear that the donor will feel pain [22].

5.2.3 Patenting life

The number of species is not evenly divided geographically. As evidence, Brazil has the greatest amount of biodiversity or species abundance in the world, followed by Indonesia [23]. However, Indonesia represents only a bit more than 1% of the world's landmass yet is the home to 17% of the globe's species [24].

In recent decades, legal, practical, and ethical issues have arisen when intellectual property (IP) ownership intersects with biology. This raises questions about how society benefits and is harmed by the search for biological entities that exist in nature but are modified by anthropocentric intervention [25]. The debate can be summarized as being between the need for incentives to encourage innovation and the national policy against monopolies [26]. The ethical debate is more nuanced, as Rebecca Eisenberg of the Michigan School of Law captured well [27]:

> Technological change often exposes unstated assumptions lurking in the law and makes them problematic, and patent law is no exception. Although the core mission of the patent system is to promote technological progress, path-breaking new technologies have not always been easily assimilated within its boundaries. The first wave of patent applications on advances in biotechnology in the 1970s illustrate some of the difficulties. Before that time, living organisms had generally been assumed to fall outside the range of patent-eligible subject matter under a time honored exclusion for "products of nature." But genetically engineered organisms, although derived from naturally occurring life forms, seemed to involve too much human intervention to be characterized as natural products. Were they eligible for patent protection? Should the default rule be protection or no protection? What are the roles of the courts, the Patent and Trademark Office (PTO), and the legislature in figuring it out?

To modulate the discomfort with the uncertainty of patenting life, the Budapest Treaty was written to streamline patent applications for inventions and other IP that involve microorganisms like bacteria and fungi [28]. This is to be accomplished by allowing deposits with the International Depository Authority (IDA), which stores the organisms and other biological materials for patent actions. Worldwide, there are 50 such authorities that accept deposits to ensure their preservation and availability for legal patent purposes. The IDA eliminates the need to deposit the same microorganisms and materials in multiple countries, with each deposit recognized by all member countries. For example, the United States has three IDAs, the United Kingdom has seven, and the Russian Federation has two. In all, 85 member states recognize deposits, but only 26 store them physically. The stored materials must be stored for at least 30 years from the date of deposit and for a minimum of five years after the most recent request for a sample.

A substantial amount of the debate about patenting life forms centers around gene patents, with ethical and legal arguments for and against the patenting of isolated DNA sequences. The debate also includes the ethics of synthetic biological and gene-editing methodologies like clustered regularly interspaced short palindromic repeats (CRISPR) and CRISPR-associated protein 9 (CRISPR-Cas9) gene-editing technologies, the customized sequence of guide RNA (gRNA) to match the specific, targeted DNA sequence in the genome. At the heart of the debate are concerns about the privatization of the fundamental building blocks of life, as well as research freedom, healthcare access, and global equity. Patent protection is an indispensable incentive for biotechnological advancement [29]; however, the ethical and scientific factors must be carefully optimized to ensure the safety, health, and welfare of the public [30].

5.2.3.1 *Bioprospecting*

Bioprospecting, the search for natural substances of medicinal value, is very divisive topic in bioethical debates. As evidence, in November of 1999, the US PTO rescinded a patent on the plant species *Banisteriopsis caapi* held by a Californian since 1986 [31]. The plant is sacred to tribal communities living in the Amazon basin and is the source of the hallucinogen *ayahuasca*, used in their religious rituals. In addition to being a harbinger of the complications of religious and cultural respect, it presages the continuation of bitter debates about the extent to which biological materials can be "owned" [32].

In fact, humankind has been bioprospecting for millennia. Like many bioethical issues, emerging technologies and research have changed the legal and biological landscapes dramatically. Powerful interests like pharmaceutical companies consider certain genetic and other natural materials to be lucrative ventures that need to be harnessed for profit. For example, the biotechnology firm, Diversa Inc., entered into an agreement with the National Park Service to find efficacious and beneficial microbes in the geysers and springs in Yellowstone National Park [33]. However, the agreement was suspended by a federal court ruling.

As controversial as patents on plant genetic material are, it pales in comparison to the bioethical debates surrounding those of animals. This is in part because patenting animals' genetic materials is linked to cloning. The larger bioethical issue is captured well by the Church of Scotland's Society, Religion and Technology Project:

> Many people would also say that knowledge of a genetic sequence itself is part of the global commons and should be for all to benefit from. To patent parts of the human genome as such, even in the form of "copy genes", would be ethically unacceptable to many in Europe. In response it is argued that patenting is the legal assessment of patent claims, and should not be confused with ethics. But patenting is already an ethical activity, firstly in that it expresses a certain set of ethical values of our society; it is a response to a question of justice, to prevent unfair exploitation of inventions. Secondly a clause excluding inventions "contrary to public order and decency" is part of most European patent legislation - an extreme case of something like a letter bomb would be excluded as immoral. But now we have brought cancerous mice and human genetic material in the potential frame of intellectual property, ethics has moved to a much more central position, where it sits uncomfortably with the patenting profession. They do not like the role of ethical adjudicator to be thrust upon them by society [34]

5.3 Environmental health: the ethics of scale and the scale of ethics

"The ultimate measure of a man is not where he stands in moments of comfort and convenience, but where he stands at times of challenge and controversy." ***Martin Luther King, Jr. (1963) [35]***

Environmental problems are often characterized by scale, a concept familiar to most engineers. In fact, we often characterize phenomena by their dimensions and by when they occur, that is, by their respective *spatial* and *temporal* scales. Engineers are comfortable with dimensional analysis, by which we measure and describe physical, chemical, and biological attributes of what we design. But can we "measure" ethics in a similar way? King's advice is that we *can* measure ethics, especially in our behavior during worst cases. How well can we stick to our principles and duties when things get tough? Philosophers and teachers of ethical philosophy at the university level frequently subscribe to one classical theory or another for the most part, but most concede the value of other models. They all agree, however, that ethics is a rational and reflective process of deciding how we ought to treat each other.

An important consideration in making bioethical decisions is the amount and type of effects that will result from an action. A particularly difficult aspect of a decision or activity is predicting the cascade of events and their future impacts. Manufacturing and commercial decisions about material use can have lasting effects for generations, such as metallic pigments in paint, lead-based fuel additives, and industrial processes that process carcinogenic byproducts. Such decisions are quite complex. A case in point is the comparison of short-term and long-term effects of using coal versus nuclear fission to generate electricity. Combusting coal releases particulate matter as well as damaging compounds like sulfur dioxide and other toxic substances, like mercury. Nuclear power presents a short-term concern about potential accidental releases of radioactive materials and long-lived (sometimes for hundreds of thousands of years) radioactive wastes. Nuclear events have also been extremely influential in our perception of pollution and threats to public health. Most notably, the cases of Three Mile Island, Dauphin County, Pennsylvania, in 1979 and the Chernobyl nuclear power-plant disaster in the Ukraine in 1986 have had an unquestionable impact on not only nuclear power but also aspects of environmental policy, such as community "right-to-know" and the importance of risk assessment, management, and communication.

Similarly, decisions regarding armed conflict must consider not only the tactical warfare but also the geopolitical changes wrought by the conflict. Furthermore, the psychological and medical effects on combatants and noncombatants must be considered. Prominent cases of these effects include the use of Agent Orange in Vietnam and decisions to prescribe drugs and to use chemicals in the Persian Gulf War in the 1990s. The World War II atomic bombings of the Japanese cities of Hiroshima and Nagasaki in August 1945 served the purpose of accelerating the end of the war in the Pacific arena but also ushered in the threat of nuclear war. In addition, the bombings were the world's first entries into the linkage of chronic illness and mortality (e.g., leukemia and radiation disease) that could be directly linked to radiation exposure. Following are cases where the effects did not become apparent until after a protracted lag period.

5.3.1 Case study: Agent Orange[1]

The defoliant "Agent Orange" was used between 1961 and 1970 during the Vietnam War. It is a glaring example of the importance of the "latency period," where possible effects may not be manifested until years or decades after the exposure to toxins [37,38]. The US military sprayed Agent Orange to remove leaves from trees behind which the enemy troops would hide. It was applied by airplanes, helicopters, trucks, and backpack sprayers. In the 1970s, years after the tours of duty in Vietnam, some veterans became concerned that exposure to Agent Orange might cause delayed health effects. One of the chemicals in Agent Orange contained small amounts of the highly toxic compound tetrachlorodibenzo-*para*-dioxin (TCDD).

The US Department of Veterans Affairs has listed several diseases that could have resulted from exposure to herbicides like Agent Orange. US law requires that some of these diseases be at least 10% disabling under VA's rating regulations within a deadline that began to run the day you left Vietnam. If there is a deadline, it is listed in parentheses after the name of the disease [37,39–41].

The combined product was mixed with kerosene or diesel fuel and dispersed by aircraft, vehicles, and hand spraying. About 80 million liters of the formulation were applied in South Vietnam during the war [40–43].

There are still many cases filed by military personnel and others exposed to Agent Orange, a reminder of the latency period of chronic diseases. For a carcinogen, for example, the effects may not appear for years or decades after exposure.

Beginning in the 1970s, military veterans expressed concern that exposure to Agent Orange might cause deferred health effects. Among the compounds in the mixture, Agent Orange contained the most highly toxic dioxin, that is, TCDD. US government and others have identified several potential adverse health outcomes associated with exposures to TCDD [39]:

- Chloracne or other acneform disease consistent with chloracne. (Must occur within one year of exposure to Agent Orange).
- Chronic Lymphocytic Leukemia
- Diabetes Mellitus, Type II
- Hodgkin's disease.
- Multiple myeloma.
- Non-Hodgkin's lymphoma.
- Acute and subacute peripheral neuropathy. (For purposes of this section, the term acute and subacute peripheral neuropathy means temporary peripheral neuropathy that appears within weeks or months of exposure to an herbicide agent and resolves within two years of the date of onset.)
- Porphyria cutanea tarda. (Must occur within one year of exposure to Agent Orange).
- Prostate cancer.
- Respiratory cancers (cancer of the lung, bronchus, larynx, or trachea).

[1] This is an update to the case discussed in: [36] D. A. Vallero and T. M. Letcher, *Unraveling environmental disasters*, Second ed. Amsterdam, The Netherlands: Elsevier Academic Press, 2023.

- Soft-tissue sarcoma (other than osteosarcoma, chondrosarcoma, Kaposi's sarcoma, or mesothelioma).

In addition to exposures to United States and allied military personnel, the government of Vietnam from prolonged exposure to dioxin are for reproductive and developmental disorders that may be occurring in the general population.

Chlorinated dioxins have 75 different forms, and there are 135 different chlorinated furans. The compounds can be separated into groups that have the same number of chlorine atoms attached to the furan or dioxin benzene rings. Each form varies in its chemical, physical, and toxicological characteristics, but all are persistent and lipophilic.

TCDD is a byproduct when certain pesticides are produced. Dioxin migrates through soil and is transported through the aquatic and soil environments. Soil and sediments are reservoirs for recalcitrant substances, leading to direct and indirect exposure pathways for human populations and wildlife. Thus, all dioxins and furans bioconcentrate and biomagnify in food webs.

Agent Orange is among several defoliant mixtures used in Vietnam, including the nine listed in Table 5.2. Most formulations included the two herbicides 2,4-D and 2,4,5-T [37,38,40,41,44].

Agent Orange illustrates the uncertainty involved in fully characterizing and linking exposure to effects in an extended type of exposure scenario, especially when many victims are affected. Consensus has been difficult to reach on linkages between Agent Orange exposures and the symptoms. Even with this rigorous and well-documented weight of evidence that appears to support cause and effect, much of the restitution has been through individual and collective lawsuits and outcry from advocacy groups.

As is the case for most persistent substances, the chemicals in Agent Orange provided a function. The US military decided that chemicals would be an efficient means of clearing vegetation, both weeds and leaves on trees, to eliminate or to reduce sites for enemy troops to hide. Agent Orange was applied by airplanes, helicopters, trucks, and backpack sprayers.

TABLE 5.2 Formulations of defoliants used in the Vietnam War.

Agent	Formulation
Purple	2,4,-D and 2,4,5,-T used between 1962 and 1964.
Green	Contained 2,4,5-T and was used 1962–64.
Pink	Contained 2,4,5-T and was used 1962–64.
Orange	A formulation of 2,4,-D and 2,4,5-T used between 1965 and 1970.
White	A formulation of Picloram and 2,4,-D.
Blue	Contained cacodylic acid.
Orange II ("Super Orange")	2,4,-D and 2,4,5-T used in 1968 and 1969.
Dinoxol	2,4,-D and 2,4,5-T. Small quantities were tested in Vietnam between 1962 and 1964.
Trinoxol	2,4,5-T. Small quantities tested in Vietnam 1962–64.

Data from: [37,38,40,41,44].

In the 1970s, years after the tours of duty in Vietnam, some veterans became concerned that exposure to Agent Orange might cause delayed health effects.

Given that the mixture was designed to be toxic to plants, a likely chemical compound linked to the health effects would be one or more of the listed active or inert ingredients in the formulation of Agent Orange. Rather, the culprit was an unintended chemical byproduct, that is, the same chemical central to the subsequent disasters at Love Canal, Times Beach, and Seveso, Italy. In these disasters, TCDD was not an intended product. In Love Canal, the TCDD migrated into homes and other buildings from an inactive hazardous waste site in Niagara Falls, New York. An estimated 21,800 tons of wastes from the former Hooker Chemical Corporation and others included halogenated solvents, chlorobenzenes, hexachlorocyclohexanes, polychlorinated biphenyls (PCBs), trichlorophenol, and halogenated dioxins and furans. In addition to reaching and infiltrating buildings, the chemicals migrated into surface waters and sediments, with concentrations of TCDD as high as 644 picograms per gram (pg g^{-1}) and 1019 pg g^{-1} total dioxins in carp (*Cyprinus carpi*) [45]. Times Beach was one of several communities and farms in Missouri where dioxin-contaminated oil was sprayed for dust control and road oiling. Seveso's dioxin release was the result of an explosion where one of the ingredients of Agent Orange was being synthesized, that is, 2,4,5-trichlorophenol [46,47]. Thus, the etiologies of the disasters were quite different, but the principal contaminant for all was TCDD.

Agent Orange contained minute amounts of the highly toxic compound TCDD. However, TCDD has a very steep cancer slope factor; meaning it is a very potent carcinogen.

TCDD has been associated with various other ailments, including increased levels of thyroid-stimulating hormone (TSH) in newborns exposed to TCDD in utero, decreased sperm concentrations, and decreased motile sperm counts in men who were one to nine years of age at the time of the Seveso accident (initial TCDD exposure event) in 1976, female reproductive effects, subtle changes in immune system components, developmental dental defects, and diabetes. Also, the US Department of Veterans Affairs has listed a number of diseases that could have resulted from exposure to herbicides like Agent Orange. The law requires that some of these diseases be at least 10% disabling under VA's rating regulations within a deadline that began the day the person left Vietnam.

Scientists from Vietnam, the United States, and 11 other countries have discussed the state of the science of research into the health effects of dioxin. In March 2002, senior scientists from the Vietnamese Ministry of Science, Technology, and the Environment, the Vietnamese Ministry of Health, the US National Institute of Environmental Health Sciences, the US Environmental Protection Agency (EPA), and the US Centers for Disease Control and Prevention met in Hanoi to establish an agreement for future research activities using the findings from the three-day conference and one-day workshop as a guide. The Vietnamese and US government agencies agreed to the following joint research plan addressing the need for direct research on human health outcomes from exposure to dioxin and research on the environmental and ecological effects of dioxin and Agent Orange. The participants identified the highest priority areas.

The primary concerns in Vietnam from prolonged exposure to dioxin are reproductive and developmental disorders that may be occurring in the general population. Chlorinated dioxins have 75 different forms, and there are 135 different chlorinated furans, depending on the number and arrangement of chlorine atoms on the molecules.

The compounds can be separated into groups that have the same number of chlorine atoms attached to the furan or dioxin ring. Each form varies in its chemical, physical, and toxicological characteristics.

Dioxins are only created unintentionally during combustion processes, that is, they have never been synthesized for any other reason than for scientific investigation, for example, to make analytical standards for testing. As mentioned, the most highly toxic form is the 2,3,7,8 configuration of TCDD. This isomer remains in the environment after certain pesticides are produced, such as those used in the Vietnam defoliants. Other isomers that may have been present in the formulations along with the 2,3,7,8 configuration are also considered to be more highly toxic than the dioxins and furans with different chlorine atom arrangements.

The key areas for research in Vietnam include miscarriage, premature birth, congenital malformations, endocrine disorders, neurological disorders, immunodeficiency, cancer, genetic damage, and diabetes mellitus. Note that a number of these conditions are transgenerational, that is, the effect is not expressed in the progeny, as opposed to the person who has been exposed to the dioxin. Preliminary discussions have suggested two areas of research that should be further developed: research on existing populations with high exposures to dioxin relative to populations with low exposures (e.g., people living near hotspots), and research on therapies to reduce dioxin body burdens in humans (such as some herbal therapies being proposed in Vietnam).

Dioxin contaminants from Agent Orange have persisted in the environment in Vietnam for over 40 years. In addition to a better understanding of outcomes of exposure, an improved understanding of residue levels and rates of migration of dioxin and other chemicals in the environment is needed. "Hot spots" containing high levels of dioxin in soil have been identified, and others are presumed to exist but have yet to be located.

Dioxin has migrated through soil and has been transported through natural processes such as wind-blown dust and erosion into the aquatic environment. Contamination of soil and sediments provides a reservoir source of dioxin for direct and indirect exposure pathways for humans and wildlife. Movement of dioxin through the food web results in bioconcentration and biomagnification with potential ecological impacts and continuing human exposure. Research is needed to develop approaches for more rapid and less expensive screening of dioxin residue levels in soil, sediment, and biological samples, which can be applied in Vietnam.

Sampling and analytical methods need to be enhanced to determine locations of highly contaminated areas, monitor remediation, and understand migration of dioxin in the natural environment more readily. Monitoring efforts need to be linked to modeling efforts to understand fate and transport of dioxin in the environment. Innovative and cost-effective approaches to environmental remediation for application in Vietnam need to be developed, tested, and applied.

Preliminary discussions have suggested two areas of research that should be further developed: ecological and restoration research on a degraded upland forest (such as the Ma Da Forest) and research on the identification, characterization, and remediation of hot spots (such as Da Nang Airport).

The Agent Orange tragedy illustrates the problem of uncertainty in characterizing and enumerating effects. There is little consensus on whether the symptoms and disorders suggested to be linked to Agent Orange are sufficiently strong and well documented, that is, they provide weight of evidence, to support cause and effect. Scientists know, for example, that even very strong weight of evidence is merely a part of cause-to-effect linkages. Medical researchers and epidemiologists employ several steps in determining causation. For example, the strength of an association between an effect and a potential cause requires sound technical judgment. In the case of cancer, characterizing the weight of evidence for carcinogenicity in humans consists of three major steps:

1. Characterization of the evidence from human studies and from animal studies, individually;
2. Combination of the characterizations of these two types of data to show the overall weight of evidence for human carcinogenicity; and
3. Evaluation of all supporting information to determine if the overall weight of evidence should be changed.

None of these steps is completely certain. All require a dose of technical judgment. For instance, an observed cluster of cancers in town may have nothing to do with the green gunk that is flowing out of the abandoned building's outfall. But in their minds, the linkage is obvious.

The challenge is to present information in a meaningful way without violating and overextending the interpretation of the data. If we assign causality when none really exists, we may suggest erroneous solutions. But if all we can say is that the variables are associated, the public is going to want to know more about what may be contributing to an adverse effect (e.g., TCDD exposure and diabetes). This was particularly problematic in early cancer research. Possible causes of cancer were being explored, and major research efforts were being directed at a myriad of physical, chemical, and biological agents. So, there needed to be some manner of sorting through findings to see what might be causal and what is more likely to be spurious results. Sir Austin Bradford Hill is credited with articulating key criteria that need to be satisfied to attribute cause and effect in medical research [48]. The factors to be considered in determining whether exposure to a chemical elicits an effect include:

- Criterion 1: Strength of association
- Criterion 2: Consistency
- Criterion 3: Specificity
- Criterion 4: Temporality
- Criterion 5: Biologic gradient
- Criterion 6: Plausibility
- Criterion 7: Coherence
- Criterion 8: Experimentation
- Criterion 9: Analogy

In assessing risks, some of Hill's criteria are more important than others. Risk assessments rely heavily on strength of association, for example, to establish dose–response relationships. Coherence is also very important. Animal and human data should not be extensions of one another and should not disagree. Biological gradient is crucial, as this is the basis for dose–response (the larger the dose, the greater the biological response).

Temporality is crucial to all scientific research, that is, the cause must precede the effect. However, this is sometimes difficult to see in some instances, such as when the exposures to suspected agents have been continuous for decades and the health data are only recently available.

The Agent Orange debate is still raging. The data themselves will never be sufficient to convince all sides. Sound technical and scientific judgment based on the best available and most reliable data is also needed for reliable risk assessments. Linking cause and effect in the aftermath of environmental disasters is often difficult. Sometimes, the best we can do is to be upfront and clear about the uncertainties and the approaches we use.

5.3.2 Case study: endocrine disruption[2]

Developmental effects from pollutants may also occur after birth. Other developmental toxins adversely affect any human life stage by altering the chemical messages between glands and receptors in cells throughout the body. These are collectively known as hormonally active agents or endocrine disruptors. These chemical compounds act in several ways at several sites in the body. The endocrine system consists of glands, that is, specialized groups of cells that manufacture, store, and release hormones directly into the bloodstream. These chemical messengers regulate bodily functions. Several pollutants appear to disrupt the activity of glands, including the adrenals, chemoreceptor organs, gonads, hypothalamus, pancreatic islets, parathyroid, pineal, pituitary, and thyroid. They also enlarge preantral and antral follicles, lower ovarian weight, decrease tertiary follicles and serum estradiol concentrations, and decrease progesterone concentrations. Hormones that are affected include semivolatile organic compounds (SVOCs), polycyclic aromatic hydrocarbons (PAHs), steroids, growth hormones, and thyroxine [50].

The mode of action can be as an agonist, an antagonist, or as an indirect effect of another endogenous process, notably changes to the nervous or immune system that alter hormonal messages. This messaging system in animals and plants employs hormones, which are chemical compounds that interact with receptor sites in the cell (usually on the membrane), which, in turn, initiate biochemical processes [51–57].

Endocrine disrupting chemicals (EDCs) provide a unique challenge since EDCs can mimic hormones, antagonize normal hormones, alter the pattern of synthesis and metabolism of natural hormones, or modify hormone receptor levels [58,59]. Anthropogenic EDCs include pesticide residues (e.g., DDT, endosulfan, and methoxychlor), PCBs, dioxins, alkylphenols (e.g., nonylphenol), plastic additives (e.g., bisphenol A, diethyl phthalate), PAHs, and pharmaceutical hormones (e.g., 17b estradiol, ethinyl estradiol) [60–62].

An endocrine-disrupting molecule may react with a receptor molecule on the cell's surface. This reaction may signal the feminine or masculine responses (e.g., hair growth, testis, or ova development) much like a hormone would do. In other words, the EDC and the natural hormone both bind to the cell's receptor. They are both ligands, that is, molecules that travel through the bloodstream as chemical messengers that will bind to a target cell's receptor. Or the new polypeptide that is formed from this receptor—contaminant interaction

[2] This is an update to the discussion of endocrine disruption in: [49] D. A. Vallero, *Fundamentals of air pollution*, 6th ed. Amsterdam, NV: Academic press, 2025.

may react with DNA in the nucleus. The former reaction is an example of an endocrine response, while the latter may lead to mutagenicity or cancer.

One of the first EDCs heavily researched was DDT [63]. Throughout the 1980s, exposure to this pesticide was associated with abnormal sexual differentiation in seagulls and thinning and cracking of bald eagle eggs [64–66]. Numerous pesticides and chemicals have been associated with endocrine-related abnormalities in wildlife, including the inducement of feminine traits, such as secretion of the egg-laying hormone, vitellogenin, in males of numerous fish downstream from wastewater treatment plants (likely the result of chemicals used in homes, e.g., household chemicals and drugs that bypassed treatment) [67–69]. Numerous species of aquatic and terrestrial animals are affected by EDCs [52,70,71]. Recently, these problems have found their way to humans, such as those exposed to halogenated compounds and pesticides [72–79]. A recent nationwide survey of pharmaceuticals in US surface water found EDCs at nanogram-per-liter levels in 139 stream sites throughout the United States. Several of these EDCs were even found at microgram per liter levels, including nonylphenol (40 $\mu g\ L^{-1}$), bisphenol A (12 $\mu g\ L^{-1}$), and ethinyl estradiol (0.831 $\mu g\ L^{-1}$) [80–82].

It is difficult to classify the chemicals that disrupt mammalian endocrine systems, given the different types of disruption, although those listed by one US agency are largely represented by relatively low vapor pressure, lipophilic compounds [51].

Many endocrine disruptors are semivolatile (vapor pressure ranging from 10–5 to 10–2 kPa at 20°C) and persistent organic pollutants (POPs). The notable exceptions include iodine (0.04 kPa at 25°C) and atrazine (vapor pressure=4 × 10–8 kPa at 20°C), given that iodine is not an organic compound, and that atrazine can be considered virtually nonvolatile. Airborne POPs are particularly dangerous since they are often hormonally active and damage unborn children and neonates, as evidenced by their routinely being detected in breast milk [83–85] and in infant cord blood [86–91]. In the United States, biomonitoring results from the National Health and Nutrition Examination Survey show widespread human exposure to industrial chemicals, many of which have not been fully evaluated for adverse health effects but a number of which are likely EDCs.

Much of the exposure to POPs is at distance from original sources and points of use. Due to their persistence, POPs can be transported to long distances. Long-range atmospheric transport can carry pollutants thousands of miles to what would otherwise be thought to be relatively pristine areas, for example, Arctic regions. POPs are a subset of the so-called "PBTs," which are compounds that are persistent, readily bioaccumulate in tissue, and are toxic at low doses. PBTs are composed of myriad compounds (see Discussion Box: The Inuit and Persistent Organic Pollutants).

The US EPA has created a more expansive list of EDCs beyond the POPs. The chemical compounds on this list are evaluated and screened for potential endocrine disruption based on biological assays and potential exposure scenarios. This is a listing of the chemicals that were first screened as part of the Endocrine Disruptor Screening Program (EDSP) based on human exposure-related factors, not the actual biological mechanisms of endocrine disruption [92–96]. The factors include whether a chemical is produced in high volumes (i.e., high production volume [HPV] chemicals). This initial list focuses on active ingredients and inert ingredients of products. Because the chemicals were selected based on exposure potential only, it was neither construed as a list of known nor likely endocrine disruptors.

After developing this first listing, the US EPA issued a revised list of chemicals [97]. Although the list is based on the ingestion pathway, the water being ingested could indeed be contaminated by air pollutants that have migrated into surface and groundwater.

Under environmental conditions, aqueous solubility of the compounds ranges eight orders of magnitude, and vapor pressures range 18 orders of magnitude. A third listing is a group of chemicals that have been associated with abnormal spermatogenesis, feminization of males, masculinization of females, dysfunction of adrenal, pineal, and thyroid glands, autoregulatory problems, and other hormonally related problems. They are diverse in molecular structure, come from a myriad of sources, and have been detected throughout the environment, that is, in food, water, air, soil, and in plant and animal tissues.

The modes of action of endocrine disruption can be similar between mammals and other vertebrates. Thus, EDCs are also threats to ecosystems. For example, dicofol was the agent of a large spill into Lake Apopka, Florida, which resulted in the endocrine disruption of wildlife because of impaired physiology of reptiles, especially alligators, brought on by changes in their hormonal systems. Understanding human endocrine disruption has also been enhanced by extrapolations from these and other "natural experiments" and other studies following spills and releases of toxic substances.

Some endocrine disruptors act directly on hormone receptors as hormone mimics or agonists. Some EDCs block hormone receptors, that is, antagonists. An estrogenic compound, for example, may lead to the same physiological outcome, that is, feminization, as an antiandrogenic compound that blocks the estrogen receptor. The former is an estrogen agonist; the latter is an androgen antagonist. Others act directly on any number of proteins that control the delivery of a hormone to its normal target cell. The affinity of an endocrine disruptor for a hormone receptor is not directly equivalent to its potency, that is, the dose–response of an EDC is one of the many factors within the endocrine system that determines the actual hormonal response. Based on in vitro and in vivo studies, most EDC dose–response curves are nonlinear. Endocrine disruptors are not necessarily specific, that is, the same chemical can disrupt the activity of more than a single androgen, estrogen, or thyroid hormone. Indeed, some EDCs have been found to interact with multiple hormone receptors simultaneously. It appears that two different EDCs may act synergistically at low doses that would not have elicited an effect had each EDC acted alone.

Endocrine disrupting compounds can be transported from mother to child occurs through the placenta during pregnancy [64,66,88,91,98–120]. Indeed, there is much interest in the linking in utero exposures to air pollutants and other environmental contaminants to developmental outcomes [86,88,90,107,118,121–127].

5.3.3 Case study: waste sites[3]

In abandoned and active waste sites, remnant chemical contaminants are either already reaching human populations or ecosystems or are a threat to be released. A specific chemical constituent of any waste stream can cause harm. In the United States, the Resource Conservation and Recovery Act (RCRA) [128] requires proper handling of wastes from their production through their transport to their ultimate disposal, that is, the so-called cradle-to-

[3] This is an update to the discussion of endocrine disruption in: [49] ibid.

grave manifest system. RCRA divides wastes into two basic types, that is, nonhazardous solid waste and hazardous waste. Subtitle D of the act covers solid waste, and Subtitle C regulates hazardous waste. Therefore, even though it is a subset of "solid" waste, a hazardous waste may be of any physical phase, that is, solid, liquid, gas, or mixtures.

The hazardous attributes of the waste are usually based on its inherent physicochemical properties, including its likelihood to ignite, explode, and react with water. A waste's hazardous inherent properties can also be biological, such as the infectious nature of medical waste, or a chemical compound that has been shown to elicit acute effects, for example, skin irritations, and chronic effects, for example, cancer, harm to the endocrine, immune, or nervous system, interference with tissue development and reproduction, or mutations, birth defects or other toxic endpoints. Persistence is important because the longer a hazardous substance remains in the environment, the more likely is the exposure and risk. For chemicals, the most important hazard is the potential for disease or death (morbidity and mortality, respectively). The hazards to human health are referred to collectively in the medical and environmental sciences as toxicity. Human toxicology is the study of these health outcomes and their potential causes. Likewise, the study of hazards to ecosystems is known as ecological toxicology or simply ecotoxicology, which is further subdivided by discipline, for example, aquatic toxicology and mammalian toxicology.

In the United States, as in many countries, the term hazardous waste is a legal term defined by the regulatory agencies. Under the RCRA hazardous waste regulations, the US EPA is principally responsible for the permitting of hazardous waste treatment, storage, and disposal facilities. The EPA, however, has authorized most states to operate portions or the entire hazardous waste program. Arguably, the most important definition is found in Section 1004(5) of the RCRA, which describes a hazardous waste as a solid waste that may pose a substantial present or potential threat to human health and the environment when improperly treated, stored, transported, or otherwise managed.

The EPA has developed standard approaches and set criteria to determine if substances exhibit hazardous characteristics. Because RCRA defines a hazardous waste as a waste that presents a threat to human health and the environment when it is improperly managed, the government identified a set of assumptions that would allow for a waste to be disposed of if it is not subject to the controls mandated by Subtitle C of RCRA. This mismanagement scenario was designed to simulate a plausible worst case, that is, the conditions where a hazardous substance is most likely to migrate.

Thus, if a compound is persistent and it is likely to move then there is a chance for an exposure. When exposed to the myriad contaminants stored in known and unknown waste sites, there will be risks to public health and the environment. These risks reflect many small disasters and a worldwide collective disaster.

5.3.4 Case study: Persistent organic pollutants and the Inuits

Indigenous peoples often subsist on traditional food for all or part of their diet. This is certainly true for the Inuit population who live in the Arctic and sub-Arctic climes. Ironically, these very remote Arctic regions have been chronically exposed to POPs, rendering these subpopulations vulnerable to the chronic health effects. Since POPs remain in the environment for long periods of time, they can be transported over great distances from their sources. The tendency of these compounds to be toxic, to bioaccumulate, and to biomagnify in food chains increases the risk to the Inuit populations.

In addition to the need for many indigenous people in the Arctic to rely on traditional diets for nourishment, the hunting and fishing are also important parts of their cultural identity. Indeed, alternative sources of food often do not exist. Exacerbating the problem, traditional diets are often high in fat. Since most POPs are lipophilic, they tend to accumulate in fatty tissue of the animals that are eaten. Due to these physicochemical properties, POPs can move many hundreds of kilometers away from their sources, either in the gas phase or attached to particles. They are generally moved by advection, that is, along with the movement of air masses.

Ironically and sadly, most northern residents have not used or directly benefited from the activities associated with the production and use of these chemicals; yet, indigenous peoples in the Arctic have some of the highest known exposures to these chemicals.

Lactating Inuit mother's breast milk, for example, contains elevated levels of PCBs, DDT, and its metabolites, chlorinated dioxins and furans, and brominated organics, such as residues from fire retardants, that is, polybrominated diphenyl ethers (PBDEs), and heavy metals [129–132]. Infants are particularly vulnerable to persistent, bioaccumulating toxic compounds (PBTs), which, as mentioned, are lipophilic and find their way to fat reserves in warm-blooded animals.

Although POPs are found in varying tissue concentrations among women in both industrially developed and developing nations, the Canadian Inuit have shown some of the highest levels of contaminants detected. Their diet consists of seal, whale, and other species high on the marine food chain, increasing the Inuit body burden of POPs [131,133–136]. These elevated exposures are associated with reports of health effects similar to the effects of persons exposed to PCBs who also had evidence of other contaminants in body fluids.

A study of Inuit women from Hudson Bay indicated very high levels of PCBs and dichlorodiphenylethene (DDE) in breast milk [136]. These results prompted an examination of the health status of Inuit newborns [137]. Correlation analysis revealed a statistically significant negative association between male birth length and levels of hexachlorobenzene, mirex, PCBs, and chlorinated dibenzodioxins (CDDs)/dibenzofurans (CDFs) in the fat of mothers' milk. No significant differences were observed between male and female newborns for birth weight, head circumference, or TSH. Immune system effects have also been detected in Inuit infants suspected of receiving elevated levels of PCBs and dioxins during lactation. These babies showed a decline in the ratio of the CD4+ (helper) to CD8+ (cytotoxic) T cells at ages 6 and 12 months (but not at three months) [137].

A lesson for developers of emerging chemicals is the obligation to fully understand the potentially harmful properties of a compound before its use and to monitor for problems after the product reaches the marketplace. For example, all other things being equal, the recalcitrance of compounds is greater in the Arctic regions compared to temperate and tropical regions, as a direct result of lower ambient temperatures. Toxicity properties are also affected by extrinsic conditions, such as whether the substances are found in the air, water, sediment, or soil, along with the conditions of these media (e.g., oxidation–reduction, pH, and grain size). For example, the metal mercury is usually more toxic in reduced and anaerobic conditions because it is more likely to form alkylated organometallic compounds, like monomethyl mercury and the extremely toxic dimethyl mercury. These reduced chemical species are likely to form when buried under layers of sediment where dissolved oxygen levels approach zero.

Ironically, engineers have unwittingly participated in increasing potential exposures to these toxic compounds. With the good intention of attempting to clean up contaminated lakes in the 1970s, engineers recommended and implemented dredging programs in the process of removing the sediment. However, the metals and other toxic chemicals that had been relatively inert and encapsulated in buried sediment were released into the lake waters. In turn, the compounds were also more likely to find their way into the atmosphere.

The problems of long-range transport of persistent pollutants begin with the manufacture and use of a product, followed by contamination of soil and surface waters. Next, the persistence allows the chemicals to reach the ground water and to be sequestered in sediment. These are the reservoirs of the pollutants, which serve as sources of air pollutants. The pollutants are then transported many kilometers before they accumulate in the food chain. Along the way, humans and ecosystems are exposed, but the most damage to human populations may well occur in people who have not benefited from the use that occurred decades earlier. Unfortunately for the Inuit, the doses that cause the most reproductive, endocrine, neurotoxic, and other chronic effects are highest among a population separated in time and space from the benefits of pest control enjoyed by a previous generation of another population. These complexities are a lesson to environmental scientists and engineers to consider systematically the many physical, chemical, and biological characteristics of the compound and the environment where it exists.

The Inuit situation demonstrates the critical ties between humans and their environment, and the importance of physical properties of contaminants (e.g., persistence, bioaccumulation, and toxicity potentials), the conditions of the environment (e.g., the lower Arctic temperatures increase the persistence of many POPs), and the complexities of human activities (e.g., diet and lifestyle) to assess risks and, ultimately, to take actions to reduce exposures. The combination of these factors leaves the Inuit in a tragic dilemma. Since they are subsistence anglers and hunters, they depend almost entirely on a tightly defined portion of the earth for food. Their lifestyle and diet dictate dependence on food sources high in POPs.

What makes this disaster even more complex and tragic is that it forces innocent people to make extremely difficult decisions. Pediatricians rightly encourage breast feeding for its many attributes, including enhancing the infant's immune system in the critical first weeks after birth. So, in terms of risk tradeoffs, it is dangerous to discourage breast feeding. This dilemma not only applies to the Inuit, or even just subsistence farmers, hunters, and anglers but also to all of us. The key is to ensure that breast milk everywhere does not contain hazardous levels of PBTs and other contaminants. The only way to do this is to consider the entire life cycle of the pollutants and find ways to prevent their entry into the environment in the first place.

5.3.5 Case study: forever chemicals

Chemical compounds that are potentially harmful and are being detected increasingly in the environment are known as emerging chemicals. These include so-called "forever chemicals" like the perfluoroalkyl and polyfluoroalkyl substances (PFAS), as well as polybrominated diphenyl diethers (PBDEs), organophosphorus flame retardants (OPFRs), and also microplastics (MPs), nanoplastics (NPs), and nanoparticles. Ever-increasing improvements in measurement technologies lead to lower levels of detection (LoD) and levels of quantitation (LoQ) that allow for finding new chemicals in the environment. However, other reasons include the buildup of persistent compounds and continuous new releases into the environment [49]. Even banned compounds will continue to be detected in the environment given their recalcitrance. Prior to this century, finding substances in parts per million (ppm) or milligrams per liter (mg L^{-1}) in the environment was considered acceptable LoDs. Recently, substances can be detected in the parts per quadrillion (ppq) or micrograms per liter (μg L^{-1}) range and even lower for the so-called "forever chemicals" like PFAS. For example, the United States recently proposed drinking water maximum contaminant levels (MCLs) for perfluorooctanoic acid (PFOA), perfluorooctane sulfonic acid (PFOS), perfluorononanoic acid (PFNA), hexafluoropropylene oxide dimer acid (HFPO-DA, known as GenX Chemicals), perfluorohexane sulfonic acid (PFHxS), and perfluorobutane sulfonic acid (PFBS) [138,139]. The recommended MCL for PFOA and PFOS is 4 ng L^{-1} [140] and a hazard index (HI) of 1 for the rest (see Discussion: Hazard Quotient and Index).

Hazard quotient (HQ) and hazard index (HI)

In risk assessments, the threshold concentrations of a pollutant are chosen to protect humans and ecosystems. For emerging contaminants, these concentrations aim to protect human health. The amount of protection is based on the potential for an unacceptable outcome, that is, its hazard. Often, there are two approaches for addressing a hazardous pollutant in water, that is, effluent limits and water quality limits. Effluent limits apply to any source that releases the contaminant to a waterbody, for example, 100 mg L^{-1} of a chemical compound. Of course, the actual level of concern for emerging contaminants varies considerably given the large number of candidate chemicals of concern. For example, the US publishes lists of contaminants that are not yet regulated but are known or anticipated to be found in the environment. If, for instance, a chemical or biological agent of concern is expected to reach public water systems and may need to be regulated under the Safe Drinking Water Act (SDWA), it is placed on the Contaminant Candidate List (CCL). The most recent draft (CCL5), published in the Federal Register (87 FR 68060) on Novermber 14, 2022, includes 66 chemicals, three chemical groups (polyfluoroalkyl substances, cyanotoxins, and disinfection byproducts), and 12 microbial contaminants [141].

Uncertainty about outcomes, timing, and effectiveness of proposed actions is high for most emerging contaminants [142]. Uncertainty about an emerging contaminant can also stem from temporal, spatial, and operational variability, and the lack of knowledge [143]. Uncertainties vary by many factors, for example, types of habitats, sources of pollution, diseases in populations, geographic diversity, and relevance to better-studied contaminants, for example, applying quantitative structure–activity relationship from known chemicals to emerging contaminants [144–149]. For example, within a

(cont'd)

square meter of soil, the microbial populations, soil texture, organic matter, and chemical makeup can be highly diverse and may differ substantially from a nearby square meter of soil. Extrapolations to the effect of dredging near a 10-hectare wetland or trying to determine the impact of a leaking underground gasoline tank propagate even greater amounts and types of uncertainty. Uncertainty in the real world, compared to highly controlled experiments, is compounded by the measurement imprecision, voids and gaps in observations, practical obstacles (e.g., no access to private property and physical barriers), lack of consistency in measured and modeled results, and simply not being able to understand what the information means, that is, structural uncertainty [150].

Safety thresholds for chemical hazards are known as reference doses (RfD) and, in air, reference concentrations (RfC), usually based on oral exposure data. RfC is calculated in the same manner as RfD, but with units that apply to air (e.g., $\mu g\ m^{-3}$):

$$\text{RfD} = \frac{\text{NOAEL}}{\text{UF}_{\text{inter}} \times \text{UF}_{\text{intra}} \times \text{UF}_{\text{iother}}} \quad (5.1)$$

The calculation of the RfC is identical:

$$\text{RfC} = \frac{\text{NOAEC}}{\text{U}_{\text{inter}} \times \text{U}_{\text{intra}} \times \text{U}_{\text{other}}} \quad (5.2)$$

Animal studies and epidemiology provide the knowledge bases for these values. Assays, that is, in vitro studies, can also inform the calculations of RfDs and RfCs. The main reasons for uncertainty include variability among the animals and human populations being tested, as well as differences in response to the compound by different species (e.g., one species may have decreased liver function, while another may show cardiovascular effects). Subpopulations, like immunocompromised individuals and children, may respond differently to exposures that are below thresholds for the general population.

Lacking long-term human studies, the hazards and risks will be extrapolated from acute or subchronic studies of the same or similar agents, usually from secondary data sources that likely had different research objectives than those needed for an emerging contaminants [115,151–158]. The routes and pathways of exposure may differ from those applications in the real world. For example, if the dose in a research study is administered orally, but the pollutant is more likely to be inhaled, this route-to-route extrapolation adds inconsistency, which translates into uncertainty [159].

The factors underlying the uncertainties are quantified as specific uncertainty factors (UFs) in the denominator. The uncertainties in the RfD are largely due to the differences between results found in animal testing and expected outcomes in human population. As in other bioengineering operations, a factor of safety must be added to calculations to account for UFs. The increase in the size of the denominator, that is, total uncertainty, means that the safe level of exposure continues to drop, so that the RfC and RfD become increasingly small, that is, assumed high toxicity given the high levels of uncertainty.

The first of the three types of uncertainty is those resulting from the difference in the species tested and that of *Homo sapiens* (UF_{inter}). Humans may differ from other species in their sensitivity to substances, so

(cont'd)

their response to be exposed to the compound will differ from that of the tested species, that is, UF_{inter}. The second uncertainty factor results from certain subpopulations being more sensitive to the effects of a compound than the general human population, known as intraspecies uncertainty factors (UF_{intra}). The third type of uncertainty (UF_{other}) results when the available data and science are lacking, such as when a lowest observed adverse effect is used rather than a no observed adverse effect level (NOAEL). This happens when data show a dose at which an effect is observed, but the "no effect" threshold must be extrapolated. Since the UFs are in the denominator, the greater the uncertainties, the closer the safe level (i.e., the RfD) is to zero, that is, the threshold is divided by these factors. The UFs are usually multiples of 10, although the UF_{other} can range from 2 to 10.

Errors are a major source of uncertainty. When an occurrence or event is not identified during an observation or measurement, but in fact the occurrence or event exists, this is a false negative. If sampling and analysis are not sufficiently sensitive and lack resolution, it can lead to a false negative [160]. Other problems, for example, LoD being too high, chain-of-custody with missing or erroneous documentation, improper interpretation of results, and poor documentation and data entry, can also be sources of false negative findings.

The other major type of error is one that leads to a false positive, that is, an occurrence or event is found but, in fact, does not exist [161]. This can be a major challenge for emerging contaminants. For example, consider the situation where previous evidence used by an agency to list a compound as a potential endocrine disruptor has subsequently found that a wealth of new information is now showing that it has no such effect [162]. The decision makers must decide whether the new information is sufficient to remove protections that were only required based on the compound's apparently false positive endocrine disruption data. Sometimes, the false positive results from scientific shortfalls, for example, based upon faulty models or models that only work well for lower organisms (i.e., interspecies uncertainty). Often, this is rectified by scientific advancement, for example, models that are subsequently developed consider newly available data that account for physical, chemical, and biological complexities of higher-level organisms, including humans.

False positives may also lead to overemphasis on less severe problems that cause public health officials to devote inordinate amounts of time and resources to address too many problems [163]. False positives may also discourage people to make use of potentially beneficial products. False positives, especially when they occur frequently, create credibility gaps between scientists and the public.

The HQ is a risk factor applied to noncarcinogenic chemicals that relates the delivered dose at the exposure point, for example, the average daily dose (ADD) or median exposure (mg kg^{-1} day^{-1}) to an adverse health endpoint. The health endpoint can be the RfD or RfC specific that that chemical:

$$HQ = \frac{\text{Exposure concentration}}{\text{RfD}} \tag{5.3}$$

(cont'd)

For example, if the exposure concentration (EC) is 5 μg L^{-1} and the RfD for a chemical is 10 μg L^{-1}, the HQ would be 0.5. Since HQ < 1, the hazard is not considered to be of concern. If the EC were to double, the HQ would be 1, indicating a hazard of concern.

When exposures include more than one chemical, the HI is simply the sum of all HQs:

$$\mathrm{HI} = \mathrm{HQ}_A + \mathrm{HQ}_B + \mathrm{HQ}_C = \frac{\mathrm{EC}_A}{\mathrm{RfD}_A} + \frac{\mathrm{EC}_B}{\mathrm{RfD}_B} + \frac{\mathrm{EC}_C}{\mathrm{RfD}_C} \tag{5.4}$$

For example, if the EC of Chemical *A* is 5 μg L^{-1} and the RfD for that chemical is 10 μg L^{-1}, the EC of Chemical *B* is 0.3 μg L^{-1} and the RfD for that chemical is 1 μg L^{-1}, and the EC for Chemical *C* is 0.05 μg L^{-1} and the RfD for that chemical is 0.5 μg L^{-1}, the HI would be:

$$\frac{5}{10} + \frac{0.3}{1} + \frac{0.05}{0.5} = 0.9$$

So, the mixture does not reach a level of concern since HI < 1.

However, if new data show that the RfD for Chemical *C* should be revised to be 0.2, HI would be recalculated to 1.05. This mixture would then be considered to be hazard of concern since HI > 1.

Plastic waste is becoming a major environmental issue. The growth of plastic production in the world results from its minimal cost, durability, malleability, low weight, and plasticity. Society would now find it very difficult to live without plastic. In 1950, plastic production was 1.5×10^6 tons, in 2002 it was 2×10^9 t(t). In 2021, 3.9×10^8 t, with a rate of increase of over 2% per year. Since 1950, 9×10^9 t has been produced [164].

Unfortunately, not all plastic is recovered, recycled, or burnt and much of the discarded plastic ends up in landfill with a sizable fraction released into the environment; the subsequent breakdown taking hundreds of years. Overall, 9% of plastic is recycled each year, with 40% ending in landfill and 25% being incinerated [165]. The remainder is lost to the environment and probably ends up in the sea via riverine transport.

It does appear that large amount of plastic ends up in the oceans, and Lebreton has reported on five large subtropical gyres. The North Pacific Ocean Gyre has been monitored, but little is known of the other gyres. Identifying the sources is necessary in order to reduce and clean up these large ocean garbage accumulations [166]. The eventual breakdown of plastic material results in microplastic particles. Although the existence of microplastics had been known for a long time, it was not until 2004 that the name was coined. Microplastic particles are defined as particles that are less than 5 mm across at its widest part.

Microplastic material has been recorded in all types of environments, from farmlands to sewage sludge and wastewater treatment plants, to rivers, lakes, and oceans; in short, microplastics are everywhere in the air and on the land. We ingest microplastics every day with apparently no symptoms, and despite their ubiquity, the impacts of microplastics on both organisms and habitats have yet to be fully understood [167,168]. However, one of the potential problems with plastic materials is that they often contain harmful chemical additives, which could have serious consequences, such as cancer, reproductive issues, and effects on the central nervous system for humans, animals, and possibly plants. Additives include plasticizers (such as bisphenol or dibutyl phthalate), flame retardants (such as organophosphates), antioxidants (e.g., amines), biocides, pigments, lubricants, and fragrances [169].

Models like quantitative structure–activity relationships (QSARs) can extend measurement data and other information to lesser-studied compounds. Characterizing potential exposures to the class of PFAS used in converted textiles lacks reliable information and knowledge related to emissions, fate, and transport for all but a small number of PFAS compounds, notably PFOA and PFOS. Other PFAS compounds lack reliable data on physical-chemical properties required to predict chemical behavior [170]. PFAS presence and concentrations may not be known in many products, articles, and exposure media [152,171–176]. The extrapolation of biological activity and biophysiochemical properties from data-rich to data-poor compounds can be accomplished using QSAR models. However, QSAR reliability increases with the structural and bioactivity similarity between the known, data-rich compound and the unknown, data-poor compound [144,147,148,177].

Two PFAS compounds, PFOA and perfluorooctanoate sulfonate (PFOS), are the two PFASs that have been produced in the largest amounts within the United States [178]. Since the production and use of certain compounds change over time, especially as their hazards become better documented given new controls and public awareness, and because additional concentration data become available, studies of emerging contaminants must be updated frequently [179–183].

Physical properties of PBDEs, which are relatively data-rich, can be a first resource in constructing workflows for other data-poor SVOC. Unfortunately, given their high electronegativity compared to PBDEs, such extrapolations are not possible for most PFAS compounds. Also, many of the chemical properties required to classify PFAS as SVOCs are simply forecasts, so the knowledge base must continue to be updated as experimental data and improved modeling methods, for example, Bayesian and distance-based (such as structure-activity space), become available [184,185].

5.4 Green chemistry and engineering

The best and most obvious way to address recalcitrant contaminants is to avoid them in the first place. Green chemistry is advancing, so safer alternatives that meet the same societal needs are slowly growing. Unfortunately, most of the compounds mentioned in this chapter were used and released decades ago. Since they are persistent, they would be expected to be around for many more decades without intervention. Mishandling of municipal and hazardous wastes can be the source of disasters, especially at the local scale when drinking water supplies are contaminated and people are exposed to recalcitrant and toxic compounds released from landfills and other waste management systems. Thus, it is worth discussing the scientific principles that underpin best practices.

Environmental scientists and engineers have noted that natural systems, such as surface waters and soil, can break down organic materials. This is a type of "biomimicry" wherein one can emulate natural processes to achieve desired results. Various genera of microbes were observed to decompose detritus on forest floors, suspended organic material in water, and organic material adsorbed onto soil particles, using these organic compounds as sources of energy needed for growth, metabolism, and reproduction. The sanitary engineers in the 20th century correctly extrapolated from these observations that a more concentrated system could be designed to do the same thing with society's organic wastes. This started with the so-called conventional wastes (human sewage and animal manure). Trickling filters, oxidation ponds, landfills, and other wastewater treatment systems are merely supercharged versions of natural systems, so it was not a major leap in thinking to apply these principles to the more persistent and toxic compounds.

5.5 Global bioethics

As mentioned, *bioethics* was coined by the biochemist Van Rensselaer Potter II. He wrote two important books. The first, *Bioethics: A Bridge to the Future*, in which he coined the term, was a call to integrate many scientific and engineering disciplines to provide for an environment that was both livable and sustainable. The term sustainable was actually coined in the 18th century by a German mining executive named Hans Carl von Carlowitz. He applied the term to "sustained use" to indicate that overharvesting could not be sustained, so the amount of wood being harvested had to equal the amount that would be replaced by new growth [186,187].

The term bioethics was subsequently coopted by the medical community and was redefined specifically to address the morality of biomedical practice, including appropriateness of treatments and technological advances.

Meanwhile, Potter [1] grew more and more respectful of the "land ethic" espoused by Aldo Leopold, and he attempted to recapture its original meaning and coined another phrase, *global bioethics*, in his 1988 book, *Global Bioethics: Building on the Leopold Legacy* [188].

The prefix "bio" applies to bioethics and global bioethics, but the medical connotation limits its use to the human, whereas global bioethics considers the moral obligation to protect all living things. In doing so, global bioethics includes not only the instrumental, anthropocentric values, but also the intrinsic, biocentric, and eco-centric values. As such, global bioethics takes the systems view of living and nonliving components of the environment.

5.6 Conclusions

Bioethics involves many professions, scientific disciplines, and the social sciences. It includes biomedical ethics, but much more. It is a key part of environmental ethics, professional ethics, and research ethics. In a time when emerging biotechnologies and other innovations are entering the marketplace and society as a whole, it is critically important to optimize decision-making with sufficient attention to bioethics along with other scientific and cultural considerations.

References

[1] V.R. Potter, Bioethics: Bridge to the Future, Prentice-Hall, New York, NY, (1971).
[2] M.H. Akram, Balancing obligations: navigating the interplay between regulations and ethics in civil engineering, in SSRN (Formerly Social Science Research Network), ed, 2024.
[3] A.S. Daar, H. Thorsteinsdóttir, D.K. Martin, A.C. Smith, S. Nast, P.A. Singer, Top ten biotechnologies for improving health in developing countries, Nature Genetics 32 (2) (2002) 229–232.
[4] E. Wood, A.D. Smith, S.P. Datta, G.H. Smith, P.N. Campbell, R. Bentley, H.A. McKenzie (Eds.), The Oxford Dictionary of Biochemistry and Molecular Biology: General, Oxford University Press, (1997), p. 740 £ 34.95," ed: No longer published by Elsevier, 1997.
[5] C. Whitbeck, Ethics as design: doing justice to moral problems, Hastings Center Report 26 (3) (1996) 9–16.
[6] R.W. Parsa-Parsi, The international code of medical ethics of the world medical association, Journal of the American Medical Association 328 (20) (2022) 2018–2021.
[7] C.M. Kim, Preparing ethical engineers for the future: integrating modern case studies and design fiction in biomedical engineering ethics, in: 2025 ASEE Annual Conference & Exposition, 2025.
[8] L.R. Kass, R. Kass, Human Cloning and Human Dignity: The Report of the President's Council on Bioethics, Public Affairs, (2002).
[9] S.J. Laurijssen, R. van der Graaf, W.B. van Dijk, E. Schuit, R.H. Groenwold, D.E. Grobbee, et al., When is it impractical to ask informed consent? A systematic review, Clinical Trials 19 (5) (2022) 545–560.
[10] D. Rees, S. Rose, The New Brain Sciences: Perils and Prospects, Cambridge University Press, (2004).
[11] S. Priest, Descartes, Kant, and self-consciousness, The Philosophical Quarterly (1950-) 31 (125) (1981) 348–351.
[12] D.J. Stump, Rationalism in science, 2005.
[13] S.G. Crowell, Rationalism in history, Diacritics 33 (1) (2003) 3–22.
[14] K. Sharlamanov, M. Kostovska, The Modern Construction and Postmodern Deconstruction of Self, Springer, (2024).
[15] S.I. Ogaba, An epistemic reflection on David Hume's notion of the self, Nigerian Journal of Philosophical Studies 4 (1) (2025).
[16] J.J. Giordano, B. Gordijn, Scientific and Philosophical Perspectives in Neuroethics, Cambridge University Press, (2010).
[17] P.J. Paul II, , Care for patients in a "permanent" vegetative state, Origins 33 (43) (2004) 737–740.
[18] R. Hamel, The CDF statement on artificial nutrition and hydration: what should we make of it? Health Care Ethics USA 15 (4) (2007) 5–7.
[19] President's Council on Bioethics, Organ Transplantation: Ethical Dilemmas and Policy Choices, Executive Office of the President, Washington, DC, (2003).
[20] M. Massoudi, A possible ethical imperative based on the entropy law, Entropy 18 (11) (2016) 389.
[21] M. Heidelberger, The emergence of psychophysics, in: The Routledge Handbook of the History of Philosophy of Science After Kant, 2025.
[22] B.d A. Roza, et al., Reasons for organ and tissue donation refusal and opposition: a scoping review, Revista Panamericana de Salud Pública 48 (2025) e115.
[23] A.W.H. Fasa, T. Hendrix, Urgency of deposit microorganism for patent biotechnology protection in Indonesia, NTUT Journal of Intellectual Property Law and Management 11 (1) (2022) 53–74.
[24] A. Kanti, M. Ilyas, A. Nurkanto, T.R. Sulistiyani, S. Meliah, Panduan pengelolaan koleksi mikroorganisme Indonesian Culture Collection InaCC, 2018.
[25] I. Nwakoby, I. Iheukwumere, C. Iheukwumere, N. Nwakoby, M. Idigo, V. Ike, Intellectual property law and microbiology: patenting microorganisms and genetic material, IPS Journal of Education 2 (1) (2025) 43–47.
[26] A. Bouchie, E. Dorey, L. Fletcher, J.L. Fox, H. Watzman, Court upholds broad plant patenting rights, Nature Biotechnology 20 (2002).
[27] R.S. Eisenberg, The story of diamond v. Chakrabarty: technological change and the subject matter boundaries of the patent system, 2006.
[28] K. Zweigert, J. Kropholler, Budapest treaty on the international recognition of the deposit of microorganisms for the purposes of patent procedure, done at Budapest on April 28, 1977, Sources of International Uniform Law, Brill Nijhoff, (1979), pp. 264–272.

[29] A.K. Mishra, M.K. Misra, Chapter 9 - Application and ethical consideration of bioprospecting in the identification of small biomolecules for the development of novel therapeutics, in: D.B. Singh, S.K. Upadhyay (Eds.), Medicinal Biotechnology, Academic Press, (2025), pp. 157–167.
[30] National Society of Professional Engineers. NSPE code of ethics for engineers. https://www.nspe.org/resources/ethics/code-ethics, 2016 (accessed 24.01.23).
[31] J. Bosse, Re-examining the "wild" story of the ayahuasca plant patent, Intellectual Property Quarterly (2) (2024) 136–162.
[32] D.A. Vallero, Biomedical ethics for engineers, The Biomedical Engineering Series, 2007.
[33] H. Doremus, Contracts for bioprospecting: the Yellowstone national Park experience, Microbial Diversity and Bioprospecting (2003) 450–457.
[34] J.M. Francis, M. Macdonald, All things new: a brief history of the first forty years of the church of Scotland society, religion and technology project.
[35] J.M.L. King, Strength to Love, Fortress Press, Minneapolis, Minnesota, (2010).
[36] D.A. Vallero, T.M. Letcher, Unraveling Environmental Disasters, second ed., Elsevier Academic Press, Amsterdam, The Netherlands, (2023).
[37] Agent Orange. Lewis Publishing. http://www.lewispublishing.com/orange.htm (accessed 05.01.12).
[38] J.M. Stellman, S.D. Stellman, Characterization of exposure to Agent Orange in Vietnam veterans as a basis for epidemiological studies, JL & Pol'y 13 (2005) 505.
[39] U.S. Department of Veterans Affairs, Veterans' diseases associated with agent orange. https://www.publichealth.va.gov/exposures/agentorange/conditions/#:~:text=Chloracne%20(or%20similar%20acneform%20disease,year%20of%20exposure%20to%20herbicides (accessed 26.01.22).
[40] E. National Academies of Sciences and Medicine, Veterans and agent orange: update 11 (2018), 2018.
[41] M. Walker, et al., Veterans and agent orange, 2012.
[42] J. Mitchell, US vets win payouts over agent orange use on Okinawa, The Asia-Pacific Journal 10 (2012).
[43] U.S. Department of Veterans Affairs, Agent Orange. https://www.publichealth.va.gov/exposures/agentorange/index.asp (accessed 26.01.22).
[44] E. National Academies of Sciences and Medicine, Epidemiologic studies: compendium of new publications, in: Veterans and Agent Orange: Update 2014. National Academies Press (US), 2016.
[45] L.C. Skinner, Dioxins and Furans in Fish below Love Canal, New York: Concentration Reduction Following Remediation, Division of Fish and Wildlife, New York State Department of Environmental Conservation, Albany, NY, (1993).
[46] R. Fanelli, C. Chiabrando, A. Bonaccorsi, TCDD contamination in the Seveso incident, Drug Metabolism Reviews 13 (3) (1982) 407–422.
[47] N. C. f. H. S. Centers for Disease Control and Prevention, Preliminary report: 2, 3, 7, 8-tetrachlorodibenzo-p-dioxin exposure to humans–Seveso, Italy, MMWR. Morbidity and Mortality Weekly Report 37 (48) (1988) 733–736.
[48] A.B. Hill, The environment and disease: association or causation? Proceedings of the Royal Society of Medicine 58 (5) (1965) 295.
[49] D.A. Vallero, Fundamentals of Air Pollution, sixth ed., Academic Press, Amsterdam, NV, (2025).
[50] E.C. Plunk, S.M. Richards, Endocrine-disrupting air pollutants and their effects on the hypothalamus-pituitary-gonadal axis, International Journal of Molecular Sciences 21 (23) (2020) 9191.
[51] Agency for Toxic Substances Disease Registry, Toxic Substances Portal: Endocrine (Glands and Hormones), U.S. Department of Health & Human Services, (2023). Available from https://wwwn.cdc.gov/TSP/substances/ToxOrganListing.aspx?toxid=4(accessed February 12.
[52] J. Annamalai, V. Namasivayam, Endocrine disrupting chemicals in the atmosphere: their effects on humans and wildlife, Environment International 76 (2015) 78–97.
[53] R.A. Ansari, S. Alfuraih, K. Shekh, Y. Omidi, S. Javed, S.A. Shakil, Endocrine disruptors: genetic, epigenetic, and related pathways, Impact of Engineered Nanomaterials in Genomics and Epigenomics (2023) 41–82.
[54] J. McLachlan, Environmental signaling: from environmental estrogens to endocrine-disrupting chemicals and beyond, Andrology 4 (4) (2016) 684–694.
[55] M.J. Piazza, A.A. Urbanetz, Environmental toxins and the impact of other endocrine disrupting chemicals in women's reproductive health, JBRA Assisted Reproduction 23 (2) (2019) 154.

[56] B. Robaire, et al., A cross-species comparative approach to assessing multi-and transgenerational effects of endocrine disrupting chemicals, Environmental Research 204 (2022) 112063.
[57] B. Yilmaz, H. Terekeci, S. Sandal, F. Kelestimur, Endocrine disrupting chemicals: exposure, effects on human health, mechanism of action, models for testing and strategies for prevention, Reviews in Endocrine and Metabolic Disorders 21 (2020) 127–147.
[58] C. Sonnenschein, A.M. Soto, An updated review of environmental estrogen and androgen mimics and antagonists, The Journal of Steroid Biochemistry and Molecular Biology 65 (1-6) (1998) 143–150.
[59] G.R. Warner, V.E. Mourikes, A.M. Neff, E. Brehm, J.A. Flaws, Mechanisms of action of agrochemicals acting as endocrine disrupting chemicals, Molecular and Cellular Endocrinology 502 (2020) 110680.
[60] D. Azizi, et al., A comprehensive review on current technologies for removal of endocrine disrupting chemicals from wastewaters, Environmental Research 207 (2022) 112196.
[61] E. P. A. U.S, Removal of Endocrine Disruptor Chemicals Using Drinking Water Treatment Processes, US Environmental Protection Agency, Washington, DC, USA, (2001).
[62] W.T. Vieira, M.B. de Farias, M.P. Spaolonzi, M.G.C. da Silva, M.G.A. Vieira, Removal of endocrine disruptors in waters by adsorption, membrane filtration and biodegradation. A review, Environmental Chemistry Letters 18 (4) (2020) 1113–1143.
[63] D.M. Fry, C.K. Toone, DDT-induced feminization of gull embryos, Science 213 (4510) (1981) 922–924.
[64] A. Buck, J. Carrillo-Hidalgo, P.R. Camarero, R. Mateo, Organochlorine pesticides and polychlorinated biphenyls in common kestrel eggs from the Canary Islands: spatiotemporal variations and effects on eggshell and reproduction, Chemosphere 261 (2020) 127722.
[65] C.M. Butt, U. Berger, R. Bossi, G.T. Tomy, Levels and trends of poly- and perfluorinated compounds in the arctic environment, Science of the Total Environment 408 (15) (2010) 2936–2965. Available from https://doi.org/10.1016/j.scitotenv.2010.03.015, (2010/07/01/).
[66] S.N. Wiemeyer, et al., Organochlorine pesticide, polychlorobiphenyl, and mercury residues in bald eagle eggs—1969–79—and their relationships to shell thinning and reproduction, Archives of Environmental Contamination and Toxicology 13 (1984) 529–549.
[67] C. Purdom, P. Hardiman, V. Bye, N. Eno, C. Tyler, J. Sumpter, Estrogenic effects of effluents from sewage treatment works, Chemistry and Ecology 8 (4) (1994) 275–285.
[68] S. Jobling, et al., Altered sexual maturation and gamete production in wild roach (*Rutilus rutilus*) living in rivers that receive treated sewage effluents, Biology of Reproduction 66 (2) (2002) 272–281.
[69] S. Jobling, J.P. Sumpter, D. Sheahan, J.A. Osborne, P. Matthiessen, Inhibition of testicular growth in rainbow trout (*Oncorhynchus mykiss*) exposed to estrogenic alkylphenolic chemicals, Environmental Toxicology and Chemistry: An International Journal 15 (2) (1996) 194–202.
[70] G.A. Fox, Effects of endocrine disrupting chemicals on wildlife in Canada: past, present and future, Water Quality Research Journal 36 (2) (2001) 233–251.
[71] P. Matthiessen, J.R. Wheeler, L. Weltje, A review of the evidence for endocrine disrupting effects of current-use chemicals on wildlife populations, Critical Reviews in Toxicology 48 (3) (2018) 195–216.
[72] P.S. Guzelian, Comparative toxicology of chlordecone (Kepone) in humans and experimental animals, Annual Review of Pharmacology and Toxicology 22 (1) (1982) 89–113.
[73] T.B. Hayes, et al., Hermaphroditic, demasculinized frogs after exposure to the herbicide atrazine at low ecologically relevant doses, Proceedings of the National Academy of sciences 99 (8) (2002) 5476–5480.
[74] J.E. López-Pérez, J.M. Goessling, C.M. Murray, Testing androgen-induced immunosuppression: environmental androgens as a model system for steroid-immune interaction, Journal of Experimental Zoology Part A: Ecological and Integrative Physiology (2024).
[75] M. Perret, Transgender frogs turn your son gay: endangered amphibians, estrogenic pollution, and male extinction, Frontiers: A Journal of Women Studies 44 (1) (2023) 27–52.
[76] T. Mitra, et al., Endocrine disrupting chemicals: gestational diabetes and beyond, Diabetology & Metabolic Syndrome 16 (1) (2024) 95.
[77] S. Ramezanifar, et al., Occupational exposure to physical and chemical risk factors: a systematic review of reproductive pathophysiological effects in women and men, Safety and Health at Work 14 (1) (2023) 17–30.
[78] D. Wang, et al., Occupational exposure to polybrominated diphenyl ethers or decabromodiphenyl ethane during chemical manufacturing: occurrence and health risk assessment, Chemosphere 231 (2019) 385–392. Available from https://doi.org/10.1016/j.chemosphere.2019.05.165, (2019/09/01/).

[79] O. Mehrpour, P. Karrari, N. Zamani, A.M. Tsatsakis, M. Abdollahi, Occupational exposure to pesticides and consequences on male semen and fertility: a review, Toxicology Letters 230 (2) (2014) 146–156.

[80] H.T. Buxton, D.W. Kolpin, Pharmaceuticals, hormones, and other organic wastewater contaminants in US streams, Water Encyclopedia 5 (2005) 605–608.

[81] D.W. Kolpin, et al., Pharmaceuticals, hormones, and other organic wastewater contaminants in US streams, 1999– 2000: a national reconnaissance, Environmental Science & Technology 36 (6) (2002) 1202–1211.

[82] D.W. Kolpin, M. Skopec, M.T. Meyer, E.T. Furlong, S.D. Zaugg, Urban contribution of pharmaceuticals and other organic wastewater contaminants to streams during differing flow conditions, Science of the Total Environment 328 (1-3) (2004) 119–130.

[83] L. de Vries, Do human milk concentrations of persistent organic chemicals really decline during lactation? Chemical concentrations during lactation and milk/serum partitioning, Breastfeeding Review 18 (2) (2010) 34–35.

[84] A. Witczak, A. Pohoryło, H. Abdel-Gawad, Endocrine-disrupting organochlorine pesticides in human breast milk: changes during lactation, Nutrients 13 (1) (2021) 229.

[85] M.S. Wolff, Occupationally derived chemicals in breast milk, American Journal of Industrial Medicine 4 (1-2) (1983) 259–281.

[86] K. Mattonet, et al., Prenatal exposure to endocrine disrupting chemicals is associated with altered DNA methylation in cord blood, Epigenetics 17 (9) (2022) 935–952.

[87] L. Montrose, et al., Maternal levels of endocrine disrupting chemicals in the first trimester of pregnancy are associated with infant cord blood DNA methylation, Epigenetics 13 (3) (2018) 301–309.

[88] T. Nakamura, K. Nakai, T. Matsumura, S. Suzuki, Y. Saito, H. Satoh, Determination of dioxins and polychlorinated biphenyls in breast milk, maternal blood and cord blood from residents of Tohoku, Japan, Science of the Total Environment 394 (1) (2008) 39–51.

[89] Y. Yan, F. Guo, K. Liu, R. Ding, Y. Wang, The effect of endocrine-disrupting chemicals on placental development, Frontiers in Endocrinology 14 (2023) 1059854.

[90] J.B. Herbstman, et al., Prenatal exposure to PBDEs and neurodevelopment, Environmental Health Perspectives 118 (5) (2010) 712–719.

[91] Y. Kawashiro, et al., Perinatal exposure to brominated flame retardants and polychlorinated biphenyls in Japan, Endocrine Journal 55 (6) (2008) 1071–1084.

[92] U.S. Environmental Protection Agency, Endocrine Disruptor Screening Program Chemical Screening and Testing Progress, USEPA, (2024). Available from https://www.epa.gov/endocrine-disruption/endocrine-disruptor-screening-program-chemical-screening-and-testing-progress, (accessed May 5).

[93] D.G. Stump, J.C. O'Connor, J.M. Lewis, M.S. Marty, Key lessons from performance of the US EPA Endocrine Disruptor Screening Program (EDSP) tier 1 male and female pubertal assays, Birth Defects Research Part B: Developmental and Reproductive Toxicology 101 (1) (2014) 43–62.

[94] F.M. Fishel, EPA's Endocrine Disruptor Screening Program (EDSP): PI227, 3/2013, EDIS 2013 (3) (2013).

[95] P. Browne, P.D. Noyes, W.M. Casey, D.J. Dix, Application of adverse outcome pathways to US EPA's endocrine disruptor screening program, Environmental Health Perspectives 125 (9) (2017) 96001.

[96] U.S. Environmental Protection Agency, EDSP Approach, USEPA, (2024). Available from https://www.epa.gov/endocrine-disruption/edsp-approach, (accessed May 5).

[97] U.S. Environmental Protection Agency, Final second list of chemicals for tier 1 screening, USEPA, (2024). Available from https://www.epa.gov/system/files/documents/2023-08/List%202.pdf, (accessed 06.05.24).

[98] S.H. Peterson, J.T. Ackerman, R.R. Holser, B.I. McDonald, D.P. Costa, D.E. Crocker, Mercury bioaccumulation and cortisol interact to influence endocrine and immune biomarkers in a free-ranging marine mammal, Environmental Science & Technology 57 (14) (2023) 5678–5692.

[99] B.J. Tornabene, et al., Broad-scale assessment of methylmercury in adult amphibians, Environmental Science & Technology 57 (45) (2023) 17511–17521.

[100] K. Dack, M. Fell, C.M. Taylor, A. Havdahl, S.J. Lewis, Mercury and prenatal growth: a systematic review, International Journal of Environmental Research and Public Health 18 (13) (2021) 7140.

[101] L. Yang, Y. Zhang, F. Wang, Z. Luo, S. Guo, U. Strähle, Toxicity of mercury: molecular evidence, Chemosphere 245 (2020) 125586.

[102] United Nations Environment Programme, Global mercury assessment 2018, in: UN Environment Programme, Chemicals and Health Branch Geneva, Switzerland, Geneva Switzerland, 2019.

[103] J. Wang, et al., Relation of prenatal low-level mercury exposure with early child neurobehavioral development and exploration of the effects of sex and DHA on it, Environment International 126 (2019) 14–23.

[104] P. de Almeida Rodrigues, R.G. Ferrari, L.N. Dos Santos, C.A.C. Junior, Mercury in aquatic fauna contamination: a systematic review on its dynamics and potential health risks, Journal of Environmental Sciences 84 (2019) 205–218.

[105] M. Sakamoto, et al., Health impacts and biomarkers of prenatal exposure to methylmercury: lessons from Minamata, Japan, Toxics 6 (3) (2018) 45.

[106] W. Dong, L. Liang, S. Brooks, G. Southworth, B. Gu, Roles of dissolved organic matter in the speciation of mercury and methylmercury in a contaminated ecosystem in Oak Ridge, Tennessee, Environmental Chemistry 7 (1) (2010) 94–102.

[107] K. Suzuki, et al., Neurobehavioral effects of prenatal exposure to methylmercury and PCBs, and seafood intake: neonatal behavioral assessment scale results of Tohoku study of child development, Environmental Research 110 (7) (2010) 699–704.

[108] S.W. Tan, J.C. Meiller, K.R. Mahaffey, The endocrine effects of mercury in humans and wildlife, Critical Reviews in Toxicology 39 (3) (2009) 228–269.

[109] S. Tan, J. Meiller, K. Mahaffey, Response to letter from Dr. Gianpaolo Guzzi related to "The endocrine effects of mercury on humans and wildlife", Critical Reviews in Toxicology 39 (6) (2009) 539 539.

[110] V. Celo, D.R. Lean, S.L. Scott, Abiotic methylation of mercury in the aquatic environment, Science of the Total Environment 368 (1) (2006) 126–137.

[111] C. o. L. S. National Research Council, Board on environmental studies, committee on the toxicological effects of methylmercury, Toxicological Effects of Methylmercury, National Academies Press, Washington, DC, (2000).

[112] B. Aberg, L. Ekman, R. Falk, U. Greitz, G. Persson, J.O. Snihs, Metabolism of methyl mercury (203Hg) compounds in man: excretion and distribution, Archives of Environmental Health: An International Journal 19 (4) (1969) 478–484.

[113] A.A. Balalian, et al., Perinatal exposure to polychlorinated biphenyls (PCBs) and child neurodevelopment: a comprehensive systematic review of outcomes and methodological approaches, Environmental Research (2024) 118912.

[114] L. Montano, et al., Polychlorinated biphenyls (PCBs) in the environment: occupational and exposure events, effects on human health and fertility, Toxics 10 (7) (2022) 365.

[115] C.A. Weitekamp, L.J. Phillips, L.M. Carlson, N.M. DeLuca, E.A.C. Hubal, G.M. Lehmann, A state-of-the-science review of polychlorinated biphenyl exposures at background levels: relative contributions of exposure routes, Science of the Total Environment (2021) 145912.

[116] G. Winneke, Developmental aspects of environmental neurotoxicology: lessons from lead and polychlorinated biphenyls, Journal of the Neurological Sciences 308 (1-2) (2011) 9–15.

[117] R. Gioia, et al., Polychlorinated biphenyls in air and water of the North Atlantic and Arctic Ocean, Journal of Geophysical Research: Atmospheres 113 (D19) (2008).

[118] J.B. Herbstman, et al., Determinants of prenatal exposure to polychlorinated biphenyls (PCBs) and polybrominated diphenyl ethers (PBDEs) in an urban population, Environmental Health Perspectives 115 (12) (2007) 1794–1800.

[119] K.R. Chowdhury, A. Singh, C. Singh, Neurotoxic environmental pollutants: a comprehensive review on lead, mercury, pesticides, and polychlorinated biphenyls.

[120] D.B. Barr, A. Bishop, L.L. Needham, Concentrations of xenobiotic chemicals in the maternal-fetal unit, Reproductive Toxicology 23 (3) (2007) 260–266.

[121] K.B. Michels, I. De Vivo, A.M. Calafat, A.M. Binder, In utero exposure to endocrine-disrupting chemicals and telomere length at birth, Environmental Research 182 (2020) 109053.

[122] B.E. Blake, S.E. Fenton, Early life exposure to per- and polyfluoroalkyl substances (PFAS) and latent health outcomes: a review including the placenta as a target tissue and possible driver of peri- and postnatal effects, (in Eng), Toxicology 443 (2020) 152565. Available from https://doi.org/10.1016/j.tox.2020.152565.

[123] A. Reddam, et al., Prenatal exposure to polybrominated diphenyl ethers and birth outcomes, Environmental Research 216 (2023) 114830.

[124] W.J. Cowell, et al., Associations between prenatal and childhood PBDE exposure and early adolescent visual, verbal and working memory, Environment International 118 (2018) 9–16.

[125] W.J. Cowell, A. Sjödin, R. Jones, Y. Wang, S. Wang, J.B. Herbstman, Determinants of prenatal exposure to polybrominated diphenyl ethers (PBDEs) among urban, minority infants born between 1998 and 2006, Environmental Pollution 233 (2018) 774–781.

[126] W.J. Cowell, et al., Prenatal exposure to polybrominated diphenyl ethers and child attention problems at 3–7 years, Neurotoxicology and Teratology 52 (2015) 143–150.

[127] J.B. Herbstman, J.K. Mall, Developmental exposure to polybrominated diphenyl ethers and neurodevelopment, Current Environmental Health Reports 1 (2014) 101–112.

[128] Public Law 94-580, Resource Conservation and Recovery Act of 1976, 42 USC 6901, et seq, Readily available, 1976.

[129] B. Ritz, M. Wilhelm, Air pollution impacts on infants and children, Southern California Environmental Report Card (2008) 2–6.

[130] B.R. Sonawane, Chemical contaminants in human milk: an overview, Environmental Health Perspectives 103 (suppl 6) (1995) 197–205.

[131] M. Long, M. Wielsøe, E.C. Bonefeld-Jørgensen, Time trend of persistent organic pollutants and metals in Greenlandic Inuit during 1994–2015, International Journal of Environmental Research and Public Health 18 (5) (2021) 2774.

[132] P.I. Bank-Nielsen, M. Long, E.C. Bonefeld-Jørgensen, Pregnant inuit women's exposure to metals and association with fetal growth outcomes: accept 2010–2015, International Journal of Environmental Research and Public Health 16 (7) (2019) 1171.

[133] S.G. Alan, P.G. Donald, Field experiments and natural experiments, in: E.G. Robert (Ed.), The Oxford Handbook of Political Science, Oxford University Press, Cambridge, UK, (2013).

[134] B.E. Johansen, The inuit (and others): if it swims, it's probably poisonous, Resource Devastation on Native American Lands: Toxic Earth, Poisoned People, Springer, (2023), pp. 115–150.

[135] A. Gagnon-Chauvin, et al., Pre-and postnatal exposure to legacy environmental contaminants and sensation seeking in Inuit adolescents from Nunavik, PLOS Global Public Health 3 (10) (2023) e0002478.

[136] É. Dewailly, P. Ayotte, S. Bruneau, C. Laliberté, D. Muir, R.J. Norstrom, Inuit exposure to organochlorines through the aquatic food chain in arctic Québec, Environmental Health Perspectives 101 (7) (1993) 618–620.

[137] E. Dewailly, Breast milk contamination by PCBs and PCDDs/PCDFs in arctic Quebec: preliminary results on the immune status of Inuit infants, Dioxin'93 13 (1993) 403–406.

[138] M.G. Evich, et al., Per-and polyfluoroalkyl substances in the environment, Science 375 (6580) (2022) eabg9065.

[139] J.C. Quezada Davalos, M.A. Michaud, L.E. Lowe, E.N. Hanson, J.E. Owens, Per- and Polyfluoroalkyl Substances (PFASs) in the fountain creek watershed, Colorado Springs, CO, USA: a yearlong investigation of PFAS levels in water, soils, and sediments, ACS ES&T Water 3 (1) (2023) 96–105. Available from https://doi.org/10.1021/acsestwater.2c00440.

[140] J.A. Cotruvo, S.B. Goldhaber, A.J. Cohen, EPA's unprecedented interim drinking water health advisories for PFOA and PFOS, Groundwater (2023).

[141] U.S. Environmental Protection Agency, Drinking water contaminant candidate list 5-draft: a proposed rule by the environmental protection agency on 07/19/2021, Federal Register (2021) vol. FR Doc. 2021-15121, Filed 7-16-21; 8:45 am https://www.federalregister.gov/documents/2021/07/19/2021-15121/drinking-water-contaminant-candidate-list-5-draft.

[142] J.D. Solomon, "ommunicating Reliability, Risk and Resiliency to Decision Makers, Duke University, October 12, 2017.

[143] M.B. Van Asselt, Perspectives on uncertainty and risk, Perspectives on Uncertainty and Risk, Springer, (2000), pp. 407–417.

[144] U.S. Environmental Protection Agency, Quantitative structure activity relationship. http://www.epa.gov/nrmrl/std/qsar/qsar.html (accessed 19.03.15).

[145] A. Lagunin, A. Zakharov, D. Filimonov, V. Poroikov, QSAR modelling of rat acute toxicity on the basis of PASS prediction, Molecular Informatics 30 (2-3) (2011) 241–250.

[146] S. Vilar, G. Cozza, S. Moro, Medicinal chemistry and the molecular operating environment (MOE): application of QSAR and molecular docking to drug discovery, Current Topics in Medicinal Chemistry 8 (18) (2008) 1555–1572.

[147] M.L. Card, et al., History of EPI suite[trade mark sign] and future perspectives on chemical property estimation in US Toxic Substances Control Act new chemical risk assessments, Environmental Science: Processes & Impacts 19 (3) (2017) 203–212. Available from https://doi.org/10.1039/C7EM00064B.

[148] (2007). *69*, Guidance Document on the Validation of (Quantitative) Structure-Activity Relationship [(Q)Sar] Models.
[149] T. Kimura, Y. Miyashita, K. Funatsu, S.-i Sasaki, Quantitative structure– activity relationships of the synthetic substrates for elastase enzyme using nonlinear partial least squares regression, Journal of Chemical Information and Computer Sciences 36 (2) (1996) 185–189.
[150] D. Vallero, Translating Diverse Environmental Data Into Reliable Information: How to Coordinate Evidence from Different Sources, Academic Press, (2017).
[151] A.M. Evans, G.E. Rice, J.M. Wright, L.K. Teuschler, Exploratory cumulative risk assessment (CRA) approaches using secondary data, Human and Ecological Risk Assessment: An International Journal 20 (3) (2014) 704–723.
[152] N.M. DeLuca, M. Angrish, A. Wilkins, K. Thayer, E.A.C. Hubal, Human exposure pathways to poly-and perfluoroalkyl substances (PFAS) from indoor media: a systematic review protocol, Environment International 146 (2021) 106308.
[153] P. Whaley, et al., Improving the quality of toxicology and environmental health systematic reviews: what journal editors can do, ALTEX-Alternatives to Animal Experimentation 38 (3) (2021) 513–522.
[154] E.A.C. Hubal, et al., Advancing systematic-review methodology in exposure science for environmental health decision making, Journal of Exposure Science & Environmental Epidemiology 30 (6) (2020) 906–916.
[155] E.A. Cohen Hubal, D.M. Reif, R. Slover, A. Mullikin, J.C. Little, Children's environmental health: a systems approach for anticipating impacts from chemicals, International Journal of Environmental Research and Public Health 17 (22) (2020) 8337.
[156] J. Mitchell, et al., A decision analytic approach to exposure-based chemical prioritization, PLoS One 8 (8) (2013) e70911. Available from https://doi.org/10.1371/journal.pone.0070911.
[157] J. Mitchell, et al., Comparison of modeling approaches to prioritize chemicals based on estimates of exposure and exposure potential, Science of the Total Environment 458 (2013) 555–567.
[158] J.F. Wambaugh, et al., High-throughput models for exposure-based chemical prioritization in the ExpoCast project, Environmental Science & Technology 47 (15) (2013) 8479–8488.
[159] D.A. Vallero, Environmental Systems Science: Theory and Practical Applications, first ed., Elsevier Academic Press, (2021).
[160] M.D. Adams, P.S. Kanaroglou, A criticality index for air pollution monitors, Atmospheric Pollution Research 7 (3) (2016) 482–487. Available from https://doi.org/10.1016/j.apr.2015.11.004.
[161] P. Boffetta, J.K. McLaughlin, C. La Vecchia, R.E. Tarone, L. Lipworth, W.J. Blot, False-positive results in cancer epidemiology: a plea for epistemological modesty, Journal of the National Cancer Institute 100 (14) (2008) 988–995.
[162] J.L. Furman, K. Jensen, F. Murray, Governing knowledge in the scientific community: exploring the role of retractions in biomedicine, Research Policy 41 (2) (2012) 276–290.
[163] D. Psaltopoulos, A.J. Wade, D. Skuras, M. Kernan, E. Tyllianakis, M. Erlandsson, False positive and false negative errors in the design and implementation of agri-environmental policies: a case study on water quality and agricultural nutrients, Science of the Total Environment 575 (2017) 1087–1099.
[164] Statista, Annual Production of Plastics Worldwide from 1950 to 2021 (Energy & Environment, no. March 28, 2023), National Academy Press, Washington, DC, (2023).
[165] H. Jones, Plastic pollution facts. https://www.clearcommunity.org/plastic-pollution-facts/?gclid=EAIaIQobChMI-8-FuoD-_QIVyu3tCh1ObgQDEAAYAiAAEgK1FvD_BwE (accessed 28.03.23).
[166] L. Lebreton, The status and fate of oceanic garbage patches, Nature Reviews Earth & Environment 3 (11) (2022) 730–732.
[167] X. Lim, Microplastics are everywhere—but are they harmful, Nature 593 (7857) (2021) 22–25.
[168] N. Welden, Microplastics: emerging contaminants requiring multilevel management, Waste, Elsevier, (2019), pp. 405–424.
[169] Y. Sarkingobir, et al., The dangers of plastics to public health: a review, NIPES Journal of Science and Technology Resaerch 2 (2) (2020) 195–200.
[170] M.I. Gomis, Z. Wang, M. Scheringer, I.T. Cousins, A modeling assessment of the physicochemical properties and environmental fate of emerging and novel per- and polyfluoroalkyl substances, (in Eng), Science of the Total Environment 505 (2015) 981–991. Available from https://doi.org/10.1016/j.scitotenv.2014.10.062.

[171] (2019). Product – chemical profile for treatments containing perfluoroalkyl or polyfluoroalkyl substances for use on converted textiles or leathers. [Online] Available from: https://dtsc.ca.gov/wp-content/uploads/sites/31/2019/11/Product-Chemical-Profile-for-Treatments-with-PFASs.pdf

[172] L. Huang, P. Fantke, K. Isaacs, O. Jolliet, Stochastic modeling of exposure to chemicals in consumer products, in: ISEE Conference Abstracts, 2018, no. 1.

[173] S.A. Csiszar, A.S. Ernstoff, P. Fantke, O. Jolliet, Stochastic modeling of near-field exposure to parabens in personal care products, Journal of Exposure Science and Environmental Epidemiology (2016).

[174] S.A. Bălan, V.C. Mathrani, D.F. Guo, A.M. Algazi, Regulating PFAS as a chemical class under the California Safer Consumer Products Program, Environmental Health Perspectives 129 (2) (2021) 25001.

[175] C.-J. Chang, et al., Serum per-and polyfluoroalkyl substance (PFAS) concentrations and predictors of exposure among pregnant African American women in the Atlanta area, Georgia, Environmental Research 198 (2021) 110445.

[176] A.O. De Silva, et al., PFAS exposure pathways for humans and wildlife: a synthesis of current knowledge and key gaps in understanding, Environmental Toxicology and Chemistry 40 (3) (2021) 631–657.

[177] D.E. Dawson, B.L. Ingle, K.A. Phillips, J.W. Nichols, J.F. Wambaugh, R. Tornero-Velez, Designing QSARs for parameters of high-throughput toxicokinetic models using open-source descriptors, Environmental Science & Technology 55 (9) (2021) 6505–6517.

[178] United States Environmental Protection Agency, Technical fact sheet–perfluorooctane sulfonate (PFOS) and perfluorooctanoic acid (PFOA), 2017.

[179] D.A. Vallero, Estimating and predicting exposure to products from emerging technologies, Synthetic Biology 2020: Frontiers in Risk Analysis and Governance, Springer, (2020), pp. 107–142.

[180] D.A. Vallero, C.K. Gunsch, Applications and implications of emerging biotechnologies in environmental engineering, Journal of Environmental Engineering 146 (6) (2020) 3120005.

[181] J.B. Roca, P. Vaishnav, M.G. Morgan, J. Mendonça, E. Fuchs, When risks cannot be seen: regulating uncertainty in emerging technologies, Research Policy 46 (7) (2017) 1215–1233.

[182] P.P. Egeghy, et al., Computational exposure science: an emerging discipline to support 21st-century risk assessment, Environmental Health Perspectives (Online) 124 (6) (2016) 697.

[183] T. Malloy, B.D. Trump, I. Linkov, Risk-Based and Prevention-Based Governance for Emerging Materials, ACS Publications, (2016) ed.

[184] D. Wang, et al., A hybrid framework for improving uncertainty quantification in deep learning-based QSAR regression modeling, Journal of Cheminformatics 13 (1) (2021) 69. Available from https://doi.org/10.1186/s13321-021-00551-x.

[185] I.A. Balabin, R.S. Judson, Exploring non-linear distance metrics in the structure–activity space: QSAR models for human estrogen receptor, Journal of Cheminformatics 10 (1) (2018) 47. Available from https://doi.org/10.1186/s13321-018-0300-0.

[186] S.C. Brink, Sustainability: a 21st century concept? Trends in Plant Science 27 (7) (2022) 619–620. Available from https://doi.org/10.1016/j.tplants.2022.05.001.

[187] R. Lozano, Sustainable development and sustainability, in: R. Lozano (Ed.), Organisational Change Management for Sustainability, Springer Nature Switzerland, Cham, (2024), pp. 9–21.

[188] V.R. Potter, Global Bioethics: Building on the Leopold Legacy, MSU Press, East Lansing, Michigan, (2012).

CHAPTER

6

Paradigms, ontologies, and systems thinking

OUTLINE

"Cogito, ergo sum." (Latin: "I think, therefore I am.") [1] ***René Descartes (1637)***

6.1 Introduction

In the broadest sense, engineering is an outgrowth of rationalism, while technology is an outgrowth of science and engineering. Engineers apply reason to solve problems. Descartes' observation can be extended to say that "I think, therefore I design." Or, "I think, therefore I invent." But on what do we base such designs? If the precedents and data are incomplete or flawed, what makes an engineer or other innovator choose another path? Often, for a new or changing technology, a new paradigm in design and thinking is needed [2]. It is not a good idea to force-fit a new design or technology into an old paradigm [3–9]. According to physicist-philosopher Thomas S. Kuhn, there are essentially two types of paradigm shifts: those that result from a discovery caused by encounters with anomalies, and those that result from the invention of new theories brought about by failures of existing theories to solve problems that the theory defines. In the case of a paradigm shift brought about by discovery, the first step in shifting to a new paradigm is the discovery of the anomaly itself.

Technical and Ethical Optimization: Challenges and Opportunities for the Modern Engineer and Scientist
DOI: https://doi.org/10.1016/B978-0-443-44925-3.00015-6

The shifting paradigms often call for new ontologies, that is, a shared set of concepts and knowledge about their properties and the relationship of these properties with each other [10]. It includes a shared vocabulary, but also explains how the terms and definitions connect to concepts needed to understand how these terms and definitions related to the new discoveries.

These may begin with an understanding of existing protocols and guidelines but must be adapted to the new paradigms. Moral tenets and codes are also useful in forming a new ontology, for example, there should not be markedly different standards of behavior for the technologist, even for a new paradigm [11]. However, the tenets and codes may need clarification of how a scientist, engineer, or inventor may approach a new technology that was not envisioned when the codes of ethics and behavior were established. For example, most codes of ethics written just a few years ago gave sufficient attention to the ethical challenges introduced by generative artificial intelligence.

Interestingly, the above definition of ontology is listed second in the dictionary, but for a book that is written from both a technical and moral perspective like this one, the first definition of ontology must also be provided. That is, ontology is the branch of metaphysics that addresses the nature of being or the philosophy of existence and reality. The distinction between the two definitions appears in how they are used. The connotation in this chapter is ontology as a set, which can be expressed with an article, for example, an engineering ontology. The philosophical connotation is more conceptual, for example, ontology tells you about what exists, but epistemology tells you about how we are able to know things [12].

Once the paradigm has been adjusted so that the anomalous becomes the expected, it is said that the paradigm change is complete. In the case of a paradigm shift that results from the invention of new theories caused by the failure of existing theory, the first step is the failure itself (when the system in place fails, the creation of a new system is necessary). Several factors can bring about this failure: observation of discrepancies between the theory and the facts, changes in the surrounding social or cultural climate, and academic criticism of the existing theory. While these problems are generally known for a long time, Kuhn noted that on numerous occasions, the scientific community has often been highly resistant to changes in paradigms. Kuhn also noted that in the early stages of a paradigm, it is easy to invent theoretical alternatives that can be placed on a given set of data. However, once the paradigm becomes more well established, these theoretical alternatives are strongly resisted [13].

Similarly, the process through which advances are made and paradigms developed may also be examined through the lens of the social sciences, especially economics. Although there are numerous examples of the differences in thinking between economists and engineers, this is commonly an area of agreement. For example, economist John Maynard Keynes wanted to understand the dynamics of economics fundamentally. At the same time, he wanted to solve the problem of economic depression. Thus Keynes was taking the engineering perspective, seeing the fusion of these goals as a joint desire to extend basic knowledge and reach applied goals.

Objective and consistent protocols and procedures are essential and inviolable qualities of sound science, engineering, and technology. The same can be said for other credible societal undertakings in law, finance, skilled labor, and myriad other human endeavors. However, science usually connotes something more than these. These protocols and procedures are codified in the scientific method, that is, the progression from observation and measurement to experimentation to the generation and testing of hypotheses.

The scientific method is introduced in elementary schools. I have observed "systems" thinking in kindergarteners, who have surprising, interdisciplinary thought processes. Unfortunately, by middle school and certainly early high school, the systems view gives way to reductionist pedagogy. Course content in biology, chemistry, and physics becomes sequestered from one another. Perhaps, to instill primary concepts, this is necessary. The specifics must be learned before integrating them, but the isolation remains throughout educational development. In my experience in the 1960s and 1970s education, "general science" courses were often perceived to be remedial or, at best, introductory. As the student progresses in college, they become more specialized and reductionist. Researchers can become so specialized that their discipline consists of so few colleagues that their knowledge base becomes increasingly esoteric. There is some truth to the humorous adage that receiving a PhD means one has learned more and more about less and less until one knows everything about almost nothing [14,15].

Within environmental circles, the scientific method may well evoke an image of an effort to link an outcome, for example, breathing unhealthy air, to a cause, for example, exhaust from automobiles. This is certainly important, but environmental quality is a concept of systems science. As evidence, a human body is a system, so there are countless steps between exposure to the exhaust and the adverse outcome. Obviously, ecosystem contains the term "system", which connotes interactions between the myriad components to the condition of a forest, wetland, lake, or estuary. Often, the word science or scientist also leads to visions of white lab coats and goggles, that is, the experiment. Of course, there are many other reasonable interpretations of what science is about.

6.2 Systems science

Much of what is called systems science entails merging and integrating various content and forms. Systems science is something more than "multi-disciplinary" or even "interdisciplinary" science. Systems thinking must be "transdisciplinary" [16]. These distinctions are subtle but important. Multidisciplinary science is merely additive, that is, it comprises separate disciplines working side by side. Interdisciplinary science is interactive, that is, it integrates multiple disciplines for a coordinated whole. Aristotle argued for interactive thinking when he postulated that the whole is greater than the sum of its parts [17]. He could have stated this arithmetically as: the integration (transdisciplinary science) creates something greater than adding all its parts (multidisciplinary science).

Transdisciplinary science applies elements of both multidisciplinary and interdisciplinary sciences but transcends bright discipline borderlines to form something entirely new, all-inclusive, and holistic [18–25]. The result is a shared ethos based on a unified framework of several traditional fields. It is a layered system of various and separate perspectives, blending viewpoints into a shared perspective, and fusing these perspectives into new, innovative, overarching knowledge [26].

6.2.1 Emergence

Much of what is called systems science entails merging various content and forms. As mentioned, systems science is something more than "multi-disciplinary"; it must be "trans-disciplinary," that is, not simply applying data, information, and knowledge from other scientific disciplines, but building lasting alliances that combine expertise and points of view from scientific and other fields that allow for completely new knowledge, that is, emergence [27].

Emergence describes how complex patterns arise from the shared, self-organized interactions of simpler parts, yielding a new collective property that differs from the properties of each individual part. It is a foundational concept in complexity science, which addresses how large systems of many interacting parts, like a biological tissue, a town, or an economy, self-organize to produce complex, often surprising patterns that are not predictable solely from the parts, that is, based on interconnectedness, adaptation, and holistic perspectives instead of analysis, that is, deconstructing phenomena.

Emergent properties can be rather simple yet profound, such as the emergent property of water's wetness that is it is not predictable from the individual properties of the hydrogen and oxygen atoms. Such weak emergence is indeed surprising but arguably can be deconstructed with enough data about the individual parts [28,29], that is, H and O, to build a model to simulate wetness. A strong emergence is so complex and surprising, like human consciousness, that it is not reducible to the parts, that is, it is not predictable from a vast knowledge of each brain cell in the central nervous system nor even from an infinite number of brain scans [28–31]. Emergence is critical for understanding systems where individual components are simple, but where their combined, networked behavior leads to complex, unpredictable, and often functional outcomes [28–33].

6.2.2 A bit of history

The fundamental understanding of any natural or physical phenomenon is based on the first principles of physics and chemistry, processes and mechanisms of biology, applications of ecology, medicine, biochemistry, toxicology, and epidemiology, and methods of risk assessment and the social sciences. The merged forms can include theory, laboratory experiments, field studies, surveys, wet chemistry, computational and numerical models, and methods development. Systems science relies on the results of any of these approaches that help lead to advancement of knowledge.

The rules of science layered over centuries precluded systems thinking. Scientific inquiry was highly analytical, so that good science mainly consisted of intellectually breaking things down into simpler parts. This is the opposite of systems thinking, which relies on synthesis and synergy. Analysis finds meaning by incrementally reducing and dissecting phenomena. Systems thinking synergizes the parts into something more than can be explained by analysis [34,35]. Expertise and scientific disciplines increasingly specialized to carry out these analyses. Ironically, the so-called "Renaissance Men," typified by Leonardo da Vinci, were the consummate polymaths, that is, persons not restricted to a single knowledge domain, but who could address the unknown from multiple bodies of knowledge [36]. However, from its beginnings, modern science had constricted studies into nice, neat compartments. This is known as reductionism. Systems science over the past couple of decades has freed up studies to be multidisciplinary, translational, and synergistic.

One might say that systems science resembles the liberal arts and pre-Renaissance views of natural philosophy. Let me be clear: systems science does not replace reductionist analysis. The methods are complementary [34]. Thus the individual disciplines and specialties from which environmental systems science draws must be even more reliable than when used independently. If not, errors will be propagated and exacerbated when applied across disciplines. The natural philosophers took bold moves to distinguish science from other ways of knowing and the manner of explaining natural phenomena [37]. Two common, yet opposing, approaches are rationalism and empiricism. Rationalism holds that knowledge is substantially acquired independently of observations from the senses, that is, *a priori*. Conversely, sensory experiences, that is, *a posteriori* knowledge, provide the major source of knowledge for empiricists.

The Ancients applied both rationalism and empiricism, but most knowledge was thought to be gained rationally. This changed when famous Renaissance scientists, Galileo Galilei, Johannes Kepler, Francis Bacon, Robert Boyle, and their contemporaries devoted most attention to *a posteriori* methods, especially the controlled experiment and field observation. This empirical paradigm codified what is now called the scientific method. These empiricists were soon joined by scientists using rational tools and models, that is, the theoretical paradigm [7,38]. Aristotle explicated the methods for observing natural phenomena via observation, discovery, demonstration, reason, and logic [39,40]. As such, one could argue he was the first "scientist" [39,41–43].

The requirement of the application of the scientific method is relatively new. In the West, its origins can be traced to the Renaissance scientists of the 16th and 17th centuries [44]. Prior to that, understanding the natural world was the province of philosophy, that is, literally, the love of knowledge. Natural philosophy was a subset of one of the several liberal arts needed to explain natural phenomena [44,45]. Rene Descartes and Francis Bacon likely referred to themselves as natural philosophers, rather than as "scientists" *per se*. Society allowed just a few of its highly respected members to define and to set rules on what science is and what distinguishes sound science from unsound and unscientific methods. Robert Boyle, leader of the British Empire's Royal Society, for example, set empirical rules for what would be considered a valid scientific study, that is, it would require an experiment, witness, and publication [46].

The paradigms have basically stayed intact for the past half-millennium. Modifying any established scientific approach or perspective meets resistance from within the scientific discipline or the scientific community as a whole, particularly if these approaches deviate from current scientific orthodoxy [13]. Even with strong evidence of the superiority of new approaches, adherents are reluctant to move to the new paradigm. In fact, scientific discovery often retains major portions of the old paradigm when shifting to the new one. This is often desirable, since many aspects of the old paradigm remain sound, but some key aspects are identified as wrong, for example, approaches to manage water resources properly in Europe have largely been sound for decades and have remained the same, but one element has recently shifted, that is,managed according to basin, watershed, or catchment levels rather than administrative boundaries [47]. Thus at least for environmental health systems and ecosystems, the shift resembles an evolution as opposed to a revolution [6,48–51].

This discussion would be incomplete and a bit misleading if I were to say that the scientific community has unanimity about the scientific method. Most scientists and philosophers of science might agree that science must be objective and based on measured evidence. Again, such elements are not the sole province of science. Much of the criticism of the scientific method lies in its rigid adherence to the stepwise procedure, not in the goal of unbiased findings. Another criticism is that what works well for the physical sciences may be limiting or even wrong for other scientific inquiries, especially the social sciences [52].

Systems science is proof of that, having to fuse the results from various experiments, models, and other methods to arrive at meaning beyond what each of these data sources could provide on its own. Thus the reality is that there is no singular scientific method in systems science, but rather an amalgamation of the standard methods applied by each discipline, from which new knowledge can emerge. As evidence, most recent Nobel prize winners have engaged in systems thinking, introducing emergent findings.

6.3 The computational paradigm

The empirical paradigm asserts that knowledge is derived from sensory experiences, observations, and evidence. As such, it is fact-driven. It challenges theories against reality using measurement instead of reason and intuition [53,54]. Based on logic, principles, and existing theories, the theoretical paradigm provides the framework within which scientists work, offering the underlying assumptions about the physical world [55]. Both paradigms were largely articulated during the Renaissance. A historic shift, that is, the computational paradigm, occurred in the mid-20th century when scientists began to simulate the physical and natural world computationally [56].

Even the strong experimental advocates of the Renaissance adhered simultaneously to both the empirical and the theoretical paradigms. In modern science, new scientific findings increasingly come from ideas generated by a type of research crowdsourced, that is, from data tools like informatics, data mining, and other data analytics. For many scientists, "data-intensive scientific discovery" is the first stage of their current paradigm for research and development [7], applying tools crafted from recent advances in computation, data storage, grid computing, and network technologies [57–59]. This paradigm also allows for transdisciplinarity, that is, joining knowledge, information, and data from diverse scientific fields to address a problem [60]. Recently, Nobel Prize winners have increasingly been transdisciplinarians [21]. This approach also encourages knowledge catalysis, that is, gathering data and information from various sources to find new understanding of a problem at hand [61,62]. Indeed, such "interoperability" across disciplines and ideas is the current trend, which increasingly draws from databases and models that are modular and open-sourced, as opposed to stand-alone and expensive proprietary systems, and should make data-intensive discovery more cost-effective, efficient, and available to larger numbers and more diverse user groups [63]. For example, users can apply middleware architecture to accept incoming messages from applications and use them by following rules for encoding into a single format, for example, extensible markup language—XML, and directory structures to support interconnecting applications. Standardized middleware architecture is expected to support future software developers by providing a single, consistent interface for both between and within application communications [64].

In a sense, data-intensive discovery generally follows the scientific method, albeit with other people's data. The ideas and insights come from observations that are extensive and wide-ranging. In fact, the biggest obstacle to scientific advancement may not be a lack of data. To the contrary, the systems scientist's major challenge will involve sorting through massive amounts of data and metadata to find and to scale what is relevant, pertinent to the hypothesis at hand, and what truly matters in this massive storehouse of big data [65–67]. In this sense, having big data is not particularly valuable unless we have the intelligence and judgment to sort through it.

6.4 Ontologies in environmental engineering

Systems science involves much more than simply applying data, information, and knowledge from more than one scientific discipline, it also entails building lasting alliances that combine expertise and points of view from scientific and other fields, allowing for completely new knowledge, that is, emergence [27]. Systems thinking sparks ideas that drive research and actions beyond the problem at hand, setting the stage to build knowledge that would be impossible within the confines and auspices of one or just a few disciplines. For example, environmental systems science allows broader ecological and public health communities to be informed and empowered to advance environmental knowledge [27].

All nations address environmental quality, albeit in different ways and with different governmental structures. Put very simply, environmental protection consists of agencies and departments that regulate environmental media, for example, air, water, food, land, products, and occupations, to protect public health and welfare. From an environmental perspective, welfare mainly addresses ecosystems at various scales, from habitats to the planet.

The environment is a highly complex system, consisting of many interconnected and sundry systems at numerous scales. Among its many connotations and characterizations, pollution is chiefly a scientific phenomenon. The term "scientific" has a fairly clear and accepted connotation, even for the very young. This was not always true. For most of the first half of the 20th century, the term had little resemblance to its current meaning. It is gratifying that pollution now engenders a deeper concept of an ethos. Most people want to avoid pollution because its presence threatens human health and environmental quality. Thus it has a moral and value component and a scientific component. This environmental value system relies on an approach that requires objectivity and attention to detail. Environmental professionals and scientists need a consistent environmental ontology, that is, a shared set of concepts and knowledge about its properties and the relationship of these properties with each other [10].

Some of the ontology can be based on the numerous laws and rules throughout the world. Faith traditions also provide some of the morals and values related to an environmental ontology. Much has been learned since the environmentalism movement of the late 1960s and the need to address the traumas that progressed and magnified from the Industrial Revolution in the 19th century into the petrochemical era of the 20th century [68]. Well before the 1960s, however, legislation was enacted to protect the air, water, and soil, but new laws were needed to address the larger quantities of toxic substances and to apply to much larger and emerging categories of pollutants and polluters.

Environmental protection is one of several ways to protect public health. For example, in the United States, the Food and Drug Administration (FDA), the Department of Agriculture (USDA), and the Environmental Protection Agency (EPA) each protect the public's health, but with differing objectives. Other agencies have more specific roles, such as the Nuclear Regulatory Commission's role in protecting the public from radiation releases, the Department of Defense's and Department of Energy's programs to clean up military and nuclear sites, the National Aeronautics and Space Administration's satellite payloads that monitor the ozone hole and carbon dioxide levels, the US Fish and Wildlife Service's protection of endangered species and wetlands (along with the EPA and the Army Corps of Engineers), and the National Oceanic and Atmospheric Administration's roles in protecting marine and coastal habitats. Protecting various aspects of public health and the environment is one of many roles for these and other US federal agencies, with one exception. The EPA's *only and principal goal* is to protect human health and the environment by providing [69]:

- Clean air, land, and water.
- Reduced environmental risks based on the best available scientific information.
- Implementing federal laws to protect human health and the environment.
- Environmental stewardship integral to US policies concerning natural resources, human health, economic growth, energy, transportation, agriculture, industry, and international trade, and these factors are similarly considered in establishing environmental policy.
- Access to accurate information for all parts of society, that is, communities, individuals, businesses, and state, local, and tribal governments, to manage human health and environmental risks.
- Cleanup of contaminated lands and toxic sites by potentially responsible parties and revitalized; and
- Reviewing chemical substances in the marketplace for safety.

Other countries have numerous programs to address public health and the environment. Canada separates health and environment, literally, between two agencies, that is, Health Canada and Environment and Climate Change Canada. The former's goals are to help Canadians maintain and improve their health, ensure that high-quality health services are accessible, and work to reduce health risks. The latter's goal is to inform Canadians about protecting and conserving our natural heritage, and ensuring a clean, safe, and sustainable environment for present and future generations [70]. Like Canada and the United States, the Government of Japan has a dedicated Ministry of the Environment, with numerous other ministries involved in environmental protection [71]; as do the United Kingdom, the European Union, and other nations [50,72–74].

Almost all nations also protect the environment through various forms of federalism or shared responsibilities among national and local governments, for example, states, provinces, prefectures, counties, parishes, cities, towns, and villages. In the United States, for example, states are sovereigns with shared authority in governance with the federal government. Certain authorities are further delegated to lower jurisdictions, for example, a state requires environmental actions by cities and counties, for example, inspections to limit vehicle pollutant emissions, surface water monitoring, and solid waste management programs [74].

Systems thinking begins with the recognition that all environmental and human health problems involve complex systems [62]. A complex system is composed of heterogeneous components that interact with one another within the system's boundaries. Also, complex systems have emergent properties, that is, those that are not explained by understanding the individual elements of the system, that persist, and that respond to changing circumstances and conditions by adaptation [75]. For example, the life cycle risk assessment (LCRA) is an increasingly useful tool [76]. Most risk assessments have been conducted for one or few pollutant events, for example, risks emanating from the release and exposure from these events, but the LCRA broadens this to address the entire chemical's life cycle.

Complex and multipathway risk models call for diverse knowledge of chemical production, product use, and the activities and demographics of users [77–79]. Exposure science, especially high-throughput screening based on exposure potential, is increasingly reliable [80–85]. However, the number of chemicals and products entering commerce is constantly increasing, presenting a challenge to understanding their physical, chemical, and biological properties and potential risks [10]. Prevention is now a key part of air quality programs. Countries are increasingly requiring manufacturers to consider the entire life cycle of a chemical compound and to search for safer alternatives to reduce risks to human health and the environment.

The interactions that lead to pollution involve physical, chemical, and biological entities. The physical laws underpinning thermodynamics and motion, respectively, drive the transformation and transport of substances in the environment. Thus the hydrodynamics and aerodynamics of environmental fluids are fundamentally determined by physical principles. Water and air are the principal fluids, but others are also crucial, notably alcohols, organic solvents, acids, and bases. Often, fluids are complex mixtures, such as blood, tissues, and sap.

The environment is composed of systems that are embedded with systems containing all environmental media, notably the atmosphere, soil horizons, groundwater, and surface water. Contaminants move and transform physically, chemically, and biologically within and between each medium. The laws of motion and thermodynamics determine the spatial and temporal distribution of substances after they are released into the environment. For example, a grain field recently treated with a pesticide will transport some of the pesticide to adjacent habitats when the pesticide becomes dissolved or suspended in water vapor and carried by winds. The pesticide-laden droplets return to the Earth as precipitation, known as wet deposition or sorbed to dry particulate, known as dry deposition. The pesticide subsequently reaches surface and groundwater.

Physical scientists agree that CO_2 is a prominent greenhouse gas (GHG). However, simply knowing its atmospheric concentration is insufficient to predict climate change. Modeling requires the interpretation of GHG concentrations but also much more. Other variables serve as drivers and constraints. As such, knowledge of climate change is emergent. Physical and chemical changes, in turn, affect biological systems, including microbial populations, and ecological structure and function [86].

Adverse environmental effects can increase and propagate the interactions among these drivers and constraints. For example, a wetland's structure will be altered as the ratio of aerobic to anaerobic bacteria changes, with attendant increases in rates of decomposition of organic matter. These changes then increase the formation and emission of methane (CH_4), another potent GHG. A product of anaerobic decomposition from myriad natural and human activities and facilities, such as landfills, CH_4 is also emitted during the combustion of fossil fuels and cutting and clearing of forests. The increased CH_4 will increase global warming if other factors remain constant, for example, no increase in cooling agents like sulfates. Of course, in the real world, other factors never remain constant [62,87].

Thus an oversimplistic, unsystematic model would predict overall temperature increases. However, depending on the presence and weights of other factors, this could be an overprediction. For example, when a system experiences greater biological activity from plants that increase rates of photosynthesis, wetland depth is decreased, and CH_4 global concentrations would fall, which could ameliorate global temperature rise. Or, heightened biological activity and photosynthesis could decrease forest floor detritus mass, so that anaerobic degradation would fall with decreased CH_4 emissions.

Chemistry is the principal concern of most environmental laws and regulations, that is, how to prevent people and other organisms from contacting certain hazardous chemical substances.[1] However, chemistry is even more foundational. For example, chemistry explains the environment where hazardous and essential substances reside. Surface and ground water always contain solutes and suspensions. Dry air within the troposphere is almost entirely made up of nitrogen and oxygen molecules. Moist air has some additional water molecules. However, the minor substances and relatively small fractions of chemical compounds make up nearly all the pollution. Even polluted water and air are still mainly water and air. I recall in the 1980s the humorous title of an environmental group's newsletter that stated the Ohio River is still "mostly water." Indeed, even highly polluted air is still about 78% N_2 and 21% O_2. The few parts per million, billion, or even trillion of certain pollutants can add to cancer and other risks in a population.

Physics, especially the laws of motion, explains how substances move and become pollutants, that is, precursors move with air, water, and other fluids carrying them, which is known as advection. There is also simultaneous mixing by turbulence from convection. Chemistry also determines interactions within the air, water, soil, sediment, and the organisms. For instance, harmless and even beneficial molecules can turn into pollutants under certain conditions. Combustion and biological processes like respiration convert organic compounds and O_2 to CO_2 in the atmosphere. Cation and anion exchange within soils determines important interactions between roots, tubers, and other subterranean plant parts, along with microbes, that determine access to water and nutrients. Anoxic conditions in river sediments can reduce, that is, lower the oxidation state of metals, making them more

[1] The laws and regulations also address exposure to biological agents, such as limits on fecal coliform and other harmful microbes in drinking water, and contamination of beaches and shores, for example, pfiesteria and red tide. Biological toxins are also regulated as medical wastes, for example, by the Resource Conservation and Recovery Act. Physical stressors are also regulated directly, for example, noise controls, and indirectly, for example, controlling chemical compounds that deplete ozone.

harmful to living creatures. A clear example is the conversion of zero- and divalent forms of mercury into highly toxic methylmercury compounds.

Thermodynamics is key to air pollution and all environmental systems. A compound with a rapid reaction rate in water may have undergone complete transformation into one or more different chemical compounds. Conversely, a slow-reacting, persistent compound may be transported long distances and over a long time period without changing before it is deposited back to the earth's surface. After emission from a source, the residence time can be expressed as the compound's atmospheric half-life ($t_{1/2}$). For example, half of the mass of the reactive compound may be gone after three days, whereas half of the mass of the persistent compound may be depleted after three months. Thus reactive compound's $t_{1/2}$=3 d, and the persistent compound's $t_{1/2}$=90 d. This indicates that the first compound is 30 times more reactive than the second.

As mentioned, the biotic and abiotic transformation of an air pollutant occurs throughout the environment and within organisms, involving kinetics and equilibria, which are in part determined by the inherent properties of a substance and in part by the characteristics of the air and other environmental media. Some substances oppose change, while others readily change. Scale matters, but most processes in the external environment are almost identical to those that take place within an organism, albeit at different temporal and spatial scales and rates. For example, after a parent compound is emitted from a stack, a fraction of its mass will break down in the air, a fraction deposits to the earth, part of which migrates into the soil. Some soil microbes gain energy from the compound via metabolism, further degrading the compounds. If the degradation continues to completion, CO_2 and water remain for aerobic metabolism, and CH_4 from anaerobic metabolism. These molecules can be used by plants in photosynthesis. Of course, this is a gross oversimplification of the physical, chemical, and biological processes, with many reactions occurring between the release of substances and their transformation and fate within the environment and within an organism. During transport, potentially toxic intermediates form, each of which undergoes its own transformation and fate.

The fate of an air pollutant is an example of the interface between ecological systems and human health systems, since humans are among the polluters and consumers within these webs and chains [88]. Environmental scientists measure and predict the transfer of mass and energy upward and downward between levels of biological organization, with apex species like eagles, alligators, and humans at the top of these transfers. Predictions must document uncertainties, since the diversity and abundance of species and interaction rates vary in time and space.

Arguably, the one feature that distinguishes systems science is emergence. An emergent property would not exist if it were not for the system [89]. An emergent property differs from any property of the individual components of the system.

Chemistry almost always involves emergence. Carbon and oxygen each have unique properties. However, carbon monoxide (CO) has different properties than either of these elements. Furthermore, carbon dioxide (CO_2) has substantially different properties than CO. These property differences are physical, chemical, and biological. Among the differences: CO_2 is essential to plants; CO is extremely toxic; CO_2 is a GHG; CO is not. The previously mentioned absorption, distribution, metabolism, elimination, and toxicity (ADMET) system has all the properties of the chemical compound, each environmental medium, and each cell of the organism, but also some new properties that none of these components have.

An ecosystem provides a prominent example of emergence. Each biotic and abiotic component of the ecosystem has its own properties. A bacterium has mass, chemical composition, biological structure, biofilm, and other properties that deem it to be a distinct organism. The same goes for a tree, a clump of grass, a house wren, a chipmunk, an alligator, and a human being. Abiotic components, like clay particles, gravel, water, and nutrients, each have unique properties. Combining all of these into an ecosystem allows them to retain their properties, but the ecosystem itself has its own emergent properties, for example, nutrient cycling rates, buffer zones, predator-prey relationships, and seasonal changes in biodiversity and species abundance.

Thus emergence is a common theme of environmental systems science. The term "emergence" is rooted in the works of great philosophers [90]. One can argue that the ancient Greeks owned the term long before 21st-century systems scientists, dating back at least to Aristotle, as mentioned, notably the whole being greater than the sum of its parts—a clear example of emergence specifically and systems thinking in general [91]. Therefore, consideration of emergence and other aspects of systems science is key to

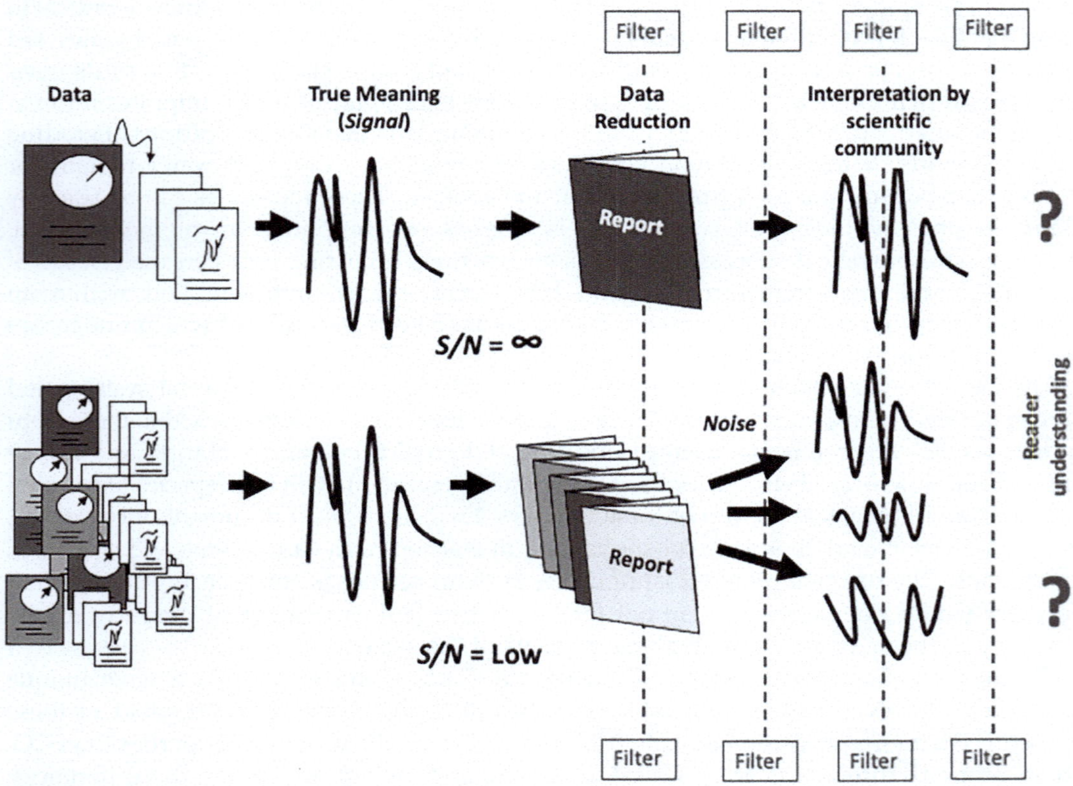

FIGURE 6.1 Scientific communication depends on high-quality data that is properly reduced and interpreted to become useful information for users. However, even data that are properly collected may become noisy if their volume is sufficiently large to override a meaningful interpretation in a timely manner. This leads to a low signal-to-noise (S/N) ratio. *Data from D.A. Vallero, Biomedical Ethics for Engineers, The Biomedical Engineering Series (2007) [93].*

understanding scale and complexity of air pollution but requires some attention to what philosophers and "pre-scientists" have written about systems. Indeed, one of the barriers to systems thinking falls under one of Saint Thomas Aquinas' philosophical principles, *Quidquid recipitur per modum recipientis recipitur*, that is, "whatever is received into something is received according to the condition of the receiver" [92]. The beginning of understanding is limited entirely by the information that the person receives. The amount of correct information taken in by the receiver is the primary rate-limiting step in decision making. No matter how eloquent or fact-based the information, if the receiver is not in the proper "condition" to receive it, the information is not received [62]. This can be likened to an electronic transmission where the signal must be sufficient to overcome the noise, that is, the signal-to-noise ratio (S/N) of the message is strong enough to overcome the filters between the sender and receiver (see Fig. 6.1).

6.5 Conclusions

The essence of a scientific or professional discipline involved in developing a technology is that the discipline follows certain norms and adheres to methods important or even unique to that discipline. Every discipline has some agreement on how to address the scale and complexities of problems they encounter. An engineer's approach may differ only somewhat from that of the physicist, but quite a bit from an economist, and likely quite substantially from that of an anthropologist. All are "disciplined" but apply their own protocols and approaches to building and accessing knowledge. Many scientific phenomena are so complex and scale-dependent that scientific rigor *and* systems thinking are needed to understand and address them.

Emergent technologies arise from the transdisciplinary sciences, where weak and strong emergences spark ideas and call of new paradigms. This also requires new and revised ontologies so that the innovations can be properly understood.

References

[1] R. Descartes, A discourse on method: meditations and principles, Dent (1912).

[2] A. Mishra, A.V. Jha, B. Appasani, A.K. Ray, D.K. Gupta, A.N. Ghazali, Emerging technologies and design aspects of next generation cyber physical system with a smart city application perspective, International Journal of System Assurance Engineering and Management 14 (Suppl 3) (2023) 699–721.

[3] J.A. Moshiri, Emerging paradigms: exploring the latest trends in science, technology, and business, International Multidisciplinary Journal of Science, Technology & Business 2 (1) (2023) 29–34.

[4] A. Das, P. Dobbidi, Self-powered devices: a new paradigm in biomedical engineering, Bioelectronics, CRC Press, (2022), pp. 323–339.

[5] A. Misuri, V. Cozzani, A paradigm shift in the assessment of Natech scenarios in chemical and process facilities, Process Safety and Environmental Protection 152 (2021) 338–351.

[6] E.G. Snyder, et al., The changing paradigm of air pollution monitoring, Environmental Science & Technology 47 (20) (2013) 11369–11377.

[7] T. Hay, eScience and the Fourth Paradigm: Data-Intensive Scientific Discovery and Digital Preservation-APA Conference, ed: London, 2011.

[8] D. Vallero, Paradigms Lost: Learning from Environmental Mistakes, Mishaps and Misdeeds, Butterworth-Heinemann, (2005).
[9] S. Macgill, Y. Siu, A new paradigm for risk analysis, Futures 37 (10) (2005) 1105–1131.
[10] D.E. Meyer, S.C. Bailin, D. Vallero, P.P. Egeghy, S.V. Liu, E.A.C. Hubal, Enhancing life cycle chemical exposure assessment through ontology modeling, Science of the Total Environment 712 (2020) 136263.
[11] M. Nuzzaco, Ethics of information: ontology, responsibility and new moral agents. Applying ethics to information computer technology's challenges. An Interview with Luciano Floridi, Politica & Società 3 (2) (2014) 325–340.
[12] G. Goertz, J. Mahoney, Concepts and measurement: ontology and epistemology, Social Science Information 51 (2) (2012) 205–216.
[13] T.S. Kuhn, D. Hawkins, The structure of scientific revolutions, American Journal of Physics 31 (7) (1963) 554–555.
[14] A.T. Schwartz, Graduate education in chemistry: more and more about less and less? Journal of Chemical Education 71 (11) (1994) 949.
[15] A. Casadevall, F.C. Fang, Specialized science, vol. 82, ed: American Society for Microbiology Washington, DC, 2014, pp. 1355–1360.
[16] A. Engel, Systems Science for Engineers and Scholars, John Wiley & Sons, (2024).
[17] J. Barnes, 3 Metaphysics, The Cambridge Companion to Aristotle (1995) 66.
[18] B.C. Choi, A.W. Pak, Multidisciplinarity, interdisciplinarity, and transdisciplinarity in health research, services, education and policy: 2. Promotors, barriers, and strategies of enhancement, Clinical and Investigative Medicine 30 (6) (2007) E224–E232.
[19] D.P. Peters, et al., An integrated view of complex landscapes: a big data-model integration approach to transdisciplinary science, BioScience 68 (9) (2018) 653–669.
[20] R.W. Scholz, G. Steiner, Process ownership in science–practice collaborations: the special role of transdisciplinary processes in sustainable transitioning, Sustainability Science 18 (3) (2023) 1501–1518.
[21] O. Descalzi, S. Curilef, L. Velazquez, V. Muñoz, Complex systems and inter/transdisciplinary research: a review, Chaos: An Interdisciplinary Journal of Nonlinear Science 34 (1) (2024).
[22] D. Ludwig, Transdisciplinary plant sciences: a review, Plants, People, Planet 8 (1) (2026) 73–86.
[23] L. Söller, et al., Addressing complex social-hydrogeological challenges through inter-and transdisciplinary knowledge co-production, Hydrogeology Journal (2026) 1–24.
[24] M. Marseglia, F. Cantini, T. Celli, E. Brunelli, G. Lotti, Transdisciplinary approach in biodesign: a case study of hybridization between design and scientific method in material design, Biotechnology Design 4 (2026) e4.
[25] D. Black, J. Gopsill, G. Rosenberg, T. Silvonen, A. Kukreja, B. Hicks, Using Natural Language Processing (NLP) to assess changes in transdisciplinary understandings across a large research consortium, 2026.
[26] P.T. Anastas, Beyond reductionist thinking in chemistry for sustainability, Trends in Chemistry 1 (2) (2019) 145–148.
[27] P.T. Anastas, Fundamental Changes to EPA's Research Enterprise: The Path Forward, ed: ACS Publications, (2012).
[28] D.J. Chalmers, Strong and weak emergence, The Re-Emergence of Emergence 675 (2006) 244–256.
[29] A. Mantini, The re-emergence of emergence, A Franciscan Theological-Metaphysical Foundation of Emergence: Theology and Science in an Enriching Dialogue, Springer, (2024), pp. 65–116.
[30] S. Johnson, Emergence: The Connected Lives of Ants, Brains, Cities, and Software, Simon and Schuster, (2002).
[31] F.E. Turkheimer, et al., Conflicting emergences. Weak vs. strong emergence for the modelling of brain function, Neuroscience & Biobehavioral Reviews 99 (2019) 3–10.
[32] D.F. Channell, The emergence of the engineering sciences: an historical analysis, Philosophy of Technology and Engineering Sciences, Elsevier, (2009), pp. 117–154.
[33] D. Cabrera, A general, measurable definition of emergence (E= ΔO): structure→ behavior as the universal mechanism underlying emergent phenomena, Journal of Systems Thinking 5 (1) (2025) 1–23.
[34] P. Corning, Synergy and the systems sciences, Institute for the Study of Complex Systems, International Society for the Systems Sciences and the Systems Sciences Primer. Online. Available from: http://www.complexsystems.org/commentaries/sept97.html, 1996 (accessed 03.11.11).
[35] J. Nielsen, et al., Engineering synergy in biotechnology, Nature Chemical Biology 10 (5) (2014) 319–322.
[36] K. Murphy, Robert Burton and the problems of polymathy, Renaissance Studies 28 (2) (2014) 279–297.
[37] T. Kuhn, What are Scientific Revolutions, MIT, Boston, (1981).

[38] A.M. Sarikhani, An Adaptive Provenance Collection Architecture in Scientific Workflow Systems, University of Adelaide, School of Computer Science, (2015).
[39] A.M. Leroi, The Lagoon: How Aristotle Invented Science, Bloomsbury Publishing, (2014).
[40] R.L. Presti, History of science: the first scientist, Nature 512 (7514) (2014) 250–251.
[41] H.S. Lang, Aristotle's science of nature, Astrid Schürmann (Hg.): Geschichte der Mathematik und der Naturwissenschaften in der Antike. Bd 3 (2005) 77–92.
[42] A. Schürmann, Physik/Mechanik, Franz Steiner Verlag, (2005).
[43] B. Beilage, Bibliographische Beilage Nr. 4, Gnomon 78 (7) (2006) 141–176.
[44] J.D. North, Universal Frame: Historical Essays in Astronomy, Natural Philosophy and Scientific Method, Bloomsbury Publishing, (1989).
[45] S. Gaukroger, J. Schuster, J. Sutton, Descartes' Natural Philosophy, Routledge, (2003).
[46] S. Shapin, S. Schaffer, Leviathan and the Air-Pump: Hobbes, Boyle, and the Experimental Life (New in Paper), Princeton University Press, (2011).
[47] P. Haastrup, J. Würtz, Environmental Data Exchange Network for Inland Water, Elsevier, (2011).
[48] R.T. Paine, Advances in ecological understanding: by Kuhnian revolution or conceptual evolution? Ecology 83 (6) (2002) 1553–1559.
[49] V. Dellarco, T. Henry, P. Sayre, J. Seed, S. Bradbury, Meeting the common needs of a more effective and efficient testing and assessment paradigm for chemical risk management, Journal of Toxicology and Environmental Health, Part B 13 (2-4) (2010) 347–360.
[50] I. Linkov, et al., Comparative, collaborative, and integrative risk governance for emerging technologies, Environment Systems and Decisions (2018). Available from https://doi.org/10.1007/s10669-018-9686-5. May 04.
[51] J. Schukoske, The evolving paradigm of laws on lead-based paint: from code violation to environmental hazard, SCL Rev. 45 (1993) 511.
[52] R.A. Brown, A. Kumar, The scientific method: reality or myth? Journal of College Science Teaching 42 (4) (2013) 10.
[53] T. Sousa, T. Domingos, S. Kooijman, From empirical patterns to theory: a formal metabolic theory of life, Philosophical Transactions of the Royal Society B: Biological Sciences 363 (1502) (2008) 2453–2464.
[54] Y. Jihai, Research paradigm issues and reflection on educational empirical research methods, Journal of East China Normal University (Educational Sciences) 35 (3) (2017) 64.
[55] B. Zheng, et al., Agent4S: the transformation of research paradigms from the perspective of large language models, arXiv preprint arXiv:2506.23692, 2025.
[56] N. Maity, S.C. Bhat, Artificial intelligence: the dawn of a new technological era, Psychological Review 65 (386) (1958).
[57] B. Allcock, et al., Data management and transfer in high-performance computational grid environments, Parallel Computing 28 (5) (2002) 749–771.
[58] G. Singh, et al., A metadata catalog service for data intensive applications, in: Supercomputing, 2003 ACM/IEEE Conference, IEEE, 2003, pp. 33–33.
[59] D. Vallero, Translating Diverse Environmental Data Into Reliable Information: How to Coordinate Evidence from Different Sources, Academic Press, (2017).
[60] B.E. Turvey, A. Coronado, K.V. Baltazar, Chapter 7 - Integrated forensic assessments: a psychosocial approach with the human rights perspective, in: B.E. Turvey (Ed.), Criminal Profiling, fifth Edition, Academic Press, (2023), pp. 205–234.
[61] G.S. Clair, D. Stanley, Knowledge services: the practical side of knowledge management, Information Outlook (2008) 1.
[62] D.A. Vallero, Environmental Systems Science: Theory and Practical Applications, first ed., Elsevier Academic Press, (2021).
[63] A.L. Washington, The interoperability of US federal government information: interoperability, in: Information Resources Management Association (Ed.), Big Data: Concepts, Methodologies, Tools, and Applications: Concepts, Methodologies, Tools, and Applications, IGI Global, Hershey, Pennsylvania, (2016), pp. 210–228.
[64] 2002-P-00017, EPA management of information technology resources under the clinger-cohen act [Online]. 2002. Available from: https://www.epa.gov/sites/production/files/2015-10/documents/clingercohen.pdf.
[65] D. Boyd, K. Crawford, Critical questions for big data: provocations for a cultural, technological, and scholarly phenomenon, Information, Communication & Society 15 (5) (2012) 662–679.
[66] J.M. Durden, J.Y. Luo, H. Alexander, A.M. Flanagan, L. Grossmann, Integrating "big data" into aquatic ecology: challenges and opportunities, Limnology and Oceanography Bulletin 26 (4) (2017) 101–108.

[67] D. Lazer, R. Kennedy, G. King, A. Vespignani, The parable of Google Flu: traps in big data analysis, Science 343 (6176) (2014) 1203–1205.
[68] D.A. Vallero, Fundamentals of Air Pollution, sixth ed., Academic Press, Amsterdam, NV, (2025).
[69] U.S. Environmental Protection Agency, Our mission and what we do: our mission. https://www.epa.gov/aboutepa/our-mission-and-what-we-do (accessed 14.02.20).
[70] Government of Canada, Health and the environment. Government of Canada. https://www.canada.ca/en/health-canada/topics/health-environment.html, 2019 (accessed 14.02.20).
[71] A. Silva, M. Rosano, L. Stocker, L. Gorissen, From waste to sustainable materials management: three case studies of the transition journey, Waste Management 61 (2017) 547–557.
[72] M. Jänicke, K. Jacob, Environmental Governance in Global Perspective: New Approaches to Ecological Modernization, Freie Universität, (2017).
[73] A. Jordan, A. Zito, and R. Wurzel, Environmental Governance in Europe, 2019.
[74] D.A. Vallero, 33. Waste governance, in: T.M.a V. Letcher, Daniel A (Eds.), Waste: A Handbook for Management, Elsevier, (2019), pp. 651–671.
[75] D.A. Luke, K.A. Stamatakis, Systems science methods in public health: dynamics, networks, and agents, Annual Review of Public Health 33 (2012) 357–376.
[76] L. Aissani, F. Jabouille, J. Bourgois, P. Rousseaux, A new methodology for risk evaluation taking into account the whole life cycle (LCRA): validation with case study, Process Safety and Environmental Protection 90 (4) (2012) 295–303.
[77] K.K. Isaacs, et al., Characterization and prediction of chemical functions and weight fractions in consumer products, Toxicology Reports (2016).
[78] K.L. Dionisio, et al., Exploring consumer exposure pathways and patterns of use for chemicals in the environment, Toxicology Reports 2 (2015) 228–237.
[79] M.-R. Goldsmith, et al., Development of a consumer product ingredient database for chemical exposure screening and prioritization, Food and Chemical Toxicology 65 (2014) 269–279.
[80] P.P. Egeghy, et al., Computational exposure science: an emerging discipline to support 21st-century risk assessment, Environmental Health Perspectives 124 (6) (2016) 697.
[81] J.E. Rager, et al., Linking high resolution mass spectrometry data with exposure and toxicity forecasts to advance high-throughput environmental monitoring, Environment International 88 (2016) 269–280.
[82] J.F. Wambaugh, et al., High throughput heuristics for prioritizing human exposure to environmental chemicals, Environmental Science & Technology 48 (21) (2014) 12760–12767.
[83] J. Mitchell, et al., Comparison of modeling approaches to prioritize chemicals based on estimates of exposure and exposure potential, Science of the Total Environment 458 (2013) 555–567.
[84] J.F. Wambaugh, et al., High-throughput models for exposure-based chemical prioritization in the ExpoCast project, Environmental Science & Technology 47 (15) (2013) 8479–8488.
[85] S. Gangwal, et al., Incorporating exposure information into the toxicological prioritization index decision support framework, Science of the Total Environment 435 (2012) 316–325.
[86] D. Vallero, Environmental Biotechnology: A Biosystems Approach, Elsevier Science, (2015).
[87] D.A. Vallero, Introduction: importance of systems sciences and first principles, Methods and Calculations in Environmental Physics, AIP Publishing LLC Melville, New York, (2022). pp. 1-1–1-20.
[88] T.E. Graedel, On the concept of industrial ecology, Annual Review of Energy and the Environment 21 (1) (1996) 69–98.
[89] M.M. Waldrop, Complexity: The Emerging Science at the Edge of Order and Chaos, Simon and Schuster, (1993).
[90] S. Burmaoglu, O. Sartenaer, A. Porter, Conceptual definition of technology emergence: a long journey from philosophy of science to science policy, Technology in Society (2019).
[91] J.C. Schmidt, Is there anything new under the sun? Emergence and Modularity in Life Sciences, Springer, (2019), pp. 3–36.
[92] A. Freddoso. New English Translation of St. Thomas Aquinas's Summa Theologiae (Summa Theologica). https://www3.nd.edu/~afreddos/summa-translation/TOC.htm.
[93] D.A. Vallero, Biomedical ethics for engineers, The Biomedical Engineering Series (2007).

CHAPTER

7

Environmental ethics

OUTLINE

7.1 Introduction

We could begin exploring environmental ethics from the perspective of a particular moral school of thought. We could certainly talk about duty ethics and Immanual Kant. We could argue and debate the instrumental and intrinsic utilities by applying the consequentialist thought and the harm principle of John Stuart Mill. We could invoke John Rawls's veil of ignorance in arguing for a biocentric or eco-centric viewpoint to protect the environment because we may indeed be the most vulnerable creatures when it comes to pollution. Of course, we could argue that we are approaching environmental and climate tipping points by employing the carrying capacity parable of Garrett Hardin's Tragedy of the Commons. We could recall the speeches of Barry Commoner or the writings of Rachel Carson. We could cite passages from Pope Francis's *Laudato Si'*. Yes, these are legitimate approaches.

Although most of the ethical content discussed in the first chapter of this book applies to the environment, there are unique aspects that deserve attention. Prominent among these aspects is that the moral questions and decisions extend beyond humans to all creatures and

Technical and Ethical Optimization: Challenges and Opportunities for the Modern Engineer and Scientist
DOI: https://doi.org/10.1016/B978-0-443-44925-3.00021-1

even the abiotic components of the planet. Environmental ethics is a two-pronged endeavor with concerns for both human health and ecosystems.

Advances in environmental ethics map directly to emerging scientific inquiry and discovery. Emerging technologies have provided ways to document environmental harm, now down to the angstrom scale, for example, mass spectrometry and computational models. As new data and assessment tools have emerged, the certainty of past and pending pollution episodes has been undeniable. The emerging environmental and human health corpus provides an ever-growing data source for environmental ontologies that supports hypothesis generation and testing that drive research that supports environmental decision-making. The corresponding growth of and confidence in the domains of the ecological and human health sciences, along with an improved understanding of environmental ethics, provides a prime example of the need to optimize decisions and actions based on both technological and moral variables. Indeed, what has driven environmental protection for more than 50 years has been as much about the immorality of pollution as it has been the result of previously undiscovered and impossibly known empirical scientific evidence. The latter has continued to improve the accuracy and precision of environmental science, but the former continues to be the driver for many environmental remediation and pollution prevention efforts.

Advances in the computational sciences have intersected with systems thinking and environmental decision making. This has supported the need to align technological advancements, for example, artificial intelligence, machine learning, and big data, with ecological sustainability, public health, and social responsibility.

From another ethical standpoint, computational models and other emerging technologies must be considered for their threats. Like other applications, the use of AI models for environmental decision-making must be transparent and unbiased and provide a balance of technological benefits with potential health and ecological harm. In a sense, it is metaphorically a Heisenberg uncertainty problem. If we become enamored with AI to understand environmental conditions, are we changing the environment in the process? For example, is the growing dependence on computational tools to document evidence of environmental quality feeding the same problems as other societal demands? Will the seemingly unstoppable growth of electricity-demanding data centers for computations be exacerbated and, in turn, increase their energy demands, proliferate their environmental footprints, and extend the very harm to the environment that we are trying to protect [1]? AI-related data processing by supercomputers is powered by the electricity grid and often requires backup with diesel-powered generators. For example, training one AI system can emit more than 113,000 kg of carbon dioxide. AI technology across all sectors produces carbon dioxide emissions that compare to those of the aviation industry [2].

Could this also hurt the provenance of data by increasing the access for hackers and other bad actors who can invade privacy and threaten businesses and governments? Emerging technologies have positive and negative ethical aspects, so these should be properly weighed when conducting optimizations [3,4].

7.2 Tipping points and the state of nature

Since many around the world share a common set of environmental values based on a seemingly intuitive ethos, it must have something to do with a collective societal need to protect Spaceship Earth [5,6]. That said, let us begin with philosopher Thomas Hobbes's concept of the social contract.

Perhaps one of the fears that drives the modern environmental ethos is the tipping point toward an irreversible trend of ecological destruction. Sudden changes have arisen over humanity's time on Earth. Gradual changes in one component of the Earth system reach a point when a part of the system crosses a critical threshold, or tipping point, after which impacts cascade through coupled biogeochemical systems [7]. These changes and trends have affected human populations in dramatic ways. After all, all macroscale land creatures, including humans, breathe from the air in a common troposphere. They need water that circulates through the hydrological cycle. Plant life stores and converts the solar energy from the sun. Humans make products derived from the same plants and other Earth resources.

The Earth's environment can be envisaged as a "common" shared by humankind. Biologist Garrett Hardin captured the concept of the tipping point in his allegorical common area of an English village. Every farmer had a single cow that freely grazed the grass of the common. The common readily provided nourishment for every cow. Village life remained stable. However, this began to change after an idea entered the mind of one of the villagers who thought that if he added a cow, the cost of the new cow would be shared by everyone, but the profit would be entirely his. The increase of one cow allowed him to prosper in comparison to his neighbors for a while. However, the other farmers noticed his financial success and correspondingly each decided to add a cow to the common. Their logical cascade continued in the villagers' minds so that if two are better than one, then three must certainly be better than two. At some point, however, the village common was no longer able to support the large number of cows, that is, the caloric demand exceeded the common's carrying capacity. As a result, the system crashed, and everyone suffered because the carrying capacity went from a sustainable amount to zero. A tipping point was reached and surpassed. The formerly sustainable system became unsustainable and irreversibly damaged.

This concept can be extended intellectually to the finite resources, which drive the carrying capacity of ecosystems. For millennia, the atmosphere, surface waters, oceans, soil, and groundwater were perceived to be endless commons from which we expected to draw life and resources in perpetuity. A similar argument can be made for the exploitation of nonrenewable resources. Treating diminishing resources like oil and minerals as capital gains will reach a point of insufficiency of the Earth's vast but finite resources [8].

The Intergovernmental Panel on Climate Change (IPCC) has applied the concept of the tipping point to large-scale discontinuities, that is, global warming exceeding 5°C above preindustrial conditions. But in recent years, some now consider the point to be at 1°C and 2°C [9–11]. Secondary and tertiary effects following the tipping points include the dissociation of ocean methane hydrates and thawing permafrost, as well as large-scale economic impacts [12].

Climate change, chemical exposures, pandemics, and many environmental disasters can be treated as irreversible trends with the fear that we are approaching or have begun to be at physical tipping points [7,10–13]. Along with planetary tipping points, social tipping points also occur, that is, socio-environmental relationships wherein small changes in underlying elements or behavior of actors trigger a large nonlinear response in the social system. This threatens socio-environmental chaos that requires societies to adapt [14]. These adaptations, hopefully before but assuredly after a social tipping point, must be feasible and socially acceptable. In this case, Hardin's parable applies to large segments of society.

Hobbes's concept of the commonwealth [15] implies a social tipping point when the collective selfishness and greed of society's members cause institutions to revert to what he called the "state of nature" or a complete abandonment of the social contract, resulting in a society that cannot perform properly [16].[1]

In *Leviathan* [15], Hobbes argued that the right of nature (*jus naturale*) is a law of survival, that is, the fundamental liberty possessed by every person to use one's own power accordingly to preserve one's own life, especially during the state of nature. When this happens, a "war of all against all" is on the horizon, in which individuals have a right to every act of survival. Though the terms sound similar, right of nature has nearly the opposite meaning of the "law of nature" (*lex naturale*), which are reason-based precepts to seek peace for self-preservation that ultimately lead people to form a social contract to create a sovereign power for security. Thus, a social contract is needed to transfer each person's state of nature to a sovereign body like a government to protect each person from the acts of the other. The societal breakdown in a "Mad Max" [17] type of dystopia results when the free will of members becomes out of balance and harms others who would otherwise be protected by a compassionate social contract.

A social contract is not inherently good or bad. To wit, when a government fails to respect the human rights of its members, the social contract becomes incarceration. The moral concept of empathy is key, that is, placing oneself in another's situation [18,19].

7.2.1 Environmental legislation as the tipping point approached

As in most moral decision-making, temperance and moderation rule the day. Indeed, Socrates, Plato, and Aristotle considered temperance to be among their cardinal or primary virtues, along with courage, wisdom, and justice [20]. It can be argued that as a tipping point nears, the members of society are more likely to relinquish some of their selfish desires for a better condition, that is, a healthier and more livable environment. This happened in the late 1960s when the fear of irreversible environmental damage motivated large numbers of people to call for legislation to protect the environment. In the United States, for example, the National Environmental Policy Act (NEPA) was symbolic of a new federal commitment to environmental stewardship. It was signed into law in 1970 after combative hearings in the US Congress. NEPA was not a technical law like the forthcoming Clean Water Act, Clean

[1] Ironically, in our discussion of an ethical obligation to nature, the choice of words of a 17th-century philosopher conveys a bit of detest for "nature" because he considered the undomesticated, that is, the natural, to be outside of civil society. Animals kill other animals, so the state of nature allows violent death at the hands of others, which obviously must be avoided in the commonwealth.

Air Act (CAA), and hazardous waste laws. Instead, NEPA created the environmental impact statement (EIS) and established the Council on Environmental Quality (CEQ) in the Office of the President. Of the two, the EIS represented a sea change in how the federal government was to conduct business related to the environment. Any federal agency was required to prepare EISs on any major action that it was considering that could significantly affect the quality of the environment. From the outset, each agency had to reconcile often competing values, that is, its mission and the protection of the environment.

The CEQ was charged with developing guidance for all federal agencies regarding NEPA compliance, especially when and how to prepare an EIS. The EIS process combines scientific assessment with public review. The process is similar for most federal agencies. Agencies often strive to receive a finding of no significant impact (FONSI), so that they would remain unfettered to pursue a mission-oriented project [21]. Whether a project either leads to a full EIS or a waiver through the FONSI process, it must undergo an evaluation of potential environmental impacts. This step is referred to as an "environmental assessment." An incomplete or inadequate assessment will lead to delays and increase the chance of an unsuccessful project. The final step is to establish a record of decision (ROD) to document the alternatives considered and the rationale for selecting the chosen alternative. The ROD also summarizes the comments received during the public reviews and how the comments have been addressed. Most states have adopted similar requirements for their RODs.

The EIS documents were intended to provide full disclosure of actual or possible problems if the project is implemented. This was accomplished by looking at all the potential impacts to the environment from any of the proposed alternatives and comparing those outcomes to a so-called "no action" alternative. At first, many agencies attempted to demonstrate that mission was inherently environmentally sound, arguing that the environment would be better off with the project than it would be without it, that is, the chosen action is better than no action. Too often, however, an EIS was written to justify the agency's mission-oriented project. One of the key advocates for the need for a national environmental policy, Lynton Caldwell, is said to have referred to this as the federal agencies using an EIS to "make an environmental silk purse from a mission-oriented sow's ear!" [22,23]. To remedy this whitewashing attempt, several courts ruled that federal agencies must take NEPA seriously. With time, the federal agencies' growing commitment to environmental protection led to a recognition of the need for sound science in assessing environmental conditions and possible impacts, and the very large role of the public in deciding on the environmental worth of a highway, airport, dam, land management program, agricultural project, electric power generating station, waterworks, treatment plant, or any other major project sponsored by or regulated by the federal government.

All federal agencies must follow the CEQ regulations [24] to "adopt procedures to ensure that decisions are made in accordance with the policies and purposes of the Act." This process must be set up in advance, early in an agency's planning stages. For example, if waste remediation or reclamation is a possible action, the NEPA process must be integrated into the remedial action planning processes, beginning with the identification of the need for and possible kinds of actions being considered. Noncompliance or inadequate compliance with NEPA rules and regulations has led to lawsuits, increased project costs, delays, and the loss of the public's trust and confidence, even if the project is designed to improve the environment, and even if the compliance problems seem to be only "procedural."

Thus, NEPA provided a major impetus in the growth of the environmental disciplines since the 1970s. Experts would be needed who conduct sound environmental studies and communicate their findings objectively to the public.

Subsequently, in 1970, the US Environmental Protection Agency (EPA) assumed the responsibility for reviewing the environmental effects of all federal agencies' actions. This authority was provided in Section 309 of the CAA. The review must be followed with the EPA's public comments concerning the environmental impacts of any matter related to the duties, responsibilities, and authorities of the EPA's Administrator, including EISs. The EPA's rating system is designed to determine whether a proposed action by a federal agency is unsatisfactory from the standpoint of public health, environmental quality, or public welfare. This determination is published in the *Federal Register* (for significant projects) and referred to the CEQ [24].

The 1970 amendments to the CAA arguably ushered in the era of environmental legislation with enforceable rules [25–27]. The previous version of the CAA was amended to provide a comprehensive set of regulations to control air emissions from area, stationary, and mobile sources. The CAA's goal, from the outset, has been to protect public health and the environment from air pollution. In the more recent versions, air pollutants are classified into categories, with a distinction between "conventional" and "toxic" pollutants: ambient concentrations of vapors and particulate matter (PM), including those to reduce photochemical oxidant smog or ozone (see Discussion Box: Smog and Ozone). Six conventional or "criteria" pollutants are covered under the National Ambient Air Quality Standards (NAAQS): PM; nitrogen dioxide; sulfur dioxide; carbon monoxide; ground-level ozone; and the metal lead [28]. The NAAQS are continuously evaluated, with the possibility of adding criteria pollutants (the metal lead was the last to be added). Existing NAAQS requirements are also revised when supported by sound science. For example, the particulate standard was changed to address finer particles, that is, particles with aerodynamic diameters <2.5 μm ($PM_{2.5}$), in response to research showing that exposure to these smaller particles is more likely to lead to health problems than exposure to larger particles. Smaller particles can penetrate further into the lungs and are likely more bioavailable than the larger PM_{10}. Recently, however, concerns about larger particles (e.g., PM_{10}) have increased, particularly as research has begun to link coarse particles to chronic diseases like asthma.

NEPA's passage was the beginning of a landslide of legislation. The 1990 Amendments to the CAA profoundly changed the law by adding new initiatives and imposing dates to meet the law's new requirements. The Clean Water Act has been amended several times to address emerging surface and groundwater challenges. Hazardous waste laws were passed that required risk management decisions to be based on the sound science provided by risk assessments conducted in scientifically sound ways. After the mid-1980s, other subsequent environmental legislation, raising a collective desire to improve air and water quality, regulate pesticides, ensure the safety of products, and to support first responders dealing with environmental emergencies and disasters.

7.2.2 Science and the tipping point

Scientific principles can be used to select actions at various points in a cause-effect event cascade leading to various outcomes, including a tipping point. The steps, that is, events, in the chain occur before larger impacts in terms of space and time. Therefore, preventing the impact calls for actions prior to larger scale problems' occurrence and well before the tipping point. After the tipping point is reached, a whole new set of events will occur, with some outcomes that are adverse, irreversible, and severe. Delay in action can mean bigger problems ahead. However, ill-conceived and poorly thought-out solutions can lead to other societal tipping points. Solving one problem, like limits on small farming to obviate the need for fossil-fuel-derived-fertilizer, must not cause an arguably worse problem, like diminishing the world's safe food supplies and preserving cultures. Nor should we blindly displace farmland with solar farms at the expense of optimal and sustainable uses of fertile farmland. Building offshore wind farms to reduce fossil fuel demand should not create sea mammal disasters. Indeed, synergies among societal needs are fundamental to avoiding such tipping point tradeoffs [29]. Solutions to complex problems are seldom "either-or" propositions, but balances among best alternatives.

Of course, a systematic review and assessment of actions related to avoiding tipping points must consider all relevant factors. For example, climate change may indeed adversely affect food supplies if they result in droughts and other outcomes that harm agriculture. In this case, actions taken to limit these effects would benefit food supplies in the future. This calls into question the quality of models used to predict such outcomes. For example, do these models consider changes in incoming solar radiation that would affect photosynthesis? How would contravening variables like this affect the predicted outcomes?

All events induced by various actions must also receive objective scientific scrutiny. An acceptable outcome balances environmental considerations with other societal values. For example, food supply is not simply evaluated in terms of balances of matter and energy per the first principle of thermodynamics. Selecting the right actions must also consider social issues like the effects on individual farmers at the expense of increasing the size of large-scale corporate farms that may be perceived to have higher levels of food production when only nonmonetizable values are considered. In fact, small farms almost always provide sustainable agriculture, especially compared to "big food" monocultures. In addition, the biodiversity and cultural problems displacement of small farms with transnational corporate operations often increases obesity and chronic disease incidence in communities [30,31].

Net primary productivity is just one factor in sustainable agriculture. In fact, plantation-type agriculture consists of one or a few crops year after year. This usually requires large investments in fossil fuels to cultivate and harvest crops, fossil fuel-based commercial fertilizers, genetically modified seeds, and pesticides. These add up to a lack of diversity and vulnerabilities to infestations and plant diseases. Indeed, small farms are often highly productive with fewer nonsustainable side effects than large-scale operations. For example, corn yields in traditional Mexican *chinampas*—that is, raised beds in shallow ponds and wetlands—range from 3.5 to 6.3 tons per hectare, which means that up to 20 people can maintain a nutritionally acceptable diet from each hectare farmed [32]. In addition to sustaining cultural values and preserving ways of life, small farms are sanctuaries for native species, which are threatened by genetically modified species. Such environmental decisions are complex, calling for an objective and rational process.

Environmental resources are not distributed evenly among populations in terms of amount and quality. Some breathe cleaner air than others, drink purer water than most, eat food that is less contaminated than the majority of the world's inhabitants, and have better tools and toys than everyone else.

7.3 Scale of values and virtues

Ethics can be viewed as a continuum of types of virtue and scales of virtue. Virtues can be those of the individual or those of increasingly larger communities. Honesty and integrity are generally possessed by the individual members of society. A person has different degrees of honesty, but a community does not. In a representative democracy, for example, political representatives can be honest, but political representatives of society can only impose honest behavior through the enforcement of and compliance with legislation and rules. Indirectly, society can impose norms, mores, and values through faith traditions, families, and collectives such as schools and professional organizations.

Environmental ethics consists of both microethical and macroethical obligations. Honesty is a microethical concept, exclusive to the individual. Ensuring justice for all members of society is a macroethical concept. Macroethics focuses on the collective, long-term responsibility of engineers and scientists to address broad societal impacts, such as environmental sustainability and risks from emerging technologies [33–36].

Protecting the environment certainly has microethical content, such as not littering even when no one is looking and no chance of being fined. Or even greater ethics appear when a person decides to clean up a stream that is not even in her own neighborhood. However, the environmental social contract resides at the far right of Fig. 7.1, that is, justice is the predominant source of environmental ethics. Care and actions to protect public health and ecosystems dwell at the macroethics scale.

Wasteful and unethical uses of the Earth's resources threaten human health and ecosystem effects, but the costs certainly affect society's welfare, threaten public safety, and lead to financial burdens. There is an abiotic aspect to environmental ethics, as pollution destroys irreplaceable materials, including monuments, historically important structures and buildings, as well as works of art and cultural artifacts [37].

Environmental resources have instrumental value, that is, they are valuable because they provide goods and services. For example, habitats are seen as valuable because they provide anthropocentric benefits of ecosystems, for example, flood control, fisheries, and pharmaceutical sources. Such resources have monetized valuation that can translate into monetary currency, for example, gross domestic product. However, most environmental and quality of life resources have inherent or intrinsic value and, as such, are not readily conducive to monetized valuation, for example, old growth forests have nonmonetized value by providing habitat and ecological diversity, but little monetized value since they are not used for timber.

The ethical problem with reliance on instrumental valuation can be demonstrated by an example. Consider the person who decides to drive from Townville, New York, to Alaska. He spends $1500 for fuel and wear and tear on his car, $2000 for lodging, $500 for food, $20 dollars for a gasoline can, and $5 for matches, for a total $4025. He arrives at an old growth forest in Alaska and sets fire to the forest, completely destroying 1000 acres of irreplaceable

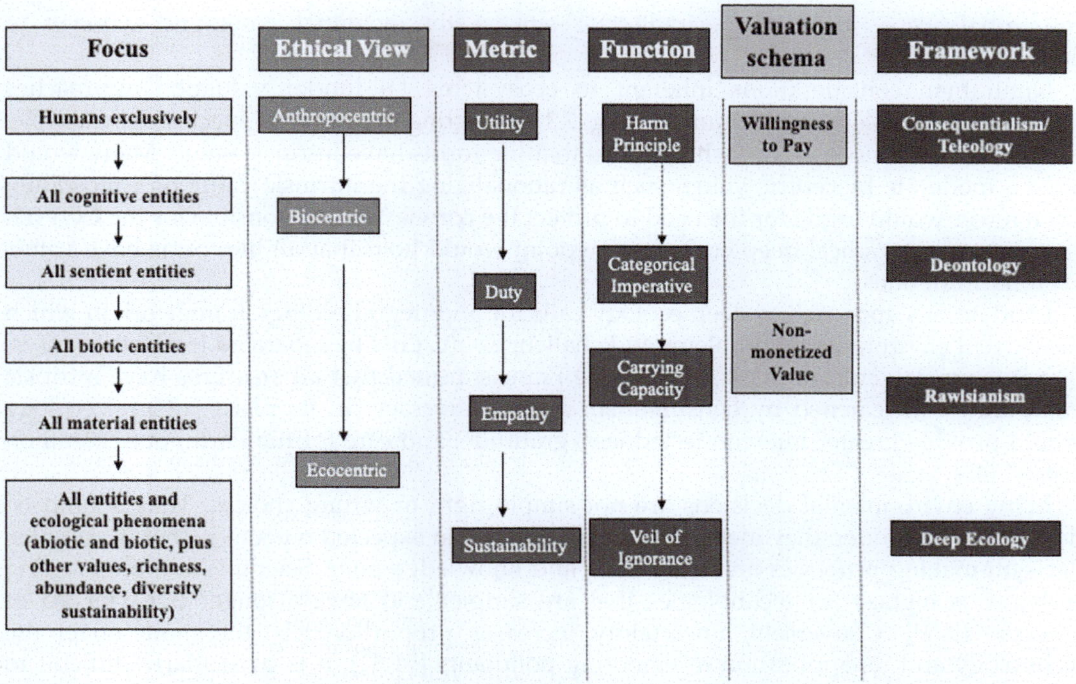

FIGURE 7.1 Environmental value systems. Source: *Data from D.A. Vallero, T.M. Letcher, Minerals, in: Unraveling environmental disasters Second ed The Netherlands: Elsevier Academic Press, Amsterdam, 2023, ch. 720; D.A. Vallero, Biomedical Ethics for Engineers, The Biomedical Engineering Series (2007).[37,38].*

old forest. An old growth has great intrinsic value but no monetized value, so he has added $4025 to the gross national product even though he has committed arson. Now, consider the same data; only on this trip he burns 1000 acres of a stand of fir trees grown by a timber company. The trees have a monetized worth of $5 million, so his action has decreased the GDP by $4,995,975.

An example of intrinsic and instrumental values is shown in Fig. 7.1.

Damage to human health is almost exclusively at the top of the diagram. The efficiency of processes employed to decrease harm can be quantified, but many environmental values are not monetizable or quantifiable. Emerging technologies are needed to improve the removal of pollutants from contaminated water, air, and other environmental resources and, preferably, before they are released into the environment. Of course, it is best to prevent the formation of the pollutant in the first place [39–46]. Sound science, however, must be paired with sound morality. Indeed, even an action that meets a scientifically based standard can still be unethical. For example, it would be an immoral act if it results in transferring the pollution and the attendant harm to another region, such as the installation of tall stacks to prevent local pollution, but, in the aggregate, leads to acid rain downwind. Mill, who is considered the founder of utilitarianism, observed this flaw in the "greatest good," and proposed that the desired action also meet a "harm principle" that can counterbalance a localized desired state against the potential costs and damage to others [47].

Nonhuman considerations inherent to ecosystems fall under valuation approaches further down the left-hand column of Fig. 7.1. Protecting endangered species, for example, is an instance of biocentrism, which states that life forms have intrinsic value. Many would have a moderate biocentric value, such as rationalizing the intrinsic value of a mosquito. Even fewer would argue for the need to protect the coronavirus responsible for the COVID-19 pandemic. A radical deep ecology viewpoint would hold that all life forms have value, even harmful ones.

Laudato Si's approach to deep ecology calls for an integral ecology framework in which the current environmental problems and challenges must not be separated from other crises, that is, spiritual, cultural, and social. Pope Francis argued that all creatures have intrinsic value that is threatened by consumerism and technocracy. In its place, integral ecology would provide greater interconnectedness, gratitude, and care for the Earth, our "common home" [48].

Many environmental decisions are not simple right or wrong choices. They should be data-driven, but often they include much uncertainty in expected outcomes, that is, complex decision-making processes are often data- and knowledge-poor. Scientific and engineering certainty is highest for technologies that are simpler and have a longer track record of success [49,50]. Conversely, uncertainty increases proportionately the more novel the technology and its application to emerging pollutants [51,52]. It is particularly difficult to predict contingent harm and downstream impacts. A decision that allows a material to corrode, for example, may appear only to be a nuisance and entail unnecessary costs, that is, an anthropocentric view of the loss of instrumental value. However, if the corrosion leads to the loss of irreplaceable resources, or worse, injury or loss of life, the decision in retrospect will have been catastrophically wrong and immoral [53–57].

7.4 Justice

Discussions of justice usually center around the collective virtues of society. We tend to talk about justice in terms of larger collections of society's members. For example, do we have justice in the dispensing of wealth in the United States? Is the engineering profession justly carrying out its duties? Are all communities justly treated in the siting of landfills?

So, it stands to reason that the environment would gain the attention of moral philosophers and ethicists. After all, ecology is the science of relationships. Individual organisms and niches make up larger associations, which, in turn, form communities.

Environmental ethics has been described as the "systematic account of moral relations between human beings and their natural environment" [58]. In a sense, then, environmental ethics extends the responsibility of human beings beyond our own species. Just allocation of resources includes the whole environment, both the biotic and abiotic components. It can be argued (and often is) that such comprehensive accountability to the natural environment is merely self-preservation and not necessarily justice. Old and New Testament admonitions to be good stewards of the environment can be viewed as a means of protecting humans, just as the Levitical rules and the Pentateuch can be viewed as a means of addressing public health concerns and of maintaining order in the Judeo-Christian culture, respectively. Whatever the motivation to dispense justice, sometimes the least appreciated species or niches are those that turn out to be the most vulnerable and most crucial parts of an ecosystem (an analogy to the responsibility to be concerned with the "least among you").

Indeed, The Gospel of Luke, Chapter 9, Verses 46 and 47 captures this well:

> "Jesus, knowing their thoughts, took a little child and made him stand beside Him. Then He said to them, 'Whoever welcomes this little child in My name welcomes Me; and whoever welcomes Me welcomes the One who sent me. For he who is least among you - he is the greatest.'"

Jesus taught that the measure of a just people is embodied in the way it treats the most downtrodden (the poor, disabled, and those of the "wrong" ethnicity or race, such as people from Samaria at that time), those with the quietest voices in society (children, women), and even those outside the mainstream of society (the "sinners"). It fits, then, that much of the environmental justice (EJ) movement has begun in churches.

The "least of these" consideration also evokes John Rawls's veil of ignorance, where the person deciding on the most ethical action is behind a metaphorical veil, without knowledge of whom is harmed the most if the action is taken. Thus, a decision must be made with an appreciation for how it may affect the most vulnerable current and future members of society, for example, the poor, the immunocompromised, those in war zones, pregnant women, the unborn, neonates, children, the infirm, and the elderly [59]. This calls for complete consideration of both the ends and the means of actions needed to protect the environment [60]. As such, decision-makers are compelled to create sustainable policies that are impartial. The decision could harm themselves or those they care about most, including future generations who will be born into an unsustainable and unhealthy future.

Given that the distribution of goods and services is so uneven, we may be tempted to assume that systems are fair simply because "most" are satisfied with the current situation. Indeed, environmental and public health decisions are seldom based on a statistical measure of central tendency, that is, a population's mean, median, or mode. To the contrary, most environmental laws and regulations aim to protect the weakest among us. The moral dictates of natural law and canons of faith traditions stress the protection of the vulnerable [61–63], for example, pregnant women, unborn, newborn and young children, immunocompromised, elderly, those with respiratory and cardiovascular problems, and those who are otherwise susceptible to exposure and effects from pollutants. Protecting individuals means overarching preservation and improvement of the ecosystems surrounding them [64]. This was recently evidenced in the societal disaster resulting from the failure of retaining structures at an abandoned mine (see Discussion: Gold King Mine disaster).

Gold King Mine disaster

The Gold King Mine near Silverton, Colorado, one of Colorado's 23,000 abandoned mines, is a source of water contamination by heavy metals. In the Upper

(cont'd)

Animas watershed in the southwestern part of the state, the mine had altered the hydrology and hydrogeology, changing the flow velocity and direction of tributaries and groundwater. The Rocky Mountains contain iron sulfide in pyrite that forms sulfuric acid when it reacts with oxygen. The acidic conditions increase the rate and quantity of metals dissolved in water. Mining at Gold King ended in 1991, and the mine was bulkheaded. Metal-laden waters were then transported [65].

While excavating atop an abandoned adit, pressurized water started to leak above the mine tunnel, over a nine-hour period, releasing over 11 million liters of water that had been stored behind the collapsed material. The contaminated water reached Cement Creek, a tributary of the Animas River and San Juan River, flowing into Lake Powell. The spill contained the metalloid, arsenic, and the metals aluminum, cadmium, copper, iron, lead, manganese, mercury and zinc [66]. EPA issued an official apology:

EPA regrets and deeply apologizes for the potential impact this is having on the health, finances, and culture of Navajo communities. EPA recognizes the San Juan River as not only an important water resource for agricultural use but also as a source of cultural significance for Diné people. We are committed to working closely with the Navajo Nation Department of Emergency Management to provide alternative drinking, agricultural, and livestock water, livestock feed, and to restore the river to prerelease conditions [67].

It is increasingly common for regulatory agencies to require that the rehabilitation of a tailings dam or other waste have operational lifetimes of 100 years or more [68–70]. The rate of erosion decreases in proportion to the surface strength of a slope; therefore, the effect of varying surface strength would move the erosion rate surface inward with increasing shear strength, or outward with decreasing shear strength, relative to that of the erosion loss axis [68,69,71]. Shielding the surface of a slope with a cover of vegetation, for example, would similarly increase the surface shear strength. Thus, slope angles must be as flat as possible, for example, 10–15 degrees is practical and approaches the ideal angle, with slope lengths of less than 25 m.

This means that long slopes should be interrupted at height intervals of about 7–8 m by horizontal berms or steps equipped with lined, erosion-proof surface drains designed to conduct the surface runoff from the slope above to ground level without causing any erosion of the lower slope surfaces. Very flat slopes and horizontal surfaces do not erode to any significant extent and hence do not require protection against erosion.

Both wind and water erosion are important. In wetter climates, water erosion predominates, and in drier climates, there is more wind erosion. In South Africa, for example, wind erosion accounts for two-thirds of total erosion, but in Iceland, wind erosion also predominates over water erosion. Unprotected tailings slopes may erode at rates of up to 1000 t ha^{-1} a^{-1}, whereas natural slopes usually erode at 50–100 t ha^{-1} a^{-1} or less. The aim of protecting tailings slopes should be such that their rate of erosion is reduced to less than 100 t ha^{-1} a^{-1}.

Interestingly, Gold King Mine is in a relatively dry location, so this is another

(cont'd)

example of the importance of finding other causal factors in root cause analysis. Indeed, many other physical features affect slope stability [72]. Sorption of soil and other unconsolidated materials is also important to slope stability, especially the cation exchange capacity (CEC). Increasing CEC often decreases the erosion rate [73–75]. Salinity also affects erosion. High salt concentrations in pore fluid increase erosion in cohesive soils [76,77]. Biota also influence erosion and slope integrity. The previously mentioned vegetative cover provides roots that increase stability and decrease erosion rates. Also, certain microbial biofilms can substantially increase soil cohesion and decrease roughness by excretion of extracellular polymeric substances [78].

After reaching the Animas River, Colorado, polluted water soon moved into communities in Colorado, New Mexico, and through Navajo communities adjacent to the San Juan River. The Native American Diné farmers were notified that water they use to irrigate their corn crops and supply to their livestock was likely contaminated and needed to be shut off [65,67]. This water is the exclusive source for Diné farmers and much of the Navajo Nation [79]. In addition to contaminated food and water and economic costs, the disaster extended into the Diné social and cultural fabric and the strong connection with the San Juan River system. Corn is much more than a source of nutrition to the Diné people; it is part of their connection to the Earth, ancestry, and essence of being.

Water is life to the Diné. Tó (water) is one of the Diné's Four Sacred Elements, joined with Kǫ' (fire/light), Niłchí (air), and Tadidiin (pollen/earth) [80]. All of these must be protected and preserved, so the mine breach was an especially abrupt and devastating insult to what the Diné refer to as Lifeway. The elements instill a state of beauty in connection with the natural world for the Diné. Harming one element disrupts this connection and creates disharmony. These Four Sacred Elements are considered indivisible and must coexist and complement one another.

For the Diné, this is an environmental injustice in its truest sense. It is a disruption of environmental quality and cultural significance. As such, EJ goes beyond cleanup and requires cultural restoration.

7.4.1 Environmental justice

Social disasters that deny or diminish human rights often result from or cooccur with environmental and public health disasters. Indeed, the only proper way to protect public health and the environment is to ensure that *all* persons are adequately protected. In the words of Reverend Martin Luther King, "Injustice anywhere is a threat to justice everywhere"[81]. By extension, if any group is disparately exposed to an unhealthy environment, then the whole community is subjected to inequity and injustice. Put in a more positive way, we can work to provide a safe and livable environment by including everyone, leaving no one behind. The term EJ is usually applied to social issues, especially as they relate to neighborhoods and communities. EJ communities possess two basic characteristics:

1. They have experienced historical (usually multigenerational) exposures to disproportionately high doses of potentially harmful substances (the *environmental* part of the definition). These communities are home to numerous pollution sources, including heavy industry and pollution control facilities, which may be obvious by their stacks and outfall structures, or may be more subtle, such as long-buried wastes with little evidence on the surface of their existence. These sites increase the likelihood of exposure to dangerous substances. Exposure is preferred to *risk*, since risk is a function of the hazard and the exposure to that hazard. "Even a substance with a very high toxicity (one type of hazard) that is, confined to a laboratory of a manufacturing operation may not pose much of a risk due to the potentially low levels of exposure." ("be confined to a laboratory | English examples in context | Ludwig")
2. EJ communities have certain specified socioeconomic and demographic characteristics. EJ communities must have a majority representation of low socioeconomic status, racial, ethnic, and historically disadvantaged people (the *justice* part of the definition).

These definitions point to the importance of an integrated response to ensure justice. The first component of this response is a sound scientific and engineering underpinning to decisions. The technical quality of designs and operations is vital to addressing the needs of any group. However, the engineering codes' call that we be "faithful agents" lends an added element of social responsibility to environmental practitioners [82]. For example, we cannot assume a "blank slate" for any design. Historic disenfranchisement and even outright bias may well have put certain neighborhoods at a disadvantage.

Thus, the responsibility of professionals cannot stop at sound science, but should consider the social milieu, especially possible disproportionate impacts. The determination of disproportionate impacts, especially pollution-related diseases and other health endpoints, is a fundamental step in ensuring EJ. But even this step relies on the application of sound physical science. Like everything else that technical professionals do, we must first assess the situation to determine what needs to be done to improve it. As a first step in assessing environmental insult, epidemiologists look at clusters and other indications of elevated exposures and effects in populations. For example, certain cancers, neurological, hormonal, and other chronic diseases have been found to be significantly higher in minority communities and in socioeconomically depressed areas. Acute diseases, as indicated by hospital admissions, may also be higher in certain segments of society, such as pesticide poisoning in migrant workers.[4] These are examples of *disparate effects*. In addition, each person responds to an environmental insult uniquely, and that person is affected differently at various life stages. For example, young children are at higher risk to neurotoxins. This is an example of *disparate susceptibility*. However, subpopulations can also respond differently to the whole population, meaning that genetic differences seem to affect people's susceptibility to contaminant exposure. Scientists are very interested in genetic variation, so that genomic techniques[5] (e.g., identifying certain polymorphisms) are a growing area of inquiry.

In a sense, historical characteristics constitute the "environmental" aspects of EJ communities, and socioeconomic characteristics entail the "justice" considerations. The two sets of criteria are mutually inclusive, so for a community to be defined as an EJ community, both sets of criteria must be present.

A recent report by the US Institute of Medicine found that numerous EJ communities experience a "certain type of double jeopardy" [83]. The communities must endure elevated levels of exposure to contaminants, while being ill-equipped to deal with these exposures because so little is known about the exposure scenarios in EJ communities. The first problem (i.e., higher concentrations of contaminants) is an example of *disparate exposure*. The latter problem is exacerbated by the disenfranchisement from the political process that is endemic to EJ community members. This is a problem of *disparate opportunity* or even *disparate protection*, which is based on the fourteenth amendment to the US Constitution's call for equal protection clause. The report also found large variability among communities regarding the type and amount of exposure to toxic substances. Each contaminant has its own type of toxicity. For example, one of the most common exposures in EJ communities is to the metal lead (Pb) and its compounds. The major health problem associated with Pb is brain, as well as central and peripheral nervous system diseases, including learning and behavioral problems. Another common contaminant in EJ communities is benzene, as well as other organic solvents. These contaminants can also be neurotoxic but have very different toxicity profiles from neurotoxic metals like Pb. For example, benzene is a potent carcinogen, having been linked to leukemia and lymphatic tumors, as well as severe types of anemia. They also have very different exposure profiles. For example, Pb exposure is often in the home and yard, while benzene exposures often result from breathing air near a source (e.g., at work or near an industry, such as an oil refinery or pesticide manufacturer). The Institute's findings point to the need for improved approaches for characterizing human exposures to toxicants in EJ communities. One of the first places to recognize the disparate exposures was in Warren County, North Carolina, but numerous other communities have experienced uneven, and arguably unjust, disparities in environmental protection. However, there is little consensus on what defines an environmental injustice and whether, in fact, an injustice has occurred in many of these communities.

Beyond the need to address traditional chemical pollutants that can be inordinately high, recent research has identified so-called nonchemical stressors like financial strain, housing instability, discrimination, and community violence [84,85]. These may manifest themselves as chronic health problems in EJ communities. Indeed, genetic predisposition and chronic exposure to psychosocial stress may act synergistically with chemical exposures to exacerbate health [84,88].

Even before the genetic engineering revolution, crop practices presented threats to sustainable food supply, notably the introduction of monocultures to what were the world's most diverse habitats. Substituting indigenous crop varieties and diverse cultivation systems with monocultures has increased vulnerability to the ravages of pests and plant diseases, loss of soil fertility, and increased application of agrochemicals. As evidence, global food supplies now depend on merely 100 or so species of food crops, rather than the thousands of species and varieties that have been used locally for millennia. This consolidation is inextricably threatening to the individual farmers in developing countries as they cannot sustain this practice, leading to more monocultures and further diminishing diversity.

Many of the farmers in developing nations and in certain demographic strata of developed nations are poor and susceptible to offers of quick fixes to food supplies. They are unlikely to complain, which is similar to other vulnerable communities. Thus, what may seem to be a scientific issue (e.g., growing crops that produce the most food) can also frequently be an issue of justice. Indeed, it is like the need for EJ, as articulated in the United

States with Executive Order 12898, "Federal Actions to Address Environmental Justice in Minority and Low-Income Populations." This order directs that federal agencies in their day-to-day operations to be sensitive to "disproportionately high and adverse human health and environmental effects of programs, policies, and activities on minority populations and low-income populations." Unsustainable food production, then, falls under the rubric of disproportionate harm to vulnerable people.

The industrial corridor along the lower Mississippi River [86], known as "Cancer Alley," is home to a predominantly low-income, minority (African-American and Latino) community that is being exposed to many pollutants. This 80-mile-long region, between Baton Rouge and New Orleans, has experienced releases of carcinogens, mutagens, teratogens (birth-defect agents), and endocrine disruptors in the atmosphere, soil, groundwater, and surface water. More than 100 oil refineries and petrochemical facilities are located in this region. It has been reported that per capita release of toxic air pollutants is about 27 kg, nine times greater than the US average of only 3 kg [87]. The US average of 260 kg of toxic air pollutants per square mile is dwarfed by the more than 7700 kg per square mile in this industrial corridor.

In the 1970s, cases of liver cancer (hepatic angiosarcoma) began to be reported in workers at polymer production facilities and other industries where vinyl chloride was present. Since then, the compound has been designated as a potent human carcinogen (inhalation slope factor=0.3 milligrams per kilogram per day).

Vinyl chloride may at first glance appear to be readily broken down by numerous natural processes, including abiotic chemical and microbial degradation (see Fig. 19.4); however, numerous studies have shown that vinyl chloride concentrations can remain elevated over long periods of time. In fact, under environmental conditions, vinyl chloride can be extremely persistent, with an anaerobic $t_{1/2}$ in soil greater than two years. It can also be difficult to treat with conventional engineering methods. For example, aerobic degradation in sewage treatment plants and surface water in an isolated bacteria culture with vinyl chloride concentrations of 20–120 mg L^{-1} needs a minimum of 35 days to completely degrade the compound. Nontraditional treatment methods, such as attack by hydroxyl radicals, can significantly reduce the half-life.[26] In heavily polluted areas like Cancer Alley, vinyl chloride repositories can remain intact for decades, serving as a continuous potential source. These repositories can be compounds other than vinyl chloride, but which break down to form the compound, for example, chloroethylene solvents degrade to vinyl chloride. With its high vapor pressure (2300 mmHg at 20°C) and high aqueous solubility (1100 mg L^{-1}), the chances of people being exposed via the air or drinking water once vinyl chloride is formed can be considerable.

7.5 Ethics and the environmental engineer

The role of the engineer is to protect safety, health, and welfare of the public [82]. This should be extended to any technologist. Most environmental protection falls under two categories, that is, public health and welfare.

Environmental engineers play a large role in preventing disease and other health problems. They do so by designing systems that provide clean water, treat wastewater, clean polluted air, decontaminate contaminated soil, prevent flooding with storm sewers, and protect wetlands and other sensitive habitats, to name a few.

The environment is a highly complex system, consisting of many interconnected and sundry systems at numerous scales. Thus, an environmental ontology must include both reductionist and systems concepts. For example, among its many connotations and characterizations, pollution is chiefly a scientific phenomenon. The term "scientific" has a generally clear and accepted connotation, even for the very young. This was not always true. For the most of the first half of the 20th century, the term had little resemblance to its current meaning. It is gratifying that pollution now engenders a deeper concept of an ethos. Most people want to avoid pollution because its presence threatens human health and environmental quality. So, it has a moral and value component and a scientific component. This environmental value system relies on an approach that requires objectivity and attention to details. Environmental professionals and scientists need a consistent environmental ontology, that is, a shared set of concepts and knowledge about its properties and the relationship of these properties with each other [89].

Some of the ontology can be based on the numerous laws and rules throughout the world. Faith traditions also provide some of the morals and values related to an environmental ontology. Much has been learned since the environmentalism movement of the late 1960s and the need to address the traumas that progressed and magnified from the Industrial Revolution in 19th century into the petrochemical era of the 20th century [28]. Well before the 1960s, however, legislation was enacted to protect the air, water, and soil, but new laws were needed to address the larger quantities of toxic substances and to apply to much larger and emerging categories of pollutants and polluters.

The need for environmental engineering is both directly and indirectly expressed in rules and regulations. For example, in the United States, the Food and Drug Administration (FDA), the Department of Agriculture (USDA), and the Environmental Protection Agency (EPA) each concerned with protecting the public's health, but with differing objectives. Other agencies have more specific roles, such as the Nuclear Regulatory Commission's role in protecting the public from radiation releases, the Department of Defense's and Department of Energy's programs to clean up military and nuclear sites, the National Aeronautics and Space Administration's satellite payloads that monitor the ozone hole and carbon dioxide levels, the US Fish and Wildlife Service's protection of endangered species and wetlands (along with the EPA and the Army Corps of Engineers), and the National Oceanic and Atmospheric Administration's roles in protecting marine and coastal habitats. Protecting various aspects of public health and the environment is one of various roles for these and other US federal agencies, with one exception. The EPA's exclusive mandate is to protect human health and the environment, by providing [90]:

- Clean air, land, and water.
- Reduced environmental risks based on the best available scientific information.
- Implementing federal laws to protect human health and the environment.
- Environmental stewardship is integral to US policies concerning natural resources, human health, economic growth, energy, transportation, agriculture, industry, and international trade, and these factors are similarly considered in establishing environmental policy. ("Our Mission and What We Do - US EPA")

- Access to accurate information for all parts of society, that is, communities, individuals, businesses, and state, local, and tribal governments, to manage human health and environmental risks.
- Cleanup of contaminated lands and toxic sites by potentially responsible parties has been revitalized; and
- Reviewing chemical substances in the marketplace for safety.

Within the United States, each state and many local governments have counterparts to the federal agencies mentioned above. Other countries have various programs to address public health and the environment. Canada separates health and environment, literally, between two agencies: Health Canada and Environment and Climate Change Canada. Health Canada's goals are to help Canadians maintain and improve their health, to ensure that high-quality health services are accessible, and to work to reduce health risks. Environment and Climate Change Canada's goal is to inform Canadians about protecting and conserving our natural heritage, and ensuring a clean, safe, and sustainable environment for present and future generations [91]. Like Canada and the United States, the Government of Japan has a dedicated Ministry of the Environment, with numerous other ministries involved in environmental protection [92]; as do the United Kingdom, the European Union, and other nations [93–96].

Almost all nations also protect the environment through various forms of federalism or shared responsibilities among national and local governments, for example, states, provinces, prefectures, counties, parishes, cities, towns, and villages. In the United States, for example, states are sovereigns with shared authority in governance with the federal government. Certain authorities are further delegated to lower jurisdictions, for example, a state requires environmental actions by cities and counties, such as inspections to limit vehicle pollutant emissions, surface water monitoring, and solid waste management programs [96].

An environmental system is comprised of heterogeneous components and has emergent properties, that is, those that are not explained by understanding the individual elements of the system, that persist, and that respond to changing circumstances and conditions by adaptation [97]. For example, the life cycle risk assessment (LCRA) is an increasingly useful tool that broadens a typical risk assessment to address the entire life cycle of a chemical substance [98]. The LCRA requires data related to chemical production, product use, and the activities and demographics of users [99–101]. Exposure science, especially high-throughput screening based on exposure potential, is becoming increasingly reliable [102–107]. However, the number of chemicals and products entering commerce is constantly increasing, presenting a challenge to understanding their physical, chemical, and biological properties and potential risks [89]. Prevention has become a key part of air quality programs. Nations are progressively requiring manufacturers to consider the entire life cycle of a chemical compound and search for safer alternatives to reduce risks to human health and the environment [28].

The interactions that form pollutants involve physical, chemical, and biological components. The physical laws underpinning thermodynamics and motion, respectively, drive the transformation and transport of substances in the environment. Thus, the hydrodynamics and aerodynamics of environmental fluids are fundamentally determined by physical principles. Water and air are the principal fluids, but others are also crucial, notably alcohols, organic solvents, acids, and bases. Often, fluids are complex mixtures, such as blood, tissue, and sap.

Engineers are relentlessly interpreting theory to provide realistic examples. Thus, engineering cannot rely on an unswerving paradigm. Instead, the competent engineer explores both promising existing technologies but always with an eye toward the anomalies.

7.6 Environmental morality

> … I not only acknowledge but insist upon the fact that morality only limits the range of morally acceptable answers, it does not always provide a unique solution to a moral problem. I hold that it is very rare that any ethical theory, including mine, can resolve any controversial ethical disagreement. ***Bernard Gert [108,109] 20th-century Ethicist***

The more complicated the problem, the less certain and unique the solution is. Why does engineering attract the best and brightest young minds? Surely, the attraction goes beyond the mathematical and scientific accolades. It extends beyond the sorting process of high analytical and quantitative scores on college admission exams. It must have something to do with a calling. That calling is an integration of a myriad of factors and variables. This integration is an uneasy one. It is simultaneously rewarding and risky, as are all great callings. It incorporates the most basic and most complex algorithms, given that it involves science and people, respectively.

Humans are quite complex. Applying the sciences to solve human problems is a complicated business. Mathematics speaks to this complexity. In 1902, the mathematician Jacques Hadamard defined a well-posed problem (un problème bien pose) as one that is uniquely solvable (déterminé). A year earlier, he defined ill-posed problems (questions mal posées) as those without a unique solution, that is, such problems depend in a discontinuous way on the measurements, so that tiny errors can create very large deviations in the solution. The modern rendition of this phenomenon is the "butterfly effect" (see Discussion Box in Chapter 1, Small Things Matter in Biotechnology). Hadamard believed anything that is physically important must be well-posed. We now know that numerous engineering, medical, and physical science problems are ill-posed. It can be argued that engineering ethics cases may often be even less uniquely solvable than some of the most complicated physical and engineering problems.

By far, most of the actual and pending environmental problems have been generally well documented, with many reaching agreement within most of the scientific community to their importance and root causes. They have served as motivation for environmental laws, regulations, and policy. Some environmental problems on the horizon, however, do not enjoy such a consensus. Examples include the hypotheses linking anthropogenic activities to changes in global climate [110], the threats posed by long-term storage of radioactive wastes [111–115], and how to generate sufficient energy to support artificial intelligence projects and data centers [1,3].

Humans are inextricably tied to the environment. Not only does it provide sustenance and shelter but also to some degree, all human enterprises are environmentally constrained. Thus, society and the environment exist as a very complex set of systems. The impact of an environmental disaster goes beyond physiological health and ecosystem condition. It affects the human condition as well.

The geologic record shows that sudden changes have arisen from slow changes in one component of the Earth system that, in due course, cross a critical threshold or tipping point, after which impacts cascade through coupled biogeochemical and social systems [7]. Humans breathe from the air in a common troposphere. Water circulates through the hydrological cycle. Plant life stores and converts the solar energy from the sun. Humans make products derived from the same Earth's resources.

7.6.1 The tragedy of the commons and the commonwealth

The Earth's environment can be visualized as a shared "common." Hardin [116] imagined an English village with a common area where every farmer's cow may graze. At first, the common readily sustains the livestock. Village life remained stable. However, over time, one of the villagers does decide that if he has two cows instead of one, the cost of the extra cow will be shared by everyone, whereas the profit will be solely his. The increase in cattle lets him prosper more than his neighbors for a while, but others notice his success and similarly want two cows. The villagers see that if two are better than one, then three must undoubtedly be better than two. Over time, however, the village common is no longer able to support the large number of cows; the system crashes, and everyone suffers. A tipping point was reached and surpassed. A sustainable system became unsustainable and irreversibly damaged.

This concept can be extended to ecological carrying capacity. Until relatively recently, the atmosphere, surface waters, oceans, soil, and groundwater were generally perceived to be boundless commons from which we expected to draw life and resources without constraint. Treating diminishing resources such as oil and minerals as capital gains in this way can also lead to scarcity and loss.

The concept of the environmental tipping point was articulated by the IPCC, after which climate systems could no longer adequately handle greenhouse gas loadings without major disruptions and problems [9–11]. Secondary and tertiary effects follow the initial tipping points, leading to a cascade of ever-increasing and even irreversible problems [12]. Climate change, chemical exposures, pandemics, and many environmental disasters display potentially severe and irreversible trends, with the fear that we are approaching or have begun to be at physical tipping points.

In addition to the physical and natural tipping points, adaptations before and after a social tipping point must be feasible and socially acceptable. Socio-environmental problems can arise as secondary tipping points, wherein small changes in underlying elements or behavior of actors trigger a large nonlinear response in the social system [28], which requires adaptive solutions [14]. That said, Hardin's Tragedy of the Commons extends to society. Thomas Hobbes's commonwealth [15] implies a social tipping point when the collective selfishness of society's members causes institutions to revert to what he called the "state of nature" or a complete abandonment of the social contract, resulting in a society that cannot perform properly [16]. Such a "Mad Max" dystopia results when the free will of members becomes out of balance and harms others who would otherwise be protected by a compassionate social contract. Conversely, when a government fails to respect the human rights of its members, the social contract becomes incarceration. The concept of empathy is key, that is, placing oneself in another's situation. The "Golden Rule" tells us to care as much about others as we do about ourselves [117]. This sentiment is at the heart of Immanuel Kant's categorical imperative, which basically holds that when deciding whether an action under consideration is virtuous, one must ask if the action or inaction would make for a better world if all others acted in the same way [18].

7.7 Conclusions

Environmental ethics is like other moral constructs discussed in this book. It consists of combinations of ethical frameworks that guide decisions and actions by an individual and my collectives. Environmental ethics differs from engineering and medical ethics in that most of the decisions and actions are to be accomplished at at any scale, from personal to global, whereas engineering and medical ethical behavior mainly occurs at the individual scale, such as a specific hazardous waste site, a neighborhood, or a person. Justice is a key aspect of environmental ethics, which is a macroethical concern. Environmental ethics does indeed also require microethical content, that is, the aggregation of individual actions is needed to protect the environment.

References

[1] J. Ferreira, G. Callou, A. Josua, P. Maciel, Estimating the environmental impact of data centers, 2018 IEEE 17th International Symposium on Network Computing and Applications (NCA), IEEE, (2018), pp. 1–4.

[2] E. Jones, B. Easterday, Artificial intelligence's environmental costs and promise. https://www.cfr.org/blog/artificial-intelligences-environmental-costs-and-promise#:~:text=Training%20a%20single%20AI%20system, comparable%20to%20the%20aviation%20industry.

[3] H. Rong, H. Zhang, S. Xiao, C. Li, C. Hu, Optimizing energy consumption for data centers, Renewable and Sustainable Energy Reviews 58 (2016) 674–691.

[4] A.A. Bashiir, Ethical considerations in emerging technologies: balancing innovation and morality, Research Invention Journal of Engineering and Physical Sciences 4 (1) (2025) 50–55. Available from https://doi.org/10.59298/RIJEP/2025/415055.

[5] R.B. Fuller, Operating Manual for Spaceship Earth, Lars Muller Publishers, Baden, Switzerland, (1969).

[6] M. McLuhan, The rise and fall of nature, Journal of Communication 27 (4) (2006) 80–81. Available from https://doi.org/10.1111/j.1460-2466.1977.tb01859.x.

[7] V. Brovkin, et al., Past abrupt changes, tipping points and cascading impacts in the Earth system, Nature Geoscience 14 (8) (2021) 550–558.

[8] D.A. Vallero, T.M. Letcher, Unraveling Environmental Disasters, Second ed, Elsevier Academic Press, Amsterdam, The Netherlands, (2023).

[9] I. Stadelmann-Steffen, C. Eder, N. Harring, G. Spilker, A. Katsanidou, A framework for social tipping in climate change mitigation: what we can learn about social tipping dynamics from the chlorofluorocarbons phase-out, Energy Research & Social Science 82 (2021) 102307.

[10] T.M. Lenton, et al., Climate tipping points—too risky to bet against, Nature 575 (7784) (2019) 592–595.

[11] D.I. Armstrong McKay, et al., Exceeding 1.5 C global warming could trigger multiple climate tipping points, Science (New York, N.Y.) 377 (6611) (2022) p. eabn7950.
[12] S. Dietz, J. Rising, T. Stoerk, G. Wagner, Economic impacts of tipping points in the climate system, Proceedings of the National Academy of Sciences 118 (34) (2021) e2103081118.
[13] Society For Risk Analysis, Global Risks @ the Tipping Point, in: SRA Annual Meeting, Tampa, Florida, 2022, Herndon, Virginia: SRA Worldwide Headquarters.
[14] D.A. Vallero, Engineering aspects of climate change, Climate Change, Elsevier, (2021), pp. 771–797.
[15] T. Hobbes, Leviathan: or, the matter, form, and power of a commonwealth ecclesiastical and civil. G. Routledge and sons, 1894.
[16] S. Shapin, S. Schaffer, Leviathan and the Air-Pump: Hobbes, Boyle, and the Experimental Life (New in Paper), Princeton University Press, (2011).
[17] G. Miller, The Road Warrior, ed, USA: Warner Brothers, Burbank, California, (1981).
[18] I. Kant, Groundwork of the Metaphysics of Morals, Translated by H.J. Paton (1953) as The Moral Law, Hutchinson, London, (1785).
[19] F.E. Sparshott, Five virtues in plato and aristotle, The Monist 54 (1) (1970) 40–65 [Online]. Available http://www.jstor.org/stable/27902161.
[20] C.M. Young, Aristotle on temperance, The Philosophical Review 97 (4) (1988) 521–542. Available from https://doi.org/10.2307/2185414.
[21] (2020). Programmatic environmental assessment - broadband deployment to rural america: finding of no significant impact. [Online] Available from: https://www.rd.usda.gov/sites/default/files/Final_FONSI_RUS_PEA_signed.pdf.
[22] L.K. Caldwell, The National Environmental Policy Act: An Agenda for the Future, Indiana University Press, Bloomington, Indiana, (1998).
[23] C.A. Moore, The 1990 clean air act amendments: silk purse or Sow's Ear? Duke Envtl. L. & Pol'y F. 2 (1992) 26.
[24] (2005). 40 CFR Parts 1500-1508, Regulations for Implementing The Procedural Provisions of the National Environmental Policy Act.
[25] M. Greenstone, The impacts of environmental regulations on industrial activity: evidence from the 1970 and 1977 clean air act amendments and the census of manufactures, Journal of Political Economy 110 (6) (2002) 1175–1219.
[26] P.G. Rogers, The clean air act of 1970, EPA Journal / United States Environmental Protection Agency, Office of Public Awareness 16 (1990) 21.
[27] U.S. Environmental Protection Agency, Principles and practices of air pollution - student manual: APTI course 452, 3rd ed. Research Triangle Park, North Carolina, 2003. [Online]. Available: http://www.4cleanair.org/APTI/452combined.pdf.
[28] D.A. Vallero, Fundamentals of Air Pollution, sixth ed, NV: Academic Press, Amsterdam, (2025).
[29] I. Vera, et al., Land use for bioenergy: synergies and trade-offs between sustainable development goals, Renewable and Sustainable Energy Reviews 161 (2022) 112409.
[30] C.A. Monteiro, G. Cannon, The impact of transnational "big food" companies on the South: a view from Brazil, PLoS Medicine 9 (7) (2012) e1001252.
[31] R. Patel, How society subsidizes big food and poor health, JAMA Internal Medicine 176 (8) (2016) 1132–1133. Available from https://doi.org/10.1001/jamainternmed.2016.3068.
[32] M.A. Altieri, Agroecology, small farms, and food sovereignty, Monthly Review 61 (3) (2009) 102–113.
[33] J.R. Herkert, Future directions in engineering ethics research: microethics, macroethics and the role of professional societies, Science and Engineering Ethics 7 (3) (2001) 403–414.
[34] J.R. Herkert, Ways of thinking about and teaching ethical problem solving: microethics and macroethics in engineering, Science and Engineering Ethics 11 (3) (2005) 373–385.
[35] D.A. Vallero, Beyond responsible conduct in research: new pedagogies to address macroethics of nanobiotechnologies, Journal of Long-Term Effects of Medical Implants 17 (1) (2007).
[36] D.A. Vallero, Macroethics and engineering leadership, Leadership and Management in Engineering 8 (4) (2008) 287–296.
[37] D.A. Vallero, T.M. Letcher, "Minerals", in Unraveling Environmental Disasters, 2, Elsevier Academic Press, Amsterdam, The Netherlands, (2023). ch. 720.
[38] D.A. Vallero, Biomedical ethics for engineers, The Biomedical Engineering Series (2007).

[39] M. Cantley, P. Laget, European responses to biotechnology: research, regulation, and dialogue, Issues in Science and Technology (no Summer) (2001).

[40] C.H. Collins, A.J. Beale, Safety in Industrial Microbiology and Biotechnology, Elsevier, (2015).

[41] M.M. Epstein, T. Vermeire, Scientific opinion on risk assessment of synthetic biology, Trends in Biotechnology 34 (8) (2016) 601–603.

[42] J. Naimon, Using expert panels to assess risks of environmental biotechnology applications: a case study of the 1986 Frostban risk assessments, in: Risk Assessment in Genetic Engineering: Environmental Release of Organisms. McGraw-Hill Environmental Biotechnology Series, series editors RM Atlas, EP Greenberg, AL Demain, BH Olson, and G. Sayler (New York: McGraw-Hill, 1991), 1991.

[43] National Research Council, Environmental effects of transgenic plants: the scope and adequacy of regulation, The National Academies Press (in English), Washington, DC, (2002), p. 342.

[44] O.V. Singh, S. Ghai, D. Paul, R.K. Jain, Genetically modified crops: success, safety assessment, and public concern, Applied Microbiology and Biotechnology 71 (5) (2006) 598–607.

[45] C.G. Turvey, E.M. Mojduszka, C.E. Pray, The precautionary principle, the law of unintended consequences, and biotechnology in: Presented at the Agricultural Biotechnology: Ten Years After, Ravello, Italy, July 6–10, 2005, 2005.

[46] D. Vallero, Environmental Biotechnology: A Biosystems Approach, Elsevier Science, (2015).

[47] J.S. Mill, On liberty. Longmans, Green, Reader, and Dyer, 1869.

[48] P. Francis, Laudato Si: on care for our common home. Our Sunday Visitor, 2015.

[49] A. Aderem, Systems biology: its practice and challenges, Cell 121 (4) (2005) 511–513. Available from https://doi.org/10.1016/j.cell.2005.04.020.

[50] J.D. Pleil, L.S. Sheldon, Adapting concepts from systems biology to develop systems exposure event networks for exposure science research, Biomarkers Biochemical Indicators of Exposure, Response, and Susceptibility to Chemicals 16 (2) (2011) 99–105.

[51] D.A. Vallero, The future: the physics of emerging technologies, Applications and Calculations in Environmental Physics, AIP Publishing LLC Melville, New York, (2022), pp. 7–20 7–1–.

[52] C. Munthe, Precaution and Ethics: Handling Risks, Uncertainties and Knowledge Gaps in the Regulation of New Biotechnologies, Federal Office for Buildings and Publications and Logistics (FOBL), (2017).

[53] A. Ghosh, H. Adams, S. Ash, M. Southard, Operators need to know how to maintain corrosion control conditions, Opflow 49 (7) (2023) 18–22.

[54] V.S. Sastri, Challenges in Corrosion: Costs, Causes, Consequences, and Control, John Wiley & Sons, (2015).

[55] M.H. Wood, A.V. Arellano, L. Van Wijk, Corrosion related accidents in petroleum refineries, European Commission Joint Research Centre, Report no. EUR 26331 (2013).

[56] B.V. Salas, et al., H2S pollution and its effect on corrosion of electronic components, Air Quality—New Perspective, IntechOpen, London, (2012), pp. 263–285.

[57] A. El-Meligi, Corrosion preventive strategies as a crucial need for decreasing environmental pollution and saving economics, Recent Patents on Corrosion Science 2 (1) (2010).

[58] J.R. Des Jardins, Environmental Ethics, Cengage Learning, (2012).

[59] P.A. Vesilind, A.S. Gunn, Hold Paramount: The Engineer's Responsibility to Society, Cengage Learning, (2010).

[60] D. Vallero, Ethical Decisions in Emergent Science, Engineering and Technologies, INTECH Open Access Publisher, (2012).

[61] E.L. Kurnat-Thoma, Laudato Si lessons: nurses recognizing our responsibility in climate science and health systems resilience, The Journal of Nursing Administration 52 (6) (2022) 319.

[62] P. Francesco, Lettera Enciclica Laudato Si, Libreria Editrice Vaticana, (2015).

[63] Catholic Daily Readings, Matthew 25: 31-46 - whatsoever you do for the least of my brothers. https://catholicreadings.org/whatsoever-you-do-to-the-least-of-my-brothers-you-do-unto-me/ (accessed 31.03.23).

[64] J. Burger, et al., Combining ecological, eco-cultural, and environmental justice parameters to create Eco-EJ indicators to monitor cultural and environmental justices for diverse communities around contaminated sites, Environmental Monitoring and Assessment 194 (3) (2022) 177.

[65] U.S. Environmental Protection Agency, Gold King Mine watershed fact sheet. https://www.epa.gov/goldkingmine/gold-king-mine-watershed-fact-sheet (accessed 20.03.18.

[66] (2017). EPA/600/R-16/296 analysis of the transport and fate of metals released from the Gold King Mine in the animas and San Juan Rivers: executive summary.
[67] U.S. Environmental Protection Agency, Emergency response to August 2015 release from Gold King Mine. https://www.epa.gov/goldkingmine (accessed 20.03.18).
[68] D.A. Vallero, G. Blight, Mine waste: a brief overview of origins, quantities, and methods of storage, Waste, Elsevier, (2019), pp. 129–151.
[69] G.E. Blight, Long-term research on amelioration of wind and water erosion of mine tailings storages in Southern Africa, Transactions of the Royal Society of South Africa 67 (2) (2012) 91–102.
[70] G. Blight, A. Fourie, Catastrophe revisited–disastrous flow failures of mine and municipal solid waste, Geotechnical & Geological Engineering 23 (3) (2005) 219–248.
[71] G. Blight, J. Troncoso, A. Fourie, W. Wolski, Issues in the geotechnics of mining wastes and tailings, ISRM International Symposium, OnePetro, (2000).
[72] U. S. B. o. R. D. O. T. S. Center, Erosion and sedimentation manual. Government Printing Office, 2006.
[73] A. Daraei, B.M. Herki, A.F.H. Sherwani, S. Zare, Slope stability in swelling soils using cement grout: a case study, International Journal of Geosynthetics and Ground Engineering 4 (1) (2018) 1–10.
[74] M. Mirnik, et al., General discussion, Discussions of the Faraday Society 42 (1966) 14–22.
[75] E.J.W. Verwey, J.T.G. Overbeek, Theory of the stability of lyophobic colloids, Journal of Colloid Science 10 (2) (1955) 224–225.
[76] Y. Zhang, Z. Qu, L. Yang, Analysis and influence of salt content on stability of saline soil subgrade slope, in: IOP Conference Series: Earth and Environmental Science, 2019, 384, 1: IOP Publishing, p. 12072.
[77] J.L. Sherard, R.S. Decker, N.L. Ryker, Piping in earth dams of dispersive clay, Performance of Earth and Earth-Supported Structures, ASCE, (1972), p. 589.
[78] D. Paterson, Biological mediation of sediment erodibility: ecology and physical dynamics, Cohesive Sediments (1994).
[79] R.J. Clausen, et al., Diné-centered research reframes the Gold King Mine Spill: understanding social and spiritual impacts across space and time, Journal of Rural Studies 97 (2023) 449–457. Available from https://doi.org/10.1016/j.jrurstud.2022.12.021.
[80] Y.O. Van Horne, et al., Impacts to diné activities with the san juan river after the Gold King Mine Spill, Journal of Exposure Science & Environmental Epidemiology 31 (5) (2021) 852–866.
[81] Martin Luther King, Jr., Letter from Birmingham jail, UC Davis L. Rev. 26 (4) (1993) 835.
[82] National Society of Professional Engineers, NSPE code of ethics for engineers. http://www.nspe.org/resources/ethics/code-ethics, 2016 (accessed 26.01.16).
[83] Institute of Medicine, Toward environmental justice: research, education, and health policy needs, National Academy Press, Washington, DC, (1999).
[84] D.C. Payne-Sturges, et al., Methods for evaluating the combined effects of chemical and nonchemical exposures for cumulative environmental health risk assessment, International Journal of Environmental Research and Public Health 15 (12) (2018) 2797.
[85] L. Racz, W. Rish, Exposure monitoring toward environmental justice, Integrated Environmental Assessment and Management 18 (4) (2022) 858–862.
[86] Advisory Commission on Regulatory Barriers to Affordable Housing (US), Not in My Backyard: Removing Barriers to Affordable Housing, Government Printing Office, Washington, DC, (1991).
[87] R.R. Kuehn, Loyola University - New Orleans, The "justice" in environmental justice, Miscellaneous Publications of the Center for Environmental Communications 1 (2002) 15–23.
[88] R. Morello-Frosch, E.D. Shenassa, The environmental "riskscape" and social inequality: implications for explaining maternal and child health disparities, Environmental Health Perspectives 114 (8) (2006) 1150–1153.
[89] D.E. Meyer, S.C. Bailin, D. Vallero, P.P. Egeghy, S.V. Liu, E.A.C. Hubal, Enhancing life cycle chemical exposure assessment through ontology modeling, Science of the Total Environment 712 (2020) 136263.
[90] U.S. Environmental Protection Agency, Our mission and what we do: our mission. https://www.epa.gov/aboutepa/our-mission-and-what-we-do (accessed 14.02.20).
[91] G.O.. Canada, Health and the environment. Government of Canada. https://www.canada.ca/en/health-canada/topics/health-environment.html, 2019 (accessed 14.02.20).
[92] A. Silva, M. Rosano, L. Stocker, L. Gorissen, From waste to sustainable materials management: three case studies of the transition journey, Waste Management 61 (2017) 547–557.

[93] M. Jänicke, K. Jacob, Environmental Governance in Global Perspective: New Approaches to Ecological Modernization, Freie Universität, (2017).

[94] A. Jordan, A. Zito, R. Wurzel, Environmental Governance in Europe, ed, 2019.

[95] I. Linkov, et al., Comparative, collaborative, and integrative risk governance for emerging technologies, Environment Systems and Decisions, Journal Article (2018). Available from https://doi.org/10.1007/s10669-018-9686-5.

[96] D.A. Vallero, 33. Waste governance, in: T.M.A.V. Letcher, A. Daniel. (Eds.), Waste: A Handbook for Management, Elsevier, (2019), pp. 651–671.

[97] D.A. Luke, K.A. Stamatakis, Systems science methods in public health: dynamics, networks, and agents, Annual Review of Public Health 33 (2012) 357–376.

[98] L. Aissani, F. Jabouille, J. Bourgois, P. Rousseaux, A new methodology for risk evaluation taking into account the whole life cycle (LCRA): validation with case study, Process Safety and Environmental Protection 90 (4) (2012) 295–303.

[99] K.K. Isaacs, et al., Characterization and prediction of chemical functions and weight fractions in consumer products, Toxicology Reports (2016).

[100] K.L. Dionisio, et al., Exploring consumer exposure pathways and patterns of use for chemicals in the environment, Toxicology Reports 2 (2015) 228–237.

[101] M.-R. Goldsmith, et al., Development of a consumer product ingredient database for chemical exposure screening and prioritization, Food and Chemical Toxicology 65 (2014) 269–279.

[102] P.P. Egeghy, et al., Computational exposure science: an emerging discipline to support 21st-century risk assessment, Environmental Health Perspectives 124 (6) (2016) 697.

[103] J.E. Rager, et al., Linking high resolution mass spectrometry data with exposure and toxicity forecasts to advance high-throughput environmental monitoring, Environment International 88 (2016) 269–280.

[104] J.F. Wambaugh, et al., High throughput heuristics for prioritizing human exposure to environmental chemicals, Environmental Science & Technology 48 (21) (2014) 12760–12767.

[105] J. Mitchell, et al., Comparison of modeling approaches to prioritize chemicals based on estimates of exposure and exposure potential, Science of the Total Environment 458 (2013) 555–567.

[106] J.F. Wambaugh, et al., High-throughput models for exposure-based chemical prioritization in the ExpoCast project, Environmental Science & Technology 47 (15) (2013) 8479–8488.

[107] S. Gangwal, et al., Incorporating exposure information into the toxicological prioritization index decision support framework, Science of the Total Environment 435 (2012) 316–325.

[108] B. Gert, Common Morality: Deciding What to Do, Oxford University Press, (2004).

[109] G. B, Letter to the Editor, The Ag Bioethics Forum - Iowa State University, Letter to the Editor 5, 2, November 1993.

[110] J. Odenbaugh, Disagreement and the environment, The Routledge Handbook of Philosophy of Disagreement, Routledge, (2024), pp. 508–517.

[111] R. Clayton, J. Kirk, A. Banford, L. Stamford, A review of radioactive waste processing and disposal from a life cycle environmental perspective, Clean Technologies and Environmental Policy 27 (2) (2025) 665–682.

[112] P. Kumar, S. Yadav, A. Yadav, Radioactive waste minimization and management, Green Chemistry Approaches to Environmental Sustainability, Elsevier, (2024), pp. 165–184.

[113] A.O. Adeola, et al., Advances in the management of radioactive wastes and radionuclide contamination in environmental compartments: a review, Environmental Geochemistry and Health 45 (6) (2023) 2663–2689.

[114] D. Cherry, E. Friedman, M. Vincent, A. Maier, The legacy of weapons grade plutonium production: health status of Hanford complex workers who manage the waste, Toxicology and Industrial Health 37 (5) (2021) 260–269. Available from https://doi.org/10.1177/0748233721996555.

[115] K. Colborn, Radioactive materials transportation, Nuclear Engineering Handbook, CRC Press, (2016), pp. 557–588.

[116] G. Hardin, The tragedy of the commons: the population problem has no technical solution; it requires a fundamental extension in morality, Science (New York, N.Y.) 162 (3859) (1968) 1243–1248.

[117] L. Wang, L. Xie, M. Jia, How can empathy be achieved?—A comparative study between the christian "golden rule" and the buddhist "five precepts and ten virtues" in China, Religions 16 (10) (2025) 1229 [Online]. Available https://www.mdpi.com/2077-1444/16/10/1229.

CHAPTER

8

Land ethics☆

OUTLINE

The land is the only thing in the world worth working for, worth fighting for, worth dying for, because it's the only thing that lasts. **Margaret Mitchell** *Gone with the Wind* [1]

8.1 Introduction

With a few exceptions, land represents a finite amount of the earth's surface. As the real estate folks say, they are not making any more of it. About 5000 square kilometers (km^2) in the Zuiderzee

☆ Daniel J. Vallero, P.E., Engineer for the City of Durham, NC, United States, provided valuable information for this chapter

Technical and Ethical Optimization: Challenges and Opportunities for the Modern Engineer and Scientist
DOI: https://doi.org/10.1016/B978-0-443-44925-3.00006-5

in the Netherlands and human-made lands and islands in China, the United Arab Emirates, Japan, and the United States represent a tiny fraction of the 149 million km^2 of land on earth.

Every hectare of this massive amount of land, however, is susceptible to harm in many ways. Its upper layers can wash away due to wind and water erosion. It can be contaminated by chemical pollutants. It can be misused, and it can be altered physically in a manner that renders it less useful and less sustainable.

The land's components of soil and water become contaminated by the chemical constituents of materials and wastes, such as leachate from landfills, hazardous waste storage and treatment facilities, and indirectly from atmospheric pollutants, for example, sulfur and nitrogen compounds that lead to acid rain. This chapter focuses on how land can be impaired by human activities, such as construction, agriculture, and transportation. These and other human activities lead to the release of chemical contaminants, but they also result in landscape damage, such as soil erosion, habitat destruction, and loss of resources, including wetlands and coastal ecosystems.

8.2 Land and freedoms

Philosopher, John Locke, argued that government needs to be subservient to the natural rights of individuals, not the other way around [2–4]. This philosophy was a key part of the debates that led to the US Declaration of Independence, which was rooted in Locke's requirement that all persons should be allowed to seek life, liberty, and property. This means that everyone has the inherent right to own things through their labor or exchange. Thomas Jefferson subsequently adapted the Declaration to say "life, liberty, and the pursuit of happiness." However, the pursuit of happiness includes the right to own property, without which a nation cannot survive socially or economically. The key property of ownership is land. Indeed, the US Constitution later recognized that owning land is the cornerstone of a person's independence and is a guardrail against government overreach [5].

A full accounting of the value of land must include its environmental, cultural, and social attributes. In the United States, for example, environmental impact statements (EISs) under the National Environmental Policy Act (NEPA) require that land-related factors be part of the decision as to whether a federal project would have a significant effect on the environment. For example, the changes in land use can lead to direct environmental damage like dynamiting landscapes to allow rails and roads to be built. It can also lead to indirect harm if the access created by these transportation systems invites urban sprawl and other secondary impacts [6–9]. These include historic preservation, economics, psychology (e.g., perceptions of open space, green areas, and crowding), aesthetics, urban renewal, and the so-called "land ethic" [8,10].

NEPA was, in many ways, symbolic of the new federal commitment to environmental stewardship and a template for proper land development. It was signed into law in 1970 after contentious hearings in the US Congress. NEPA was not really a technical law. It did two main things. In addition to requiring EISs, it established the Council on Environmental Quality (CEQ) in the Office of the President. Of the two, the EIS represented a sea change in how the federal government was to conduct business. This new way of doing business has saved many square miles of land that would have been misused. It supported a growing appreciation of why and how land can be conserved and protected [11–13].

More recently, the Omnibus Public Lands Management Act of 2009 [14] directs federal land management agencies to develop proposals that restore sensitive land, like old-growth forest land, to regain "pre-fire suppression old growth conditions characteristic of the forest type" using "a collaborative process that includes multiple interested persons representing diverse interests and is transparent and nonexclusive" [15]. It continues to require federal agencies to meet the land use planning requirements of NEPA and other federal statutes, including the Clean Air Act, the Clean Water Act, the Threatened and Endangered Species Act, the National Forest Management Act, and the Antiquities Act [15].

8.3 The land ethic

Aldo Leopold, in his famous essays published posthumously as *A Sand County Almanac* [16,17], argued for a holistic and assertive approach toward the protection and use of land [18]:

> A thing is right when it tends to preserve the integrity, stability and of the biotic community. It is wrong when it tends otherwise.

The title of this chapter is Land Ethics, which is semantically similar but intellectually different from Leopold's land ethic. The plural form represents a broader, mutable, and individual set of codes, norms, and duties that a person or group holds toward the land. Leopold's land ethic expands this morality to include care and respect for the living and nonliving land components, for example, the soil, water, plants, microbes, and animals. The land ethic specifies that something can only be right if it tends to "the preserve the integrity, stability, and beauty of the biotic community" and it is not right "when it tends otherwise" [18].

8.4 Land misuse

Leopold's concept of the land ethic is a form of value or axiological ethics, where inherent worth is the principal guide to decision-making and optimization. This differs from deontology or duty ethics, and it also differs from consequentialism or utilitarian ethics. From an axiological standpoint, land is intrinsically valuable. Its value goes beyond what is monetizable and extrinsic or instrumental. Unfortunately, the massive loss of farmland to urbanization is an example of where the instrumental, extrinsic value of land has overridden the intrinsic value. The value of the land as fertile and biodiverse treasures is rapidly and unremittingly lost to impervious surfaces of buildings and roads, habitats are fragmented so that natural migrations are interrupted, and the farming culture is being decimated. As shown in Fig. 8.1, the spectrum of values ranges from purely instrumental to humans, that is, anthropocentric, to purely intrinsic, with concern for other creatures and the abiotic parts of nature.

It is tempting to think that land is a "blank slate" or that buildings are simply three-dimensional structures ready to be built, changed, or demolished as means to engineering ends. In fact, land and structures are human enterprises that will affect people's lives

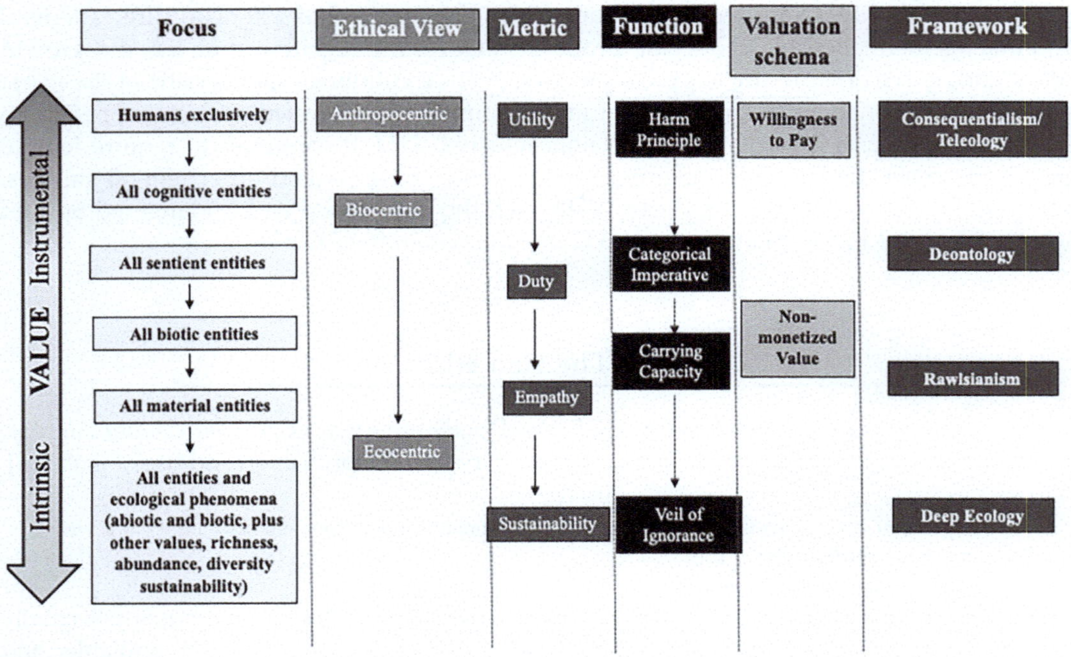

FIGURE 8.1 Value spectrum. *Adapted from D.A. Vallero, "Biomedical Ethics for Engineers," The Biomedical Engineering Series, 2007 [19]; R.B. Meyers, "Environmental Values, Ethics and Support for Environmental Policy: A Heuristic, and Psychometric Instruments to Measure Their Prevalence and Relationships," 2004 [20].*

directly and indirectly. Regrettably, it is not difficult to find ample examples of improper development of land and other waste, no matter where one resides. Land waste can result from subdivisions with improper erosion and sediment control systems, clear-cutting trees, and complete disregard for, and elimination of, sensitive ecosystems, as discussed later. Misuses of land are examples of not heeding Leopold's admonition of the need for biotic integrity, stability, and beauty, that is, sustainability.

One type of sustainability is the need for ongoing support of the human community. Improper land development has societal and cultural importance. One horrific example of land misuse was the Pruitt-Igoe public housing project. In this case, the wrongness was the failure to include sufficient integrity, stability, and beauty in the planning, design, construction, and maintenance of the building-land-people enterprise. When it was designed, this housing development in St. Louis, Missouri, was touted to be emblematic of advances in fair housing and progress in the war on poverty. Instead, the demolition of the Pruitt-Igoe buildings long before their expected lifespan has become an icon of failure of imagination, especially imagination that properly accounts for the human condition.

Although we may immediately think of public housing projects in terms of housing, they often represent elements of environmental justice. Contemporary understanding of environmental quality is often associated with physical, chemical, and biological contaminants, but in the formative years of the environmental movement, aesthetics and other "quality of life" considerations were essential parts of environmental quality.

The project was designed before environmental policies, and the land ethic was yet to be articulated following the passage of NEPA. The Pruitt-Igoe designers did not benefit from the insights of Leopold and his contemporaries. However, the problems that led to the premature demolition of this costly housing experiment arguably should have been anticipated intuitively if the designers had taken the time to understand what people needed and expected.

Blame and excuses for the failure have been plentiful. Some blame the inability of the modern architectural style to create livable environments for people living in poverty [21], largely because they "are not the nuanced and sophisticated 'readers' of architectural space the educated architects were [22]." This is a revealing observation and an important lesson for anyone developing land. Care must be taken to ensure that the use and operation of whatever is designed are sufficiently understood by those living with it, long after the designers have been paid and have left [23,24].

In retrospect, it was not logical to expect a high-rise building design to be compatible with the basic needs of low-income families with children. Indeed, high-rise configurations are less satisfactory than most other housing designs for children. Often, social interactions are more impersonal, helping behavior is less common, crime and the fear of crime are greater, and other antisocial factors are more prevalent than in most other forms of housing [25].

Socioeconomics plays a large role in the success or failure of housing developments, whether public or private. Most large cities have large populations of families with children living in such environments. Indeed, at the time, the St. Louis metropolitan area had successful luxury townhomes near Pruitt Igoe, sharing many of the same design features. Part of the difference can be traced to racism and classism in the era when the project was conceived [22,26]. As evidence, when originally inhabited, the Pruitt section was restricted to African Americans and Igoe to whites.

Costs are a factor in land misuse. The building contractors' bids in the Pruitt-Igoe project were increased to a level where the project construction costs in St. Louis exceeded the national average by 60%. The response to the local housing authority's refusal to raise unit cost ceilings to accommodate the elevated bids was to reduce room sizes, eliminate amenities, and raise densities [27]. As originally designed, the buildings were to become "vertical neighborhoods" with nearby playgrounds, open-air hallways, porches, laundries, and storage areas. The compromises eliminated these features. Some of the removal of so-called "amenities" led to dangerous situations. Elevators were undersized and stopped only every third floor, and lighting was inadequate in the stairwells. Indeed, safe access and adequate lighting are not amenities, they are essentials. So, another lesson must be to know the difference between desirable and essential design elements. No self-respecting structural engineer involved in the building design would have short-cut the factors of safety built into load-bearing. However, human elements essential to a vibrant community were eliminated without much, if any, accommodation [28].

A final lesson for the designer is the need for a near-perfect match between the design and the client's needs, whether those needs are spoken or unspoken. The Pruitt-Igoe project was mismatched to the people who would live there. Many were relocated from single-family residences. They were a displaced population. They were moved to a very large, imposing project with 2800 units and almost 11,000 people living there. This quadrupled the size of the next largest project of the time. When the failure of the project became crushingly clear, the only reasonable decision was to demolish it, and this spectacular implosion became a lesson in failure for planners, architects, and engineers. In the designer's own words [27,29]:

> I never thought people were that destructive. As an architect, I doubt if I would think about it now. I suppose we should have quit the job. It's a job I wish I hadn't done.

8.4.1 Waste sites

Love Canal, New York, was the site that led to passage of groundbreaking environmental legislation, that is, the Resource Conservation and Recovery Act of 1976, to ensure that wastes are tracked from "cradle to grave" for active hazardous waste facilities, that is, those that are still operational and where an owner/operator is identified [30]. Later, the Comprehensive Environmental Response, Compensation and Liability Act of 1980, aka "Superfund," was passed to address and clean up abandoned waste sites [31]. These two laws authorized regulations and guidance that define when pollution thresholds are crossed.

Numerous other waste sites were discovered in the late 1970s, including several sites contaminated with chlorinated dioxins in Times Beach, Missouri [32,33], and a massive area of abandoned barrels of chemicals in the Valley of the Drums in West Point, Kentucky [34]. These and several other similar discoveries prompted regulators to approach pollutants from the perspective of risk.

People living near hazardous waste sites, for good reason, want to be assured that they will be "safe." But safety is a relative term. Designating something "safe" incorporates a value judgment that is invariably accompanied by uncertainties. The safety of a product or process can be described in objective and quantitative terms. Factors of safety are part of every waste management decision (see Chapter 15).

When a contaminated site is discovered, one of the queries is how to rehabilitate or restore native habitats. For example, the Environmental Protection Agency's (EPA) Land Revitalization Program has been designed to restore land and other natural resources into sustainable community assets that maximize beneficial economic, ecological, and social uses and ensure protection of human health and the environment [35].

Projects have included sustainable cleanup and redevelopment, green building, and sustainable development alternatives for affordable housing, incorporating the reuse of construction and demolition debris into redevelopment projects, and comprehensive "hands on" training for local government officials on assessing, remediating, and revitalizing contaminated properties in a sustainable manner. When a previously contaminated site is remediated to the point that it has regained a commercial or other beneficial use, this is known as a brownfield. Several sites now exist throughout the United States.

8.5 The complaint paradigm

In many places, environmental response is often precipitated first by a complaint. A common land-related protest is NIMBY, "not in my backyard." Often, those who proclaim this really mean "in someone else's backyard."

The underlying assumption of fairness is that everyone has a voice. Voice is never equal, however, there is a modicum of fairness if everyone's voice is loud enough to be heard. However, if a certain group of people has had little or no voice in the past, they are likely to feel and be disenfranchised. For example, for centuries, African-American communities were not successful in voicing concerns about environmentally unacceptable conditions in their neighborhoods. Presently, Hispanic Americans may have even less voice in environmental and public health matters since the government may not be seen as a viable place to lodge a complaint, and the perceived risks of complaining (e.g., questions about citizenship and possible deportation) may outweigh the expected benefit (e.g., a cleaner environment).

Aldo Leopold's "land ethic" is a reminder that the use of land is dependent on the values placed on it. One means of determining land's value is a scientific tabulation of the land's resources and threats to those resources. As mentioned, since NEPA's passage, land use decisions in the United States must follow from an environmental assessment. Most environmental impact assessment handbooks prior to the late 1990s contained little information and guidelines related to fairness issues in terms of housing and development. They were usually concerned about open space, wetland, and farmland preservation, housing density, ratios of single versus multiple-family residences, owner-occupied housing versus rental housing, building height, signage, and other restrictions, designated land for public facilities like landfills and treatment works, and institutional land uses for religious, health care, police, and fire protection.

An environmental response is often precipitated first by a complaint. But to complain, one must have a "voice." If a certain group of people has had little or no voice in the past, they are likely to feel and be disenfranchised. For most of Western history, African-American communities have had little success in voicing concerns about environmentally unacceptable conditions in their neighborhoods. Hispanic Americans may have even less voice in environmental matters since their perception of government, the final arbiter in many environmental disagreements, is one of skepticism and outright fear of reprisal in the form of being deported or being "profiled." Many of the most adversely affected communities are not likely to complain.

Land use is a crucial part of an environmental assessment. Usually, however, land justice issues are not necessarily part of these assessments. Most environmental impact assessment handbooks prior to the late 1990s contained little information and guidelines related to fairness issues in terms of housing and development. They were usually concerned about open space, wetland and farmland preservation, housing density, ratios of single versus multiple-family residences, owner-occupied housing versus rental housing, building height, signage, and other restrictions, designated land for public facilities like landfills and treatment works, and institutional land uses for religious, health care, police, and fire protection.

When land uses change (usually to become more urbanized), the environmental impacts may be direct or indirect. Examples of direct land use effects include eminent domain, which allows land to be taken with just compensation for the public good. Easements are another direct form of land use impacts, such as a 100 m right-of-way for a highway project that converts any existing land use (e.g., farming, housing, or commercial enterprises) to a transportation use. Land use change may also come about indirectly, such as so-called secondary effects of a project that extend, in time and space, the influence of a project. For example, a wastewater treatment plant and its connected sewer lines will create accessibility that spawns suburban growth [36]. People living in very expensive homes may not even realize that their building lots were once farmland or open space and, had it not been for some expenditure of public funds and the use of public powers like eminent domain, there would be no subdivision [37].

Environmentalists are generally concerned about increased population densities, but housing advocates may be concerned that once land use has been changed, environmental and zoning regulations may work against affordable housing. Even worse, environmental protection can be used as an excuse for some elitist and exclusionary decisions. In the name of environmental protection, certain classes of people are economically restricted from living in specific areas. This problem first appeared in the United States in the 1960s and 1970s during a search for ways to preserve open spaces and green areas. One measure was the minimum lot size. The idea was that rather than having the public sector secure land through easements or outright purchases (i.e., fee simple) to preserve open spaces, developers could either set aside open areas or require large lots to have their subdivisions approved. Thus green areas would exist without the requisite costs and O&M entailed by public parks and recreational areas. Such areas have numerous environmental benefits, such as wetland protection, flood management, and aesthetic appeal. However, minimum lot size translates into higher costs for residences. The local rules for large lots that result in less affordable housing are called exclusionary zoning. One value (open space and green areas) is pitted against another (affordable housing). In some cases, it could be argued that preserving open spaces is simply a tool for excluding people of lesser means or even people of minority races [37,38].

8.5.1 Case study: habitat for humanity[1]

About 20 years ago, the housing advocacy group, Habitat for Humanity, proposed a development of affordable houses in Chapel Hill, North Carolina (NC), which has one of the most expensive real estate markets in the Southeast United States. Being a college town (home to the University of North Carolina), numerous groups, including churches and student coalitions, are calling for "livable housing" [37]. However, the cost of housing in this town is well above the state average, so several advocates supported the habitat model, where potential homeowners invest in their own homes through "sweat equity" and receive help from volunteers building the home. But some groups opposed the plan. In an early meeting, one neighbor stated a desire that the homes be like a nearby high-cost subdivision, that is, houses costing much more than even the already expensive town average. She recommended that they be "single family homes with a nice, friendly, college town look and feel."[2] Another later said that "From day one, we have said that that parcel is not suited for a high-density project." These statements may have been factual, but the result of such thinking is, in the end, exclusionary. People are very passionate and protective about their neighborhoods, well beyond concern about property values. Such cries of not in my

[1] This case is an updated version of the original in [37] D. A. Vallero and P. A. Vesilind, *Socially responsible engineering: Justice in risk management*. Hoboken, New Jersey: John Wiley & Sons, 2007.

[2] The quotes have been changed to protect anonymity, but the meanings are maintained.

backyard (NIMBY) are so common that the designers and engineers must be prepared to balance science and social needs in residential areas, much like their professional responsibilities to site unpopular facilities like landfills and treatment plants.

Planners and engineers must be sensitive to protecting the quality of neighborhoods, but simultaneously, they must take great care that their ends (in this case, environmental protection) are not used as a rationale for unjust means (in this case, less than fair development practices). Environmental laws and policies, like zoning ordinances and subdivision regulations, should not be used to prevent lower socio-economic groups from living in or near more affluent neighborhoods.

How can the engineering profession help people who have historically had no voice or have not been taught how to make it heard? For environmental fairness, everyone potentially affected needs a voice and a place at the table from the earliest planning stages of a project. The default seems to be changing, but some argue that environmental quality is still being used, knowingly or innocently, to work against fairness. Additionally, people who are likely to be exposed to the hazards brought about by land use decisions need to be aware of options as they are considered and well before decisions are made. This should be applied not only to land development and civic decisions, but to everything engineers do that may have an impact on health, safety, and welfare of the public.

Sometimes we need to remind ourselves of just how far we have come in battling pollution in the past 50 years. We also need to realize that the general perception of environmental quality and the public's expectations have grown substantially in a relatively short time. The revolution in thinking and the public acceptance of strong measures to regulate the actions of private industry has been phenomenal. So, what may have previously been simply the "cost of doing business" (brown haze, smelly urban areas, obviously polluted water, and disposal of pollutants in pits, ponds, lagoons, and by land burial) is now considered to be inappropriate and even immoral activities. However, it is not automatic that all private and public entities have received that message. Engineers can help continue to raise their clients' appreciation of fairness and justice, as well as the improvement to the "bottom line" that can result from strong environmental programs.

8.5.2 Case study: Carver Terrace[3]

In the 1970s, the citizens of Carver Terrace, in Texarkana, Texas, a predominantly African-American community, began to see dark, vile-smelling "gunk" oozing out of their lawns [39,40]. They could not interest the local authorities in the problem, even when they started to believe that their community was experiencing a higher-than-usual number of medical issues. As mentioned, the problems at Love Canal made hazardous waste a national issue. A year later, after Congress ordered large chemical companies to identify what chemicals they had disposed of and at what locations, the discovery was made that Koppers Company of Pittsburgh had operated a creosote plant in the area now known as Carver Terrace, and when Koppers closed the plant, they bulldozed everything into the ground, including the vats and pond holding creosote, a known human carcinogen. Because the land was

[3] This case is an updated version of the original in [37] D. A. Vallero and P. A. Vesilind, *Socially responsible engineering: Justice in risk management*. Hoboken, New Jersey: John Wiley & Sons, 2007.

inexpensive, poorer families eagerly bought lots and built homes in what became known as Carver Terrace. About 25,000 people, 85% racial minority, lived within 4 miles of the former creosote plant.

When it became known that the ground contained large quantities of creosote, the US EPA sent a team of hazardous waste experts who nosed around in their "moon suits" and concluded that there was no problem. The citizens knew better and soon learned that the EPA had conducted other studies and reports that clearly showed this area to be a candidate for the Superfund cleanup, but that these were not made available to the citizens of Carver Terrace. The extent of pollution was large; 45 million gallons of shallow groundwater were contaminated, along with 2150 cubic yards of soil, contaminated to a depth of one foot. The situation was so bad that the citizens' group urged the government to buy out the residents of Carver Terrace just as they were buying out the people who lived around Love Canal and Times Beach, Missouri. The Carver Terrace neighbors could not see why they were being treated differently. The only obvious difference between the Love Canal residents and the Carver Terrace people was that the former were mostly white, whereas Carver Terrace residents were predominantly African American. Eventually, through energetic and determined activism, the residents of Carver Terrace were also bought out.

8.6 Land justice

Certainly, environmental injustice is caused by other factors in addition to racial bias. It also occurs when profit motive drives corporate decisions to site environmentally hazardous facilities where workers living near a facility fear losing their jobs, so they are less likely to complain. Public decisions have also brought lower socio-economic communities into environmental harm's way. Although public agencies, such as housing authorities and public works administrations, do not have a profit motive, *per se*, they do need to address budgetary and policy considerations. If open space is cheaper and certain neighborhoods are less likely to complain (or by extension, vote against elected officials), the "default" for unpopular facilities, such as landfills and hazardous waste sites, may be to locate them in lower-income neighborhoods. Also, elected and appointed officials and bureaucrats may be more likely to site other types of unpopular projects, such as public housing projects, in areas where complaints are less likely to be put forth or where land is cheaper.

8.6.1 Case study: West Dallas Lead Smelter[4]

In 1954, the Dallas, Texas, Housing Authority built a large public housing project on land immediately adjacent to a lead smelter. The project had 3500 living units and became a predominantly African-American community [41]. During the 1960s, the lead smelter stacks emitted over 200 tons of lead annually into the air. Recycling companies owned and operated the smelter to recover lead from as many as 10,000 car batteries per day. The lead

[4] This case is an updated version of the original in [37] ibid.

emissions were associated with blood lead levels in the housing project's children, which were 35% higher than in children from comparable areas.

Lead is a particularly insidious pollutant because it can result in developmental damage. Study after study showed that the children in this project were in danger of higher lead levels, but nothing was done for over 20 years. Finally, in the early 1980s, the city sued the lead smelter, and the smelter immediately initiated control measures that reduced its emissions to allowable standards. The smelter also agreed to clean up the contaminated soil around it and to pay compensation to people who had been harmed.

This case illustrates two issues of environmental injustice. First, the housing units should never have been built next to a lead smelter. The reason for locating the units there was justified based on economics. The land was inexpensive, and thus this saved the government money. The second issue was the procrastination by the city in insisting that the smelter clean up the emissions. Once the case had been made, within two years, the plant met the emission standards. As of 2003, blood lead levels in West Dallas were below the national average.

The key question: Why did it take 20 years for the city to do the right thing?

When land uses evolve (usually to become more urbanized), the environmental impacts may be direct or indirect. Examples of direct land use effects include eminent domain, which allows land to be taken with just compensation for the public good. Easements are another direct form of land use impacts, such as a 100 m right-of-way for a highway project that converts any existing land use (e.g., farming, housing, or commercial enterprises) to a transportation use. Land use change may also come about indirectly, such as so-called secondary effects of a project that extend, in time and space, the influence of a project. For example, a wastewater treatment plant and its connected sewer lines will create accessibility that spawns suburban growth [36]. People living in very expensive homes may not even realize that their building lots were once farmland or open space and had it not been for some expenditure of public funds and the use of public powers like eminent domain, there would be no subdivision.

8.6.2 Competing values: affordable housing versus conservation

Land use plans must somehow balance the need to preserve the economic value of homes and to enhance the quality of our neighborhoods with fair and affordable housing. Thus engineers and land use planners must take great care that their ends (environmental protection) are not used as a rationale for unjust means (unfair development practices). Environmental laws and policies, like zoning ordinances and subdivision regulations, should not be used to keep lower socio-economic groups out of privileged neighborhoods.

Environmental quality continues to be used, knowingly or innocently, to compete with fairness. This may be most dramatically demonstrated when people who benefit little bear too large a burden in terms of pollution and waste. Often, exposures to the hazards brought about by land use decisions are set in place well before the actual land use decisions are made. In fact, since everything engineers do may have an impact on health, safety, and welfare, inclusiveness and affordability should be factors in all standard operating procedures for all designs that potentially affect the public. Engineers and planners must find ways to ensure plans are fair and just, often looking beyond property value "bottom lines." Sustainable development, therefore, is the result of applications of both physical and social sciences.

Environmental injustice can result from factors other than the profit motive and its driving corporate decisions to site environmentally hazardous facilities where people are less likely to complain. Public decisions have also brought lower socio-economic communities into environmental harm's way. Although public agencies, such as housing authorities and public works administrations do not have a profit motive, *per se*, they do need to address budgetary and policy considerations. If open space is cheaper and certain neighborhoods are less likely to complain (or by extension, vote against elected officials) the "default" for unpopular facilities such as landfills and hazardous waste sites may be to locate them in lower-income neighborhoods.

It may be easier to cite other types of unpopular projects, such as public housing projects, in areas where complaints are less likely to be put forth or where land is cheaper, such as occurred in the West Dallas smelter case. There are similar cases throughout the US where minority neighborhoods have be chosen for hazardous sites owing to the lack of voice from residents and the relatively cheap available land that has resulted from decades of neglect. In addition, it has often been politically expedient to choose these sites since resistance has been low compared to richer and more politically connected communities. As a result, such sites are known as "sacrifice zones" [42] because the greater number of society members and individual developers receive the benefit of inexpensive means to carry out needed, albeit environmentally unacceptable, practices at the expense of a relatively few members of society.

8.7 Land pollution

Land and water are intricately connected. In addition to the damage caused to the land when topsoil is lost after a precipitation event, the stormwater transports both suspended material (e.g., soil, debris) and soluble matter (e.g., organic and inorganic chemicals). These materials either directly enter surface waters or reach a storm sewer system. The material reaching the water is harmful in two ways. First, it harms the integrity of the receiving stream, increasing turbidity, suspended and dissolved solids content, and the organic load. Second, the material can be hazardous.

The material in the runoff is collectively referred to as "sediment," which can substantially decrease the ability of light to penetrate the water column, that is, increased turbidity. Sediment loading diminishes the growth capacity of aquatic plants and disrupts normal processes, such as seasonal turnover and nutrient cycling in the surface waters.

Household hazardous wastes, such as insecticides, pesticides, paint, solvents, used motor oil, and other auto fluids, are also frequently part of the sediment load. These are often highly toxic to aquatic life and human populations. These wastes are discussed in detail in Chapter 13.

Pathogenic microbes suspended or sorbed to suspended particles. Notably, *Escherichia coli* can cause major health problems and indicate the presence of enteric bacteria, that is, those that grow in the intestines of animals, so they are also known as fecal coliform bacteria. This can be an indication of agricultural and human waste sources, for example, septic tanks or even poorly treated effluent from municipal wastewater treatment systems.

Sediment loading can change the species composition and diversity of lakes and other surface waters. For example, the increased input of nitrogen, phosphorus, and other nutrients encourages the growth of algae, leading to algal blooms. When these algae die, they fall to the lake bottom. As they continue to decompose, more oxygen is demanded by the bacteria and abiotic processes, so the concentrations of dissolved oxygen (DO) decline. As a result, fish and other organisms at higher trophic levels in the food chain are stressed. This can result in fish kills and complete alterations to aquatic habitats.

8.7.1 Land-based stress to water bodies: biochemical oxygen demand

The biochemical oxygen demand (BOD) is the amount of oxygen that bacteria will consume in the process of decomposing organic matter under aerobic conditions. The BOD is measured by incubating a sealed sample of water for five days and measuring the loss of oxygen by comparing the O_2 concentration of the sample at time=0 (just before the sample is sealed) to the concentration when time equals five days (known specifically as BOD_5). Samples are commonly diluted before incubation to prevent the bacteria from depleting all of the oxygen in the sample before the test is complete [43]. BOD_5 is merely the measured DO at the beginning time, that is, the initial DO ($D_)$), measured immediately after it is taken from the source, minus the DO of the same water measured exactly five days after D_1, that is, D_5:

$$\text{BOD} = \frac{D_1 - D_5}{P} \tag{8.1}$$

where P=decimal volumetric fraction of water utilized. D units are in mg L^{-1}. If the dilution water is seeded, the calculation becomes:

$$\text{BOD} = \frac{(D_1 - D_5) - (B_1 - B_5)f}{P} \tag{8.2}$$

where B_1= initial DO of seed control; B_5= final DO of seed control; and f=the ratio of seed in sample to seed in control $=\frac{\%\text{ seed in }D_1}{\%\text{ seed in }B_1}$. B units are in mg L^{-1}.

For example, to find the BOD_5 value for a 10 mL water sample added to 300 mL of dilution water with a measured DO of 7 mg L^{-1} and a measured DO of 4 mg L^{-1} five days later:

$$P = \frac{10}{300} = 0.03$$

$$\text{BOD}_5 = \frac{7-4}{0.03} = 100.\ \text{mg L}^{-1}$$

Thus the microbial population in this water is demanding 100 mg L^{-1} DO over the five-day period. So, if a conventional municipal wastewater treatment system is achieving 95% treatment efficiency, the effluent discharged from this plant would be 5 mg L^{-1}.

Chemical oxygen demand (COD) does not differentiate between biologically available and inert organic matter, and it is a measure of the total quantity of oxygen required to oxidize all organic material completely to carbon dioxide and water. COD values always exceed BOD values for the same sample. COD (mg L^{-1}) is measured by oxidation using potassium dichromate ($K_2Cr_2O_7$) in the presence of sulfuric acid (H_2SO_4) and silver. By convention, 1 g of carbohydrate or 1 g of protein accounts for about 1 g of COD. On average, the ratio BOD:COD is 0.5. If the ratio is <0.3, the water sample likely contains elevated concentrations of *recalcitrant* organic compounds, that is, compounds that resist biodegradation [44]. That is, there are numerous carbon-based compounds in the sample, but the microbial populations are not efficiently using them for carbon and energy sources. This is the advantage of having both BOD and COD measurements. Sometimes, however, COD measurements are conducted simply because they require only a few hours compared to the five days for BOD.

Since available carbon is a limiting factor, the carbonaceous BOD reaches a plateau, that is, the ultimate carbonaceous BOD (see Fig. 8.2). However, carbonaceous compounds are the only substances demanding oxygen. Microbial populations will continue to demand O_2 from the water to degrade other compounds, especially nitrogenous compounds, which account for the bump in the BOD curve. Thus in addition to serving as an indication of the amount of molecular oxygen needed for the biological treatment of organic matter, BOD also provides a guide to sizing a treatment process, assigning its efficiency, and giving operators and regulators information about whether the facility is meeting its design criteria and complying with pollution control permits.

If effluent with high BOD concentrations reaches surface waters, it may diminish DO to levels lethal to some fish and many aquatic insects. As the water body re-aerates as a result of mixing with the atmosphere and by algal photosynthesis, O_2 is added to the water, and the oxygen levels will slowly increase downstream. The drop and rise in DO concentrations downstream from a source of BOD is known as the DO sag curve, because the concentration

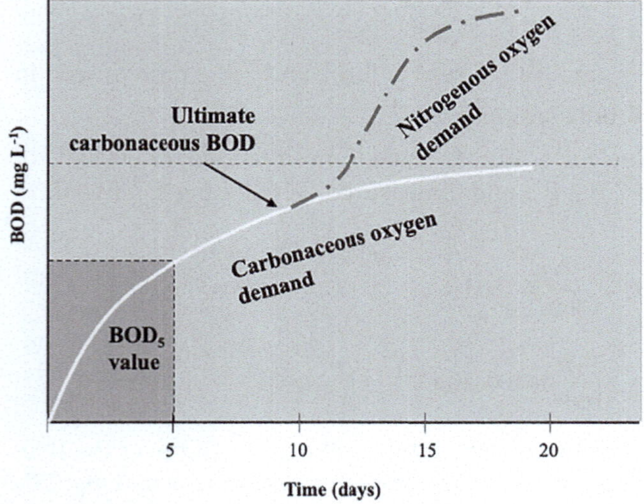

FIGURE 8.2 Biochemical oxygen demand (BOD) curve showing ultimate carbonaceous BOD and nitrogenous BOD [44,45]. *Data from C.P. Gerba and I.L. Pepper, "Wastewater treatment and biosolids reuse," Environmental microbiology, second edn. Elsevier Inc., USA, pp. 503–530, 2009; D. Vallero, Fundamentals of Water Pollution: Quantifying Pollutant Formation, Transport, Transformation, Fate and Risks, first ed. Amsterdam, NV: Elsevier, 2024.*

of DO "sags" as the microbes deplete it. Thus the falling O_2 concentrations decrease with both time and distance from the point where the high BOD substances enter the water.

8.7.2 Predicting sediment-induced stress

The stress from decreasing DO is usually indicated by the BOD. Like most environmental systems, the water bodies that receive sediment loads are complex in their response to increased input of materials. The DO will respond both positively and negatively to increased nutrient levels. Since the biota have unique optimal ranges of growth and metabolism that vary among species (e.g., algae will add some O_2 with photosynthesis but use some O_2 for metabolism, whereas bacteria will generally be net consumers of molecular oxygen).

The most widely applied environmental indices are those that follow the framework of an index of biological integrity. In biological systems, *integrity* is the capacity of a system to sustain a balanced and healthy community. This means the community of organisms in that system meets certain criteria for species composition, diversity, and adaptability, often compared to a reference site, that is, a benchmark for integrity. As such, biological integrity indices are designed to integrate the relationships of chemical and physical parameters with each other and across various levels of biological organization. They are now used to evaluate the integrity of environmental systems using a range of metrics to describe the system.

Thus environmental indices combine attributes to determine a system's biodiversity, productivity, and other conditions and to estimate stresses. The original index of biotic integrity developed by Karr [46,47] was based on fish fauna attributes and has provided predictions of how well a system will respond to a combination of stresses. In fact, the index is completely biological, with no direct chemical measurements. However, the metrics (see Table 8.1) are indirect indicators of physicochemical factors (e.g., the abundance of game fish is directly related to DO concentrations). The metrics provide descriptions of a system's structure and function.

An example of the data that is gathered to characterize a system is provided in Table 8.2. The information gleaned from this data is tailored to the physical, chemical, and biological conditions of an area, for example, large spatial regions. The information from a biologically based index can be used to evaluate a system, as shown in Fig. 8.3.

Indices include several biological and chemical inputs. For example, a fish's direct aqueous exposure (*AE* in μg day^{-1}) is the product of the organism's ventilation volume, that is, the flow Q (in mL day^{-1}), and the compound's aqueous concentration, C_w (μg mL^{-1}). The fish's exposure through its diet (*DE*, in μg day^{-1}) is the product of its feeding rate, F_w (g wet weight day^{-1}), and the compound's concentration in the fish's prey, C_p (μg g^{-1} wet wt). If the fish's food consists of a single type of prey that is at equilibrium with the water, the fish's aqueous and dietary exposures and the bioconcentration factor (*BC*) can be calculated when they are equal:

$$AE = DE; \; QC_w = F_w C_p; \; BCF = \frac{Q}{F_w} \tag{8.3}$$

TABLE 8.1 Biological metrics used in the original index of biotic integrity (*IBI*).

Integrity aspect	Biological metric
Species richness and composition	Total number of fish species (total taxa)
	Number of *Catostomidae* species (suckers)
	Number of darter species
	Number of sunfish species
	Number of intolerant or sensitive species
Indicator species metrics	Percent of individuals that are *Lepomis cyanellus* (Centrarchidae)
	Percent of individuals that are omnivores
Trophic function metrics	Percent of individuals that are insectivorous Cyprinidae
	Percent of individuals that are top carnivores or piscivores
	Percent of individuals that are hybrids
Reproductive function metrics	Abundance or catch per effort of fish
Abundance and condition metrics	Percent of individuals that are diseased, deformed, or have eroded fins, lesions, or tumors

Data from J.R. Karr, "Assessment of biotic integrity using fish communities," Fisheries, 6, 6, pp. 21–27, 1981.

The ventilation-to-feeding ratio for a 1 kg trout has been found [49,50] to be on the order of $10^{4.3}$ mL g^{-1}. Assuming the quantitative structure-activity relationship (QSAR) for the trout's prey is BCF=0.048 times the octanol-water coefficient (K_{ow}), it appears that the trout's predominant route of exposure for any chemical has a $K_{ow} > 10^{5.6}$. Exposure must also account for the organism's assimilation of compounds in food, which for very lipophilic compounds, will probably account for most of the exposure compared to that from the water. Even though chemical exchange occurs from both food and water via passive diffusion, the uptake from food, unlike direct uptake from water, does not necessarily relax the diffusion gradient into the fish. The difference between digestion and assimilation of food can result in higher contaminant concentrations in the fish's gut.

Predicting expected uptake where the principal route of exchange is dietary can be further complicated by the fact that most fish species exhibit well-defined size-dependent, taxonomic, and temporal trends regarding their prey. Thus a single bioaccumulation factor (BAF) may not universally be useful for risk assessments for all fish species. Indeed, the BAF may not even apply to different sizes of the same species.

The systematic biological exchange of materials between the organism, in this case various species of fishes, is known as uptake, which can be expressed by the following three differential equations for each age class or cohort of fish [51]:

$$\frac{dB_f}{d_t} = J_g + J_i + J_{bt} \tag{8.4}$$

TABLE 8.2 Biological metrics that apply to various regions of North America.[a]

Alternative IBI metrics	Midwestern United States	Central Appalachians	Sacramento-San Joaquin	Colorado Front Range	Western Oregon,	Ohio	Ohio Headwater Sites	Northeastern United States,	Ontario	Central Corn Belt Plain	Wisconsin-warm water	Wisconsin-coldwater	Maryland Coastal Plain	Maryland nontidal
1. Total number of species	X	X	X	X				X		x			X	X
# native fish species					X	X	X		X		X			
# salmonid age classes[c]				X	X									
2. Number of darter species	X	X		X		X				X	X			
# sculpin species					X									
# benthic insectivore species								X						
# darter and sculpin species							X							
# darter, saupin, and madtom species											X			
# salmonid juveniles (individuals)[c]			X		X									
% round-bodied suckers						X[b]								
# sculpius. (individuals)			X											
# benthic species													X	X
3. Number of sunfish species	X			X		X				X	X			
# cyprinid species					X									
# water column species								X						
# sunfish and trout species									X					
# salmonid species			X							X				
# headwater species							X							
% headwater species							X			X				

(*Continued*)

TABLE 8.2 (Continued)

Alternative IBI metrics	Midwestern United States	Central Appalachians	Sacramento-San Joaquin	Colorado Front Range	Western Oregon, Ohio		Ohio Headwater Sites	Northeastern United States, Ontario		Central Corn Belt Plain	Wisconsin-warm water	Wisconsin-coldwater	Maryland Coastal Plain	Maryland nontidal
4. Number of sucker species	X				X	X		X			X			
# adult trout species[c]			X		X									
# minnow species				X			X			X				
# sucker and catfish species									X					
5. Number of intolerant species	X			X	X	X		X			X	X	X	X
# sensitive species							X			X				
# amphibian species			X											
presence of brook trout									X					
% stenothermal cool and cold-water species												X		
% **salmonid individuals as brook trout** (*Salvelinus fontinalis*)												X		
6. % Green sunfish	X													
% common carp					X									
% white sucker				X				X						
% tolerant species						X	X			X	X	X	X	X
% creek chub		X												
% dace species									X					
% eastern mudminnow													X	
7.% Omnivores	X			X		X	x	x	x	x	x			

TABLE 8.2 (Continued)

Alternative IBI metrics	Midwestern United States	Central Appalachians	Sacramento-San Joaquin	Colorado Front Range	Western Oregon, Ohio		Ohio Headwater Sites	Northeastern United States, Ontario		Central Corn Belt Plain	Wisconsin-warm water	Wisconsin-coldwater	Maryland Coastal Plain	Maryland nontidal
% generalis feeders		X												
% generalizes, omnivores, and invertivores														X
8. % Insectivorous cyprinids	X												X	
% insectivores					X			X			X		X	x[d]
% specialized insectivores		X		X										
# juvenile trout			X											
% insectivorous species						X	X							
9. % Top carnivores	X					X		x	x	x	x	x		
% catchable salmonids					X									
% catchable trout			X											
% pioneering species							X			X			X	
Density of catchable wild trout			X											
10. Number of individuals (or catch per effort)	X	X	x	X	X	X	X[e]	X[e]		X	X	x[e]	X	
Density of individuals								X						X
% abundance of dominant species													X	X
Biomass (per m^2)														X[f]
11. % Hybrids	X							X						
% introduced species				X	X									

(*Continued*)

TABLE 8.2 (Continued)

Alternative IBI metrics	Midwestern United States	Central Appalachians	Sacramento-San Joaquin	Colorado Front Range	Western Oregon, Ohio	Ohio Headwater Sites	Northeastern United States, Ontario	Central Corn Belt Plain	Wisconsin-warm water	Wisconsin-coldwater	Maryland Coastal Plain	Maryland nontidal
% simple lithophills					X			X	X			X
% simple lithophills species						X						
% native species			X									
% native wild individuals			X									
% silt-intolerant spanners											X	
12. % Diseased individuals (deformities, eroded fins, lesions, and tumors)	X	X		X	X	X	X	X	X	X	X	X

Note: X=metric used in region. Many of these variations are applicable elsewhere.
[a] *Taken from numerous references cited by [48].*
[b] *Boat sampling methods only (i.e., larger streams/rivers).*
[c] *Metric is a provisional replacement metric in small western salmonid streams.*
[d] *Noncoastal Plain streams only.*
[e] *Excluding individuals of tolerant species.*
[f] *Coastal Plain streams only.*
Data: M.T. Barbour, Rapid bioassessment protocols for use in wadeable streams and rivers: periphyton, benthic macroinvertebrates and fish. US Environmental Protection Agency, Office of Water, 1999.

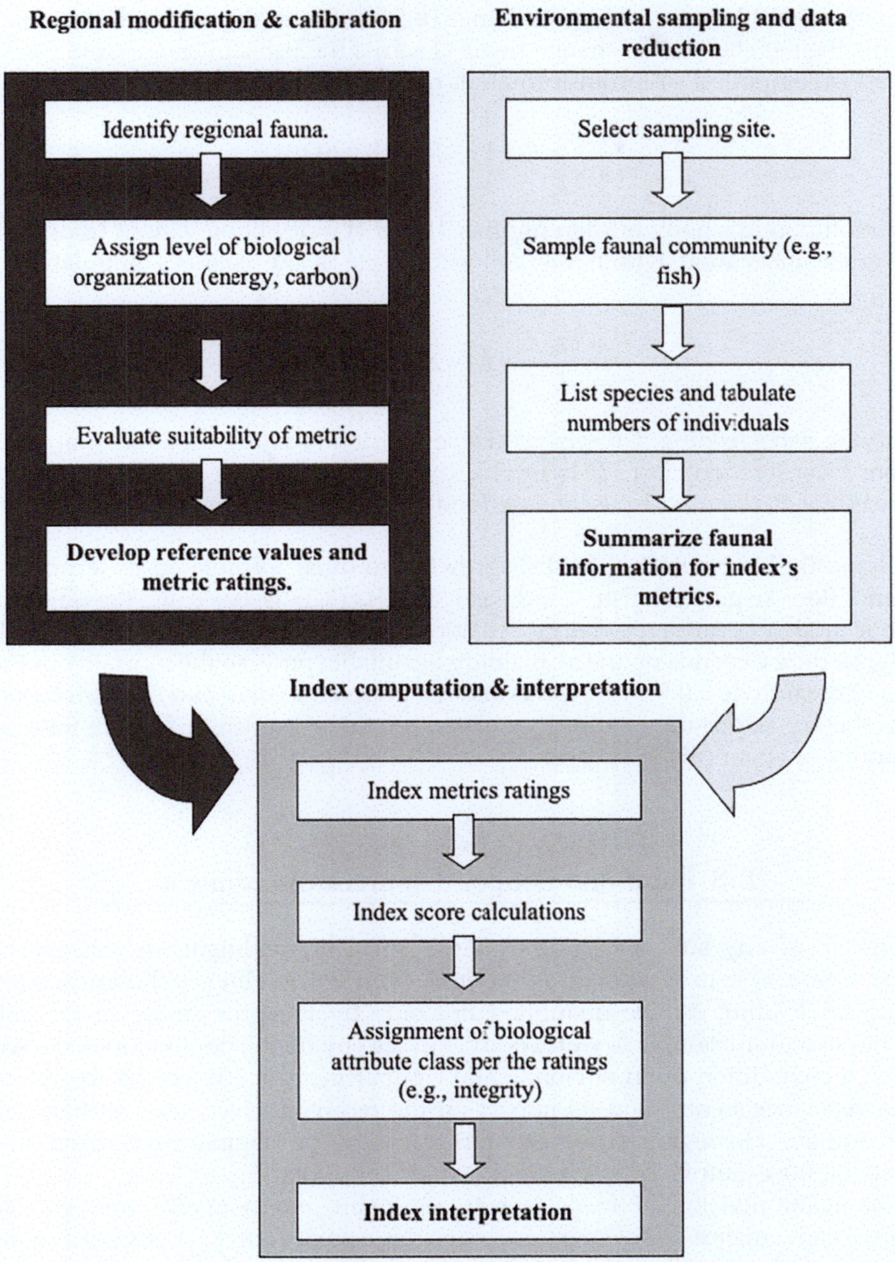

FIGURE 8.3 Sequence of activities involved in calculating and interpreting an index of biotic integrity (*IBI*). *Data from [46–48].*

Where, J_g represents the net chemical exchange (μg d^{-1}) across the fish's gills from the water; J_i represents the net chemical exchange (μg d^{-1}) across the fish's intestine from food; and J_{bt} represents the compound's biotransformation rate (μg d^{-1}).

$$\frac{dW_d}{d_t} = F_d - E_d - R - EX - SDA \tag{8.5}$$

Where, B_f=compound's body burden (μg $fish^{-1}$) and W_d=dry body weight (g dry wt $fish^{-1}$) of the average individual within the cohort; and N is the cohort's population density (fish ha^{-1}).

$$\frac{dN}{d_t} = EM - NM - PM \tag{8.6}$$

Where, Fd=the fish's feeding; Ed=*egestion* (i.e., expulsion of undigested material); R=routine respiration; EX=excretion; and SDA=specific dynamic action (i.e., the respiratory expenditure in excess of R required to assimilate food). All these parameters have units of g dry wt day^{-1}.

Physiologically based models for fish growth are often formulated in terms of energy content and flow (e.g., kcal $fish^{-1}$ and kcal day^{-1}). (8.3) is basically the same as such bioenergetic models because the energy densities of fish depend on their dry weight [52–54]. Obviously, feeding depends on the availability of suitable prey, so the mortality of the fish is a function of the individual feeding levels and the population densities of its predators. Thus the fish's dietary exposure is directly related to the organism's feeding rate and the concentrations of chemicals in its prey.

8.8 Point and nonpoint sources of sediment

The amount of any substance, for example, nutrient, pollutant, or microbes, that is discharged into a system is known as the *load*. Nutrient loading, pollutant loading, and microorganism loading can be quantified not only by the gross mass of the substance entering a system, for example, a water body, but also by the response of that system to the load. Thus, the condition of an envionmental compartment is affected by the relationship between a substance's entry and its impact on the receiving river, lake, wetland, estuary, ocean, or aquifer. However, these receptors can also be compartmentalized into their various media, for example, soil, sediment, water, and biota.

The first major division of loads is between point and nonpoint sources. This is a distinction of convenience. What some may consider a point source, others see as a nonpoint source. For example, if a 1-hectare (ha) field is near a small creek, this may be a relatively large and dispersed source of pesticide being released into the creek. However, a 1-ha trickling filter system near a large river may be classified as a point source since much of the discharge to the river comes through a conduit (i.e., the outfall structure) from the system. However, it is highly likely that pollutants are being released from the treatment facility from sources other than the outfall structure.

Thus nonpoint sources are generally associated with runoff, that is, multiple sources in a given two-dimensional space that leave these sources and are transported toward a receptor system. These receiving bodies can be above the ground surface, for example, streams and lakes, or below the ground surface, that is, aquifers. They are usually cascades of systems, such as a contaminant moving atop the soil, with some infiltration. The infiltrated fraction may find its way to the aquifer and move much more slowly than the runoff above the ground. However, the contaminated ground water may recharge stream flow. The same stream may have already been contaminated by the overland flow, so the stream is receiving a continuous flow from the ground water source to the stream and a plug or episodic flow from overland runoff. Material in the flow runs off, infiltrates through the soil column, follows ground water flow lines, and eventually reaches the receiving body. Along the way, a fraction of the material may be chemically transformed and physically held (e.g., sorbed to soil particles). For volatile compounds, a fraction will also be evaded to the atmosphere. Thus the net amount that reaches the water body is a function of numerous factors, including vegetative cover, soil porosity and permeability, ground water flow rates, soil texture, and Henry's Law and other equilibrium coefficients for the released material [55].

For biota, especially bacteria, the nonpoint and point contributions are even more complex than those of chemical compounds. Microbes may be directly loaded into a stream or other receiving body. In addition, they may accumulate as nonpoint source loads from a combination of sources. As they are transported overland and beneath the soil surface, the microbes encounter hostile and accommodating conditions, which will decrease and increase their numbers, respectively. However, the hostile conditions may lead to the formation of cysts and spores, which are more likely than the bacteria themselves to be transported by advection in the air and water or carried by fauna.

When the conditions change to become more accommodating, the spores will generate increased microbial populations. Thus the actual loading to the body consists of a complicated set of transport and fate processes. In a point source, the growth, metabolism, and formation of cysts and spores depend on the conditions of the receiving water body, which vary over time but not in space. In a nonpoint source scenario, the growth, metabolism, and formation of spores and cysts are a cascade in time and space. In accommodating environments, the microbes will undergo metabolism and growth, which may allow for larger microbial populations to be delivered to other compartments and eventually to the receiving body. However, various compartments in the flow may be hostile (insufficient nutrients, water, pH, temperature, etc.). These may induce the formation of spores, which can be advectively transported in the atmosphere or by the flow of ground and surface waters [55].

Loading calculations must account for background sources. The hydrologic and atmospheric systems join at the interfaces between terrestrial and water bodies and the atmosphere, that is, terrestrial, hydrological, and tropospheric fluxes. For example, atmospheric deposition is an important source of nutrient loading to ecosystems. Ecosystems like wetlands exchange nitrogen, phosphorus, and other nutrients with the atmosphere, with transformation of different forms of the nutrients. Many of these reactions are biotic and are mediated by microbes and plants [55].

The exchanges between biota and the land are influenced by the availability and forms of nutrients. When rain containing nitrates falls on the wetland water and land surfaces, for instance, plant roots take it up and metabolize it into organic forms of nitrogen. Meanwhile, bacteria in sediments reduce it to ammonia. In the opposite direction, the reduced forms are oxidized by other bacteria (e.g., *Nitrosomonas* converts ammonium ions [NH_4^+] to nitrite [NO_2^-], and *Nitrobacter* converts the NO_2^- to nitrate [NO_3^-]). Thus two opposite reactions are constantly at play in loading scenarios. In this instance, these are *ammonification* (or *deamination*) and *nitification:*

$$\text{Ammonification: } NH_4^+ + OH^- \Leftrightarrow NH_3 + H_2O \tag{8.7}$$

$$\text{Nitrification: } NH_4^+ + 2O_2 \rightarrow NO_3^- + 2H^+ + H_2O \tag{8.8}$$

Ammonification is an equilibrium reaction, and nitrification is an oxidation reaction. However, the loading depends on numerous environmental conditions, including pH. Thus under acidic conditions, most of the ammonia nitrogen will ionize to ammonium, and under basic conditions, the nonionized ammonia concentrations will increase in proportion to the ammonium. This is important since nonionized ammonia is very toxic to aquatic biota, whereas ionized species are nutrients needed for plants and algae. Add to this, the decomposition of organic matter, and the loading becomes a complicated mix of reactions:

$$\begin{aligned}\text{Organic nitrogen} + O_2 &\rightarrow NH_3\,\text{nitrogen} + O_2 \rightarrow NO_2\,\text{nitrogen} + O_2 \\ &\rightarrow NO_3\,\text{nitrogen} + O_2\end{aligned} \tag{8.9}$$

Even this is an oversimplification, since the oxidation and reduction depend on the types of bacteria present. For example, the O_2 required for nitrification is theoretically 4.56 mg O_2 $mg^{-1}NH_4^+$, but this is an autotrophic reaction. Thus O_2 is produced by the nitrifying bacteria, decreasing the amount needed. However, the growth rate of nitrifying microbes is far less than that of heterotrophic microbes decomposing organic wastes. When large amounts of organic matter are degraded, the nitrifiers' growth rate will be sharply limited by the heterotrophs, which means the rate of nitrification will commensurately decrease [56–58]. Furthermore, these same responses to conditions exist for all other nutrients in the system, for example, organic P will be oxidized and oxidized P species reduced, etc. Thus the kinetics of nutrient loading are quite complex.

8.8.1 Total maximum daily loading

The United States and other nations restrict the total maximum daily load (TMDL) of pollutants released to water bodies that have been deemed impaired, where technology-based and other engineering controls have been inadequate to achieve water quality standards. Thus the TMDL reflects the amount of a pollutant that can be discharged from point, nonpoint, and natural background sources, including a margin of safety (MOS) for any water quality-limited water body. The TMDL process consists of five steps:

1. Selection of the pollutant in need of consideration.
2. Estimation of the water body's assimilative capacity (i.e., loading capacity).
3. Estimation of the pollutant loading from all sources to the water body.
4. Analysis of the current pollutant load and determination of needed reductions to meet assimilative capacity; and,
5. Allocation, including a MOS, of the allowable pollutant load among the different pollutant sources in a manner that achieves water-quality standards.

A beneficial use that has been impaired is a function of a change in the chemical, physical, or biological integrity of surface waters, such as:

- restrictions on fish and wildlife consumption
- tainting of fish and wildlife flavor
- degradation of fish and wildlife populations
- fish tumors or other deformities
- bird or animal deformities or reproduction problems
- degradation of benthos
- restrictions on dredging activities
- eutrophication or undesirable algae
- restrictions on drinking water consumption, or taste and odor problems
- beach closings
- degradation of aesthetics
- added costs to agriculture or industry
- degradation of phytoplankton and zooplankton populations
- loss of fish and wildlife habitats.

Meeting water quality standards is a way to protect beneficial uses, including recreational activities, aquatic life, and water supply. For example, recreational beneficial use can be based on protection against fecal coliform, including *E. coli* bacteriological criteria. For example, if a lake is designated for primary contact recreation water, potential exposures to bacteria and a baseline level of disinfection will be required. This can be more straightforward for point sources, for example, building wastewater treatment systems, than for nonpoint sources. The latter often requires management approaches, such as measures to prevent soil erosion, since soil may carry coliform bacteria.

8.9 Land use planning

Terrestrial organisms depend on both the quantity and quality of land for their survival. Indeed, all creatures, to some extent, are affected by land habitats; some because a life stage is terrestrial, for example, amphibians, and some because the land is the source of stress on the water.

Land use plans must include measures to complement the natural and cultural environment and to avoid destruction or adverse alteration of the land. This means that good practice should always consider approaches to minimize the impacts of construction, development, and facility operation on the surrounding watershed by preventing water pollution and actions that disrupt natural drainage patterns.

In addition, land use planning must directly address habitat destruction by recommending ways that any development conserves land that serves as habitat for native flora and fauna, both on and off site, directly and indirectly. An example of indirect destruction or secondary impacts from infrastructure development is the location of new sewer lines or roads in previously undeveloped or low-density land. This invites new residential and commercial development, which in turn calls for more roads, structures, and impervious surfaces. These changes then lead to less soil infiltration and increase the risks of flooding on site and downstream, as the holding capacity for water during rain events decreases.

Another consideration for land development is the presence of threatened or endangered species, and the potential that their habitats may be affected by developing a site.

A question crucial to any land management decision is how well the plan fits with land development practices, codes, and design criteria. Land use planning is actually "applied social science" because it is based on not only the geographic and engineering sciences, but also sociological, economic, and political theories to guide the development and use of land for the benefit of society [59]. It bridges social needs with physical space by involving public participation in land use decision-making.

Proper land development calls for a set of criteria that draws from the ample reservoir of safety criteria, especially these four rules [60]:

1. The design must comply with applicable laws.
2. The design must adhere to "acceptable engineering practice."
3. Alternative designs must be sought to see if better practices are available.
4. Possible misuse of the design must be foreseen.

Land use decisions in the United States are predominantly made at the local level. Durham, NC's development plan [61], for example, incorporates several land use considerations, including:

1. Intensity/density of the proposed development, that is, any human-caused change to improved or unimproved real estate that requires a permit or approval from any agency of the City or County of Durham, including but not limited to constructing or changing buildings or other structures, mining, dredging, filling, grading, paving, excavation or drilling operations, and storage of equipment or materials.
2. Sensitive areas and related protection.
3. Any limitations on the number, type, or range of uses, that is, the purpose for which a building, structure, or area of land may be arranged or occupied, or the activity conducted or proposed in a building, structure, or on an area of land.
4. Dedications, that is, the transfer without payment of ownership or other interest in real property from a private entity to a public agency or reservations.
5. Design elements if required or otherwise provided.
6. Development phasing if required or otherwise provided.

Each of these considerations requires that land use decisions be based on thorough and scientifically sound information, and that the planners, engineers, and others involved in implementing the plan use sound judgment and follow best practices.

8.10 Addressing land pollution

Local governments address land pollution using several tools, including land use plans, zoning ordinances, and development rules, such as subdivision regulations, which must undergo significant public review. An illustrative example as that of Durham, NC. The City of Durham Public Works mission statement is based on the City of Durham's *Customer Bill of Rights*, which is a citizen-focused initiative "to provide excellent service at all times by determining the needs of our citizens and customers and satisfying those needs beyond their expectations" [61,62]. Human health and environmental quality are top priorities in fulfilling the city's mission. The chief instrument in the environmentally responsible approach is achieved by the city's growth adhering to the Unified Development Ordinance, which is designed to promote the health, safety, and general welfare of the residents of Durham City and County. It does so by:

A. Protecting existing neighborhoods, preventing their decline, and promoting their livability.
B. Addressing future needs, growth, and change in the jurisdiction.
C. Conserving land and water resources.
D. Preserving groundwater quality and supply.
E. Recognizing geologic features, soil, and topography.
F. Improving air quality.
G. Minimizing congestion in the streets and reducing reliance on automobiles by providing options for walking, bicycling, and transit use.
H. Securing safety from fire and other dangers.
I. Providing adequate light and air.
J. Preventing overcrowding of land and undue concentrations of population.
K. Providing adequate transportation, water supplies, sewer service, schools, parks, open space, and public facilities.
L. Conserving the value of buildings.
M. Examining the most appropriate use of the land.
N. Regulating the locations of businesses and industries.
O. Regulating the height and bulk of buildings.
P. Protecting the capacity of floodways and nonencroachment areas to prevent loss or damage to homes or property.
Q. Regulating the area of yards and open spaces for buildings.
R. Providing for the needs of agriculture.
S. Protecting historic sites and areas.
T. Encouraging an aesthetically attractive community; and
U. Preventing secondary effects from land uses that could negatively impact nearby land uses, consistent with prior ordinances restricting such uses and with evidence supporting those restrictions.

This list of expectations provides a yardstick for land use. If an activity adheres to all these factors, it is likely to be environmentally acceptable. The ordinance provides for the orderly, efficient, and economic development of the City and County by coordinating streets, highways, and other public facilities within proposed subdivisions with existing or planned streets and highways or other public facilities. It also provides for the dedication and reservation of rights-of-way, easements, or sites for streets, utilities, open space, recreation areas, and other public facilities, the protection of historic resources and the natural environment, and the distribution of population and traffic to avoid congestion and overcrowding, which shall create conditions essential to public health, safety, and the general welfare.

8.11 Conclusions

As real estate professionals are fond of saying about land, "they aren't making any more of it." With the small exception of the Zuiderzee and a few other sedimentation projects, this is generally true. So, wasting land by either losing it through erosion, diminishing its beneficial uses, and contaminating it is wholly unacceptable.

Protecting land not only relies on the sound application of the physical and natural sciences but also on the social sciences. Where people choose to live is based on environmental quality, as well as sociological and psychological factors. Land use planners, engineers, and waste managers each have vital roles in preventing and addressing land pollution. They are key leaders in adhering to a land ethic that preserves the land for future generations.

References

[1] M. Mitchell, Gone with the Wind, Random House, (2020).
[2] J. Locke, Natural rights, Moral Reasoning: A Philosophic Approach to Applied Ethics, *Dryden Press, London*, (1990), pp. 133–135.
[3] L. Strauss, On Locke s doctrine of natural right, The Philosophical Review 61 (4) (1952) 475–502.
[4] M. Matulovic, John Locke's Theory of Natural Rights, Zbornik Pravnog Fakulteta Sveučilišta u Rijeci 16 (1995) 179.
[5] P.J. Larkin Jr, , The original understanding of property in the constitution, Marquette Law Review 100 (2016) 1.
[6] M. Welde, E. Tveter, The wider local impacts of new roads: a case study of 10 projects, Transport Policy 115 (2022) 164–180. Available from https://doi.org/10.1016/j.tranpol.2021.11.012.
[7] M.L. York, Dealing with secondary environmental impacts of transportation, Land Use Law & Zoning Digest 51 (3) (1999) 3–10.
[8] P.H. Guldberg, F.H. Benesh, T. McCurdy, Secondary impacts of major land use projects, Journal of the American Institute of Planners 43 (3) (1977) 260–270.
[9] S. Bascom, Secondary impacts of transportation and wastewater investments, Office of Research and Development, US Environmental Protection Agency, (1975).
[10] J.B. Shopley, nvironmental impact analysis: the identification of secondary impacts, 1981.
[11] J.R. Rasband, M. Garrett, A new era in public land policy? The shift toward reacquisition of land and natural resources, Rocky Mountain Mineral Law Institute 53 (2007).
[12] L.K. Caldwell, The National Environmental Policy Act: an Agenda for the Future, Indiana University Press, Bloomington, Indiana, (1998).
[13] B. Anderson, Lynton Keith caldwell: an environmental visionary and the national environmental policy act, Electronic Green Journal 1 (37) (2014).
[14] B. Obama, Obama Signs the Omnibus Public Lands Management Act of 2009, Diakses dari https://www. nytimes. com/2009/03/30/us/politics/30lands-text. html pada, vol. 1, 2021.
[15] D. Brewer, Planning and NEPA, Breaking Barriers Building Bridges (2012) 82.
[16] A. Leopold, A Sand County Almanac, and Sketches Here and There, Oxford University Press, USA, (1989).
[17] A. Leopold, A sand county Almanac 1949, *Ballantine, New York*, (1970).
[18] A. Leopold, From A Sand County Almanac (1949). Oxford, 1968.
[19] D.A. Vallero, Biomedical ethics for engineers, The Biomedical Engineering Series (2007).
[20] R.B. Meyers, Environmental values, ethics and support for environmental policy: a heuristic, and psychometric instruments to measure their prevalence and relationships, 2004.
[21] C.A. Jencks, The language of post-modern architecture, Journal of Aesthetics and Art Criticism 37 (2) (1978).
[22] E. Birmingham, Reframing the ruins: Pruitt-Igoe, structural racism, and African American rhetoric as a space for cultural critique. 1998.
[23] D.J. Vallero, D.A. Vallero, Land Pollution, Waste, Elsevier, (2011), pp. 445–466.

[24] D.A. Vallero, T.M. Letcher, Society, Unraveling Environmental Disasters, Second ed, Elsevier Academic Press, Amsterdam, The Netherlands, (2023), p. 720 ch.
[25] R. Gifford, The consequences of living in high-rise buildings, Architectural Science Review 50 (1) (2007) 2–17. Available from https://doi.org/10.3763/asre.2007.5002.
[26] J. Heathcott, Planning note: pruitt-igoe and the critique of public housing, Journal of the American Planning Association 78 (4) (2012) 450–451.
[27] A. Von Hoffman, Why they Built Pruitt-Igoe (1995).
[28] J. Bailey, The case history of a failure, in Architectural Forum 123 (5) (1965) 22–25.
[29] A. Von Hoffman, Why They Built the Pruitt-Igoe Project, Joint Center for Housing Studies, Harvard University, http://www. soc. iastate. edu/sapp/PruittIgoe. html (10/05/04), 2005.
[30] Public Law 94-580, Resource Conservation and Recovery Act of 1976, 42 USC 6901, et seq, Readily available, 1976.
[31] Comprehensive Environmental Response, Compensation, and Liability Act (Superfund), in 42 U.S.C. §9601 et seq. (1980), ed. United States, 1980.
[32] M. Sun, Missouri's costly dioxin lesson: up to 100 sites in the state were contaminated with dioxin 12 years ago; a wrong assumption that the chemical would break down delayed action, Science (New York, N.Y.) 219 (4583) (1983) 367–369.
[33] D. Vallero, Paradigms Lost: Learning from Environmental Mistakes, Mishaps and Misdeeds, Butterworth-Heinemann, (2005).
[34] M.R. Greenberg, Stigma as a multispatial-scale process: revisiting the worst US superfund sites, Risk Analysis 43 (9) (2023) 1811–1823.
[35] U.S. Environmental Protection Agency. "Land revitalization." https://www.epa.gov/land-revitalization (accessed.
[36] B.B. Marriott, Environmental Impact Assessment: A Practical Guide, McGraw-Hill Inc, (1997).
[37] D.A. Vallero, P.A. Vesilind, Socially Responsible Engineering: Justice in Risk Management, John Wiley & Sons, Hoboken, New Jersey, (2007).
[38] M. Ritzdorf, Locked out of paradise: contemporary exclusionary zoning, the Supreme Court, and African Americans, 1970 to the present, Urban Planning and the African American Community in the Shadows (1997).
[39] K. Shrader-Frechette, Ethics and environmental advocacy, Environmental Ethics and Forestry: a Reader (2000) 209.
[40] K. Shrader-Frechette, Environmental Justice: Creating Equality, Reclaiming Democracy, Oxford University Press, (2002).
[41] D.E. Newton, Environmental justice, 2009.
[42] J. Erickson, 2016. Targeting minority, low-income neighborhoods for hazardous waste sites. University of Michigan News., (January 19, 2016).
[43] Georgia Department of Natural Resources, Development and Implementation of Aucilla River Watershed Management Plan, 2018. Accessed: January 18, 2024. [Online]. Available from: file:///Users/daniela.vallero/Downloads/AucillaRiver_WMP_0311010301_Y2018.pdf.
[44] C.P. Gerba, I.L. Pepper, Wastewater treatment and biosolids reuse, Environmental MIcrobiology, *2nd edn., Elsevier Inc., USA*, (2009), pp. 503–530.
[45] D. Vallero, Fundamentals of Water Pollution: Quantifying Pollutant Formation, Transport, Transformation, Fate and Risks, 1st ed., Elsevier, Amsterdam, NV, (2024).
[46] Assessment of Biological Integrity in Running Waters: A Method and Its Rationale, 1986.
[47] J.R. Karr, Assessment of biotic integrity using fish communities, Fisheries 6 (6) (1981) 21–27.
[48] M.T. Barbour, Rapid Bioassessment Protocols for Use in Wadeable Streams and Rivers: Periphyton, Benthic Macroinvertebrates and Fish, US Environmental Protection Agency, Office of Water, (1999).
[49] R.J. Erickson, J.M. McKim, A model for exchange of organic chemicals at fish gills: flow and diffusion limitations, Aquatic Toxicology 18 (4) (1990) 175–197.
[50] D.J. Stewart, D. Weininger, D.V. Rottiers, T.A. Edsall, An energetics model for lake trout, Salvelinus namaycush: application to the Lake Michigan population, Canadian Journal of Fisheries and Aquatic Sciences 40 (6) (1983) 681–698.
[51] M.C. Barber, Bioaccumulation and Aquatic System Simulator (BASS) User's Manual Beta Test Version 2.1, *US Environmental Protection Agency, Athens, GA*, (2001), p. 76.

[52] J.A. Kushlan, S.A. Voorhees, W.F. Loftus, and P. Frohring, Length, mass, and calorific relationships of Everglades animals, Florida scientist (USA), 1986.
[53] K.J. Hartman, S.B. Brandt, Estimating energy density of fish, Transactions of the American Fisheries Society 124 (3) (1995) 347–355.
[54] K. Schreckenbach, R. Knosche, K. Ebert, Nutrient and energy content of freshwater fishes, Journal of Applied Ichthyology 17 (3) (2001) 142–144.
[55] D. Vallero, Environmental Biotechnology: A Biosystems Approach, Elsevier Science, (2015).
[56] V. Novotny, G. Chesters, Delivery of sediment and pollutants from nonpoint sources: a water quality perspective, Journal of Soil and Water Conservation 44 (6) (1989) 568–576.
[57] V. Novotny, Water Quality. Diffuse Pollution and Watershed Management, John Willey & Sons, Inc, Hoboken, New Jersey, (2003).
[58] V. Novotny, Integrated water quality management, Water Science and Technology 33 (4-5) (1996) 1–7.
[59] J.B. Cullingworth, J.B. Cullingworth, The Political Culture of Planning: American Land Use Planning in Comparative Perspective, Routledge, (2002).
[60] C.B. Fleddermann, Engineering ethics, Prentice Hall, Upper Saddle River, NJ, (1999).
[61] Unified Development Ordinance. Section 3.5.6. [Online], 2010. Available from: http://www.durhamnc.gov/udo/mainPage.asp.
[62] B.A. Mohammad, Assessing Municipal Operations as a Stormwater Pollution Prevention Management Strategy: A Possible Approach for the City of Durham, North Carolina, School of the Environment Master's Project submitted in partial fulfillment …, 2009.

CHAPTER

9

Optimization methods☆

OUTLINE

9.1 Introduction

Making sound, systems-based environmental decisions requires rigorous attention not only to physical and biological sciences but also to the feasibility of any recommendations to address an emerging technology. There are several techniques for optimizing the development and operation of an emerging technology based on its feasibility to complement the technical and scientific rigor and ethics. This includes attention to project planning, communicating decisions, and working with stakeholders. For example, designing a pollution control system must be underpinned by sound science, but it must also be realistic in terms of economics. Indeed, every engineering solution incorporates aspects of what is known as engineering economy.

☆ This chapter includes excerpts from and is an update to Chapter 11 in [1] D. A. Vallero, *Environmental Systems Science: Theory and Practical Applications*, First ed. Amsterdam, NV: Elsevier, 2020.

Technical and Ethical Optimization: Challenges and Opportunities for the Modern Engineer and Scientist
DOI: https://doi.org/10.1016/B978-0-443-44925-3.00014-4

Ultimately, the basis for acceptability must integrate environmental and health risk-based factors with economics and feasibility. It must be both environmentally and economically sustainable. It must be reliable in its design and operation, resilient to external stresses, and reasonable in the expectations that the project can be properly managed. Furthermore, any systems-based project must provide for the appropriate means of communicating these decisions to and working with various stakeholders.

Optimization consists of methodical, structured, and logical sequence of steps to ensure thoroughness, accuracy, and reproducibility of a process that follows defined techniques as opposed to reliance on chance. Optimization makes use of objective analysis, critical evaluation, and a well-ordered progression to minimize errors and gain deeper understanding. For example, the scientific method of observation, experimentation, and objectivity is a type of optimization for credible research. There are many texts that cover the computational and mathematical steps in optimization [2–18]. However, this discussion will apply the broadest, nonmathematical connotation of the term for optimization because our goal here is to merge the emerging technology development process with the ethical optimization process. Indeed, any optimization must be rational, orderly, and logical. It must be structured, following a framework that is constructed and defined before a project or process begins. It must be comprehensive, using and vetting all available data and information that are relevant to the task at hand. It must be transparent and reproducible so as to be generalized and withstand peer review. Finally, it must identify potential errors as the optimization effort proceeds. Any optimization relies on getting the best output(s) from all the inputs.

Of course, the scientific aspects of a technology must adhere to the tenets of engineering optimization, that is, finding the design by maximizing desired outcomes. For example, the most efficient, reliable, profitable, and useful technology must meet specific constraints like size, multiple applications, safety, material types and limits, budget, and cost of the final technology. To this list, we must add moral permissibility. Engineering optimization applies mathematical models and algorithms to explore various alternatives, transforming complex engineering challenges into quantifiable, optimal results for better performance and resource use.

Combining technological drivers with ethical considerations provides a balance among numerous motivations and mandates, whether for an individual company's financial profit or for the budgets of every governmental jurisdiction. For example, using or adapting a technology by a public works or engineering department should not view the new technology's value separately from other civic or corporate functions like environmental sustainability, transportation, waste collection, hazardous materials handling, treatment, engineering, and remediation activities. The technology is merely part of the entire system needed to provide an efficient and effective function. As such, public agencies and other enterprises must search for economies of scale that solve and prevent multiple environmental, public health, and other societal problems.

9.2 Engineering and environmental sustainability

Engineering economics is a good place to start to introduce approaches for evaluating the acceptability of proposed new technologies, upgrades, and enhancements to existing technologies.

Though they share many commonalities, it is important to note that technical and cost efficiencies are calculated differently, usually by standards and methods of dissimilar professional disciplines. For example, the first requirement of technical efficiency involves optimizing among scientific variables to achieve the best alternative that protects public health and the environment *before* cost estimates. Thus a less costly alternative is only acceptable if it also meets the essential technical specifications necessary to meet a standard or other quality metrics. For example, environmental legislation and regulations require steps to be taken to limit the concentrations of pollutants released from a facility. To do so, management plans are written and implemented to address environmental quality. These plans include measures to prevent and to control pollution of the air, surface waters, ground water, as well as to preserve soil and land resources, including solid and hazardous waste management. Government jurisdictions at all levels engage in prevention and control strategies to protect and enhance environmental quality and public health. These plans are periodically reviewed for effectiveness in both technical success and cost efficiencies.

Most environmental programs establish baselines and conduct inventories of pollutants from sources. The inventories are tools for targeting control strategies for various pollution sources, including air emissions, water discharges from industrial facilities and other stationary sources, emissions from mobile sources, waste site contamination, and potential threats to land and soil [19]. For example, the United Kingdom requires annual reporting by all entities that [20]:

1. operate a sewage treatment works with a capacity greater than or equal to 100,000 population equivalents;
2. dispose of radioactive waste to air, water, or sewers covered by a permit issued under the Environmental Permitting Regulations; or
3. operate an opencast mine or quarry with a surface area of over 25 hectares or any underground mine and related operation.

One section of the reporting form is dedicated to resource efficiency. Many countries and other jurisdictions require similar reporting.

9.2.1 Feasibility of a project

A public agency, those who contract with public entities, or those who have projects affected by public agencies require a grasp of how these projects are evaluated, especially in terms of benefits and costs.

The benefit-cost analysis (BCA) is a tool for maximizing benefits for a certain amount of costs as detailed in budgets. The cost-effectiveness analysis aims to maximize net benefits and, similarly, to minimize costs to get a desired amount of benefits.

The BCA assumes:

1. Equivalence, that is, benefits or costs are in the same units in all projects under review. Most often, this is represented by dollars.
2. The public is the "user" or "client" who will benefit from the selected project.
3. The "sponsor" is the government (e.g., town, city, county, state, nation, or even multinational organizations).

The BCA:

1. Identifies all user benefits (positive outcomes) and "disbenefits" (negative outcomes);
2. Quantifies and compares these benefits and disbenefits;
3. Identifies and quantifies the sponsor's costs;
4. Finds the equivalent net benefits and net costs at the base period, with a discount rate specific to the project; and
5. Considers the project to be "acceptable" if the equivalent user's net benefits exceed the equivalent sponsor's net costs. For a public process, it is preferable to have a rule for exceedance before the project is accepted, for example, some percentage above a measure of central tendency like a mean net cost for a state or local area.

Essentially, the goal is to determine whether the benefits are worth the economic investment. This can be viewed at various points in the design process. For example:

- Project planning: From an economic perspective, are the benefits of the project costs sufficiently better than present conditions or expected conditions without the project?
- Design and environmental study: From an economic perspective, for example, are the benefits of location "A" worth the project costs? How does location "A" compare to "B" or "C"?
- Construction planning: From an economic perspective, for example, are the benefits of closing certain lanes during construction worth the traffic delays and diversion costs compared to keeping some lanes open?

These comparisons may be conducted using linear programming (LP), a mathematical optimization technique that is used to find the best possible outcome within a specific set of constraints, that is, linear inequalities or equalities. A linear function, known as the objective function, is subjected to various constraints. Ideally, the BCA would project and evaluate all possibilities, but this is not possible or even desirable due to the large uncertainties. So, expert judgment is needed to decide the type and amount of information that can be eliminated or abbreviated, and what information to include as realistic options to compare them to some base case, which is not necessarily the "no action" alternative. However, the other options usually have substantial improvements over the base case.

Feasibility includes numerous perspectives, for example, systems engineering, optimization, and economics. Often, the three are mutually inclusive. For example, optimization techniques, like LP, can be employed to consider economic feasibility of a project with a lifecycle/systems perspective. However, public projects must differentiate economic costs and benefits from other societal costs and benefits. Thus the BCA differs from an economic impact analysis (EIA) [21]:

> A BCA measures the value of a project's benefits and costs to society, while an economic impact analysis measures the impact of increased economic activity within a region. Common metrics for measuring economic impacts include retail spending, business activity, tax revenues, jobs, and property values. Economic impact analyses often take a strictly positive view, (i.e., increased jobs, spending) and do not examine how the resources used for a project might have benefitted alternative societal uses of the resources (i.e., they do not assess the net effect on society).

An economic impact analysis, in this view, considers the initial investment in infrastructure as a stimulus to the local economy, as opposed to a cost to the local government.

Also, economic impact analyses are smaller scale than the usual BCA, which usually consider broad, economy-wide, and "societal" benefits and costs. Effects in one region may be offset by losses in a neighboring region, reflecting a transfer of spending or jobs that may be a net neutral summation. Thus the two local jurisdictions would have different economic impact results, but the BCA would not change, since its focus is larger.

Benefits of a public investment are the direct, positive effects of that project, that is, the desirable outcome the public receives by directly investing in the project. For example, a road improvement may reduce the number or severity of crashes, eliminate long delays during peak hours, or reduce circuitry of travel by offering a shorter route. In highway BCA, benefits are often first estimated in physical terms and then valued in economic terms. Thus the analyst must first estimate the number of crashes eliminated, the travel time saved, and/or vehicle-miles reduced before assigning or calculating monetary values. As mentioned, any public project will require a benefit-cost comparison. The projects vary in the amount and kind of information needed, but all require a quantification of benefits and costs. In economic terms, the cost of a public investment is the value of the resources that must be consumed to bring the project about. The total value of construction and any additional maintenance costs must be estimated. It is important to note that the analysis does not emphasize who incurs the cost but rather aims to include all costs that are involved in bringing about the project.

We will continue with the highway example.

Capital costs comprise the total investment required for an improvement in service. When possible, capital costs should be grouped into similar life-cycle categories. These include engineering, right of way, major structures, grading and drainage, subbase and base, surfacing, and miscellaneous items. ("Benefit-Cost Analysis - MnDOT") These life-cycle groupings make it easier to calculate remaining capital value. Estimates of capital costs, ranging from detailed engineers' estimates to planning-level cost estimates, should be refined and updated during project development process. Within a BCA period, future investments may be needed to maintain the serviceability of a major environmental facility. ("Benefit-Cost Analysis - MnDOT") For example, with a new or reconstructed treatment plant, complete tank replacements may be required 8, 12, or 15 years after the initial construction year. The cost of tank replacement and ancillary work should be included in the analysis and allocated to the year when they are anticipated to occur. When evaluating investments, all future operating and maintenance costs of the facility must be incorporated into the feasibility study and BCA. As mentioned, treatment facilities require preventive maintenance, control technologies must be updated and replaced, and roadway lanes must be plowed and patched each year. In the case of an upgrade of an existing facility or service, it is necessary to estimate the marginal or additional maintenance costs that would be required for the alternative as compared to the base case. For a new facility, the entire additional maintenance costs should be included as the incremental increase in costs. Many components of a project retain some residual useful life beyond the BCA period. At the end of the analysis period, the still-useful parts of infrastructure that will continue to provide benefits into the future should be included as a value in the analysis. The remaining capital value is calculated by determining the percentage of useful life remaining beyond the analysis period and multiplying that percentage by the construction cost for that component [22]. The estimate of the remaining capital value at the end of the analysis period is then converted to a present value and subtracted from the initial capital cost. For many public investments, costs are incurred in the initial years, whereas the benefits from the investment accrue over many years into the future. When assessing the costs and benefits of a project, it is necessary to consider the time value of money by converting the costs and benefits that take place in different years into a common year [22]. Such discounting converts future costs and benefits that occur in different years into a value for a common year, that is, present value.

9.3 Economic efficiency

Certain governmental, environmental regulations apply to the construction of a new facility that will release a certain amount of a specific pollutant to an area that is presently above ambient standards for that pollutant. These regulations require that the applicant document a reduction of at least that amount of the pollutant from another source in the area. For example, US air pollutant sources are allowed to effectuate this reduction in another plant that air permit applicant owns in the area, or the applicant can shut down that plant. The Clean Air Act refers to this reduction as an "offset." However, if the applicant does not own another such plant or prefers not to shut down or effect such reduction in a plant they own, they can seek the required reduction or offset from another owner. Thus such offsets are marketable credits that can be bought, sold, traded, or stockpiled ("banked") if the state or local regulatory agency legitimizes, records, and certifies these transactions. ("Transactional certificates - Big Chemical Encyclopedia") The new source will still have to meet other air quality standards when applicable [23].

Emissions trading can be accomplished by netting, offsets, bubbles, and banking. Netting allows large new sources and major modifications of existing sources to be exempted from certain review procedures if existing emissions elsewhere in the same facility are reduced sufficiently. An offset allows a major new source to locate in an area that does not attain a given air quality standard, that is, a nonattainment area, if emissions from an existing source are decreased by at least as much as the new source would contribute after installation of stringent controls. A bubble allows a regulated entity to combine the limits for several different sources into one combined limit and to determine compliance based on that aggregate limit instead of from each source individually. ("Emissions Trading - an overview | ScienceDirect Topics") The name alludes to an imaginary "bubble" placed over the several sources. Banking allows a regulated entity to accumulate emission credits for future use or sale to take actions to lower emissions below the relevant standard. ("Emissions Trading - an overview | ScienceDirect Topics") These approaches all have a common objective of flexibility to comply with traditional source-specific command-and-control standards, but achieving or providing even less emissions to improve local air quality [24]. Under the more recent "bubble concept" (Fig. 9.1) adopted by some US states with the approval of the Environmental Protection Agency (EPA), the addition of the new source is allowed, provided the total emission of the relevant regulated pollutants from the total complex is decreased. This can be accomplished by obtaining the required offset from one or more of the other sources within the complex, by shutdown, or by improvement in control efficiency.

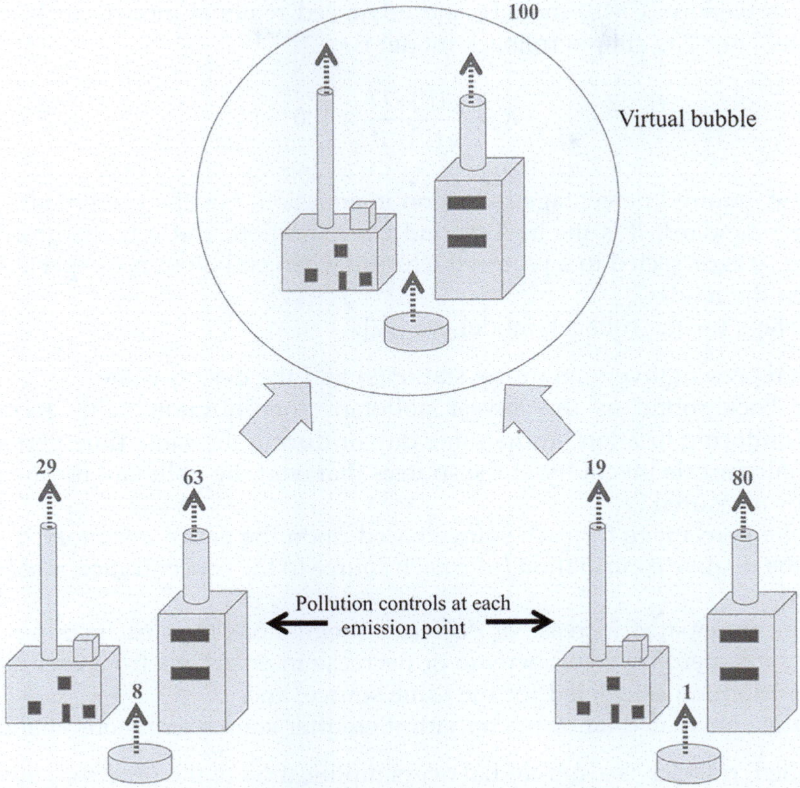

FIGURE 9.1 The so-called "bubble" pollution control concept places an imaginary three-dimensional boundary around an entire industrial plant, evaluating emissions from the facility as a whole instead of requiring control point-by-point on emission sources. Numbers represent emissions from individual sources, some of which can be fugitive sources, and from the entire industrial plant. Information from the Chemical Manufacturing Association. *Data from U.S. Environmental Protection Agency. "How Do Emissions Trading Programs Work?" USEPA. https://www.epa.gov/emissions-trading/how-do-emissions-trading-programs-work (accessed March 25, 2026).*

The bubble concept has been subject to litigation and may require a ruling or challenge to make it acceptable [25].

9.3.1 Proportional emission predictions

The proportional modeling or rollback approach assumes that emissions and atmospheric concentrations are proportional and linearly related, that is, a specified percentage reduction in emissions will result in a similar percentage reduction in atmospheric concentrations. This is most likely a valid assumption for a nonreactive gas such as carbon monoxide, whose principal source is vehicular but is not recommended for many pollutants and/or those with more complicated and complex physics and chemistry, for example, ground-level ozone. Estimation methods include dispersion modeling, source/receptor models such as the

Chemical Mass Balance (CMB) models, and advanced regional models such as the Urban Airshed Model [26]. The simple rollback model is,

$$R = \frac{g(P) - D}{g(P) - B} \times 100 \tag{9.1}$$

where, R is the required percentage reduction in emission, P is the present air quality, D is the air quality standard, B is the background concentration, and g is the growth factor in emissions (e.g., g is projected to a year in the future when emissions are expected to apply to all vehicles on the road).

The basic steps for the rollback model are [27]:

1. Determine representative monitoring station(s) and the design value.
2. Determine background as the lowest pollutant concentration value recorded at an upwind monitoring location on the same day or during the same time period.
3. Prepare a micro-scale inventory of the sources that emit the pollutant for the time period the monitor measured.
4. Calculate the percentage for each source based upon the entire inventory.
5. Calculate the relative contribution from each source to the concentration measured for the time period.
6. Estimate the anticipated increase or decrease in emissions from each source.
7. Apply the same percentage of increase or decrease in emissions from each source to the relative contribution calculated for the same source; and
8. Calculate the anticipated ambient concentration after source emissions change.

As mentioned, rollback is a substantial oversimplification of the processes involved in air pollution, so basing emission reduction on rollback is fraught with error. Indeed, proportional modeling is mentioned in US air quality modeling only for particulate matter (PM) and lead (Pb). Proportional models are not recommended for PM screening analysis unless such techniques are used in conjunction with receptor modeling. Refined models are recommended for PM_{10}. However, where possible, particle size, gas-to-particle formation, and their effect on ambient concentrations may be considered. For point sources of small particles and for source-specific analyses of complicated sources, the appropriate recommended steady-state plume dispersion model should be used. Receptor models have proven useful for helping validate emission inventories and for corroborating source-specific impacts estimated by dispersion models. The CMB model is useful for apportioning impacts from local sources. Other receptor models, for example, the Positive Matrix Factorization (PMF) model and Unmix, which do not have some of CMB's constraints, have also been applied. In regulatory applications, dispersion models have been used in conjunction with receptor models to attribute source (or source category) contributions [28].

For lead (Pb), an entire major urban area or model areas without significant sources of Pb emissions may use a proportional model for air quality analysis at a minimum. However, urban or other dispersion models are encouraged in circumstances where the use of such models is feasible. In modeling the effect of traditional line sources (such as a specific roadway or highway) on lead air quality, dispersion models applied for other pollutants can be used [28].

9.4 Engineering economics: a tool for environmental optimization

Once the decision is made to address an environmental pollution problem, engineering economics, also known as engineering economy, is used to determine whether a specific project is most financially worthwhile compared to other candidates. Engineering economics is also used to prioritize various projects that may serve the same purpose or function, and to assign value or importance to design specifications. For example, engineering economics may help the city engineer determine whether to invest in new control technologies or keep repairing the old ones, which commercial equipment to purchase, which improvements to public assets and operations should be funded, and whether the added cost of materials, for example, fabric filters, is worth the price.

A principal economic concept is that time represents money, that is, the so-called "time value of money." Environmental risk management is an ongoing endeavor, so funding and other resources are viewed over several years. Expensive equipment may be purchased and immediately installed, yet paid for years, even if the benefit provided by the equipment ends before payoff. Therefore, the first step in evaluating a project is the cash flow, which represents the time value of money. This requires that the cash receipts and disbursements (i.e., expenses) for each project be described over a specified time period.

The costs include initial costs (e.g., purchase price, set-up, and installation), operation and maintenance (O&M), overhaul, and end-of-project life costs, for example, decommissioning, disassembly, and recycling. The costs are somewhat fluid and interrelated. For example, a fly ash or scrubber sludge pit may have major decommissioning costs if hazardous substances have accumulated, such as in the cases of water contamination from power plant ash ponds in North Carolina. Therefore, O&M and surveillance costs upfront can prevent major costs over time. The disbursements include salvage value, that is, what the equipment or property is worth after its operational life, and revenues, such as selling compost or recycled materials.

The perils of decommissioning: ash pits [29]

Coal combustion residuals (CCRs) represent the largest volume of industrial wastes generated in the United States. The 10,000 tons of CCRs generated per year from 470 electric generating plants consist of bottom ash, fly ash, boiler slag, and byproducts from fluidized gas flue desulfurization (FGD). Surface impoundments (ash ponds) and landfills hold half of the CCRs [30,31]. If the effluent from ash is released into a surface waterbody in the United States, the amount and type of effluent are limited by the National Pollutant Discharge Elimination System (NPDES) and require an NPDES permit. Since this effluent very likely contains highly toxic metals and other contaminants, the permit process is complicated beyond a single effluent limitation threshold and requires that other factors, including background concentrations and characteristics of the receiving waterbody. For example, the US Environmental Protection Agency (EPA) applies a formula to calculate waste load allocations and to find the in-stream pollutant concentration (C_r) in mg L^{-1} in the stream reach after complete mixing occurs [32]:

$$C_r = \frac{Q_d C_d + \quad Q_s C_s}{Q_r} \tag{9.2}$$

(cont'd)

where: Q_d=waste discharge flow in units commonly used by US environmental engineers [31], that is, million gallons per day (MGD) or cubic feet per second (cfs); C_d=pollutant concentration in waste discharge in milligrams per liter (mg L^{-1}); Q_s=background stream flow in MGD or cfs above point of discharge; C_s=background in-stream pollutant concentration in mg L^{-1}; and, Q_r=resultant in-stream flow after discharge in MGD or cfs).

This and similar equations are used by permitting authorities as part of the process of calculating water quality-based limits. If a value of zero is used for the ambient concentration of a pollutant (C_s) in the equation, the permitting authority can establish an entire limit on discharge. The EPA is aware that the resulting limit would not account for upstream background concentrations of the pollutant [32]. Depending on the pollutant, the background concentration is almost never zero, so the limit would not prevent an in-stream exceedance of the water quality criteria. In addition, some of the released pollutant is likely to come from nonpoint and other sources besides the direct discharge of the effluent, especially seepage from the ash storage system [32]. Thus, handling CCRs requires coordination among water quality, air quality, and waste management authorities. ("Coal Waste Streams - ScienceDirect") Indeed, failures of ash storage systems can involve direct and indirect releases into surface and groundwater.

On February 2, 2014, an estimated 39,000 tons of coal ash spilled from Duke Energy's Dan River Steam Station's ash pond into the Dan River in Eden, N.C. The ash remained visible in a reservoir 80 miles downstream from the spill for days after the coal ash spill. The day after the spill, the North Carolina Department of Environment and Natural Resources, the Virginia Department of Environmental Quality, the US Fish and Wildlife Service, and the US EPA, along with the utility, Duke Energy, initiated a unified command to address the spill, including technical assistance, oversight, data management, and environmental sampling and analysis. Other agencies eventually became involved in a massive effort to protect human health and the environment and guide the cleanup. These include the US Army Corps of Engineers, North Carolina Department of Public Health, Virginia Department of Health, as well as county and local agencies from communities along the Dan River and within the Kerr Reservoir watershed [31].

An extensive breach like this threatens both public health and ecosystems. Metals and other contaminants can reach drinking water supply intakes. The Dan River basin is a recreation area for fishing, boating, and wildlife viewing. The river is a vital habitat area for fish and wildlife, which include the endangered Roanoke logperch (*Percina rex*) and the James spiny mussel (*Pleurobema collina*), along with a species currently under consideration for protection, that is, the green floater (*Lasmigona subviridis*), a freshwater mussel [33].

On May 22, 2014, EPA and Duke Energy entered into an Administrative Order on Consent (AOC). Under the requirements of the AOC, the utility removed coal ash that had accumulated at the Schoolfield Dam, Town Creek Sand Bar, and at both the Danville and South Boston water treatment facilities. The removal action was completed

(cont'd)

in July 2014, and an estimated 4000 cubic yards of coal ash were removed.

"Placing the wastes in a pit or other storage system can allow mercury (Hg), lead (Pb), arsenic (As), and other toxic metals and metalloids to migrate." ("Coal Refining - an overview I ScienceDirect Topics") For example, they can reach groundwater, a major source of drinking water for much of the world. An ash pit that is upgradient from a well can contaminate personal and municipal well water. Water stands in a well at a particular level, known as hydraulic head (*h*). Groundwater flows in the direction of decreasing head. The change in head measurement is the hydraulic gradient, $\frac{\partial h}{\partial x}$.

The capability of a fluid to move through underground materials is known as hydraulic conductivity (*K*). The various strata under the earth's surface may be consolidated rock or unconsolidated materials, like soil, sand, or gravel. Most commonly, however, the movement of groundwater occurs through the unconsolidated materials, that is, porous media. Darcy's Law is an empirically derived equation for the flow of fluids through a cross-section of porous media (*A*). It states that the specific discharge, *Q*, is directly proportional to *K* and the hydraulic gradient:

$$\text{Darcy's Law: } Q =\_\ KA\frac{\partial h}{\partial x} \quad (9.3)$$

The law assumes that flow is laminar and that inertia can be neglected. Thus if K=0.5 m second^{-1}, dh/dx=0.0085 m.m^{-1} and A= 500 m^2, then Q=2.125 m^3 second^{-1}. Thus if a leak has produced a plume, the aquifer is being contaminated at a rate of 2.125 m^3 second^{-1} (see Fig. 9.2).

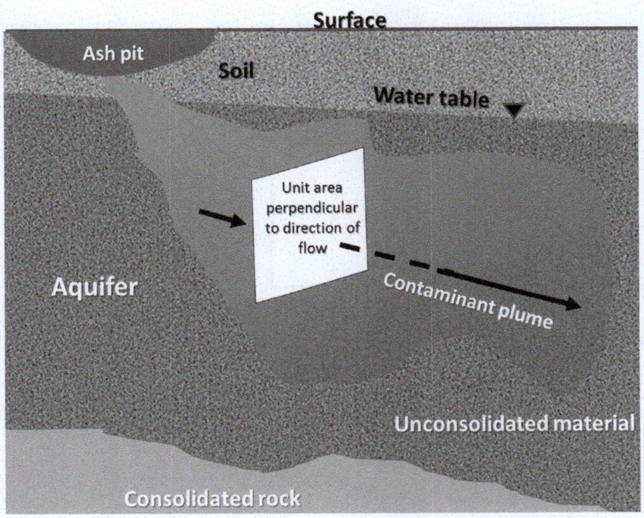

FIGURE 9.2 Plume of pollutants flowing from an ash pit into a drinking water aquifer. The plane perpendicular to groundwater flow is used to calculate the movement, that is, flux, of the plume.

9.4.1 Cash flow

The cost flow diagram (CFD) is used to evaluate costs and disbursements (see Fig. 9.3). The CFD shows the amount and timing of individual cash flows and forms the basis for engineering economic analysis [34]. The CFD is developed by first drawing a segmented, time-based horizontal line divided into appropriate time units. At each cash flow event, a vertical arrow points down for costs and up for revenues or benefits. For clarity, the CFD segments should be drawn to a relative scale.

The end of period t is the same as the beginning of period (t+1). The beginning of period cash flows includes costs for rent, lease, and insurance payments, whereas the end-of-period cash flows are O&M, salvages, revenues, and overhauls. Deciding where t=0 is arbitrary, for example, when a project is analyzed, when funding is approved, or when construction begins. Note that one group's cash outflow, represented as a negative value, is another group's inflow, represented as a positive value. Incidentally, two or more cash flows in the same year should be diagrammed individually to clearly indicate the connection from the problem statement to each cash flow in the diagram.

Example: A city purchases a sludge combustion facility, including air and water pollution control technologies, for $2 million, with O&M costs of $30,000 in year 1 and increasing by $5000 each year. Some of the sludge is converted to commercial-grade fertilizer with equipment that has a local salvage company and a plastic company, providing an annual revenue of $50,000. After year 7, the converter is salvaged, that is, sold to another jurisdiction, for $110,000. To simplify, let us apply the year-end convention, that is, all expenses and revenue are assumed to occur at the end of the year. This makes the math discrete. The CFD is shown in Fig. 9.4.

9.4.2 Discounting

Engineering economics dictates equivalence so that all values can be compared consistently. The present worth (P) is the present amount at t=0. Future worth (F) is the

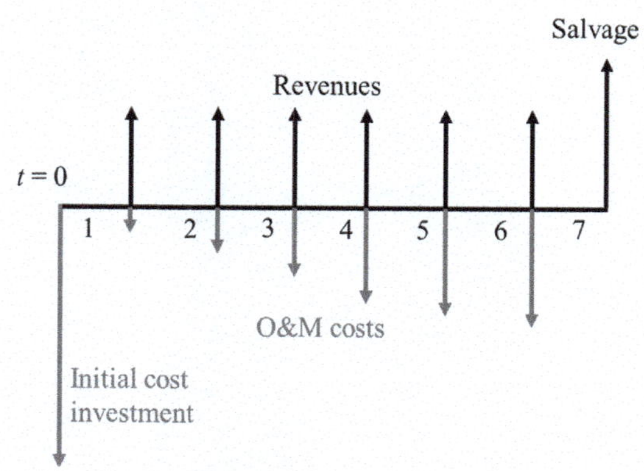

FIGURE 9.3 Simple cash flow diagram. In this case, after the initial investment, revenues are constant, whereas operation and maintenance costs are increasing each year. After year 5, O&M costs exceed revenues. After year 7, the project ends, and salvage is recovered.

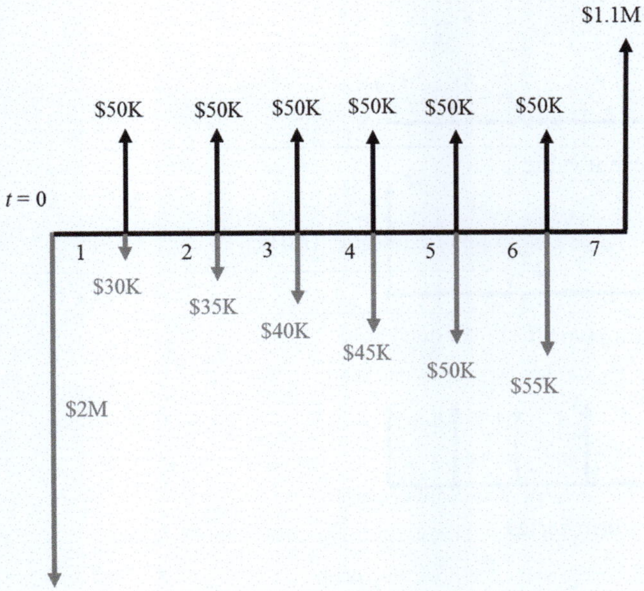

FIGURE 9.4 Example cash flow diagram showing an initial $2 million expenditure. Following the initial investment, revenues are constant, whereas operation and maintenance costs are increasing each year. After year 5, O&M costs exceed revenues, so the project is ended, and salvage is recovered.

equivalent value that will have accumulated after n periods (e.g., years) from P. The annual amount (A) is the uniform amount of money that repeats at the end of each period for n periods. If we do not assume a discrete payout, we also need to have a uniform gradient amount (G), which is the amount of money that repeats at the end of each period, beginning at the end of period 2 and ending at the end of period n.

Therefore, the CFDs can represent present value, future worth, uniform series, uniform gradient series, and exponential gradient (EG) series (see Fig. 9.5). Present value and future worth CFDs are used when determining whether a purchase is acceptable and feasible. Uniform series CFDs can be part of multiyear budgets. The two bottom CFDs in Fig. 9.5 may best apply to equipment and systems that demand greater maintenance and replacement. For example, if the reliability of a system declines with a linear slope, the uniform gradient series is used, but if the failure rates accelerate with time, the exponential (E) gradient CFD may be preferable.

Another important factor is the discount rate (i) that needs to be paid to borrow money, that is, the effective rate per time period. The annual i can be calculated as:

$$i = \left(1 + \frac{r}{k}\right)^{\frac{k}{p}} - 1 \tag{9.4}$$

where, r=the annual interest; k=number of compounding periods per year, and p=the number of periods per year corresponding to the basis for n. Thus if payments are made monthly, p=12; if daily p=365.

The discounting principle states that investing P at t=0 will pay F at t=n. Thus the discount formula is:

$$P = F(P/F,\ i\%,\ n) \tag{9.5}$$

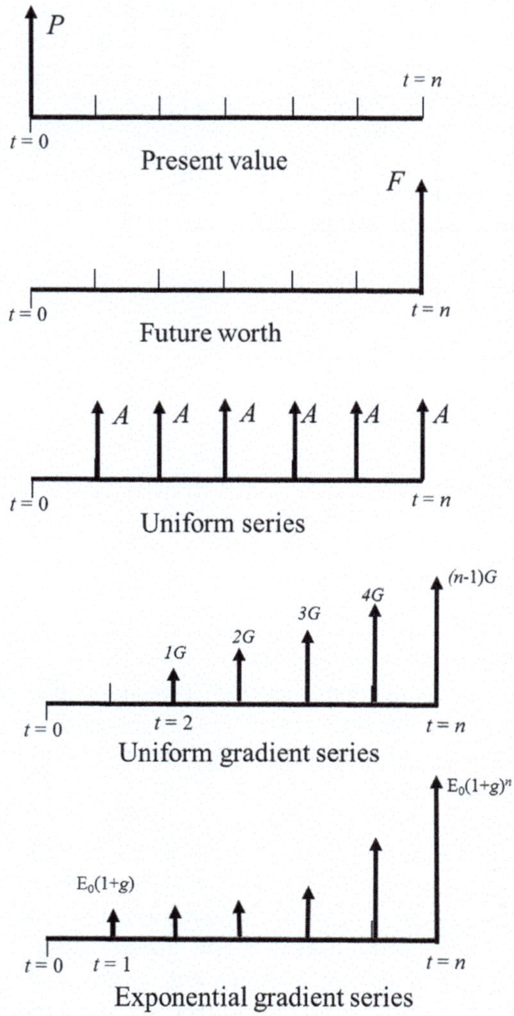

FIGURE 9.5 Commonly used cash flow diagrams for publicly and privately funded projects.

where (*P/F,i%, n*) is the discount factor.

Table 9.1 provides the discount factor conversion formulas for the most common conversions for *P*, *F*, *A*, *G*, and *E* used by environmental managers and engineers. Numerous online calculators are available for these conversions, as well as spreadsheet programs, especially Microsoft Excel (see far-right column of Table 9.1). Microsoft [36], for example, provides support for calculating, *P*, *F*, and *i* at:

- *P=PV(i,N,A,F,Type)*: https://support.office.com/en-us/article/PV-function-23879d31-0e02–4321-be01-da16e8168cbd
- *F=FV(i,N,A,P,Type)*:https://support.office.com/en-us/article/FV-function-2eef9f44-a084-4c61-bdd8-4fe4bb1b71b3

TABLE 9.1 Conversions for environmental systems economics of present value (P), future worth (F), annual amount (A), gradient (G), and exponential gradient (EG). Note: Excel formulas for ($F/G,i\%,n$) and ($A/G,i\%,n$) are based on the algebraic equivalence of $F/G=(P/G)\times(F/P)$ and $A/G=(P/G)\times(A/P)$. The discount factor symbols ($P/G,i\%,n$), ($F/P,i\%,n$) and ($A/P,i\%,n$) need to be replaced with the appropriate discount factor formula listed in the table. ("Calculating Discount Factors in Excel - Discount Factor Table") [35].

Conversion	Symbol	Discount factor formula	Discount factor formula in Excel
Discrete compounding			
P to F	$(F/P,i\%,n)$	$(1+i)^n$	$=FV(i,n,0,-1)$
F to P	$(P/F,i\%,n)$	$(1+i)^{-n}$	$=PV(i,n,0,-1)$
F to A	$(A/F,i\%,n)$	$i/((1+i)^n-1)$	$=\mathrm{PMT}(i,n,0,-1)$
P to A	$(A/P,i\%,n)$	$i\times(1+i)^n/((1+i)^n-1)$	$=\mathrm{PMT}(i,n,-1)$
A to F	$(F/A,i\%,n)$	$((1+i)^n-1)/i$	$=FV(i,n,-1)$
A to P	$(P/A,i\%,n)$	$((1+i)^n-1)/(i\times(1+i)^n)$	$=PV(i,n,-1)$
G to P	$(P/G,i\%,n)$	$((1+i)^n-1)/(i^2\times(1+i)^n)-n/(i\times(1+i)^n)$	{=NPV(i,(ROW(INDIRECT("1:"&n))−1))}
G to F	$(F/G,i\%,n)$	$((1+i)^n-1)/i^2-(n/i)$	$\{=(P/G,i\%,n)\times(F/P,i\%,n)\}$
G to A	$(A/G,i\%,n)$	$(1/i)-n/((1+i)^n-1)$	$\{=(P/G,i\%,n)\times(A/P,i\%,n)\}$
EG to P	$(P/EG,z-1,n)$	$(z^n-1)/(z^n(z-1))$, $z=(1+i)/(1+g)$	$=PV(z-1,n,-1)$
Discrete compounding			
F to P	$(P/F,r\%,n)$	$e^{-r\times n}=1/e^{r\times n}$	$=1/\mathrm{EXP}(r\times n)$
P to F	$(F/P,r\%,n)$	$e^{r\times n}$	$=\mathrm{EXP}(r\times n)$
F to A	$(A/F,r\%,n)$	$(e^r-1)/(e^{r\times n}-1)$	$=(\mathrm{EXP}(r)-1)/(\mathrm{EXP}(r\times n)-1)$
A to F	$(F/A,r\%,n)$	$(e^{r\times n}-1)/(e^r-1)$	$=(\mathrm{EXP}(r\times n)-1)/(\mathrm{EXP}(r)-1)$
P to A	$(A/P,r\%,n)$	$(e^r-1)/(1-e^{-r\times}n)$	$=(\mathrm{EXP}(r)-1)/(1-1/\mathrm{EXP}(r\times n))$
A to P	$(P/A,r\%,n)$	$(1-e^{-r\times n})/(e^r-1)$	$=(1-1/\mathrm{EXP}(r\times n))/(\mathrm{EXP}(r)-1)$

Adapted from: J. Wittwer. "Discount Factor Table for Excel: Summary of Discount Factor Formulas for TVM Calculations in Excel®." https://www.vertex42.com/ExcelArticles/discount-factors.html (accessed May 17, 2018).

- *i=RATE(N,A,P,F,Type,guess)*:https://support.office.com/en-us/article/RATE-function-9f665657-4a7e-4bb7-a030–83fc59e748ce
- Where, "Type=0" means end-of-period cash payments; "Type=1" means beginning-of-period payments; and "guess" is the value of the interest rate.

The symbols and nomenclature are not used consistently in engineering and environmental systems economics. For example, public agencies may also represent present value as PV:

$$PV = \frac{B_{y_i}}{(1 + r)^{y_i - y_0}} \tag{9.6}$$

where, B=the annual benefit, r=the discount rate, y_i=the year the benefit occurs, and y_0=the year to which future dollars are discounted.

Present value is similarly determined for annual cost (C):

$$PV = \frac{C_{y_i}}{(1 + r)^{y_i - y_0}} \tag{9.7}$$

where, y_i=the year the cost occurs.

Discounting

Assume an electric power generating station must change to a higher sulfur-content coal and must install a scrubber to meet sulfur dioxide standards. The scrubber is estimated to cost $5,000,000. Construction and installation will occur in 2031. The year of analysis is 2026. If the discount rate is 3.6% for the year 2026, what is the present value of the construction cost?

Given: $AC = \$5,000,000$; $r \rightarrow =0.036$, $n = 2025\text{–}2020 = 5$

Then, $PV = 5,000,000/(1 + 0.036)^5$

Therefore, $PV = \frac{5000000}{(1+0.036)^5} = \$4,189,587$ (see Fig. 9.6).

The $5 million is the present value. If the project can wait until 2031, the money could be put into a bank where it earns interest or put into other projects where it generates immediate benefits. The amount that needs to be deposited into the bank to have $5 million in 2031, using a 3.6% discount rate, is only $4.2 million—the present value of the cost of the project in 2026. The same logic applies to benefits.

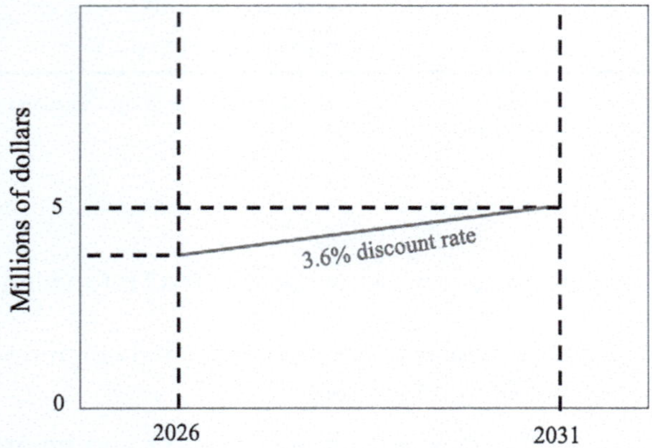

FIGURE 9.6 Discount rate for hypothetical SO_2 scrubber. *Source of funding information (2012). Benefit cost analysis for transportation projects. [Online] Available: https://www.dot.state.mn.us/planning/program/benefitcost.html.*

Thus if the only thing of importance is money, the scrubber in this example would not be installed. However, large public projects or, in this case, highly regulated utilities, must fill societal needs, including increasing the capacity of existing facilities or systems, and/or improving public health, environmental quality, safety, and aesthetics of neighborhoods, and enhancing the performance of other existing facilities or systems. In this case, large amounts of SO_2 would be emitted leading to adverse effects on human health and environmental quality, such as increased incidents of asthma and acid rain, respecitively. In fact, the benefit-cost and other alternative selection techniques (see next section) must include fines, penalties, and other monetized and nonmonetized costs of not removing the SO_2.

The analysis of a large public project is conducted in stages:

1. Planning the analysis and defining its scope.
2. Performing engineering analyses of the alternatives.
3. Calculating the present value of project costs and benefits.
4. Evaluating the results—BCA. ("Benefit-Cost Analysis - MnDOT")

Planning the BCA and performing the engineering analysis (the first two stages) require careful thought. The analysis should capture the appropriate benefits and cost differences between the base case and the identified alternatives [22]. ("Benefit-Cost Analysis - MnDOT") These first two stages are the most complicated and require the most time and effort. "The economic calculation stage is a relatively short and straightforward process." ("Benefit-Cost Analysis - MnDOT") Evaluation and interpretation of the results require judgment and experience. The process of conducting a BCA can be iterative: it may require going back to a previous stage to verify results and explore subalternatives.

9.4.3 Comparing alternatives

In most jurisdictions, the company, utility, and other regulated entities have latitude in the strategies that they employ to meet air quality standards, especially for standards that are performance oriented. Some standards are more prescriptive and technical, but even these are often open to more than one approach on how to comply [37]. This means that expenditures are seldom a one-time expense, usually requiring the comparison of alternatives with expenditures varying in amounts and spending occurring at multiple times over a project's or equipment's useful life. Not only may the expenditures vary, but also the timing of expenditures may also be different. This section addresses the selection of appropriate implementation and funding timeframes, addressing the time value of money, adjusting for prices over time, and selection of the appropriate measure of cost ("Chapter 2 - Cost Estimation: Concepts and Methodology").

9.4.3.1 Social costs

It is important to keep in mind that the calculations in this chapter generally apply to monetizable costs and benefits, often the costs borne by a private or quasipublic entity. For instance, if the entity pays for the installation and operation of an environmental pollution control technology, all these costs are considered private. Thus selection of alternatives usually rests on minimizing these private costs. Conversely, social costs are more the province of macroeconomics:

> From a regulatory standpoint, social cost represents the total burden a regulation will impose on the economy. It can be defined as the sum of all opportunity costs incurred as a result of the regulation. These opportunity costs consist of the value lost to society of all the goods and services that will not be produced and consumed if firms comply with the regulation and reallocate resources away from production activities and towards pollution abatement. To be complete, an estimate of social cost should include both the opportunity costs of current consumption that will be foregone as a result of the regulation, and also the losses that may result if the regulation reduces capital investment and thus future consumption [38].

Comparing private costs of various pollution control alternatives is more straightforward than social cost assessment because private cost comparison generally quantifies expenses incurred by a facility to purchase, finance, and operate the equipment. For example, if a governmental jurisdiction wished to encourage pollution control for a power generation industrial sector, it could provide monetary support, for example, grants, for 25% of the costs of a treatment system, for example, scrubber. The private cost for the industry would be reduced to 75% of the cost of a scrubber. Other less direct monetary support strategies have been employed, for example, eliminating or rebating sales and other taxes, which means that the private cost to the firm is lessened by the total that would otherwise need to include the tax.

The following sections stress private costs, but it is important to keep in mind that the basis for many of the costs is to provide a societal, macroeconomic value, that is, cleaner air and water. Thus the private costs lead to incremental improvements or, if the air is already clean, the maintenance of existing air quality, that is, nondeterioration [37].

Uncertainty is highest during the conceptual phases of a project. Cost estimate accuracy depends on the decision being made. Simply knowing that emissions must be reduced may begin with order-of-magnitude cost estimates for air pollution control projects. For example, chemical engineers and air pollution control engineers [39] consider the order-of-magnitude to have an error between +50% and −30%.

9.5 Proposal review

Rather than simply deciding whether to accept a proposal, it is better to decide on a functional objective and then identify ways to provide that function in the most efficient and effective manner. Both benefits and costs vary, so this analysis must find alternative means of maximizing net benefits and, similarly, minimizing costs to achieve a desired benefit. As mentioned, this is known as the BCA, which accounts for equivalent units, along with the identification and characterization of affected members of the public, This will vary by jurisdictional needs. For example, the national government may be interested in different benefits, e.g., protection of air quality in national parks, than that of state or local jurisdiction. e.g., conserving aesthetics and preserving historical values of neighborhoods.

The proposal will be reviewed to compare its expected positive and negative outcomes. Of course the benefits must substantially exceed the costs and risks. As mentioned, the review should be quantifiable wherever possible so that the proposed approach can be compared to other alternatives objectively.

Basically, the BCA used in the proposal review seeks to show whether the benefits are worth the economic investment. This can be viewed at various points in the design process.

The BCA is most valuable after a decision is made, based on risk and reliability. It is more valuable as a tool for comparing and evaluating alternatives than for deciding whether to provide a service. For example, if a risk assessment indicates that water is being contaminated, the BCA would not override the need to clean the water but would seek the best and most feasible means of doing so. The water treatment and cleanup would have to meet a risk-based or other environmental standard. Thus the project would have to provide water that meets this standard, and the risk manager would decide which of the options for doing this is most cost effective.

The BCA is a tool for maximizing benefits for a certain amount of costs, for example, the basis for a risk management or public works budget. So, the BCA needs to be more than a financial comparison. Indeed, various methods are used for comparing benefits and costs for a public project. After deciding which alternatives are feasible and viable, different techniques can be applied to compare them, including:

Net present value (NPV): NPV is the value obtained by subtracting the expected costs from the expected benefits. Projects having an NPV of less than zero should be avoided.

NPV is an expression of economic efficiency, which is useful for private cost comparisons and for estimating social costs. In the latter case, economic efficiency is the maximization of social welfare. An efficient market allows society to maximize the NPV of benefits, that is, it is the difference between social benefits and social costs over time. The efficient level of production is "Pareto optimal," that is, it is not possible to rearrange production or to reallocate goods so that someone is better off without making someone else worse off in the process. Economic efficiency merely requires that net benefits be maximized, irrespective of to whom those net benefits accrue. It does not guarantee an "equitable" or "fair" distribution of benefits among consumers and producers, or between subgroups of consumers or producers [38]. This is an example of a societal zero-sum game. It is also an indication that engineering optimization without adequate consideration of moral permissibility and justice is an incomplete optimization.

Internal rate of return (IRR): The IRR is the discount rate at which the present value of costs equals the present value of benefits. ("Mastering the Internal Rate of Return (IRR): A Comprehensive Guide to …") Alternatively, IRR is the discount rate at which NPV=0 and benefit-cost ratio (BCR)=1.0. If IRR exceeds the opportunity cost of the capital, the project is economically sound and worth pursuing [40].

9.5.1 Benefit-cost ratio

The BCR is the ratio of the present value of benefits to the present value of costs [41]. A BCR is a gross tool that may serve as a screen and is part of a larger analysis. In its simplest form, it is a ratio of benefits to cost of a project. For example, the city's air quality department is asked to participate in an analysis of a proposed landfill. The project receives the following scores:

	Score	Weight	Total
Benefits			
Water quality	3	5	15
Gas sales	1	3	3
Removal of open dumps	5	5	25
Customer satisfaction	3	4	12
Costs			
Expenses	5	3	15
Fugitive dust	2	2	4
Odors	3	2	6

The scores are weighted according to importance, so the most important factors in this case are water quality and closing dumps. Thus the benefits equal: 15 + 3 + 25 + 12=55. The costs equal: 15 + 4 + 6=25. Therefore, this benefit-to-cost ratio is 55/25=2.2. This may indicate that the landfill is acceptable. However, the scores and weightings are highly subjective. Indeed, the engineer from the air quality department was overruled on the scoring and weighting of fugitive dust (she thought it should be 5 and 5, respectively, increasing the cost side of the ratio by 21, i.e., B/C=1.2). Such ratios are notorious for justifying previously ordained projects.

As mentioned, the BCA would project and evaluate even remote possibilities. Therefore, project alternatives need to be screened. Expert judgment can be a first screen to differentiate infeasible alternatives from realistic options and compare them to a base case. The base case is not necessarily the "no action" alternative, although, in most BCAs, one or more of the other options have substantial improvements over the base case. LP and other optimization techniques can be used to quantify the economic feasibility of a project from a life-cycle/systems perspective. Public projects must also differentiate economic costs and benefits from other societal costs and benefits. As mentioned, the BCA differs from an EIA. The US Department of Transportation, for example, differentiates the two processes [42,43]. Recall that the EIA is mainly focused on economic activity, whereas the BCA must consider all relevant benefits and costs, notwithstanding whether they are monetizable. This presents a challenge, since many environmental values cannot easily be assigned values, so environmental costs may be underrepresented in either an EIA or BCA.

> In this sense, an economic impact analysis would consider the initial investment in infrastructure as a stimulus to the local economy, as opposed to a cost to the local government. Also, economic impact analyses are smaller scale than the usual BCA, which usually consider broad, economy-wide and "societal" benefits and costs. Effects in one region may be offset by losses in a neighboring region, reflecting a transfer of spending or jobs that may be a net neutral summation. Thus, the two local jurisdictions would have different economic impact results, but the BCA would not change, since its focus is larger.

9.6 Replacement cost analysis

As any manager can attest, things wear out. Replacement and repair will result; it is merely a matter of time. Often, replacement applies to a "supply," that is, something that is expected to be used and changed frequently. Repair is often seen as returning "capital equipment" to its designed efficiency.

These are semantical distinctions. Of course, any piece of a system needs to be replaced, including capital equipment. This is not always due to wear. Sometimes, the function has changed, for example, labor-intensive sorting equipment may be fine until a town's population reaches a threshold where the equipment needs to be replaced by higher-capacity equipment. Sometimes, there is a safety issue, for example, the older equipment may have met previous worker safety regulations but no longer does. Often, however, economics drives the decision, for example, the mean time to repair (MTTR) may indicate it is time to replace equipment. Any type of asset eventually loses value and must be replaced due to:

- Obsolescence: fast for information technology, slower for mechanical equipment.
- Depletion: supplies.
- Deterioration: mechanical equipment.

Replacement cost analysis (RCA)[1] uses specific terminology. The "defender" is the existing piece. The "challenger" is the best available replacement equipment. Current market value is the selling price of the defender in the marketplace.

There are a few ways to compare useful life, but a popular one is the equivalent uniform annual cost (EUAC).

The RCA is done in steps:

1. Identify alternatives (reasonable and available), at least: Defender, ... Best challenger.
2. Determine whether good defender marginal cost data are available.
3. If data are available, check whether the marginal costs are increasing.
4. If increasing, then compare defender's next-year marginal cost with challenger's EUAC. (Use Analysis Technique 1, described below)
5. If not increasing, then must find lowest EUAC for defender and compare it to the challenger's EUAC at its minimum cost life. (Use Analysis Technique 2, discussed below)
6. If defender marginal cost data not available, must find EUAC over given equipment life. (Use Analysis Technique 3; described below)
7. Compare defender's EUAC over its remaining life to the challenger's EUAC at its minimum cost life.
8. Calculate the EUAC for each value of the useful life (e.g., n=1, n=2, n=3, etc.).
9. Time period at which the EUAC is minimized is the minimum cost life (economic useful life).

[1] Unfortunately, some initialisms and acronyms are not unique in certain areas of environmental systems science. Recall that RCA is also the initialism-abbreviation for root cause analysis.

Replacement cost analysis

Assuming the conditions in Table 9.2, the engineer must decide how to address failing equipment.

Goal: Narrow replacement down to one challenger.

Approach: Evaluate each candidate challenger at its minimum cost of life.

This can be done using the various alternative evaluation approaches, that is, net present worth (NPW), as well as EUAC, IRR, or BCR.

Next, find minimum cost life. An asset can have several possible lives, depending on frequency of asset replacement. The life that minimizes the asset's EUAC is known as the asset's minimum cost life.

Example:

If initial cost (P) is \$7500, arithmetic gradient maintenance cost (G) is \$900, uniform cost (A) is \$500, and arithmetic gradient operating cost (G) is \$400, at an 8% interest rate.

There is no salvage value.

The "n" is changing depending on the year, on an arithmetic gradient, that is, it is changing each year (n=1; n=2, ...).

The 15-year spreadsheet is shown in Table 9.3:

The second column can be constructed in Excel as \$7500(A/P,8%,n). Third column: \$900(A/P,8%,n). Fourth column: \$500+400(A/P,8%,n). The fifth column is the sum of the values in columns 2, 3, and 4.

Thus the lowest total EUAC is in year 4, so this is the economic useful life, and minimum cost life is when n=4 years (see Fig. 9.7):

Next, find the year-by-year costs for keeping an asset (i.e., marginal costs). This includes loss in value of the asset for keeping it (usually per year) and the lost interest from tying up funds in this asset. The costs are directly related to the asset, for example, expenses like insurance and O&M.

TABLE 9.2 Defender and challenger example from P.R. Rosenkrantz, Cal Poly Pomona.

	Where		Compare	
	Defender marginal cost	**Defender marginal cost**	**Defender**	**Best challenger**
1	Available	Increasing	Next year's marginal cost	Equivalent uniform annual cost (EUAC) at minimum cost life
2	Available	Not increasing	EUAC at minimum cost life	EUAC at minimum cost life
3	Not available		EUAC over remaining useful life	EUAC at minimum cost life

P.R. Rosenkrantz. "Chapter 12 - Replacement." https://www.cpp.edu/~rosenkrantz/egr403site/egr403online/egr403_sv15_chapter12.ppt (accessed May 17, 2018).

(cont'd)

TABLE 9.3 Spreadsheet for equivalent uniform annual cost (*EUAC*) example from P.R. Rosenkrantz, Cal Poly Pomona.

Year	EUAC of capital recovery costs	EUAC of maintenance and repair costs	EUAC of operating costs	EUAC total	Interest rate
Initial year	–7500	0	–500		8%
Arithmetic gradient		–900	–400		
1	$8100.00	$0.00	$500.00	$8600.00	
2	$4205.77	$432.69	$692.31	$5330.77	
3	$2910.25	$853.87	$879.50	$4643.62	
4	$2264.41	$1263.56	$1061.58	$4589.55	<—MIN
5	$1878.42	$1661.82	$1238.59	$4778.84	
6	$1622.37	$2048.71	$1410.54	$5081.62	
7	$1440.54	$2424.30	$1577.47	$5442.31	
8	$1305.11	$2788.67	$1739.41	$5833.19	
9	$1200.60	$3141.93	$1896.41	$6238.94	
10	$1117.72	$3484.18	$2048.53	$6650.43	
11	$1050.57	$3815.55	$2195.80	$7061.93	
12	$995.21	$4136.17	$2338.30	$7469.68	
13	$948.91	$4446.19	$2476.08	$7871.18	
14	$909.73	$4745.75	$2609.22	$8264.69	
15	$876.22	$5035.01	$2737.78	$8649.02	

P.R. Rosenkrantz. "Chapter 12 - Replacement." https://www.cpp.edu/~rosenkrantz/egr403site/egr403online/egr403_sv15_chapter12.ppt (accessed May 17, 2018).

(cont'd)

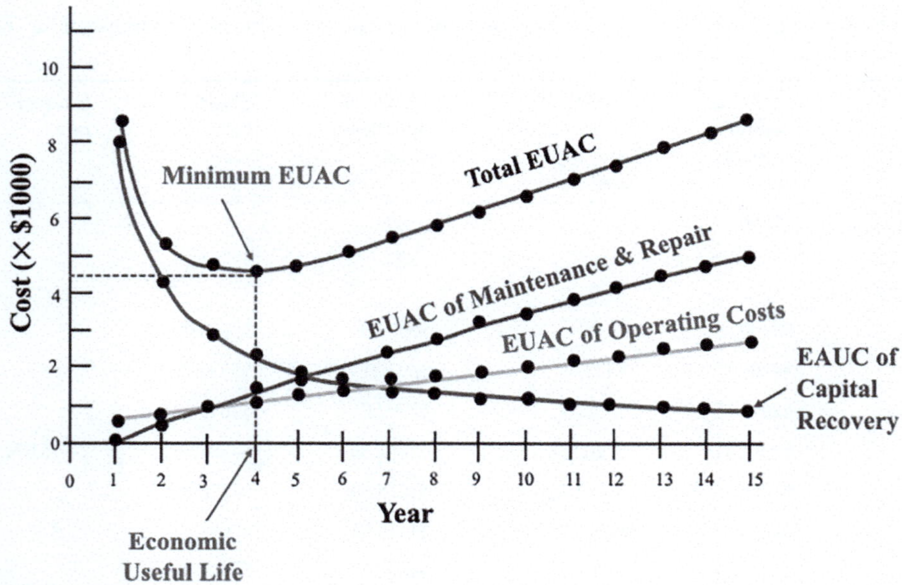

FIGURE 9.7 Minimum equivalent uniform annual cost (*EUAC*) used to find the economic useful life in example from P.R. Rosenkrantz, California Polytechnic—Pomona. *P.R. Rosenkrantz. "Chapter 12 - Replacement." https://www.cpp.edu/~rosenkrantz/egr403site/egr403online/egr403_sv15_chapter12.ppt (accessed May 17, 2018).*

Example of marginal cost comparison:
New item:

Expense	Cost	Years	Incremental cost thereafter
Investment cost	$25,000.00	0	$ -
Annual O&M cost	$2000.00	1	$500.00
Annual insurance cost	$5000.00	3	$1500.00
Useful life (years)	7		
MARR	15%		

(cont'd)

Year	Market value	Loss in market value	Foregone interest	O&M cost	Insurance cost	Total marginal cost
0	$25,000.00					
1	$18,000.00	$7000.00	$3750.00	$2000.00	$5000.00	$17,750.00
2	$13,000.00	$5000.00	$2700.00	$2500.00	$5000.00	$15,200.00
3	$9000.00	$4000.00	$1950.00	$3000.00	$5000.00	$13,950.00
4	$6000.00	$3000.00	$1350.00	$3500.00	$6500.00	$14,350.00
5	$4000.00	$2000.00	$900.00	$4000.00	$8000.00	$14,900.00
6	$3000.00	$1000.00	$600.00	$4500.00	$9500.00	$15,600.00
7	$2500.00	$500.00	$450.00	$5000.00	$11,000.00	$16,950.00

Defender:

Expense	Cost	Years	Incremental cost thereafter
Investment cost	$15,000.00	0	$ -
Annual O&M cost	$10,000.00	1	$1500.00
Annual insurance cost	$ -		$ -
Useful life (years)	5		
MARR	15%		

Year	Market value	Loss in market value	Foregone interest	O&M cost	Insurance cost	Total marginal cost
0	$15,000.00					
1	$14,000.00	$1000.00	$2250.00	$10,000.00	$ -	$13,250.00
2	$13,000.00	$1000.00	$2100.00	$11,500.00	$ -	$14,600.00
3	$12,000.00	$1000.00	$1950.00	$13,000.00	$ -	$15,950.00
4	$11,000.00	$1000.00	$1800.00	$14,500.00	$ -	$17,300.00
5	$10,000.00	$1000.00	$1650.00	$16,000.00	$ -	$18,650.00
6						
7						

(cont'd)

In this example, the marginal costs of the defender are constantly increasing, but the marginal costs of the challenger decrease for three years, then start to increase. By year 3, the challenger's marginal costs become less than the defender's and stay that way until the end of the defender's useful life (year 5).

9.6.1 Replacement cost analysis techniques

Technique 1: If the defender's marginal costs are increasing, replace when the marginal costs of ownership exceed the EUAC of the challenger. This is an appropriate approach when replacement repeatability assumptions hold:

- Best challenger is available for all subsequent years, and its costs will remain unchanged.
- The time period of service is considered to be infinite.
- If these are too restrictive, can assume a more realistic, known, and limited project life.

Technique 2: If the replacement repeatability holds, compare the EUAC of the defender asset at its minimum cost life against the EUAC of the challenger at its minimum cost life.

Technique 3: Compare the defender's EUAC over the stated life against the minimum EUAC of the challenger. There may be a problem in defining first costs of the defender and challenger. Trade-in value is not suitable; market value is appropriate.

Any cost analysis is incomplete if it neglects tax. Thus all relevant taxes will affect the remaining life of the defender, the economic life of the challenger, and defender versus challenger comparisons. After-tax replacement depends on the depreciation under the Modified Accelerated Cost Recovery System (MACRS) and assets' market over time [44].

9.7 Life-cycle comparisons

The project evaluation techniques so far have helped the engineer select from finite alternatives. Another aspect of evaluation is improving existing systems to provide the needed function or to eliminate problems that weaken that alternative. In life cycle assessment (LCA), alternatives assessment is a formal technique to characterize the hazards of substances based on a full range of human health and environmental information [45]. Changes in the life cycle made based on these assessments ameliorate the potential for unintended consequences *a priori*.

The current approach for alternatives assessment shown in Table 9.4 goes beyond hazard alternatives to address benefits and utility provided by a risk management option, that is, steps in LCA stages may require helium (He), which is becoming less available. A thorough alternative assessment may identify steps where a different, more renewable chemical or even use a nonchemical technique to perform the same function. Thus the steps in Table 35.1 may be applied to other substitutions and replacements, for example, energy use, labor and skill mix, and social justice.

TABLE 9.4 Key steps in alternatives assessment for substituting toxic chemicals in the life cycle of a process.

Process	Action	Application to finite resources
Step 1: Determine feasibility	Consider whether alternatives: **1.** are commercially available and cost effective; **2.** have the potential for an improved health and environmental profile; **3.** are likely to result in lasting change; and **4.** to what extent stakeholders are interested.	Consider whether the goods and services can be provided with lesser amounts of these resources.
Step 2: Collect information	Find data about and consider: **1.** how well characterized the possible alternatives are; **2.** how much is known about the chemical manufacturing process; **3.** the range of functional uses that the chemical serves; **4.** the feedstock of contaminants and residuals from the production process; and **5.** the work of others who are exploring alternatives for the chemical of concern, similar chemicals, and functional uses. Based on analysis of this information and preliminary stakeholder consultations, develop a proposed project scope and an approach for developing the alternatives assessment.	Same, except collect market and production information.
Step 3: Convene stakeholders	Stakeholders are drawn from entire supply-chain and all life-cycle stages of the chemical of concern. Involvement throughout the project helps to ensure that stakeholders contribute to, understand, and support the outcome, enhancing credibility and promoting adoption of the safer alternatives. Typical stakeholders include: chemical manufacturers; product manufacturers; nongovernmental organizations; government agencies; academics; retailers; consumers; and waste and recycling companies. Chemical and technology innovators are critical members of the group.	Same process, but include additional stakeholders, for example, from similar industries.
Step 4: Identify viable alternatives	Collect information about viability of a range of potential alternatives. The focus is on finding alternatives that are functional with minimum disruption to the manufacturing process. To identify the most likely alternatives, may also include viability demonstrations by chemical and product manufacturers.	Same.
Step 5: Conduct hazard assessment	Based on the best data that are available from the literature or can be modeled, assign an adjective descriptor of hazard concern level of high, moderate, or low for each alternative across a range of endpoints, including acute and repeated dose toxicity, carcinogenicity and mutagenicity, reproductive and developmental toxicity, neurotoxicity, sensitization and irritation, acute and chronic aquatic toxicity, and persistence and bioaccumulation. In addition, we provide a qualitative description of potential endocrine activity.	Rather than an assessment solely of potential hazards, include potential substitutions and utility of these substitutions in terms of providing the same goods and services as current processes.

From [46]; data for two left columns from: [47]; right column [48,49].

Alternative assessment and other life-cycle approaches could supplant many of the single-approach, unsustainable methods with more systematic, sustainable solutions. Note that Table 9.4 explicitly calls for stakeholder involvement and risk communication. For example, to go beyond treating a water pollution problem, inquiring the expertise of researchers and practitioners in governments, universities, research institutes, and other public and private knowledge centers will help to identify ways to eliminate the use of toxic chemicals in manufacturing so that they never reach the waste stream [50].

Although expert judgment is almost always applied at the outset of project evaluation, for example, deciding which alternatives are not worthy of further scientific investigation, it is also useful as the evaluation continues. Project evaluation can be a protracted, lengthy, and iterative process. Indeed, there may come a point where decision makers ask the project engineer to "settle" on a less-than-optimal approach. In addition, after substantial time and psychic investment in narrowing down the alternative list to a few, well-understood options, it may be difficult to be completely objective. This may be a good time to bring in someone "cold" who can view the options more systematically and advise the risk manager as to whether the original functions and objectives are being met.

Comparing environmental projects usually occurs in two distinct domains: scientific credibility and economic feasibility. The chapters up to this point have addressed the former. This chapter has considered the ways to do the latter and to incorporate economic efficiencies with considerations of risk, reliability, and resilience. Economic acceptability is useless if a project fails to meet standards of performance based on public health, environmental quality, and societal acceptance.

9.10 Conclusions

Scientific soundness is an inviolable aspect of any emerging technology. However, the technology must also be feasible financially and managerially. This chapter introduced several tools and techniques for evaluating the technical and financial feasibility of existing and proposed projects.

References

[1] D.A. Vallero, Environmental Systems Science: Theory and Practical Applications, first ed, Elsevier, Amsterdam, NV, (2020).

[2] U. Sadana, A. Chenreddy, E. Delage, A. Forel, E. Frejinger, T. Vidal, A survey of contextual optimization methods for decision-making under uncertainty, European Journal of Operational Research 320 (2) (2025) 271–289.

[3] J. Noh, T. Badloe, C. Lee, J. Yun, S. So, J. Rho, Inverse design meets nanophotonics: from computational optimization to artificial neural network, Intelligent Nanotechnology (2023) 3–32.

[4] D.A. Vallero, Chapter 11 - Economics and feasibility assessments, in: D.A. Vallero (Ed.), Environmental Systems Science, Elsevier, (2021), pp. 509–542.

[5] M.M. Bugge, A.M. Fevolden, A. Klitkou, Governance for system optimization and system change: the case of urban waste, Research Policy 48 (4) (2019) 1076–1090.

[6] R. Tan, K. Aviso, D. Ng, Optimization models for financing innovations in green energy technologies, Renewable and Sustainable Energy Reviews 113 (2019) 109258.

[7] M.J. Kochenderfer, T.A. Wheeler, Algorithms for Optimization, Mit Press, (2019).
[8] P.E. Gill, W. Murray, M.H. Wright, Practical Optimization. SIAM, 2019.
[9] S.S. Rao, Engineering Optimization: Theory and Practice, John Wiley & Sons, (2019).
[10] S.B. Rajulapati, Optimization of Process Parameters for Amylase Production from Vegetable Wastes, 2014.
[11] D.P. German, M.N. Weintraub, A.S. Grandy, C.L. Lauber, Z.L. Rinkes, S.D. Allison, Optimization of hydrolytic and oxidative enzyme methods for ecosystem studies, Soil Biology and Biochemistry 43 (7) (2011) 1387–1397.
[12] S. Koziel, X.-S. Yang, Computational Optimization, Methods and Algorithms, Springer, (2011).
[13] X.-S. Yang, S. Koziel, Computational Optimization and Applications in Engineering and Industry, Springer Science & Business Media, (2011).
[14] J.D. Neelin, A. Bracco, H. Luo, J.C. McWilliams, J.E. Meyerson, Considerations for parameter optimization and sensitivity in climate models, Proceedings of the National Academy of Sciences 107 (50) (2010) 21349–21354.
[15] J. Liu, J.-X Chang, W.-G Zhang, Groundwater level dynamic prediction based on chaos optimization and support vector machine, in: 2009 Third International Conference on Genetic and Evolutionary Computing, IEEE, 2009, pp. 39–43.
[16] D.J. Brus, G.B. Heuvelink, Optimization of sample patterns for universal kriging of environmental variables, Geoderma 138 (1) (2007) 86–95.
[17] G. Zeng, R. Jiang, G. Huang, M. Xu, J. Li, Optimization of wastewater treatment alternative selection by hierarchy grey relational analysis, Journal of Environmental Management 82 (2) (2007) 250–259.
[18] D.C. McKinney, M.-D. Lin, Pump-and-treat ground-water remediation system optimization, Journal of Water Resources Planning and Management 122 (2) (1996) 128–136.
[19] State implementation plans. [Online]. 2009. Available from: http://www.dec.ny.gov/chemical/8403.
[20] Uniited Kingdom Enviornment Agency, Pollution Inventory reporting form in: Yearly Releases to Air, Controlled Waters, Land and Off-Site. Transfers in Wastewater and Waste, London, UK, December 2012.. [Online]. Available from: https://assets.publishing.service.gov.uk/government/uploads/system/uploads/attachment_data/file/296868/LIT_7654_2a3780.pdf. [Accessed 13th October 2020].
[21] Benefit-cost analysis guidance for TIGER and INFRA applications. [Online]. 2017. Available from: https://www.transportation.gov/sites/dot.gov/files/docs/mission/office-policy/transportation-policy/284031/benefit-cost-analysis-guidance-2017_1.pdf.
[22] Benefit cost analysis for transportation projects. [Online] 2012. Available from: https://www.dot.state.mn.us/planning/program/benefitcost.html.
[23] R. Schmalensee, R.N. Stavins, Policy evolution under the clean air act, Journal of Economic Perspectives 33 (4) (2019) 27–50.
[24] A.D. Ellerman, P. Joskow, D. Harrison, Emissions Trading in the US Experience, Lessons, and Considerations for Greenhouse Gases, Pew Center on Global Climate Change, Arlington, VA, (2003).
[25] U.S. Environmental Protection Agency, How do emissions trading programs work? USEPA. Available from: https://www.epa.gov/emissions-trading/how-do-emissions-trading-programs-work (accessed 25.03.26).
[26] R. Scheffe, Urban Airshed Model study of five cities: volume 1. Summary report. Environmental Protection Agency Research, Triangle Park, NC (USA). Office of ..., 1990.
[27] N. D. o. A. Q. Clark County, State implementation plan for PM. Appendix K: Rollback methodology, 2001. [Online]. Available from: http://www.clarkcountynv.gov/Depts/AirQuality/Documents/Planning/SIP/PM10/App_K–Rollback_Methodology.pdf (accessed 26.11.13).
[28] 70(216) 40 CFR Part 51, Revision to the guideline on air quality models: adoption of a preferred general purpose (flat and complex terrain) dispersion model and other revisions: final rule. November 9, 2005.
[29] T.M. Letcher, D. Vallero, Excerpt from: Waste: A Handbook for Management, second ed., Academic Press, (2019).
[30] J.S. Harkness, B. Sulkin, A. Vengosh, Evidence for coal ash ponds leaking in the southeastern United States, Environmental Science & Technology 50 (12) (2016) 6583–6592.
[31] U.S. Environmental Protection Agency, Coal ash (coal combustion residuals, or CCR). Available from: https://www.epa.gov/coalash (accessed 08.03.18).
[32] U.S. Environmental Protection Agency, "Attachment B: Water quality-based effluent limits - coal combustion waste impoundments, "[{U.S. Environmental Protection Agency, 2017 #978]} (accessed 07.03.2018).

[33] U.S. Fish and Wildlife Service, Natural resource damage assessment and restoration for the Dan River coal ash spill. Available from: https://files.nc.gov/ncdeq/Coal%20Ash/documents/Coal%20Ash/20140609%20FINAL%20Dan%20River%20cooperative%20NRDAR%20factsheet.pdf (accessed 08.03.18).
[34] T. Brown, Engineering Economics and Economic Design for Process Engineers, CRC Press, (2016).
[35] J. Wittwer, Discount factor table for excel: summary of discount factor formulas for TVM calculations in excel®. AVailable from: https://www.vertex42.com/ExcelArticles/discount-factors.html (accessed 17.05.18).
[36] Microsoft Support Office, Financial functions (reference). Available from: https://support.office.com/en-us/article/financial-functions-reference-5658d81e-6035-4f24-89c1-fbf124c2b1d8?ui=en-US&rs=en-US&ad=US (accessed 17.05.18).
[37] J.L. Sorrels, T.G. Walton, Cost estimation: concepts and methodology, in: EPA Air Pollution Control Cost Manual, 27711. Research Triangle Park, North Carolina: Office of Air Quality Planning and Standards, U.S. Environmental Protection Agency, 2017, ch. 2.
[38] E. P. A. U.S, "Guidelines for preparing economic analyses," EPA 240-R-00-003. Washington, DC, US Environmental Protection Agency, 2014. Accessed: August 27, 2018. [Online]. Available from: https://www.epa.gov/sites/production/files/2017-08/documents/ee-0568-50.pdf (accessed 27.08.18)..
[39] K.B. Schnelle Jr, R.F. Dunn, M.E. Ternes, Air Pollution Control Technology Handbook, CRC Press, Boca Raton, Florida, (2015).
[40] D.A. Vallero, Air Pollution Calculations: Quantifying Pollutant Formation, Transport, Transformation, Fate and Risks, Elsevier, (2019).
[41] FasterCapital, Benefit Cost Ratio: BCR: BCR: How to Compare the Benefits and Costs of Your Investments, ed. Nad Al Sheba, Dubai, 2025.
[42] Benefit-cost analysis guidance for TIGER and INFRA applications. [Online] 2017. Available from: https://www.transportation.gov/sites/dot.gov/files/docs/mission/office-policy/transportation-policy/284031/benefit-cost-analysis-guidance-2017_1.pdf.
[43] U. Federal Railroad Administration and D. o. Transportation, Benefit-Cost Analysis Guidance for Rail Projects, ed: US Department of Transportation, Washington DC, 2016.
[44] P.R. Rosenkrantz. Chapter 12 - Replacement. Available from: https://www.cpp.edu/~rosenkrantz/egr403site/egr403online/egr403_sv15_chapter12.ppt (accessed 17.05.18).
[45] C.P. Weis, The value of alternatives assessment, Environmental Health Perspectives 124 (2016) A40.
[46] D. Vallero, Translating Diverse Environmental Data Into Reliable Information: How to Coordinate Evidence from Different Sources, Academic Press, (2017).
[47] U.S. Environmental Protection Agency, Design for the Environment Alternatives Assessments. Available from: https://www.epa.gov/saferchoice/design-environment-alternatives-assessments (accessed 15.05.17).
[48] D.A. Vallero, T.M. Letcher, Engineering risks and failures: lessons learned from environmental disasters, Leadership and Management in Engineering 12 (4) (2012) 199–209.
[49] D.A. Vallero, T.M. Letcher, Unraveling Environmental Disasters, Newnes, (2012).
[50] M.D. Wood, K. Plourde, S. Larkin, P.P. Egeghy, A.J. Williams, V. Zemba, et al., Advances on a decision analytic approach to exposure-based chemical prioritization, Risk Analysis 40 (1) (2020) 83–96.

CHAPTER

10

Risk-based optimization☆

OUTLINE

Rapid development and deployment of powerful generative AI models comes with environmental consequences, including increased electricity demand and water consumption. **MIT News (2025) [2].**

10.1 Introduction

Determining whether a new technology or product is acceptable and worthwhile is based on its likely benefits, costs, and risks. Thus, any optimization must include drivers and

☆ This chapter draws heavily from Chapter 17 of [1] D. Vallero, *Fundamentals of Water Pollution: Quantifying Pollutant Formation, Transport, Transformation, Fate and Risks*, 1st ed. Amsterdam, NV: Elsevier, 2024.

Technical and Ethical Optimization: Challenges and Opportunities for the Modern Engineer and Scientist
DOI: https://doi.org/10.1016/B978-0-443-44925-3.00020-X

constraints regarding risks to human health and the environment [3–5]. A good way to discuss risk is to consider the introduction of a new chemical compound, such as a pesticide or nanomaterial, into the marketplace.

The likelihood of an adverse outcome is expressed as a risk that depends on both the likelihood of a hazard and the likelihood of an exposure. From an environmental perspective, the substances of greatest concern are those with high potential toxicity and high potential exposure profiles. Indeed, even when the hazard remains the same, threats with higher exposure potential will lead to higher risk (see far right of Fig. 10.1). When hazard concentrations and exposure potential intersect, populations are exposed at concentrations most likely to lead to adverse outcomes. Similarly, exposure calculations are linked to risk estimates by metrics based on toxicity thresholds that are modified by uncertainty factors (UFs). For chemical compounds, the uncertainties are principally due to the differences between results found in animal testing and expected outcomes in human population [1].

To assess the impact of a pollution event on environmental quality, the event is characterized according to its potential to evoke an adverse outcome. Evidence-based risk assessments provide estimates of the amount of harm that could result from an exposure to a pollutant. A retrospective risk assessment, as part of the root cause analysis following an adverse event, includes every potential stressor and outcome that can be determined from data. A prospective risk assessment follows the root cause analysis to identify actual and

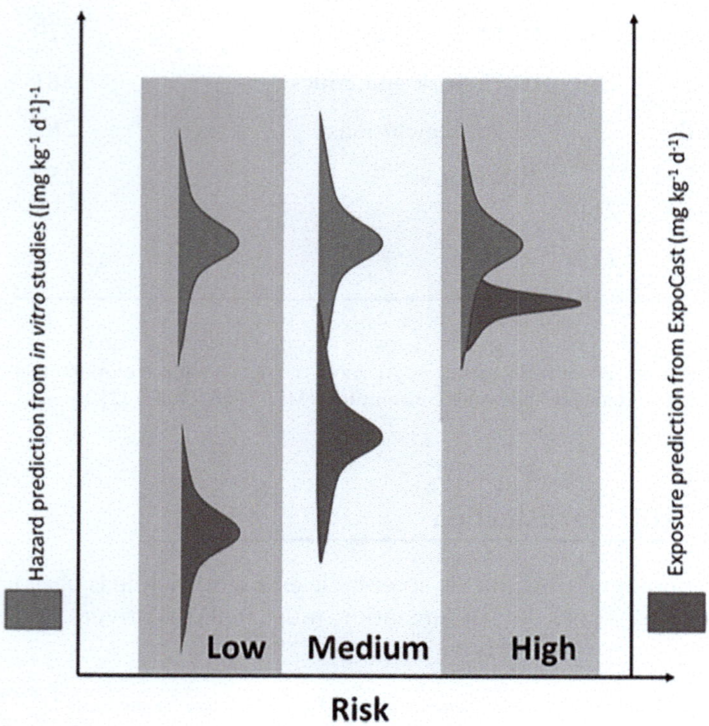

FIGURE 10.1 Combining hazard and exposure yields risk. In this case, the potential for human health risk is extrapolated from constant toxicity and increasing exposure potential. *Based on R.S. Judson et al., Estimating toxicity-related biological pathway altering doses for high-throughput chemical risk assessment, Chemical Research in Toxicology 24 (4) (2011) 451–462 [6].*

potential causes of the adverse outcome. Mathematically, the risk (R) is commonly expressed as the product of a stressor's hazard (H) and exposure (E):

$$R = H \times E \tag{10.1}$$

Hazard is an inherent trait of a substance or microbe that expresses a potential adverse outcome. The determination of this inherency is based on peer-reviewed articles, government reports, and other scientific studies. The outcome can occur relatively rapidly, for example, an outcome that occurs within one year of exposure is known as an acute hazard. Examples include exposure to high concentrations of carbon monoxide (CO), leading to cardiovascular and neurological failures after a few minutes [7–9], or drinking water contaminated with the cryptosporidium protozoan that infects humans and other vertebrates, causing acute gastroenteritis a few days after exposure. Hazard may also occur long after exposure or after continuous exposure over time, which is known as a chronic hazard, such as neurotoxicity from exposure to the metal lead or carcinogenicity from exposure to inorganic substances like asbestos or organic substances like benzene. The same hazard agent may be both an acute and a chronic hazard, depending on the concentration and exposure. For example, dimethyl mercury is acutely toxic at very low concentrations [10] and lethal at doses of approximately 5 mg kg^{-1} body weight [11,12]. However, all forms of mercury (Hg) are also chronically toxic at even lower concentrations over longer durations of exposure [13–20].

Also, as mentioned, CO at high concentrations is an acute hazard, but it is also a chronic hazard at much lower concentrations due to long-term exposure, such as tobacco smoking or side-stream (environmental) exposure to tobacco smoke for several years [21–23].

Although most environmental scientists would agree that Eq. (10.1) captures the general sense of risk, there is no unanimously accepted definition of risk within the broader scientific community. Most definitions include an adverse outcome and the probability of it occurring. Most definitions include both the probability of an event and the consequences that could result from that event, such as nuclear accidents or terrorist attacks. As evidence [24,25]:

> In most formal discussions of risk assessment, risk is defined as the product of the probability that a certain event might occur … and the consequences that could result from such an event. The probability side of the equation is basically a combination of threats and vulnerabilities.

There is variation within engineering and technical professions regarding the definition of risk, with greater emphasis recently placed on the magnitude of the adverse consequence proposed [26]:

> Different events … can have adverse effects on human life, health, property, or the environment, and consequently constitute a risk. The concept of risk can be … considered as the product between the magnitude of such adverse effects, expressed numerically as the number of deaths, percentage increase in cancer cases, property value loss, and so on, and the likelihood that the event causing them will occur or, alternatively, the subsequent likelihood that the consequences themselves will occur…. Risk r will simply be expressed as the product of magnitude m and likelihood l, so with the data coming from risk analysis n risks $r_i = m_i l_i$ can be calculated."

This connotation of risk requires a choice as to whether to include the probability of the event or the consequence of the event. In many environmental risk assessments, it is the consequence, such as the added cancer cases, the number of deaths (mortality rates), or falling below a threshold number of an endangered species that is included in the risk equation. Also, the shorthand in this text and numerous other environmental risk documents is to present risk as a unitless value, for example, one in a million or 10^{-6} cases out of a population of one million. Even in these cases, the consequence is understood. Although the risk is stated as a unitless fraction, such as a cancer risk of 10^{-6}, that numerical expression implies that the units, in fact, are the number of cancer cases, that is, one added case per million people exposed. Similarly, the National Research Council's (NRC) Committee on Risk Perception and Communication has defined risk as [27]:

> ... the product of a measure of the size of the hazard and its probability of occurrence. Regardless of how numerical estimates are made the essence of the distinction between hazard and risk is that "risk" takes probability explicitly into account. ("Understanding Hazards and Risks")

This definition adds the hazard and its magnitude to the probability of harm [27]. Furthermore, the NRC's Committee on Risk Characterization defines risk as [28]:

> A concept used to give meaning to things, forces, or circumstances that pose danger to people or to what they value. Descriptions of risk are typically stated in terms of the likelihood of harm or loss from a *hazard* and usually include: an identification of what is "at risk" and may be harmed or lost (e.g., health of human beings or an ecosystem, personal property, quality of life, ability to carry on an economic activity); the hazard that may occasion this loss; and a judgment about the likelihood that harm will occur.

Both the type of adverse outcome and its probability of occurrence determine the level of risk [29]:

> The term risk usually expresses not only the potential for an undesired consequence but also how probable it is that such a consequence will occur.... A mathematical definition of risk commonly found in the literature is:

$$\text{Risk (Consequence/unit time)} = \text{Frequency(event/unit time)} \times \text{Magnitude(Consequence/event)}. \tag{10.2}$$

Thus there are several quantitative expressions of risk. All express a probability and either explicitly or implicitly convey a consequence.

To ascertain possible risks posed during and after an event, the first step in a retrospective risk assessment is to identify a general hazard and then determine the exposure scenarios that occurred. As mentioned, this can be followed by a prospective risk assessment,whichassesses scenarios of every possible hazard and exposure that were identified in the retrospective assess that could lead to the effect in the future. Risk, then, is estimated based on the severity of the effect and its likelihood of occurring in that scenario. This combination of each hazard and the exposure likelihood represents the harm to ecological conditions and human health in the root cause analysis, and the cumulative risk of pending future adverse events in the prospective risk assessment [30,31].

The relationship between the severity and probability of a risk follows a general equation:

$$R = f(S,\ P) \tag{10.3}$$

where risk (R) is a function (f) of the severity (S) and the probability (P) of harm. The risk equation can be simplified to be a product of severity and probability:

$$R = S \times P \tag{10.4}$$

Environmental and biomedical ethics can converge and be scaled to address not only the technical aspects of environmental pollution but also the ethical aspects. For example, physicians, engineers, and research scientists each address the problems posed by lead and mercury pollution, but at different scales and focuses. The physician must give primacy to the patient's wellbeing. The engineer must hold paramount the safety, health, and welfare of the public. The researcher must respect and ensure the consent of any subject. The professional and research ethical constructs can be interwoven within the environmental and biomedical scientific communities, which must advance their individual sciences while increasingly taking a systems view to incorporate numerous perspectives. As the "bright lines" between disciplines are blurring, so must the ethical canons, that is, the need to move toward commensurate "systems ethics" and "life cycle ethics" to address environmental pollution and its impacts, which range from nanoparticles to global climate change.

Ethical decision-making in medicine, public health, and environmental protection must combine technical and ethical factors. It utilizes multiplex optimization or benchmarking, where only certain outcomes are acceptable. A technically acceptable outcome may be ethically unacceptable, and an ethically acceptable outcome may be technically and fiscally infeasible. The tools needed to evaluate the health and environmental benefits and risks share aspects of most decision-support tools. This means that new decision-support tools must be employed to consider risks and costs over the life of the technology and beyond.

10.2 Risk metrics

One metric shared by medicine and environmental professionals is whether a chemical, biological, or physical agent poses or could pose an unacceptable risk. Risk is the likelihood of negative outcomes. Indeed, the concept of environmental pollution is one of risk. The same substance can be essential in one scenario but a pollutant in another, depending on the risk. For example, there has been much recent debate in the United States about whether carbon dioxide (CO_2) should be designated a criteria pollutant under the Clean Air Act (CAA). A criteria pollutant is one considered harmful to public health and the environment. The CAA identifies two types of national ambient air quality standards:

1. Primary standards provide public health protection, including the health of "sensitive" populations such as asthmatics, children, and the elderly.
2. Secondary standards provide public welfare protection, including protection against decreased visibility and damage to animals, crops, vegetation, and buildings [32].

Unlike lead (Pb), CO, sulfur dioxide, and the other criteria pollutants, CO_2 is necessary for photosynthesis and is the product of respiration. Those opposed to including CO_2 on the list of criteria pollutants argue that it is not "inherently" toxic and is indeed essential.

Those in favor of including CO_2 on the criteria pollutant list argue that it is the product of complete combustion and, as such, is a surrogate for fossil fuel combustion, which represents the anthropogenic contribution of global greenhouse gases to the atmosphere.

The risks are displaced temporally and can seem isolated from their causative agents. The prototypical cause-and-effect paradigm of environmental pollution is violated. That is, risk is a function of exposure and hazard. Typically, society recognizes certain agents as inherently toxic, that is, carcinogens, endocrine disruptors, neurotoxins, and the like. However, the risk of many substances can only be defined within a given scenario. Chromium, for example, is inherently toxic only in certain oxidation states, for example, Cr^{6+} is a carcinogen, while Cr^{3+} is essential and might have been in your breakfast cereal this morning.

Complex, large-scale, and potentially irreversible scenarios like climate change may lead to environmental and health effects due to a series of events and outcomes that may, in themselves, be unrecognizable without sophisticated equipment or, if identified, may appear to be innocuous.

An increase in atmospheric concentrations of chemical compounds with relatively low toxicity is unknowingly acting as a greenhouse gas warning. The warming from an absolute perspective may be less problematic than the contingent outcomes its event tree elicits. One of the considerations of global climate change is the extent to which the release of environmental pollutants may create extreme heat events. Heat waves are generally characterized by several days of warm, stable air masses, with daytime temperatures >32°C and consecutive nighttime temperatures above normal. If climate change does indeed increase the number and severity of heat waves in the temperate zones, this will result in greater cardiovascular effects.

Heat waves are currently the most deadly, weather-related exposure in the United States, leading to a greater number of deaths annually than those attributed to hurricanes, tornadoes, floods, and earthquakes combined. Like direct environmental pollutant exposures, heat-sensitive subpopulations include children and the elderly. Climate change will bring more heat waves to the United States. Increases in the number of people living in cities, as well as population aging, will further increase heat-related health risks. Studies suggest that if current emissions hold steady, excess heat-related deaths in the United States may increase from the present average of about 700 each year to as many as 5000 per year by 2050 [33]. Similar conditions are likely to occur globally.

Thus, among the important considerations of global climate change is the extent to which the release of environmental pollutants is increasing the frequency and severity of heat events. If climate change does indeed increase the number and severity of heat waves in the temperate zones, this will be met with greater cardiovascular effects.

The gradual buildup leads to heat islands in urban areas that increase ambient temperatures. When temperatures pass a threshold, the incidence of cardiovascular and respiratory stress results, especially in sensitive groups, like the elderly, pregnant women and their unborn children, and people with asthma. These events lead to increased hospitalizations and/or mortality. In addition, these complex problems often include psychosocial and ethical considerations. For example, any increase in the number and severity of other events brought on by climate change, such as flooding, may also increase psychological stress and concomitant cardiovascular response.

Often, it is tempting to focus entirely on the few variables that are obvious and expected. In systems sciences, attention is also given to the more ambiguous variables that are difficult to quantify are missed or dismissed but are indeed working in the background. The above example is merely a snapshot of the contingencies of the heat island effect. Indeed, not all events and outcomes move in the same direction. For example, respiratory diseases are often lumped together as indicators of adverse health outcomes, for example, asthma and chronic obstructive pulmonary disease (COPD). In fact, COPD exacerbations tend to be highest in the winter [34], so although the heat island effects may indeed harm COPD patients, other factors could moderate the response, for example, if influenza and other viral and bacterial disease rates fall in warmer weather. The increase in exacerbations with cold temperatures is complex and likely due to numerous confounders and third variables, for example, there are strong associations with increases in viral infections that act as a trigger for COPD problems [36]. Thus, although the heat stress may indeed affect COPD-related problems, these may be obscured by other seasonal factors.

Contingent events and outcomes can be envisioned as chains of events and outcomes. For example, natural and anthropogenic factors first lead to warmer atmospheric and oceanic temperatures. This, in turn, alters the hydrological cycle. These changes over time can precipitate changes in ecosystems, such as decreased water retention capacity due to loss or changes in the root systems of vegetation. This capacity deficiency affects human safety with increased flooding frequencies and severity, as well as indirectly affects human health with the contingent loss of agricultural productivity and diminished food supply. Climate change can alter seasonality of some allergenic species and disease vectors. Insect and other vector populations could lead to increases in the number and severity of vector-borne diseases. These include changes in the spatial and temporal distributions of malaria, dengue, tick-borne diseases, cholera, and some other diarrheal diseases [35]. These diseases can be expected to increase cardiovascular and other stress in human populations. Thus, opportunistic diseases brought on by environmental change that is induced by environmental pollutants (e.g., greenhouse gases), can result in increased cardiovascular effects in populations [36].

The indirect health effects demonstrate the complexity of causal links between emissions of environmental pollutants and health and ecosystem wellbeing. Of course, this applies not only to cardiovascular effects but also to most other adverse health and ecological outcomes [37].

10.3 Acceptable risk

Societal expectations of acceptable risk are mandated by codified regulations, rules, standards, and specifications, such as health codes and regulations, zoning and building codes and regulations, principles of professional engineering and medical practice, criteria in design guidebooks, and standards promulgated by international agencies, for example, ISO, or the International Organization for Standardization, and national standard-setting bodies, for example, the National Institute for Standards and Testing [38]. These directives are enforced using quantitative or descriptive data. They may be ordinal, for example, from survey data, or ratio-interval, for example, from actual measurements.

The data available for scientific workflows, dashboards, and other decision tools may be retrieved from differing venues but may be equally useful for informatics. The same agent may be regulated differently in a biomedical setting than in an environmental setting. For example, genetically modified organisms (GMOs) are controlled to varying degrees by institutes of biomedical sciences, such as the American Medical Association and regulatory agencies, whereas food safety and environmental agencies, such as the US Food and Drug Administration, the US Department of Agriculture, and the US Environmental Protection Agency (EPA), and their respective state counterpart agencies are responsible for new biotechnologies in their respective areas. The biomedical research concern is often whether an agent can be contained within the research facility, whereas the environmental concern is whether the GMO, for example, an oil-eating bacterium, can be contained within the service area (e.g., the extent of an oil spill). Thus, if the focus of concern is GMOs in the environment, all these data sources should be considered candidates for data for informatic review according to the quality of the data and metadata.

Emerging biotechnologies and data[1]

The growing interest in so-called bioaerosols and gene flow by environmental scientists illustrates the need for an open-minded yet careful approach to data acquisition. Bioaerosols include any airborne organisms or their genetic material-bearing substances, such as cysts, seeds, spores, and pollen. In addition, there is concern about the atmospheric transport of these materials that can lead to horizontal diffusion of the genetic materials from genetically modified organisms (GMOs), that is, gene flow. Gene flow is an example of the biotechnological downstream risks, as evidenced by proteins from the common soil bacteria *Bacillus thuringiensis* (Bt). In Europe and North America, regulators generally require reporting of a GMO before it is introduced into the marketplace. Interestingly, in the United States, a newly modified organism is something akin to a new chemical under the Toxic Substances Control Act. In the United States, the data requirements include product characterization, mammalian toxicity, allergenicity potential, effects on nontarget organisms, environmental fate, and for the Bt products, insect resistance management to prevent losing use of both the microbial sprays and the Bt plant-incorporated protectants (PIPs). A transgene is an exogenous gene that has been introduced into the genome of another organism, and a transgenic species is one whose genome has been genetically altered. For instance, if biotechnology is used as a PIP, the movement of transgenes from a host plant into weeds and other crops presents a concern that new types of exposures will occur. To date, *Bt* corn and potato PIPs that have been registered have been expressed in agronomic plant species that, for the most part, do not have a reasonable possibility of passing their

[1] Source: [1] ibid.

(cont'd)

traits to wild native plants. Most of the wild species in the United States cannot be pollinated by these crops (corn and potato) due to differences in chromosome number, phenology, and habitat. There is a possibility, however, of gene transfer from Bt cotton to wild or feral cotton relatives in Hawaii, Florida, Puerto Rico, and the United States Virgin Islands. Where feral populations of cotton species like cultivated cotton exist, regulators have prohibited the sale or distribution of *Bt* cotton in these areas. These containment measures prevent the movement of the registered Bt endotoxin from Bt cotton to wild or feral cotton relatives.

Although based on data submitted by the registrant and a review of the scientific literature, US regulators have concluded that there is no foreseeable risk for the specific case of unplanned pesticide production through gene capture and expression of the Colorado potato beetle control protein (Cry3A) in wild potato relatives in the United States. Tuber-bearing *Solanum* species, including *S. tuberosum*, cannot hybridize naturally with the nontuber-bearing *Solanum* species in the United States. Three species of tuber-bearing wild *Solanum* occur in the United States: *Solanum fendleri*, *Solanum jamesii*, and *Solanum pinnatisectum*. However, successful gene introgression into these tuber-bearing *Solanum* species is virtually excluded due to constraints of geographical isolation and other biological barriers to natural hybridization [39]. In this case, the regulators are providing a very specific answer to a specific question, that is, will there be gene flow problems associated with this specific GMO used in this way? Unfortunately, this could open the door for other uses that may indeed cause downstream problems, for example, gene flow of similar GMOs into nonpotato crops. The altered genetic material is no longer contained.

This is a weakness of premanufacture notification, not just for GMOs, but for any new substance, either chemical or biological. That is, the notification is tightly and specifically defined, but the potential use of the substance can be large once it reaches the marketplace. Also, the decision for approval must be made in relatively short period of time, for example, 90 days. Thus the onus falls on the regulators to decide whether it is safe, rather than on the manufacturers to prove it is safe. This is the reason that many nations are moving from the evidence-based risk model to one of precaution, especially for decisions that may lead to severe and potentially irreversible impacts. Although the United States continues to hold a risk-based approach for pharmaceuticals, environmental pollutants, and other substances, a coalition of 13 states has released their own principles asserting that chemical and product manufacturers should be required to develop and provide sufficient toxicity, exposure, and use information to regulators to conclude that chemicals and products in commerce do not endanger the public or the environment; they also call for protocols to evaluate potential alternatives to chemicals of concern. The states maintain that emerging chemicals of concern should be assessed for public and environmental safety before they go into widespread commerce and use. States acknowledge the need for a strong federal chemical regulation system, while expressly preserving the authority of

(cont'd)

states and localities to implement measures to manage chemicals of concern [40].

These cases set the stage for distinguishing between research ethics and professional ethics. For example, if the researcher establishes a study design to evaluate the difference in risks between an exposed population (e.g., people too poor to leave a region experiencing increasing temperatures or to control their diet to avoid GMOs) and an unexposed population (e.g., people who left the region or choose to avoid GMOs), this is more or less a "natural experiment," albeit unfortunate for those being exposed. However, if the researcher is also an engineer who has found a way for people to leave or to avoid GMOs but withholds this knowledge from the experimental ("exposed") group, this is arguably unethical. If the group is assigned blindly to the control or experimental group, the scientific integrity of the study may well be satisfied, but likely not its moral integrity. Indeed, if the engineer-researcher had gained the confidence of the population, this would be a form of the therapeutic misconception experienced by physician-researchers.

The examples of climate change and biotechnologies illustrate that an environmental pollutant's concentration must be defined by what society considers to be an acceptable and reasonable risk. The question is where to draw the line between unacceptable and acceptable, and between unreasonable and reasonable. There is little trouble building a consensus that exposing people to unacceptable risk is unethical. However, the acceptability of a pollutant concentration has both inherent and instrumental aspects. For example, dimethylmercury is inherently hazardous and is used in strictly controlled laboratory settings. However, if none of the mercury ever leaves the lab hood, its actual risk is zero, notwithstanding its inherent properties. Thus, the use scenario drives an environmental pollutant's level of acceptability. As such, acceptability is value-laden. A device that destroys a tumor and advances biomedical and environmental research may be worth the exposure to its inherently hazardous properties, so long as the risks are described accurately, the decision-making is transparent, and the ethics are considered by an objective collection of reviewers, for example, the institutional review boards that must ensure respect for persons, justice, and beneficence [41].

A useful measure of risk acceptability is whether it is "as low as reasonably practical" (ALARP), a concept coined by the UK Health and Safety Commission and published by the UK Health and Safety Executive [42]. The range of possibilities fostered by this standard can be envisioned to fall within three domains. Actions falling in one domain are associated with risk that is clearly unacceptable. At the other end of the risk spectrum are actions that are associated with unquestionably acceptable risk. In the middle are those actions that are more ambiguous from a risk perspective. The size of these domains varies considerably, so reaching consensus about acceptable risk is difficult, and there is almost never unanimous agreement on the differences between acceptability and unacceptability.

Risks in the ALARP region need to be managed scientifically and ethically to

(cont'd)

produce an acceptable outcome. Thus, this process may prove useful in deciding actions related to the emission of an environmental pollutant. For example, it may identify which actions provide the greatest good compared to the potential harm they may cause. For example, consider a propellant used for an asthma inhaler that can only be delivered efficiently by a gas that, if present in the stratosphere, would destroy ozone. It may be decided that this use should continue if it is the only reliable means of delivering the asthma medication to a particularly sensitive group. However, such single-variable assessments are uncommon and can lead to erroneous predictions of outcomes. For example, biomedical research may find alternative gases that do not destroy ozone or a mechanism that does not use a propellant at all, for example, a mechanical air pump. The key is that the decision must be systematic and must consider the entire life cycle of the potential pollutant.

The ALARP depends on a defensible margin of safety that is both protective and reasonable. Hence, reaching ALARP necessitates qualitative and/or quantitative measures of the amount of risk reduced and costs incurred with the design decisions. The ALARP principle assumes that it is possible to compare marginal improvements in safety (marginal risk decreases) with the marginal costs of the increases in reliability [43,44].

Environmental pollution risk assessment has numerous sources of uncertainty. For example, the dose-response relationships are difficult to extrapolate from animal studies to human populations. Also, the likely exposure can be variable and highly uncertain. Even the physical properties of an air pollutant are difficult to characterize. As evidence, engineering materials are being created with dimensions <100 nm (i.e., nanoparticles), which behave very differently after being emitted. In fact, the air pollution community is concerned about the health effects of ultrafine particles (also with aerodynamic diameters <1 nm), while the biomedical, chemical, and other communities are trying to develop nanoparticles with special features. Indeed, the medical and environmental pollution communities are working at opposite sides of the same coin. Drug delivery researchers are looking for ways to enhance dose, while environmental pollution experts are seeking means to decrease biologically effective dose. Since risk is a function of hazard and the exposure to that hazard, reliable assessment of that risk depends on sound physical characterization of the hazard. However, if even the physics is not well understood due to the scale and complexity of the research, the expected hazards and potential for exposure to humans and ecosystems are even less well understood.

Indeed, the ethical uncertainty of environmental pollution is propagated in time and space. Nascent areas of research include ways to link protein engineering with cellular and tissue biomedical engineering applications (e.g., drug delivery and new devices); ultra-dense computer memory; nonlinear dynamics and the mechanisms governing emergent phenomena in complex systems; and state-of-the-art nano-scale sensors, including those based on photonics. Complicating the potential societal risks, much of this research frequently employs

(cont'd)

biological materials and self-assembly devices to design and build some strikingly different kinds of devices. Among the worst-case scenarios is the replication of the "nano-machines." Advancing the state of the science to improve the quality of life (e.g., treating cancer, Parkinson's disease, Alzheimer's disease, and improving life expectancies, or cleaning up contaminated hazardous wastes) can introduce different risks [45].

Rarely is there a simple answer to the questions "How healthy is healthy enough?" and "How protected is protected enough?" Managing risks consists of balancing among alternatives. Usually, no single solution to an environmental problem is available. Whether a risk is acceptable is determined by a process of making decisions and implementing actions that flow from these decisions to reduce the adverse outcomes or, at least, to lower the chance that negative consequences will occur [46].

The social contract with the scientific and engineering communities expects that environmental pollution risks will be low, all other things being equal. Derby and Keeney [47] have stated that "acceptable risk is the risk associated with the best of the available alternatives, not with the best of the alternatives that we would hope to have available." Calculating the risks associated with these alternatives is inherently constrained by three conditions [48]

1. The actual values of all important variables cannot be known completely, and thus cannot be projected into the future with complete certainty.
2. The physical and biological sciences of the processes leading to the risk can never be fully understood, so the physical, chemical, and biological algorithms written into predictive models will propagate errors in the models.
3. Risk prediction using models depends on probabilistic and highly complex processes that make it infeasible to predict numerous outcomes.

10.4 Hazard

A compound that causes cancer or elicits respiratory problems in rats would be suspected of doing the same in humans, albeit at lower doses and over longer time frames than what laboratory rats have been exposed to. Additionally, the toxicity data from other species are used in hazard identification, but there are certainly interspecies differences. For example, some species are sensitive to substances that may have relatively low human toxicity, for example, copper compounds can be highly toxic to fish [49].

Ecosystem toxicity, that is, ecotoxicity, is the hazard that a pollutant may pose to various species within an ecosystem. For example, a substance may have low toxicity to a particular fish species but high toxicity to an organism on which the fish feeds. This would not kill the fish directly but would diminish the abundance and diversity of fish species in the ecosystem.

Besides toxicity, chemical hazards include other potential harms, including fire hazards, explosiveness, corrosivity, and chemical reactivity. Hazards may also be biological (biohazards) and physical (e.g., radioactive). Air pollution experts are primarily interested in chemical hazards, but they must be aware of these other hazards, particularly in first response and emergency situations. Indeed, spills and accidents may initially be physical hazards, for example, a highly reactive compound leaking from a ruptured tank. The reactions may form highly toxic vapors and gases that threaten human populations. In addition to stopping the leak, the immediate actions may include evacuating nearby residences to prevent exposure to these toxic gases.

10.4.1 Thresholds

Health officials and occupational, environmental scientists, and engineers often refer to *thresholds* of chemical compounds. The threshold level is the lowest amount needed to induce harmful effects in an organism. Establishing a toxicity threshold begins with a substance's inherent properties, for example, cyanide is highly acutely toxic, whereas the metal iron is usually only acutely toxic in high doses. Next, the influence of factors such as a chemical's mobility in environmental media and tissues should be considered, along with its persistence in the organism and the environment. For example, a chlorinated compound is inherently harder to break down than the unsubstituted compound. It is also much more likely to accumulate in living tissue.

The traditional risk assessment paradigm for chemical compounds in the environment (Fig. 10.1) is a stepwise process, starting with the identification of a hazard. Identification comprises a summary of an agent's physicochemical properties, routes and patterns of exposure, and a review of toxic effects. The tools for hazard identification consider the chemical structures associated with toxicity, metabolic and pharmacokinetic properties, short-term animal and cell tests, long-term animal (in vivo) testing, and human studies (e.g., epidemiology, such as longitudinal and case-control studies). These comprise the core components of hazard identification; however, additional hazard identification methods have been emerging that increasingly provide improved reliability of characterization and prediction.

Biomarkers are useful in identifying hazards since many exposure biomarkers appear before tissue damage. Hazard identification is also being advanced by computational chemistry by enhancing the understanding of molecular structure. Even similar substitutions on molecules (e.g., one versus two methyl groups, methyl versus ethyl substitutes, or halogenations [Cl versus F]) can lead to quite different levels of hazards. As mentioned, computational tools like quantitative structure-activity relationship (QSAR) models are increasingly used to identify and quantify hazards. They provide clues to subtle and presymptomatic hazard information akin to precision medicine's focus on a person's entire life rather than the presence or absence of diseases [50]. New methods are constantly being developed to identify and characterize pollutant hazards.

Regulatory agencies differentiate cancer risk from noncancer risk. Carcinogens are classified according to the soundness of science that supports or rejects a substance's link to cancer (see Table 10.1). Noncancer hazard is generally based on studies that show thresholds above which a dose leads to an adverse outcome (e.g., neurological damage, reproductive and developmental disorders, and endocrine disruption). This threshold is known as the no observed adverse effect level (NOAEL). For dermal, ingestion, and other exposure pathways, the NOAEL is commonly used as the threshold. The NOAEC is used for inhalation, so it is most appropriate for air pollutants.

Cancer risk policies often assume that there is no threshold below which an exposure is acceptable. Another threshold, the "lowest observed adverse effect level" (LOAEL), is the area of the dose-response curve where studies have linked a dose to an effect. Cancer slope factors (CSFs, also known as q1*) are used to calculate the estimated probability of increased cancer incidence over a person's lifetime, known as the excess lifetime cancer risk (ELCR) [54]. Like the reference doses (RfDs), slope factors relate to specific exposure pathways, that is, water and food ingestion, inhalation, and dermal contact [55–58].

As mentioned, for cancer risk assessments, the hazard is generally assumed to be the product of the CSF and the long-term exposure (E_{lt}):

$$\text{Cancer risk} = \text{CSF} \times E_{lt} \tag{10.5}$$

Since cancer is a chronic illness, with symptoms appearing after many years of exposure or at least after a latency period following episodic exposures, the corresponding E must also be chronic, for example, lifetime average daily dose (LADD). In this case, risk will be the product of LADD and potency represented by the CSF. To demonstrate, we can calculate the direct risk to the people living near a source of vinyl chloride, a potent liver carcinogen. Assume that the mean LADD in the neighborhood is 100 ng kg^{-1} d^{-1} and, from EPA's Chemistry Dashboard [58], the vinyl chloride CSF of 7.2×10^{-1} (mg kg^{-1} d^{-1})$^{-1}$, then the mean lifetime cancer risk to the neighborhood exposed for an entire lifetime (exposure duration [ED]=70 years) gives:

$$1 \times 10^2 \text{ ng kg}^{-1}\text{d}^{-1} \times 10^{-6} \text{ mg ng}^{-1} \times 7.2 \times 10^{-1}(\text{mg kg}^{-1}\text{d}^{-1})^{-1} \approx 4.3 \times 10^{-4}.$$

TABLE 10.1 Cancer classes in the United States, based on weight-of-evidence.

Cancer classification	Description
A	The chemical is considered to be a human carcinogen
B	The chemical is a probable human carcinogen, with two subclasses
B1	Chemical compounds that have limited human data from epidemiological studies supporting their carcinogenicity
B2	Chemical compounds for which there is sufficient evidence from in vivo studies but inadequate or no evidence from human epidemiological studies
C	The chemical is a possible human carcinogen
D	The chemical is not classifiable as to human carcinogenicity
E	There is evidence that the chemical does not induce cancer in humans

Data from: [51–53].

The neighborhood lifetime risk from exposures to vinyl chloride is about 4 per 10,000. Often, in cleanups, the threshold is that the risk should not exceed one in a million. For simple removals, the cleanup risk target is not to exceed 1 per 10,000. Both risk levels are exceeded in this hypothetical example. The combination of a steep slope factor and very high lifetime exposures leads to an elevated cancer risk.

The relatively high aqueous solubility and soil sorption rates of vinyl chloride should be considered in soil exposure dose and risk calculations. Plausible exposure scenarios would include the ingestion of drinking water for people on private wells drawing water from groundwater that has been contaminated, and dermal exposures, for example, children playing in the soil are also plausible scenarios. Total risk from a single contaminant like vinyl chloride equals the sum of risks from all pathways, for example, vinyl chloride in the air, water, and soil:

$$\text{Total Risk} = \sum \text{risks from all exposure pathways} \tag{10.6}$$

Cancer risk varies by the potency of the carcinogen and the exposure to that carcinogen. As with any air pollution risk, cancer risk is a function of hazard and exposure. As mentioned, since cancer is a chronic disease, the exposure must be long-term or at least must account for the latency between exposure and cancer outcome. Thus, cancer risk assessments generally express hazard potency for the dose-response curves from animal research and human epidemiology. The quasilinear segment of these curves provides what is known as the CSF, and the long-term exposure is the LADD.

An exposure concentration (*EC*) is a time-weighted average concentration derived from measured or modeled concentrations of a contaminant in the air but adjusted according to the characteristics of the exposure scenario. For an acute drinking water exposure, the *EC* and water concentration (C_W) values will be almost the same. However, estimating cancer risks from chronic exposure involves both the C_W measured at a point of exposure as well as scenario-specific parameters, such as the ED and EF:

$$EC = \frac{C_W \times ET \times EF \times ED}{AT} \tag{10.7}$$

where EF is the exposure frequency, ED is the exposure duration, and AT is the averaging time.

The value for AT for chronic exposures is ED in years × 365 days year^{-1} × 24 hours day^{-1}. The AT units are hours, so AT here is multiplied by 24 to provide an AT in days. Thus, *EC* is usually the same as C_W [53].

Total ingestion exposures occur when an individual comes into contact, intentionally or inadvertently with a substance that then enters the gastrointestinal (GI) tract. Depending on the properties of the substance, absorption can occur at various locations throughout the GI tract. A substance may directly reach the GI tissue or subsequently enter the bloodstream. Ingestion exposure can be estimated as:

$$E_{ing} = C_{ing} \times IR \tag{10.8}$$

where: E_{ing} is ingestion exposure (mass per time), C_{ing} is the concentration of the substance in food, water, or other exposure media (mass of substance per mass of medium or mass of substance per volume of medium), and *IR*=ingestion rate (mass of medium ingested during the exposure per time). If more than one medium is being studied for ingestion exposure, the exposure from each medium is calculated separately and then summed. An exposure scenario is usually expressed as a frequency of ingestion events times the intake per event [60].

For hazardous pollutants that are inhaled, the unit risk estimate (URE) is often used instead of equations based on the CSF. The URE is an upper-bound estimate of a person's chance of contracting cancer over a lifetime of exposure when exposed to $1\ \mu g\ L^{-1}$ of drinking water or $1\ \mu g\ m^{-3}$ of air. An upper-bound risk is a likely limit to the true probability that a person will contract cancer over a 70-year lifetime due to a toxic substance [61–63].

Risks can be calculated by multiplying the actual concentration to which a person is exposed by the URE. For example, a URE of 3×10^{-6} per $1\ \mu g\ L^{-1}$ of drinking water means that as many as three additional persons in an exposed population of one million would be expected to develop cancer during a lifetime due to 1 μg of the carcinogen in 1 L of water consumed. Since the URE is an upper-bound estimate, it is conservative, that is, it avoids underestimating the cancer risk. The actual risk may be lower but is unlikely to be higher.

The EC is often combined with the inhalation-related URE to provide the inhalation unit risk (IUR). The IUR is an estimate of the increased cancer risk via inhalation exposure to a concentration of $1\ \mu g\ m^{-3}$ for a lifetime. Like the CSF, the IUR is a predictive cancer risk potency estimate. The IUR provides an upper-bound ELCR estimated to result from continuous exposure to an agent at a concentration of $1\ \mu g\ m^{-3}$ in air. For example, if the IUR for formaldehyde in a metropolitan area is $2 \times 10^{-6}\ \mu g\ m^{-3}$ predicts two excess cancer cases as an upper-bound estimate to develop per 1,000,000 people if exposed daily for a lifetime to 1 μg of the chemical in 1 cubic meter of air.

10.5 Biological hazard classification

Emerging technologies that adapt and otherwise utilize biological agents can introduce risks. The most common organisms in emerging technologies are microbes, for example, viruses, bacteria, and fungi. They may also include larger organisms or their parts, for example, pollen and spores. The concern has recently grown to include GMOs. Risks and hazards from biological agents are often different from those posed by chemical compounds, so the risk assessment process is commensurately different. As evidence, the Safety in Biotechnology Working Party of the European Federation of Biotechnology has identified four risk classes for GMOs [64,65]:

Risk class 1. No adverse effect, or very unlikely to produce an adverse effect. Organisms in this class are considered safe.

Risk class 2. Adverse effects are possible but are unlikely to represent a serious hazard with respect to the value to be protected. Local adverse effects are possible, which can either revert spontaneously (e.g., due to environmental elasticity and resilience) or be controlled by available treatment or preventive measures. Spread beyond the application area is highly unlikely.

Risk class 3. Serious adverse local effects are likely concerning the value to be protected but spread beyond the area of application is unlikely. Treatment and/or preventive measures are available.

Risk class 4. Serious adverse effects are expected with respect to the value to be protected, both locally and outside the area of application. No treatment or preventive measures are available.

These classes indicate that even the safest genetic modifications carry some risk and that with more uncertainty about an organism, one cannot assume it to be safe. Thus, risk scenarios include not only the effects resulting from the intended purpose of the environmental application, but also downstream and side effects that are not part of the desired purpose.

The biological agent risk classes are analogous to the hazard classes of various human health risk paradigms, for example, carcinogen listings or extremely hazardous designations. Thus, a microbe can be assigned to the appropriate risk class based on its physical, chemical, and biological properties, notwithstanding the way it was selected or, in the case of genetic modification and synthetic biology, how a specific variant was generated. After which, the potential risk is scored against a set of values to be protected [39]. As in the hazard identification process for chemical compounds, the specific microbe is classified according to its inherent properties. The next stage addresses the environmental conditions by characterizing aspects like the administered dose to various responses in different populations. Both the hazard identification and dose-response data are based on research that is the basis of the risk analysis. The virulence of strains of the same species or that of closely related species differs depending on the genetic makeup. For example, the bacteria *Salmonella enteritidis* and *S. pullorum* elicit various types of pathogenicity in different humans, which can vary by orders of magnitude [66,67]. Similarly, specific Escherichia *coli* variants cause enteric diseases in humans, but although most *E. coli* isolates are not considered to be pathogenic. Even two isolates with all the same virulence characteristics can lead to substantially dissimilar pathogenicity due to differences in the genetic expression of the toxicity or other characteristics [66].

Hazard classification of microbes is like that of chemical hazard identification processes in that available knowledge about other microbes with similar characteristics can be extrapolated, although this information is not sufficiently tested for the microbe of concern but has been observed in similar environmental conditions. A field study's results in one type of field may be extrapolated to an environmental remediation setting if the microbe in question has been observed in sufficient agricultural settings. Similarly, in chemical hazard identification, a chemical compound that has been sufficiently studied and has a similar chemical structure to a less-studied compound may employ a similarity model, like a QSAR [39,68–71]. Such similarity models may be applied to microbial growth and pathogenicity [72].

10.6 Toxicokinetics

So-called "emerging" chemical contaminants have added uncertainties regarding adverse outcomes, ED, timing, and predictions of the effectiveness of proposed actions. There is much greater uncertainty about the ability of regulatory agencies to meet an objective for substances like microplastics (see Discussion Box: Microplastics), nanoparticles [73–91], and perfluoroalkyl and polyfluoroalkyl substances (PFAS) [55,92,93], which initially have a scarcity of reliable data and peer-reviewed studies from which to make decisions.

Applying ADMET to microplastics

Estimating the potential for ecotoxicity associated with microplastics is fraught with uncertainties. One approach is to predict endogenous activity within an organism, such as a sentinel species [94–97]. The biological activity within the organism is a first step toward characterizing potential adverse ecological outcomes. This is based on the specific rates and amounts of absorption, distribution, metabolism, elimination, and toxicity (ADMET) of various species exposed to different classes of microplastics [98,99]. Such models begin with the mass of a specific physicochemical form of a microplastic taken up by the organism. The inherent physicochemical properties and the potential for the microplastic to induce biological activity are key inputs to ADMET models [100,101].

The characteristics of nearly one million chemical compounds can be found on websites and dashboards, such as EPA's Chemistry Dashboard [102–104], so the building blocks of plastics, for example, the base chemicals and monomers, are often well represented. Unfortunately, there is a paucity of such data on microplastics themselves. Thus, such tools are often limited to a bottom-up approach to microplastics ADMET, that is, estimating endogenous activities of simpler chemical structures that make up the plastic and using models to predict synergistic and antagonistic effects of the whole plastic. Such a screening approach may help to identify data gaps and steps toward closing knowledge gaps associated with data-poor microplastics by incrementally ranking hazard and exposure potentials [105,106]. Such screening tools can also aid in the development of a hypothetical portfolio of products for an individual consumer, for example, personal care, cleaning products, fuels, etc. A portfolio of products and an accompanying set of their chemical ingredients can allow decision makers to rank products according to potential risk, including the likelihood of the formation and transformation of pollutants [107]. These datasets are updated frequently as new studies and data become available.

In addition to the input on the chemical's characteristics, the model needs information about the organism. For example, the ADMET for a chemical needs information on the amount that is expected to be metabolized and the amount that will exit the human body. Toxicokinetic (TK) models often simulate based on estimates of the rate of clearance along with the unbound fraction of the absorbed chemical. For example, hepatic clearance is expressed as the rate at which liver hepatocytes remove a substance. These models have mainly targeted human subpopulations but can also apply to other species. In fact, data used as input to human models often begin with lower species, that is, they are derived from in vitro to in vivo extrapolations (IVIVE). The fraction unbound in plasma is the soluble portion of the chemical.

The volume of distribution (Vd) is the volume needed to remove the chemical

(cont'd)

compound if the organism's tissue concentration throughout the body were the same as the plasma concentration. Thus, Vd indicates the distribution rate of an organism's ADMET. The human steady-state plasma concentration is the time during which the plasma concentration of the chemical in the body remains consistent. A rule of thumb for a chemical to reach steady state is four to five half-lives with dosing at regular intervals, irrespective of the number of doses, the dose size, or the dosing interval. This dynamic equilibrium maintains chemical concentrations within defined limits for extended time periods. If these rates are similar to human ADMET, at a dose of 1 mg kg^{-1} day^{-1} and at the 95th percentile of distribution, which would roughly be expected for a healthy population of organisms, the Vd can be scaled linearly to estimate the blood plasma steady-state concentration for any dose [108].

The days to steady state are assumed to be stable and relatively unchanged over time, that is, the systemic concentration of the chemical. The pharmacokinetic (PK) half-life is the time needed for the body to eliminate half of the plasma concentration. Again, this is for the chemicals that make up the plastic substance, so it is very likely that the half-lives will differ for the whole plastic compared to the constituent chemicals. In addition, factors like sorption to aerosols in air, soil particles, and suspended solids in water will change the overall ADMET for an organism.

The TK model requires a variety of species-specific physiological parameters, such as cardiac output and lung ventilation rate. These estimates vary by subpopulation characteristics, for example, altered TK in older adults changes the specific information needed to estimate metabolic clearances, chemical half-lives, body burden, and volume of distribution [109]. Many of these values are altered in older adults compared to young adults for a variety of chemicals [110].

If a substance can cause damage with only a few doses in a short period of time, it is acutely toxic. Conversely, chronic toxicity occurs when a person or population is exposed over a protracted period, with recurring exposures [111]. The steepness of dose-response curves represents the contaminant's toxic potency or severity. The solid line curve is less steep than the dashed curve in Fig. 10.4 because the response elicited by the agent in the dashed curve is more potent than that of the agent in the solid curve. Potency is but one factor in articulating a hazard. A very potent agent that elicits a relatively innocuous effect, for example, temporary loss of appetite or headache, is less worrisome than an agent with a gentle slope that causes cancer or birth defects. Thus, dose-response is best when comparing curves of different agents that elicit the same effect, such as comparing the dose-response curves for three different chemicals that cause a specific type of cancer.

The first of the three types of uncertainty is inherent biological differences between the species tested and that of humans (UF_{inter}) [112]. For example, how can the results of a study that shows toxicity in a bird's gizzard be translated to humans, who have no such organ? The UF_{intra} results from the differing sensitivities of human subpopulations when exposed to a substance compared to those of the general human population. The UF_{other} occurs because available data and science are lacking, including the uncertainty introduced when the data may indicate a lowest observed adverse effect concentration (LOAEC), but not a NOAEC, that is, the LOAEC-to-NOAEC (U_L). That is, data show a dose at which an effect is observed, but the "no effect" threshold cannot be delineated.

Often, the first consideration of chemical, biological, or physical agent is the dose expected to elicit an adverse response, that is, the mass or volume that encounters an organism. Dose is expressed in many ways, such as the applied dose, absorbed dose, or the biologically effective dose, which is the amount that reaches a particular organ (see Fig. 10.2). For example, a neurotoxin harms the central nervous system (CNS), which means only the mass or volume of that neurotoxin that reaches the brain and nerve cells will be the biologically effective dose. The remaining absorbed dose is metabolized, stored, or, in a best-case scenario, eliminated from the body before causing any harm.

Increasing the concentration of a chemical or pathogen that contacts an organism will increase the biologically effective dose and the probability that a response will occur, that is, the expected adverse outcome. This gradient is known as the "dose-response" curve (see Fig. 10.3). Increasing the amount of the substance encountering the receptor increases the absorbed and biologically effective dose and, consequently, the probability of an adverse outcome [39,106,119].

An essential compound curve or substances like vitamins and certain metallic compounds is U-shaped. The left-hand side represents the low-dose region, that is, deficiency.

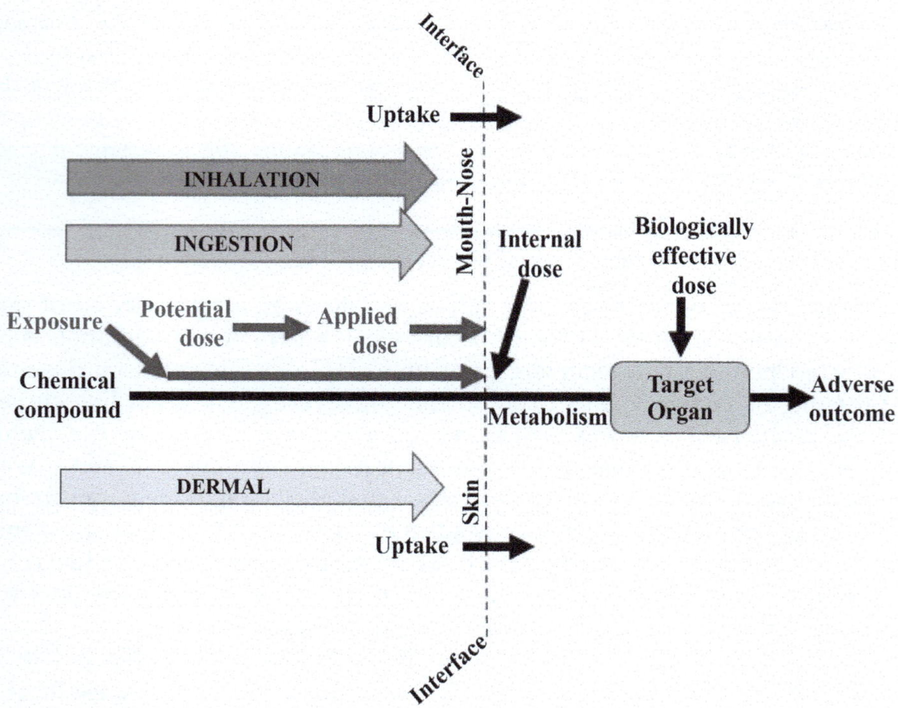

FIGURE 10.2 Continuum of dose types from potential to biologically effective via inhalation, ingestion, and dermal exposure pathways. Source: *Based on information from [113–118].*

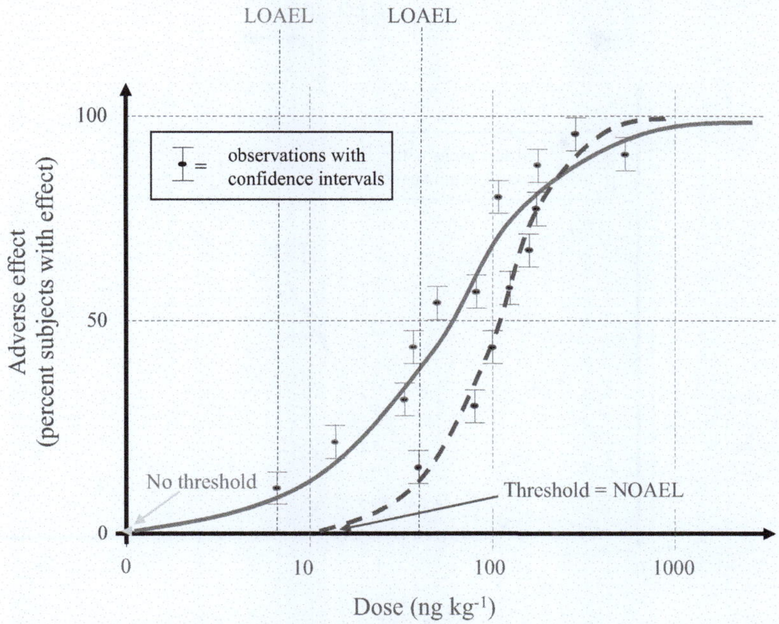

FIGURE 10.3 Hypothetical dose-response curves. The solid line curve represents the "no-threshold" curve, which predicts an adverse response no matter how low the dose. Here, the low dose end of the curve, that is, below which observational data are available, is assumed to be linear and crosses through the origin, that is, zero effect-zero dose. Thus the solid line curve represents a linearized multistage model. The dashed curve represents an adverse effect only above a certain threshold, that is, the no observed adverse effect level (NOAEL) or no observed effect concentration (NOAEC) if inhaled. Below these thresholds, there is a dose that is not expected to elicit an adverse response; in this case, the model predicts a NOAEL at about 10 ng kg^{-1}. The lowest observed adverse effect levels (LOAELs) are shown for both curves. Note that both curves have a part of the curve at doses below the LOAEL, so both curves require a model to extrapolate the unobserved curve below this dose. Source: *Data from D. Vallero, Environmental contaminants: assessment and control, Academic Press, 2010. [120].*

The right-hand side is the high-dose region, that is, toxicity or another hazard. The optimal region lies between these two adverse responses [121,122]. Between the low-dose and high-dose regions lies the optimal range, that is, sufficiently high to avoid deficiency and sufficiently low to avoid toxicity. This applies to nutrients and vitamins. For metals, the curve must be drawn for specific valence states, for example, trivalent chromium (Cr^{3+}) has a U-shaped curve, but hexavalent chromium (Cr^{6+}) has a dose-response curve resembling the solid line curve in Fig. 10.4 because it is carcinogenic [122].

Assembling a dose-response curve consists of a sequence [52]. The dose-response data from in vitro, in vivo (in living organisms other than humans), or human epidemiological studies are fitted with a mathematical model. In vitro studies include bacterial mutagenicity studies [123,124], which can be a first screen for cancer and other adverse effects. The Ames *Salmonella*/microsome mutagenicity assay is a common in vitro method that displays the short-term reverse mutation in histidine-dependent *Salmonella* strains of bacteria [125,126]. The culture allows only those bacteria with genes that revert to histidine independence to form colonies. The mutagen population growth is a first dose-response curve, that is,

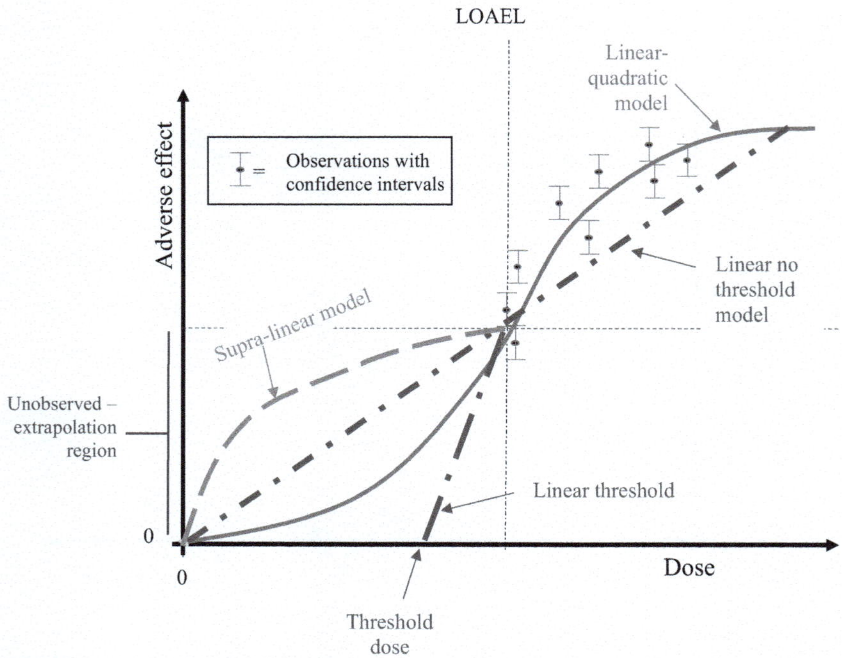

FIGURE 10.4 Possible approaches to extrapolate dose-response from observed measurements. Exposure to the pollutant is represented on the x-axis, and the corresponding dose is represented on the y-axis. A "no threshold" model assumes that there is no safe level of exposure to that substance, so the curve intercepts the x-axis at 0. Source: *Data from K.M. Seong et al., Is the Linear No-Threshold Dose-Response Paradigm Still Necessary for the Assessment of Health Effects of Low Dose Radiation?, Journal of Korean Medical Science 31 (2016) S10–S23 [139].*

increasing the amount of chemical is associated with an increase in the number of mutagens and colony size on the plate. The test is arguably biologically plausible, given that cancer is a genetic disease and a manifestation of mutations [123].

From these experimental and observational data, the upper confidence limit (e.g., 95%) line equation is expressed, and the confidence limit line is then extrapolated to a response point immediately below the lowest measured response point from the experimental data, that is, the "point of departure" (POD). For example, for carcinogens, the response is assumed to be a linear or nonlinear function of dose from the POD to zero response at zero dose. The last step is to calculate the dose on the line that is estimated to produce the response.

The sigmoidal-shaped curves in Fig. 10.3 are common for toxic substances, since increasing the dose at one point cannot increase the effect, that is, the mortality or morbidity rate cannot exceed the 100% asymptote [127]. There is often a distinction between physical and chemical agents that are suspected to cause cancer and those that are not expected to cause cancer. For suspected carcinogens, a linear no-threshold (LNT) model has been employed that resembles the solid line curve in Fig. 10.4. This applies to both ionizing radiation and chemicals. There is debate on what to do about the unobserved region, often regarding the shape of the curve below the LOAEL [111,128–132]. For example, this can be linear as suggested by the LNT, but it may also be nonlinear (see Fig. 10.5). Thus, the

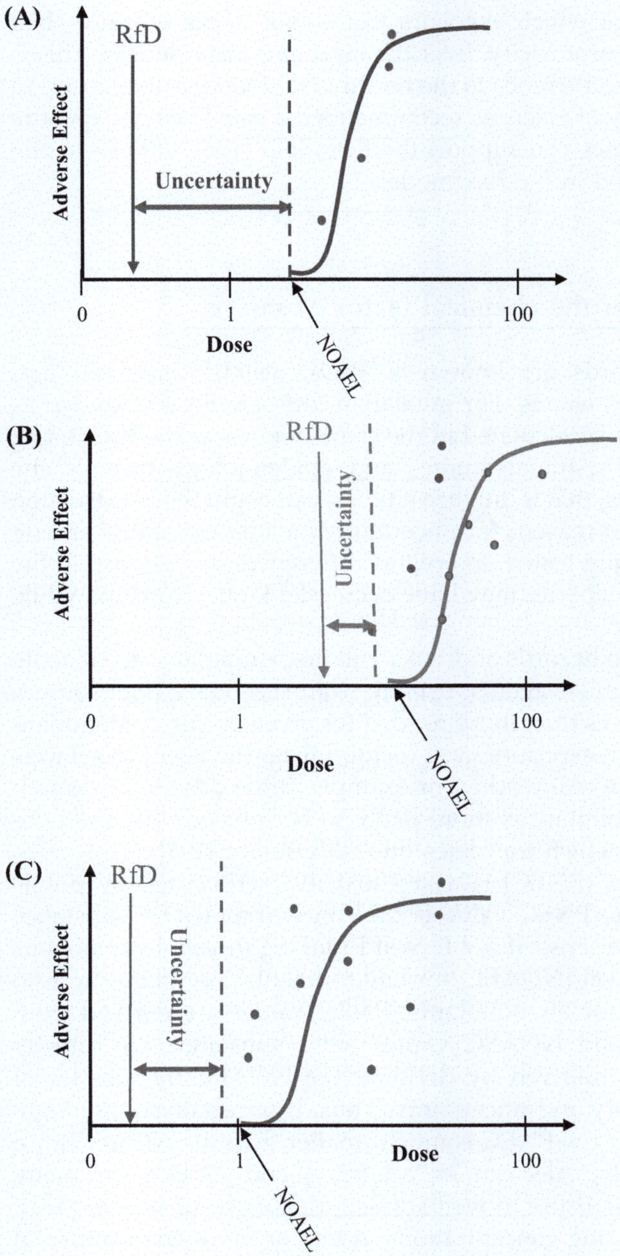

FIGURE 10.5 Constructing a reference dose (RfD) for a chemical compound with available new data. Curve (A) shows that, based on fewer reliability, the high level of uncertainty moves the RfD far from the dose-response curve's threshold, that is, no observed adverse effect level (NOAEL) as a factor of safety. Curve (B) shows that new data improve the dose-response curve and indicate that the adverse effects are not as potent as assumed from the scarce measurements, so the curve shape changes with the NOAEL moving to the right and the decreasing uncertainty moving the RfD closer to the NOAEL. Curve (C) shows what happens when new data indicate that the adverse effects are more severe than assumed from the original dataset, so the curve changes shape, with both the RfD and the NOAEL moving to the left. However, since additional data decrease the uncertainty, the distance between the RfD and the NOAEL shortens. Note that RfD could move to the right while the NOAEL moves to the left if the new data are so reliable that the uncertainty decreases more than the new data move the threshold.

arguments are not whether a single dose of radiation or a single molecule of a chemical carcinogen can, in fact, cause cancer, but how to arrive at a conservative, protective dose-response curve that may indeed have no threshold [52,133–137]. On the other hand, there are also arguments against assuming a threshold exists for every noncarcinogenic outcome, that is, the assumption of a safe level at which exposure to that substance will not elicit serious, but noncancer outcomes like neurotoxicity, liver damage, and birth defects. This is the opposite argument against exclusive adherence to the no-threshold models for cancer. In the United States, the National Academy of Sciences recommended a safe level of exposure for noncarcinogens only if strong evidence can support the threshold [138]. Otherwise, no safe level of exposure should be assumed in the risk model.

10.7 Reference dose: the chemical factor of safety

Safety thresholds for chemical hazards are known as RfDs, usually based on oral exposure data and applied to dermal exposures. For inhalation, the safe dose is known as the reference concentration (RfC), which is calculated in the same manner as the RfD, using units that apply to air (e.g., $\mu g\ m^{-3}$). Animal studies and epidemiology provide the knowledge bases for these values. Assays, that is, in vitro studies, can begin the construction of RfDs and RfCs [102,140,141]. The main reasons for uncertainty include variability among the animals and human populations being tested, as well as differences in response to the compound by different species (e.g., one species may have decreased kidney function while another may show neurological effects).

Lacking long-term human studies, the hazards and risks will be extrapolated from acute or subchronic studies of the same or similar agents, usually from secondary data sources that likely had different research objectives than those needed for an emerging contaminant [55,142–149]. The routes and pathways of exposure used to administer the agent to subjects may differ from those applications in the real world. For example, if the dose in a research study is administered orally, but the pollutant is more likely to be inhaled, this route-to-route extrapolation adds inconsistency, which translates into uncertainty [106].

The predicted no-effect concentration (PNEC) is a key toxicity metric used by public health and environmental scientists. The PNEC value is the concentration of a substance below which no effect is expected. In this sense, it is a forward-looking threshold analogous to the retrospective no adverse effect level (NOAEL) for environmental pollutants and no adverse effect concentration (NOAEC) for air pollutants. PNEC values are derived from acute toxicity data [150]. The PNEC and NOAEC values for a substance are directly proportional because the PNEC can be derived by dividing the NOAEC by a series of assessment factors (AFs) to address safety and uncertainty. Thus, if uncertainties are high, very large safety factors are used, so that the PNEC is much smaller than the NOAEC for a substance (see Fig. 10.5). Unfortunately, AFs can be arbitrary, and PNECs for many substances are not quantified, limiting their use to well-studied, data-rich substances [151].

Often, the PNEC is compared to existing concentrations in one or more environmental media or to the predicted environmental concentration (PEC). Ideally, the risk quotient (RQ) $\frac{PNEC}{PEC}$ should be much less than 1.

The uncertainties in the RfD are often caused by the differences between results found in animal testing and expected outcomes in human population. As in other bioengineering operations, a factor of safety must be added to calculations to account for UFs. So, for environmental risk analyses and assessments, the safe level is expressed in the RfD, or in air, the

RfC. This is the dose or concentration below which regulatory agencies do not expect a specific unacceptable outcome. Thus, all UFs move the measured and modeled levels of threshold values, for example, NOAELs and LOAELs, toward zero concentration. This is calculated as:

$$RfD = \frac{NOAEL}{UF_{inter} \times UF_{intra} \times UF_{iother}} \quad (10.9)$$

The calculation of the RfC is identical:

$$RfC = \frac{NOAEC}{U_{inter} \times \quad U_{intra} \times U_{other}} \quad (10.10)$$

The increase in the size of the denominator, that is, total uncertainty, means that the safe dose continues to drop, so the RfC and RfD continually decrease based on the assumed toxicity, given the high levels of uncertainty.

The first of the three types of uncertainty results from differences in the species tested and compared to *Homo sapiens* (UF_{inter}). Species differ in their sensitivity to substances, so their response to a toxic substance differs from that of the tested species, that is, UF_{inter}. The second UF is based on varying sensitivities among human subpopulations to the effects of a compound compared to the general human population, that is, intraspecies UF_{intra}. The third class of uncertainties (UF_{other}) results when the reliable data are lacking, for example, a lowest observed adverse effect is used rather than a NOAEL because a threshold cannot be properly extrapolated considering data scarcity. The UFs are in the denominator, which means that the greater the uncertainties, the closer the protective concentration is to zero. The UFs are commonly multiples of 10, although UF_{other} can range from 2 to 10 [152].

Reference dose example[2]

A reference concentration (RfC) can be constructed by using the animal data from Landsiedel et al. [153] for titanium oxide (TiO_2) NP, which shows no NOAEC but has a lowest observed adverse effect concentration (LOAEC) of 2 mg m^{-3}, while a silicate (SiO_2) NP has a NOAEC of 10 mg m^{-3}. Using the EPA's [152] most protective recommended U_L of 10 and assuming the remaining uncertainty factors $(UF)_{inter}$=10 and UF_{intra}=100 for both NPs, then:

$$RfC - TiO2 = \frac{2}{10 \times 100 \times 10} = 0.0002 \ mgm^3$$

$$RfC - SiO2 = \frac{10}{10 \times 100} = \ 0.01 \ mgm^3$$

Thus, although the thresholds may indicate that the toxicity is similar for the two NPs, given the lack of data to delineate a NOAEC, the TiO_2 NP requires a 50-times larger factor of safety. Since the UFs are in the denominator, the greater the uncertainties, the closer the safe level (i.e., the RfC) is to zero, that is, the threshold is divided by these factors. The UFs are usually multiples of 10, although the UF_{other} can range from 2 to 10.

[2] Source: [1] ibid.

Reference dose example 2[3]

The US Environmental Protection Agency (EPA) was asked to re-register the use of the organophosphate pesticide, chlorpyrifos. The acute dietary scenario had a NOAEL of 0.5 mg kg^{-1} d^{-1}, and the three uncertainty factors (UF) values equaled 100. Thus, the acute reference dose (RfD)=5 × 10^{-3} mg kg^{-1} d^{-1} but the more protective acute population adjusted dose (PAD) is 5 × 10^{-4} mg $kg^{-1}d^{-1}$. The PAD is a risk assessment value used to evaluate human health risks from a chemical exposure. It acts as a modified RfD that includes additional safety factors, especially to protect infants and children. The chronic dietary scenario is even more protective since the exposure is long-term. The chronic NOAEL was found to be 0.03 mg $kg^{-1}d^{-1}$. Thus, the chronic RfD for chlorpyrifos=3 × 10^{-4} mg $kg^{-1}d^{-1}$ and the more protective acute PAD is 5 × 10^{-5} mg $kg^{-1}d^{-1}$. This means that the allowable exposure would be three orders of magnitude higher if the NOAEL threshold had been used alone without the RfD [17].

10.8 Estimating potential exposures

Hazard is one component of risk. The other is exposure. No matter the steepness of the hazard curve of a compound, it will only elicit risk when it encounters the receptor. Humans can contact a substance in various ways in numerous settings, so-called microenvironments. Thus exposure to a pollutant varies by activities. For example, the exposure potential is relatively high for workers who work with a chemical, compared to the lower exposure potential for workers who may not work with it but are nearby and breathe the vapors, and near zero for the rest of the population. Usually, worker exposure is based on a five-workweek and 8–10 hours of exposure per day. Often, the worker risk calculations include personal protective equipment (PPE), so that exposure is decreased accordingly.

Conversely, environmental exposures are usually based on average lifetimes and continuous 24-hour daily exposure, which assumes no PPE is in use. Lifetime durations of exposure are especially needed for chronic disease risk estimates, like cancer. Thus, environmental regulations are often more stringent than occupational regulations when aimed at reducing exposure to a substance [30].

Chronic exposure may be less than the entire lifetime if the pollutant exists only for a specified time for the exposed population. For example, if it is known that the source of a pollutant only began manufacturing a chemical compound in 1990, went out of business in 2010, and cleanup of the source was documented in 2012, the risk calculation would include an ED of about 22 years. However, if the exposure route was drinking water from contaminated groundwater, even though remediation was successful in 2012, there may be residual contamination, especially if the aquifer had very low flow and the chemical

[3] Source: [1] ibid.

compound is persistent (see Chapter 11). Ecosystem exposures assume the ED is as long as the pollutant is deposited on and stored in the ecosystem's soil, water, sediment, and biota [154].

Estimates of the likelihood of exposure are part of the previously discussed UFs. The uncertainties are largely due to the differences between results found in animal testing and expected outcomes in human population. For environmental risk analyses and assessments, the safe level is expressed in the RfD for ingestion and dermal exposure, and RfC for inhalation exposures. This is the dose or concentration below which regulatory agencies do not expect a specific unacceptable outcome. Thus, all the UFs adjust the actual measured levels of no effect (i.e., the threshold values, e.g., NOAELs and LOAELs) in the direction of a zero concentration.

Varying amounts of the parent substance are absorbed and distributed within the body. Prior to metabolism, varying amounts of the parent substance, such a zero-valent metals, salts and ions formed, and other chemical species are absorbed and distributed within the body. However, within certain organic solvents, metallic substances like nanoparticle additives in diesel fuel, distribute among metallic ultrafine particles, organometallic compounds and larger arger suspended particles prior to combustion [79]. In addition to chemical factors, physical properties affect toxicokinetics, for example, the hydrophobic surface and large relative surface area of microplastic particles mean that they are prone to absorb toxic substances like heavy metals and persistent organic compounds [155]. The low mass also means that the particle is likely to be colloidal and remain suspended for a long time. Such colloidal and nano-suspensions in surface waters mean that even lipophilic chemical compounds remain in the water column rather than settle onto the surface. As a result, they are more likely to be oxidized than to be reduced, compared to the hypoxic and anoxic conditions of the sediment [156]. In the air, these characteristics mean that they are more likely to remain airborne for longer time periods sufficient for some of the organic compounds to undergo chemical transformation [78,86,156,157], and some of the metals to change their valence state, which could make them more bioavailable and toxic [158,159].

Surface charge, surface chemistry, particle shape, particle mass, and particle density of the particles that make up a matrix like a product, solid waste, or soil can mean that bulk materials will behave differently in an organism than their individual constituents in terms of the potential to be absorbed, distributed, metabolized, and excreted, so that the bulk material toxicity will differ from the sum of the constituents' toxicities [160,161]. The composite fraction of a chemical compound or its transformation products that accumulates in lipids and other tissue substrates could be higher, and the amount excreted could be decreased, so that the difference results in bioaccumulation and increased body burden [162]. The same physical and chemical principles that work outside of the organism occur within the organism but, for example, at different rates, using different catalysts, and under various endogenous redox conditions [37].

From a high-throughput, high-tier screening perspective, the first-order problem occurs when the likelihood of exposure *and* the likelihood of hazard coexist. No matter the steepness of the hazard curve of a substance, it will only elicit risk when it encounters the receptor. For example, in the EPA's risk-based chemical prioritization efforts, the substances of greatest concern are those with high potential toxicity and high potential exposure profiles. Indeed, even when toxicity remains the same, those substances with higher exposure potential will lead to higher risk (see far right of Fig. 10.1). The listing of contaminants of concern by this approach will differ from one based solely on toxicity, since even moderately toxic substances can have high risk potentials. Similarly, when hazard concentrations (e.g., greater than PNEC) and exposure potential intersect, there is the greatest concern, since populations are being exposed at concentrations most likely to lead to adverse outcomes.

10.9 Characterizing risk

Fig. 10.6 shows the traditional chemical risk assessment paradigm [163]. It begins with the identification of a hazard, which is a summary of a chemical agent's physicochemical properties, routes, and patterns of exposure, and a review of toxic and other adverse effects. Hazard identification accounts for metabolic and pharmacokinetic properties based on short-term animal and cell tests, long-term animal (in vivo) testing, and human studies (mainly epidemiological, such as longitudinal and case-control studies). Risk assessors can apply biomarkers of genetic damage (i.e., toxicogenomics), QSAR models based on

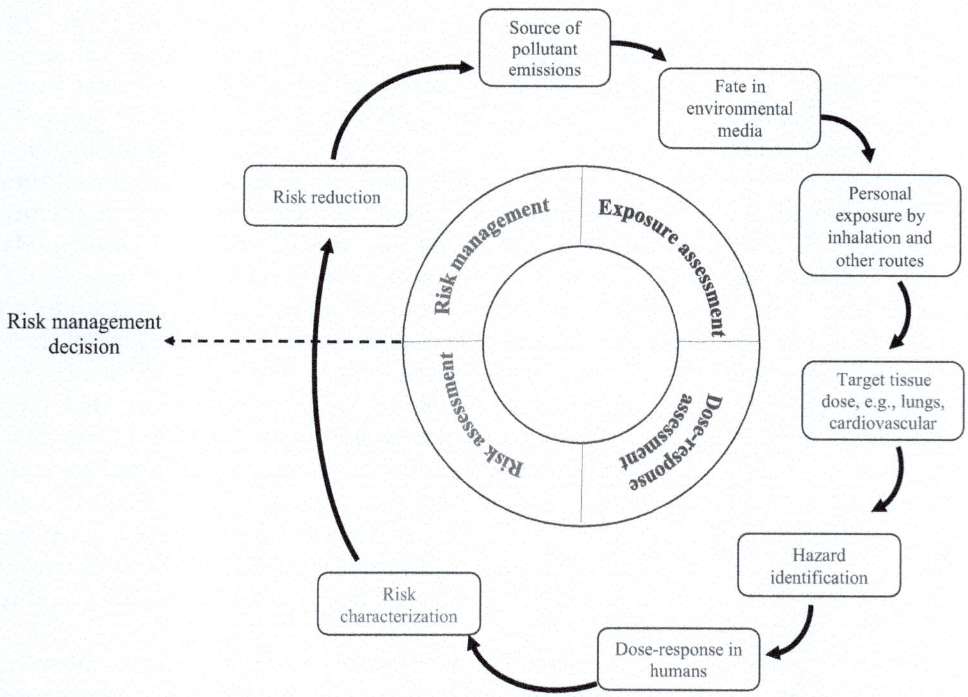

FIGURE 10.6 Risk assessment and management paradigm as employed by environmental agencies in the United States. The inner circle includes the steps recommended by the National Research Council. The outer circle indicates the activities that are currently used by regulatory agencies to meet these required steps. Source: *Data from National Research Council, Risk assessment in the federal government: managing the process, National Academy Press, Washington, DC, 1983; N. R. Council, Science and decisions: advancing risk assessment, 2009 [163,195].*

stereochemistry, and other chemical descriptions of well-studied compounds to compare with lesser-understood substances [69,164–170]. New methods such as cheminformatics [64,104,166,170–188] and computational chemistry are also increasingly applied [119,189–194].

Risk metrics are used to determine success or failure in environmental pollution control and prevention efforts. The metrics relate to environmental and health hazards, exposures, and risk characterizations that are input into environmental pollution calculations to identify and select actions to address risks [196–201].

Like risk, the reliability of a system is a metric of accountability (see Chapter 18). A reliable system is one that will not fail during its designed operating period. Thus, reliability is the mathematical expression of success, whereas risk is the expression of the likelihood of failure. Reliability expresses the probability that the system will perform its intended function for a specified time and under specified conditions [202,203]. For example, as the reliability of a pollution control system increases, the better that system will be at decreasing effluent concentrations of environmental pollutants. This, in turn, means lower risks to humans and the environment. Like risk, reliability is a probability. Risk is the likelihood of an adverse and undesirable outcome, but reliability is the likelihood of a desired outcome. As probabilities, both have values between zero or greater and less than or equal to one. They are often expressed as a percentage ranging from 0% to 100%. Complete reliability, that is, 100%, and conversely, 0% risk are the ideal targets for environmental pollution control systems [203–207].

Systems are reliable when they perform as designed under the specified conditions for the intended purpose. For example, each sampling and analytical device discussed in Chapter 13 requires a level of reliability to be useful for environmental quality measurement. Items within a system will indeed fail at some rate, for example, x number of failures per hour of operation. For example, a pump that is expected to fail no more frequently than once every 1000 hours of operation, and a water testing group study using 10 pumps in the field for 11,000 hours, they may not be sufficiently reliable even though the average pump lasted 1100 hours, if the range of endurance is 900–1200 hours. With at least one pump failing more frequently than designed, at least one extra replacement pump is likely needed.

Reliability is the probability that something that is in operation at time *0* (t_0) will still be operating until the designed life (time t=(t_t)). As such, it is also a measure of the engineer's social accountability. People want to be assured that they have water that is safe to drink and healthy for swimming and fishing, which includes assurances that water and wastewater treatment systems are working as designed [203–207].

The probability of a failure per unit time is the hazard rate or failure density [$f(t)$]. This is a function of the likelihood that an adverse outcome will occur. It is not a function of the severity of the outcome [208]. The $f(t)$ is not affected by whether the outcome is very severe, for example, lung cancer and loss of an entire species in an ecosystem, or relatively benign, for example, slight odors or minor leaf damage. The likelihood that an item will fail at a given time interval is found by integrating $f(t)$ rate over a defined time interval:

$$P\{t_1 \le T_f \le t_1\} = \int_{t_1}^{t_2} f(t)dt \quad (10.11)$$

where T_f=time of failure.

Thus, the reliability function $R(t)$ of a system at time t is the cumulative probability that the system will perform without failure in the time interval from t_0 to t_t:

$$R(t) = P\{t_1 \le T_f \le t_1\} = 1 - \int_0^t f(x)dx \qquad (10.12)$$

It is not a question of "if," but "when" an engineered system will fail. An engineer or manager may be able to extend the optimal time of a pollution control or prevention technology, that is, increasing t_t to make the system more resilient under conditions that would otherwise lead to failure. For example, proper engineering design of a water treatment system should work well not only during normal conditions but also resist failure during hostile conditions, for example, a time period of highly corrosive loading. A design flaw occurs if the wrong material is used, for example, piping not matched to worst-case corrosivity conditions. However, even when the proper materials are used, failure is not eliminated entirely. Selecting the right materials simply protracts the time before failure occurs, that is, increases T_f [39]. This means that the likelihood of the adverse outcome is reduced, which means that risk has also been decreased.

In addition to reliability, risk is also related to resilience. Reliable systems are generally resilient systems. Like many other environmental systems, engineers can learn from nature. For example, ecosystems must be resilient. Ecosystem resilience and engineering resilience are mirror images of one another. The less resilient a system is, the more the engineer and manager must do to make it reliable. A very resilient natural system can withstand a wide array of insults. A system that lacks resilience is a sensitive and vulnerable system, needing controls to remain operational when conditions change ("Ecosystem Resilience - an overview I ScienceDirect Topics").

Resilient systems have wide tolerances, whereas a sensitive system is easily harmed outside of very narrow tolerance ranges. Thus, an ecologically resilient system is in less need of engineering controls against insult than a nonresilient system. ("Ecosystem Resilience - an overview I ScienceDirect Topics"). For example, atmospheric deposition of a pollutant onto a resilient system may cause little impact and/or require a relatively short time to recover. Conversely, the same deposition onto a nonresilient system could be devastating.

Unfortunately, human activities can transform resilient systems into nonresilient ones. For example, a wild, biodiverse area may be quite resilient in the face of drought and insect infestation. Human activities, for example, building roads and infrastructure around and near this area, can reduce biodiversity and change biological processes, making it less able to withstand these insults. Thus, a natural system may have been able to withstand lower pH rainfall caused by sulfur dioxide (SO_2) emissions, but other human activities may have made the system more vulnerable to acid rain.

Engineered systems, including pollution control technologies, must also be resilient. They must be able to return to the original performance after being stressed [206,209,210]. When such complete hysteresis exists, risk levels remain stable [209,211,212]. When a system receives a sever shock, hysteresis can prevent it from returning to its original equilibrium, meaning that low resilience will occur and a new, weakened steady state is established. Ecological risks can similarly be decreased when they are biodiverse, productive, and provide ecological resilience, which increases the ability of habitat or other system to overcome hysteresis [213–218].

10.10 Unit risk estimates

Cancer models assume linearity in the extrapolation zone of the dose-response curve. The US EPA often applies a linearized multistage model (LMS), which yields a slope factor known as q1*, to predict cancer risk at a particular dose. The LMS assumes a linear extrapolation with the zero dose threshold from the upper confidence level of the lowest dose that produces cancer in observed human populations using epidemiological methods or in animal studies [219].

Models vary in their estimates of cancer risk outcomes. For example, the concentration of organochlorine pesticide chlordane ($C_{10}H_6C_{18}$) in drinking water needed to cause a lifetime risk of 10^{-6} persons in a population is estimated to be [220]:

- 0.03 μg L^{-1} using the one-hit model;
- 2 μg L^{-1} using the multihit model;
- 50 μg L^{-1} using the probit model; and,
- 0.07 μg L^{-1} using LMS.

Cancer risk can also be expressed in terms of risk per unit concentration of a substance in an exposure route, that is, unit risks. The drinking water unit risk is based on a water consumption rate of 2 L day^{-1}. The air unit risk is calculated by dividing the slope factor by a mean body weight, usually 70 kg, and multiplying by the inhalation rate (20 m^3 day^{-1}). For a substance that is not completely absorbed, that is, absorption fraction is < 1.0, when deriving the slope factor, an additional conversion factor is needed so that the unit risk is based on an administered dose. The exposure in a unit risk is assumed to be continuous over a lifetime. Thus, when no absorption conversion is required:

$$\text{Water unit risk} = \text{risk per}\,\mu\text{g}L^{-1} = \text{slope factor} \times \frac{1}{70}\text{kg} \times \frac{2L}{\text{day}} \times 10^{-3} \quad (10.13)$$

$$\text{Air unit risk} = \text{risk per}\,\mu\text{g}m^{-3} = \text{slope factor} \times \frac{1}{70}\text{kg} \times \frac{20m^3}{\text{day}} \times 10^{-3} \quad (10.14)$$

The multiplication by 10^{-3} is necessary to convert from mg to μg when the slope factor, or q1*, is given in the typical units of mg kg^{-1} day^{-1}.

10.11 Risk uncertainties

Hazard-related data for chemical substances must often be extrapolated from small-scale laboratory experiments and field studies to larger environmental exposure scenarios. As such, they rely on quantitative tools to extrapolate information from a data-rich substance to these data-poor substances [68,165,166,171,221,222]. Uncertainty in this case is tied to any deviation from the behavior of the known to the unknown, since toxic agents always have some degree of uniqueness related to transport and transformation after they enter the environment or an organism's body [206].

Chemometrics is a systematic approach to using mathematical and statistical methods to evaluate chemical data, going beyond univariate analysis to describe chemical compounds quantitatively using existing data that begins with the physicochemical structure [122,201,223,224]. For example, if a data-poor compound has a similar chemical structure to a known compound, for example, only one different functional group on a carbon ring, then statistical and mathematical structure-activity relationships (SARs) provide a first step in screening for potential hazards and exposure scenarios [68,164–166,168].

10.11.1 Error

Errors are a major source of potential risk assessment uncertainty [225]. Errors are classified as either Type 1 or Type 2. A Type I error occurs when experimental findings lead to the rejection of a null hypothesis that is, in fact, true. Conversely, a Type 2 error occurs if the findings lead to the acceptance of a null hypothesis that is, in fact, false. When an occurrence or event, such as a human disease outbreak or the loss of a species due to an environmental exposure, is not identified during an observation or measurement, but, in fact, the occurrence or event exists, this is a false negative. If sampling and analysis are not sufficiently sensitive and lack resolution, it can lead to a false negative [226]. Other problems, for example, level of detection (LoD) being too high, chain-of-custody issues with missing or erroneous documentation, improper interpretation of results, and poor documentation and data entry can also be sources of false negative findings.

The other major type of error is one that leads to a false positive, that is, an occurrence or event is found but, in fact, does not exist [227]. This can be a major challenge for emerging contaminants. For example, consider the situation where previous evidence used by an agency to list a compound as a potential endocrine disruptor has subsequently found that a wealth of new information now shows that it has no such effect [228]. The decision makers must decide whether the new information is sufficient to remove protections that were only required based on the compound's apparently false positive endocrine disruption data. Sometimes, the false positive results from scientific shortfalls, for example, based upon faulty models or models that only work well for lower organisms (i.e., interspecies uncertainty). Often, this is rectified by scientific advancement, for example, models that are subsequently developed consider newly available data that account for physical, chemical, and biological complexities of higher-level organisms, including humans.

False positives may also lead to overemphasis on less severe problems that cause public health officials to devote inordinate amounts of time and resources to address too many problems [229]. False positives may also discourage people from making use of potentially beneficial products. False positives, especially when they occur frequently, create credibility gaps between scientists and the public. In addition to physical agents like ionizing radiation and chemical agents, pathogenic microbes introduce threats to human health. Wastewater, agricultural products, and other media contain populations of microorganisms that can reach food and drinking water, exposing people to pathogens [230–234]. Pathogens can also enter the food chain, where they can reside at various trophic levels and remain in reservoirs, for example, fish and other aquatic biota that are consumed.

Biological hazards are part of the overall human health risks in populations. Cumulative risk is a function of all agents that cause an adverse outcome. Tools like life cycle analyses

(LCAs) of chemical and physical hazards must be combined with techniques that assess biological stressors to gauge the cumulative risk to a population. For example, a quantitative microbial risk assessment (QMRA) can be conducted for waterborne pathogens and combined with risk assessments of chemical stressors. Pathogenic risk is often expressed as a burden of disease, for example, disability-adjusted life years (DALYs) [231]. Thus, the DALYs can be combined with information about physical and chemical agents to depict cumulative risks [235–237].

10.12 Hazard index and hazard quotient

Recently, substances can be detected in the lower range of parts per quadrillion (ppq) for emerging chemical compounds like the so-called "forever chemicals" such as PFAS. For example, the United States recently proposed drinking water maximum contaminant levels (MCLs) for perfluorooctanoic acid (PFOA), perfluorooctane sulfonic acid (PFOS), perfluorononanoic acid (PFNA), hexafluoropropylene oxide dimer acid (HFPO-DA, known as GenX Chemicals), perfluorohexane sulfonic acid (PFHxS), and perfluorobutane sulfonic acid (PFBS) [238,239]. The recommended MCL for PFOA and PFOS is $4\,ng\,L^{-1}$ [240] and a hazard index (HI) of 1 for the rest (see Discussion: Hazard Quotient and Index).

In risk assessments, the threshold concentrations of a pollutant are chosen to protect humans and ecosystems. For emerging contaminants, these concentrations aim to protect human health. The amount of protection is based on the potential for an unacceptable outcome. Often, there are two approaches for addressing a pollutant in water, that is, effluent limits and water quality limits. Effluent limits apply to any source that releases the contaminant to a waterbody, for example, $100\,mg\,L^{-1}$ of a chemical compound. Of course, the actual level of concern (LoC) for emerging contaminants varies considerably given the large number of candidate chemicals of concern. For example, the US EPA publishes lists of contaminants that are not yet regulated but are known or anticipated to be found in the environment. If, for instance, a chemical or biological agent of concern is expected to reach public water systems and may need to be regulated under the Safe Drinking Water Act (SDWA), it is placed on the Contaminant Candidate List (CCL). The most recent draft (CCL5) includes 66 chemicals, 3 chemical groups (PFAS, cyanotoxins, and disinfection byproducts), and 12 microbial contaminants [241].

10.13 Assumptions

Risk assumptions can seriously affect risk estimates [208]. Cancer risk, for instance, is the theoretical probability of a person or population developing a form of cancer when continually or episodically exposed to a carcinogen. Often, the probability used in this estimate is the upper confidence limit. The maximum risk may be the odds in a million of contracting cancer. There are two general ways that such risk characterizations are used to support environmental and public health decisions, that is, direct risk assessments and risk-based treatment targets.

Population risk is the annual excess number of adverse events resulting from exposure to an agent, for example, cancers in an exposed population after being exposed to a carcinogen. The agent may be physical, for example, ionizing radiation, chemical, for example, a compound or class of compounds like coke oven emissions, or biological, for example, aflatoxins produced by fungi or a variant of the COVID-19 virus. For chemical compounds, the population risk is often the additive risk over some general population, for example, a "residual" or increased probability above background or expected rates.

Risk can also be individual, that is, the probability of one person having an adverse effect due to exposure to a physical, chemical, or biological agent. As a worst-case scenario, individual risk might be calculated from exposure estimates based on a maximum-exposed individual (MEI). The hypothetical MEI resides outdoors for an entire lifetime at the point where pollutant concentrations are highest. Such assumptions about exposure affect the risk estimates. For example, the cancer risk from power plants in the United States has been estimated to be 100–1000-fold lower for an average exposed individual than that calculated for the MEI [61].

A common default approach for predicting cancer risk is a linear extrapolation from exposures observed in animal or human occupational studies. A nonlinear approach is recommended when there are sufficient supporting data to establish the carcinogen's mode of action (MOA) to conclude that it is not linear at low doses and that the carcinogen does not demonstrate mutagenic or other activity consistent with linearity at low doses [242]. Thus, a straight line connects the beginning point where reliable data are not available, known as the POD, to the origin on the dose-response curve. The POD can be the lower boundary of the dose for an estimated incidence or a change in response level from a dose-response model, such as the benchmark dose (BMD), or a NOAEC or LOAEC. The default linear extrapolation approach is generally considered to be conservatively protective of public health, including sensitive subpopulations [52]. The slope of this line is the IUR, so the IUR can be presented as a linear extrapolation from a POD of 10% response (LEC_{10}):

$$IUR = \frac{0.1}{LEC_{10(HEC)}} \tag{10.15}$$

where, the $LEC_{10(HEC)}$ is the lowest effective dose using a 10% response concentration, dosimetrically adjusted to a human equivalent concentration.

The air concentration must include all physical phases, that is, both the pollutant in vapor phase and those in aerosol phase. The aerosol phase can simply be the same compound in liquid and solid phases, as well as any of the pollutants sorbed to particulates. At equilibrium, a volatile or semivolatile compound will exist in gaseous and liquid phases. A fraction can also be adsorbed on particle surfaces or absorbed within the particle matrix. The concentrations from the three phases are often normalized and expressed in units of mass per volume (e.g., mg m^{-3}) or mass per mass (e.g., mg kg^{-1}).

Volatile organic compounds (VOCs) are almost completely in the gas phase in ambient air because their vapor pressures under most environmental conditions exceed 10^{-2} kilopascals [243]. For VOCs, then, carbonic anhydrase (CA) is nearly equal to vapor-phase concentrations. Conversely, relatively nonvolatile organic compounds (NVOCs) with vapor pressures $<10^{-5}$ kilopascals are predominantly found in and on particles, unless significant energy is added to increase their volatility. For NVOCs, estimates do not expect a substance amount to reach the gas phase, and the air concentration will be entirely as the amount in and on particulate matter. Between VOCs and NVOCs are the semivolatile organic compounds (SVOCs), which, under environmental conditions, have vapor pressures between 10^{-2} and 10^{-5} kilopascals and can exist in both the gas and particle phases in the ambient air [243].

The hazard quotient (HQ) represents the risk of an adverse outcome other than cancer, that is, noncancer risk [244]. The HQ is calculated by dividing the maximum daily dose (MDD) by the acceptable daily intake (ADI):

$$\text{Noncancer risk} = \text{HQ} = \frac{\text{MDD}}{\text{ADI}} = \frac{\text{Exposure}}{\text{Reference dose or concentraton}} \quad (10.16)$$

Recall that an MDD can be equal to or less than the LADD for a noncarcinogen. HQ is an index, not a probability, which means it indicates relative risk. "If the noncancer risk is greater than 1, the potential risk may be significant, and if the noncancer risk is less than 1, the noncancer risk may be insignificant." ("Air Toxics - an overview | ScienceDirect Topics") Thus, the RfD and RfC are types of ADI.

For example, assuming the chromic acid (Cr^{6+}) mist RfC of 8×10^{-6} mg kg^{-1} day^{-1}, the actual respiratory exposure of people living near a metal processing plant was calculated to be 4×10^{-5} mg kg^{-1} day^{-1}. The HQ for the noncancer risk of the chromic acid mist to the neighborhood can be calculated from Eq. (10.16):

$$\text{HQ} = \frac{\text{Exposure}}{\text{RfC}} = \frac{4 \times 10^{-5}}{8 \times 10^{-6}} = 5.$$

Since the HQ is greater than 1, one would expect people chronically exposed at this level to show adverse effects from inhalation. However, at this same chronic exposure, the RfD for hexavalent chromic acid mists differs for dermal and oral routes, so additional adverse noncancer effects may be even larger than indicated by a single route of exposure.

The HQ value is the ratio of a potential exposure to the applicable toxicity-based benchmark.

An HQ value <1 indicates that the specific contaminant is not likely to cause adverse ecological effects. If more than one potential ecological contaminant is present at the site, the HQs for receptors that could be simultaneously exposed to the contaminants should be summed and presented as the HQ [245].

A population that is exposed to more than one contaminant requires that the HIindicate the cumulative noncancer risk from pollutants 1 through n: ("Air Toxics - an overview | ScienceDirect Topics")

$$\text{HI} = \sum_{i}^{n} \text{HQ} \quad (10.17)$$

The HI is useful in comparing risks at various locations, for example, benzene risks in Denver, St. Louis, Chicago, and Los Angeles. It can also provide an indication of the cumulative and additive risk in a single population exposed to more than one contaminant.

For example, if the HQ for toluene is 0.2 (not significant), xylene is 0.5 (not significant), and tetrachloromethane is 0.4 (not significant), the cumulative risk of the three contaminants is 1.1 (potentially significant).

Estimations of both hazard and exposure are often worst-case scenarios because the risk calculations can have large uncertainties. Effects can occur at very low doses. Human data gathered from even very high-quality epidemiological studies contain some amount of error and do not completely address variability. Humans are highly diverse in their activities with many confounding variables. Thus, uncertainty exists in every phase of risk assessment.

The HQ is a risk factor applied to noncarcinogenic chemicals that relates the delivered dose at the exposure point, for example, the average daily dose (ADD) or median exposure (mg kg^{-1} day^{-1}), to an adverse health endpoint [246]:

$$\text{Risk} = \frac{\text{ADD}}{\text{toxicity}} \tag{10.18}$$

Therefore:

$$\text{HQ} = \frac{\text{D}}{\text{MRL}} \tag{10.19}$$

Or,

$$\text{HQ} = \frac{\text{D}}{\text{RfD}} \tag{10.20}$$

where, D is the exposure dose (mg kg^{-1} d^{-1}), and MRL is the minimum risk level (mg kg^{-1} d^{-1}). MRLs represent estimates of the daily human exposure to a contaminant that are not expected to cause noncancer health effects during a specified ED. MRLs are set below concentrations that may cause adverse health effects in most humans, including sensitive populations. They are derived for acute (1–14 days), intermediate (15–364 days), and chronic (365 days and longer) EDs, and are available for oral and inhalation exposures [246].

The endpoint can be the RfD or RfC specific to that chemical compound. For example, if the EC is 5 μg L^{-1} and the RfD for a chemical is 10 μg L^{-1}, the HQ would be 0.5. Since HQ < 1, the hazard is not considered a major concern. If the EC were to double, the HQ would be 1, indicating a hazard of concern.

Exposure factors can be based on a central tendency exposure (CTE), that is, average or typical intake factors. CTE calculations use mean body weight. Reasonable maximum exposure (RME) refers to exposure factors in the upper end of the exposure distribution (approximately the 95th percentile). RME scenarios assess exposures higher than average but within a realistic exposure range [246].

If HQ < 1.0, the person is on average ingesting a safe dose. If HQ > 1.0, the person is ingesting a dose that is not safe. Neither of these are risk statements, nor do they have large amounts of uncertainty. Often, the safe dose is not a clean bill of health and likely requires more thorough analysis. HQ alone is not a justification for a cleanup action.

When exposures include more than one chemical, the HI is the sum of all HQs:

$$\text{HI} = \text{HQ}_A + \text{HQ}_B + \text{HQ}_C = \frac{\text{EC}_A}{\text{RfD}_A} + \frac{\text{EC}_B}{\text{RfD}_B} + \frac{\text{EC}_C}{\text{RfD}_C} \tag{10.21}$$

For example, if the EC of Chemical A is 5 μg L^{-1} and the RfD for that chemical is 10 μg L^{-1}, the EC of Chemical B is 0.3 μg L^{-1} and the RfD for that chemical is 1 μg L^{-1}, and the EC for Chemical C is 0.05 μg L^{-1} and the RfD for that chemical is 0.5 μg L^{-1}, the HI would be:

$$\frac{5}{10} + \frac{0.3}{1} + \frac{0.05}{0.5} = 0.9$$

So, the mixture does not reach a LoC since HI < 1. However, if new data were to indicate that the RfD for Chemical C should be revised to 0.2, the HI must be recalculated to 1.05. This mixture would then be a hazard of concern since HI > 1.

The HQ and HI method is also used at the screening level for ecological risk assessments [247], especially to estimate risk to fauna in and around contaminated sites. This often requires UFs associated with potential wildlife responses to contaminant exposure based on laboratory-derived toxicity data [248].

10.14 Ecological risks

Ecological risk assessment paradigms have differed from human health risk assessment paradigms. Ecological risk assessment is based mainly on characterizing exposure and ecological effects. Both exposure and effects are considered during problem formulation [249].

Compared to human risk assessments, the ecological risk assessment process focuses more intensely on the coherence of risk assessment and risk management than on the technical analysis of risk [249]. An ecological risk assessment differs from human health risk assessment in that the ecological assessment requires all adverse outcomes to be characterized, whereas the health risk assessment's first step is to identify the hazards of a specific chemical compound, especially its toxicity. This is because, in health risk assessments, hazard connotes either the intrinsic effect of a stressor, that is, chemical inherency, or a margin of safety by comparing an effect with an estimate of EC. Margin of safety means that the hazard is at a concentration in the environment that could be expected to induce an adverse outcome.

Eco-risk example

Peer-reviewed scientific evidence shows that a hypothetical compound, when exposed at 10 ppm, diminishes the ability of a species to reproduce, which in turn changes the predator-prey aspects of the ecosystem, leading to diminished biodiversity. This is an example of an ecological hazard. However, the threshold concentration denoting the hazard would not simply be 10 ppm due to uncertainties in the data and science (e.g., measurement and modeling errors and variability) underlying this hazard threshold. With an ample margin of safety, the hazard threshold is adjusted to a lower concentration, for example, to 5 ppm, to account for these uncertainties. The actual correction depends on the quality and representativeness of the available data, with the correction increasing with increasing uncertainty. For example, if the data are less certain in this case, the margin of safety may be 10 times more protective, that is, 0.5 ppm.

Another aspect of ecological hazard is that of the food chain, which incorporates human health and ecosystem condition. A hazard to an ecosystem must be considered at multiple levels of biological organization [250]. The food chain or web is, therefore, a means by which an ecological hazard becomes a human health hazard. It is also how a hazard can change physicochemical forms to become more hazardous, such as the methylation of mercury in sediment and aquatic fauna tissue. The environmental stressor travels both within and among levels of biological organization, reaching various receptors, thereby influencing and inducing outcomes. The outcome can be explained by physical, chemical, and biological processes [250].

After transport and transformation in the food web, the fate is mammalian tissue. Various modeling and decision tools are available to characterize the movement, transformation, uptake, and fate of the compound. From these, biochemodynamic paradigms can be constructed for individual chemical compounds and mixtures, as well as for microorganisms [39]. Although biological hazards are more complicated than purely chemical hazards, similar thresholds could exist for biota. For example, an introduced species may overwhelm another species in an ecosystem if it is introduced to the ecosystem or habitat. Seeds, spores, or cysts can be carried long distances by wind and water, as well as by attached migrating organisms. In such cases, the biological hazard would be identical to chemical hazard approach. Other biota, such as algae, would require a model that includes both direct and indirect risk factors, such as the increase in nutrient concentrations, changes in sunlight intensity and temperature, and decreases in flow rates that lead to algal blooms. These also result in decreasing dissolved oxygen concentrations and ultimately diminished health of the aquatic ecosystem and adverse outcomes like fish kills. In this case, the risk assessment had to include succession of interim outcomes to reach the ultimate outcome of concern, that is, the fish kill. Thus, there are layers of hazards and exposures that are step functions leading to the risk outcome [251].

Ecosystem hazard and exposure calculations differ from human health risk calculations. The RQ is one tool used in eco-risk assessment. It is deterministic in that the RQ is found by dividing a point estimate of exposure by a point estimate of effects to compare risks [252]:

$$RQ = \frac{\text{Exposure}}{\text{Ecotoxicity}} \tag{10.22}$$

Ecotoxicity is an effect level or endpoint that is derived from toxicity tests that yield values such as lethal concentrations, for example, the concentration at which 50% of the exposed fish die (LC_{50}). The bioassays may also indicate toxicity thresholds, for example, no observed effect concentration (NOEC) or no observed effect level (NOEL). The ecotoxicity effect level is often the median effective concentration, that is, EC_{50}, the concentration of the test substance that decreases biological activity by 50%, for example, algal growth, growth rate, or Daphnia immobilization [253]. Since these organisms are waterborne, the EC_{50} units are mg L^{-1}. The estimated environmental concentration (EEC) is found by:

$$RQ = \frac{EEC}{EC_{50}} \tag{10.23}$$

An ecological health criterion is based on the HQ for an ecosystem. This differs from the human health version of HQ. For ecosystems, an EEC of the contaminant at the site, for example, ng L^{-1} in groundwater or ng kg^{-1} in soil, is divided by the threshold (e.g., no observed adverse effect level – NOAEL) in units that match the EEC:

$$HQ = \frac{EES}{NOAEL} \tag{10.24}$$

The purpose of the RQ is to associate the exposure and toxicity ratio with a LoC. After the RQ is calculated, it is compared to the LoC. An LoC is a relative concept used to interpret the RQ in terms of potential risk to a species. For example, the LoC is used to predict potential impacts on so-called nontarget organisms, that is, those not intended to be killed by a pesticide. Decisions regarding environmental regulatory action are informed by the RQ and LoC [252]. Some representative LoCs are listed in Table 10.2

Ecosystem risks may share the same or slightly adjusted parameters as those applied to human risks. The PNEC used to assess harm to an ecosystem can be modified to resemble the derived no-effect level (DNEL) for humans. Both express the concentration of a substance below which no effect is expected. Thus, both the DNEL and PNEC are forward-looking thresholds analogous to the retrospective human threshold concentrations (NOAEL or NOAEC). As discussed in Chapter 10, DNEL is calculated by dividing a toxicological dose, like the LD_{50} or NOAEL, by an AF. For example, studies may show the incidence of adverse developmental effects in an animal species that would provide each an intraspecies AF=10, an interspecies AF=5, a duration AF=2, and route extrapolation and data quality AF=2. The product of the AF values (200) is the denominator of the DNEL equation.

PNEC values are derived from acute toxicity data from studies exposing bacteria, for example, *Bacillus subtilis* and *E. coli*, to a substance to obtain threshold concentrations for each species. For any species, the PNEC is directly proportional to the NOAEC because the PNEC can be derived by dividing the NOAEC by a series of AFs to address uncertainty and to be protective. Thus if uncertainties are high, very large safety factors are applied, so that the PNEC becomes much smaller than the NOAEC for a substance. AFs may be arbitrary, and PNECs for many substances are not quantified, which limits their applications to species and substances that have been well-studied and for which reliable data are available [151].

The PNEC can be compared to existing concentrations in one or more environmental media or to the PEC. For acceptable risk, $\frac{PNEC}{PEC}$ should be much less than 1.

10.15 Risk and ethics

The risk equation (Eq. 10.10) can also be expressed as the probability of loss (L):

$$R = p(L) \tag{10.25}$$

This indicates the uncertainty of the ethical impact of a technological decision. If $p(L)$ is the likelihood that the loss (harm) is greater than the good outcome, that is, $p(G)$, it serves as

TABLE 10.2 Risk quotients (*RQ*) and levels of concern (*LoC*) for various organisms.

Risk presumptions for birds

Risk presumption	**RQ**	**LoC**
Acute risk	Estimated environmental concentration (EEC)/LC_{50} or LD_{50}/ft^2 or LD_{50}/day	0.5
Acute restricted use	EEC/LC_{50} or LD_{50}/ft^2 or LD_{50}/day or $LD_{50} < 50$ mg kg^{-1}	0.2
Acute endangered species	EEC/LC_{50} or LD_{50}/ft^2 or LD_{50}/day	0.1
Chronic risk	EEC/no observed effect concentration (NOEC)	1.0

Risk presumptions for wild mammals

Risk presumption	**RQ**	**LoC**
Acute high risk	EEC/LC_{50} or LD_{50}/ft^2 or LD_{50}/day	0.5
Acute restricted use	EEC/LC_{50} or LD_{50}/ft^2 or LD_{50}/day or $LD_{50} < 50$ mg kg^{-1}	0.2
Acute endangered species	EEC/LC_{50} or LD_{50}/ft^2 or LD_{50}/day	0.1
Chronic risk	EEC/NOEC	1.0

Risk presumptions for aquatic animals

Risk presumption	**RQ**	**LoC**
Acute high risk	EEC/LC_{50} or EC_{50}	0.5
Acute restricted use	EEC/LC_{50} or EC_{50}	0.1
Acute endangered species	EEC/LC_{50} or EC_{50}	0.05
Chronic risk	EEC/NOAEC	1.0
Acute endangered species	EEC/EC_{50} or NOEC	1.0

Risk presumptions for aquatic plants

Risk presumption	**RQ**	**LoC**
Acute high risk	EEC/EC_{50}	1.0
Acute endangered species	EEC/EC_{05} or NOEC	1.0

Risk presumptions for terrestrial and semiaquatic plants

Risk presumption	**RQ**	**LoC**
Acute high risk	EEC/EC_{25}	1.0
Acute endangered species	EEC/EC_{05} or NOEC	1.0

Data from: U.S. Environmental Protection Agency. Technical Overview of Ecological Risk Assessment: Risk Characterization. https://www.epa.gov/pesticide-science-and-assessing-pesticide-risks/technical-overview-ecological-risk-assessment-risk. (accessed September 5, 2018).

a first screen that the decision is likely not to be morally permissible. This can be a construct under utilitarianism, especially when applying the harm principle.

This is, at best, a semiquantitative calculation. The expected harm can be estimated based on the probability and the amount of harm. For example, if the expected outcome has a high probability and a low degree of harm, for example, as a 95% chance of delaying a road repair by one day, this may be acceptable if it is the best alternative that meets all factors of safety and is the most acceptable in terms of public welfare. However, a low-probability, high-harm decision may be immoral even at 10^{-6} if the potential outcome would be catastrophic, for example, creating a cancer cluster in a town as a result of the decision to allow a plant to release a few kilograms of a toxic substance.

Of course, ethical harm is often difficult to quantify. Loss of an aspect of quality of life, for example, less green space in a city, is not merely a calculation of the acreage lost, but also the type of vegetation, the prior usage rates, and the psychosocial perception of the green space's nonmonetized value [254–257].

10.16 Conclusions

This chapter has discussed the risk paradigm, beginning with identifying the hazard, followed by dose-response curves and interpretations, exposure assessment, and risk characterization. The chapter's focus is on chemical pollutants but also discusses biological and physical agents. The risk assessment and management tools are essential for optimizing the selection of new and adaptive technology and ethical decision-making.

References

[1] D. Vallero, Fundamentals of Water Pollution: Quantifying Pollutant Formation, Transport, Transformation, Fate and Risks, 1st ed., Elsevier, Amsterdam, NV, (2024).

[2] A. Zewe, Explained: Generative AI's environmental impact, MIT News. Luettavissa: Available from https://news.mit.edu/2025/explained-generative-ai-environmental-impact-0117. Luettu, 16, p. 2025 (2025).

[3] Y. Cai, J. Cai, D. Chen, Y. Xie, J. Feng, Development of an integrated prediction-optimization modeling approach for coupled risk management of water and energy nexus systems, Science of the Total Environment 781 (2021) 146744.

[4] H. Braakhuis, W. Fransman (Ed.), The future of chemicals is Safe and Sustainable by Design (2024).

[5] J. Jaworska, S. Gabbert, T. Aldenberg, Towards optimization of chemical testing under REACH: a Bayesian network approach to Integrated Testing Strategies, Regulatory Toxicology and Pharmacology 57 (2-3) (2010) 157–167.

[6] R.S. Judson, et al., Estimating toxicity-related biological pathway altering doses for high-throughput chemical risk assessment, Chemical Research in Toxicology 24 (4) (2011) 451–462.

[7] B.S. Wankhade, et al., Neurological Sequelae After Acute Carbon Monoxide Poisoning, Cureus 16 (1) (2024).

[8] A. Dorey, P. Scheerlinck, H. Nguyen, T. Albertson, Acute and chronic carbon monoxide toxicity from tobacco smoking, Military Medicine 185 (1-2) (2020) e61–e67.

[9] M.T. Kleinman, Carbon monoxide, Environmental Toxicants: Human Exposures and Their Health Effects (2020) 455–486.

[10] D. HaMai, S.C. Bondy, Dimethylmercury, in: P. Wexler (Ed.), Encyclopedia of Toxicology (Second Edition), Elsevier, New York, (2005), pp. 54–57.

[11] R.E. Gosselin, R.P. Smith, H.C. Hodge, J.E. Braddock, Mercury, fifth ed., Clinical toxicology of commercial products 1085 Williams & Wilkins, Baltimore, Maryland, (1984), pp. 262–275.
[12] D.W. Nierenberg, et al., Delayed cerebellar disease and death after accidental exposure to dimethylmercury, New England Journal of Medicine 338 (23) (1998) 1672–1676.
[13] S. Saavedra, Á. Fernández-Recamales, A. Sayago, A. Cervera-Barajas, R. González-Domínguez, J.D. Gonzalez-Sanz, Impact of dietary mercury intake during pregnancy on the health of neonates and children: A systematic review, Nutrition Reviews 80 (2) (2022) 317–328.
[14] K. Dack, M. Fell, C.M. Taylor, A. Havdahl, S.J. Lewis, Mercury and Prenatal Growth: A Systematic Review, International Journal of Environmental Research and Public Health 18 (13) (2021) 7140.
[15] J. Wang, et al., Relation of prenatal low-level mercury exposure with early child neurobehavioral development and exploration of the effects of sex and DHA on it, Environment International 126 (2019) 14–23.
[16] P. de Almeida Rodrigues, R.G. Ferrari, L.N. Dos Santos, C.A.C. Junior, Mercury in aquatic fauna contamination: a systematic review on its dynamics and potential health risks, Journal of Environmental Sciences 84 (2019) 205–218.
[17] M. Sakamoto, et al., Health impacts and biomarkers of prenatal exposure to methylmercury: lessons from Minamata, Japan, Toxics 6 (3) (2018) 45.
[18] C.-B. Jiang, et al., Mercury concentration in meconium and risk assessment of fish consumption among pregnant women in Taiwan, Science of the Total Environment 408 (3) (2010) 518–523.
[19] Northeast Waste Management Officials' Association. Indoor Air Mercury. Available from http://www.newmoa.org/prevention/mercury/MercuryIndoor.pdf (accessed July 26, 2018.
[20] B. Aberg, L. Ekman, R. Falk, U. Greitz, G. Persson, J.O. Snihs, Metabolism of methyl mercury (203Hg) compounds in man: Excretion and distribution, Archives of Environmental Health: An International Journal 19 (4) (1969) 478–484.
[21] M. Leventer-Roberts, et al., Environmental tobacco smoke exposure among children by urinary biomarkers and parent report, Academic Pediatrics 21 (4) (2021) 663–669.
[22] N.L. Benowitz, Biomarkers of environmental tobacco smoke exposure, Environmental Health Perspectives 107 (suppl 2) (1999) 349–355.
[23] N.L. Benowitz, Cotinine as a biomarker of environmental tobacco smoke exposure, Epidemiologic Reviews 18 (2) (1996) 188–204.
[24] C. Wormuth, Homeland security risk assessments: Key issues and challenges. Testimony before the Subcommittee on, Intelligence, Information Sharing and Terrorism Risk Assessment Committee on Homeland Security United States House of Representatives, Center for Strategic and International Studies, (2005).
[25] U.S. House of Representatives. Serial No. 109-58, (2005). Homeland security risk assessments: key issues and challenges. Testimony before the Subcommittee on Intelligence, Information Sharing and Terrorism Risk Assessment Committee on Homeland Security United States House of Representatives. [Online] Available from: https://www.govinfo.gov/content/pkg/CHRG-109hhrg35939/html/CHRG-109hhrg35939.htm
[26] E. Cameron, G.F. Peloso, Risk management and the precautionary principle: a fuzzy logic model, Risk Analysis: An International Journal 25 (4) (2005) 901–911.
[27] National Research Council, Improving Risk Communication, National Academy Press, Waashington, DC, (1989).
[28] National Research Council, Understanding Risk: Informing Decisions in a Democratic Society, National Academies Press, Washington, DC, (1996).
[29] N.C. Rasmussen, The application of probabilistic risk assessment techniques to energy technologies, Readings in Risk, RFF Press, Washington, DC, (1990), pp. 195–206.
[30] D.A. Vallero, Hazardous wastes, Encyclopedia of Science and Technology, AccessScience, McGraw-Hill Education, New York, NY, (2017).
[31] D.A. Vallero, T.M. Letcher, Unraveling Environmental Disasters, second ed., Elsevier Academic Press, Amsterdam, The Netherlands, (2023).
[32] U.S. Environmental Protection Agency, NAAQS table. Available from: https://www.epa.gov/criteria-air-pollutants/naaqs-table (accessed 25.08.18).
[33] Centers for Disease Control and Prevention, Temperature Extremes, March 2, 2024. Available from https://www.cdc.gov/climate-health/php/effects/temperature-extremes.html; accessed on May 8, 2026.

[34] G.I. Otero, A.M. Blanco, M.C. Montero, L.P. Valiño, H.H. Verea, The epidemiology of COPD and asthma exacerbations in a general hospital, Archivos de Bronconeumologia 38 (6) (2002) 256.

[35] *Intergovernmental Panel of Climate Change. 2013,. Climate Change 2013: The Physical Science Basis. Contribution of Working Group I to the Fifth Assessment Report of the Intergovernmental Panel on Climate Change* [Stocker, T.F., et al. (eds.)]. Cambridge University Press. Available from https://doi.org/10.1017/CBO9781107415324.

[36] P. Almagro, C. Hernandez, P. Martinez-Cambor, R. Tresserras, J. Escarrabill, Seasonality, ambient temperatures and hospitalizations for acute exacerbation of COPD: a population-based study in a metropolitan area, International Journal of Chronic Obstructive Pulmonary Disease 10 (2015) 899.

[37] D.A. Vallero, Fundamentals of Air Pollution, fifth ed., Elsevier Academic Press, Waltham, MA, (2014), p. 999 pages cm.

[38] D. Vallero, Ethical Decisions in Emergent Science, Engineering and Technologies, INTECH Open Access Publisher, (2012).

[39] National Research Council, Biosolids Applied to Land: Advancing Standards and Practices, National Academies Press, (2002).

[40] P.P. Egeghy, D.A. Vallero, E.A.C. Hubal, Exposure-based prioritization of chemicals for risk assessment, Environmental Science & Policy 14 (8) (2011) 950–964.

[41] C. Grady, Institutional Review Boards: Purpose and Challenges, Chest 146 (2015) 1148–1155. Available from https://doi.org/10.1378/chest.15-0706.

[42] D. Vallero, Environmental Biotechnology: A Biosystems Approach, Elsevier Science, (2015).

[43] Health and Safety Executive, Guidance on ALARP decisions in COMAH. HSE. Available from http://www.hse.gov.uk/foi/internalops/hid_circs/permissioning/spc_perm_37/ (accessed 14.05.17).

[44] R. Bell, D. Reinert, Risk and system integrity concepts for safety-related control systems, in: Computer Assurance, 1993. COMPASS'93, Practical Paths to Assurance. Proceedings of the Eighth Annual Conference on, 1993, pp. 15–34. IEEE.

[45] D.A. Vallero, P.A. Vesilind, Socially Responsible Engineering: Justice in Risk Management, John Wiley & Sons, Hoboken, New Jersey, (2007).

[46] R. S. S. Group, Risk: analysis, perception and management (1992).

[47] S.L. Derby, R.L. Keeney, Risk analysis: understanding "how safe is safe enough?", Risk Analysis 1 (3) (1981) 217–224.

[48] M.G. Morgan, Risk assessment: choosing and managing technology-induced risk: how much risk should we choose to live with? How should we assess and manage the risks we face? IEEE Spectrum 18 (12) (1981) 53–60.

[49] New Jersey Department of Health and Senior Services, Hazardous substance fact sheet: copper acetoarsenate. Available from https://nj.gov/health/eoh/rtkweb/documents/fs/0529.pdf (accessed 26.07.18).

[50] N.R. Council, Biosolids Applied to Land: Advancing Standards and Practices, National Academies Press, (2002).

[51] U.S. Environmental Protection Agency, Quantitative estimate of carcinogenic risk from oral exposure: aniline. U.S. EPA. Available from https://iris.epa.gov/ChemicalLanding/&substance_nmbr=350 (accessed 10.02.23).

[52] U.S. Environmental Protection Agency, EPA's approach for assessing the risks associated with chronic exposure to carcinogens: background document 2 - January 17, 1992. U.S. EPA. Available from https://www.epa.gov/iris/epas-approach-assessing-risks-associated-chronic-exposure-carcinogens, 2020 (accessed 29.09.20).

[53] Proposed guidelines for carcinogen risk assessment. [Online] Available from http://cfpub.epa.gov/ncea/cfm/recordisplay.cfm?deid=2833, 2015.

[54] U.S. Environmental Protection Agency, Risk assessment for carcinogenic effects. U.S. EPA. Available from https://www.epa.gov/fera/risk-assessment-carcinogenic-effects (accessed 30.01.24).

[55] D. Dawson, et al., Assessment of non-occupational 1, 4-dioxane exposure pathways from drinking water and product use, Environmental Science & Technology 56 (8) (2022) 5266–5275.

[56] N.M. DeLuca, M. Angrish, A. Wilkins, K. Thayer, E.A.C. Hubal, Human exposure pathways to poly-and perfluoroalkyl substances (PFAS) from indoor media: a systematic review protocol, Environment International 146 (2021) 106308.

[57] C.L. Ring, et al., Consensus modeling of median chemical intake for the US population based on predictions of exposure pathways, Environmental Science & Technology 53 (2) (2018) 719–732.

[58] K.L. Dionisio, et al., Exploring consumer exposure pathways and patterns of use for chemicals in the environment, Toxicology Reports 2 (2015) 228–237.
[59] U.S. Environmental Protection Agency, Vinyl chloride: 75-01-4 | DTXSID8021434. Available from https://comptox.epa.gov/dashboard/chemical/executive-summary/DTXSID8021434?list=ATSDRLST, 2020 (accessed 18.01.22).
[60] U.S. Environmental Protection Agency, Guidelines for human exposure assessment risk assessment (no. EPA/100/B-19/001), 2019, Washington, DC: Risk Assessment Forum Washington, DC. [Online]. Available from https://www.epa.gov/sites/default/files/2020-01/documents/guidelines_for_human_exposure_assessment_final2019.pdf. [Online]; https://www.epa.gov/sites/default/files/2020-01/documents/guidelines_for_human_exposure_assessment_final2019.pdf
[61] K. Huang, S.A. Sanchez, M.J. Eckelman, Using the US National Air Toxics Assessment to benchmark the USEtox inhalation-mediated carcinogenic impacts of air emissions, The International Journal of Life Cycle Assessment 26 (7) (2021) 1417–1430.
[62] National-scale air toxics assessment for 2005: fact sheet. [Online] Available from http://www.epa.gov/ttn/atw/nata2005/05pdf/nata2005_factsheet.pdf, 2013.
[63] M. Strum et al., The 2011 National Air Toxics Assessment (2011 NATA): Emissions and Modeling Approach.
[64] ENV/JM/MONO(2012)10/PART1, The Adverse Outcome Pathway for Skin Sensitisation Initiated by Covalent Binding to Proteins: Part 1: Scientific Evidence, 2012.
[65] ENV/JM/MONO(2011)8, Report of the Workshop on Using Mechanistic Information in Forming Chemical Categories, 2011.
[66] R.L. Buchanan, J.L. Smith, W. Long, Microbial risk assessment: dose-response relations and risk characterization, International Journal of Food Microbiology 58 (3) (2000) 159–172. Available from https://doi.org/10.1016/S0168-1605(00)00270-1/07/15/ 2000.
[67] M. Coleman, H. Marks, Topics in dose-response modeling, Journal of Food Protection 61 (11) (1998) 1550–1559.
[68] U.S. Environmental Protection Agency, Quantitative structure activity relationship. Available from http://www.epa.gov/nrmrl/std/qsar/qsar.html (accessed 19.03.15).
[69] U.S. Environmental Protection Agency (2015). Quantitative Structure Activity Relationship. [Online] Available from http://www.epa.gov/nrmrl/std/qsar/qsar.html
[70] D.E. Dawson, B.L. Ingle, K.A. Phillips, J.W. Nichols, J.F. Wambaugh, R. Tornero-Velez, Designing QSARs for parameters of high-throughput toxicokinetic models using open-source descriptors, Environmental Science & Technology 55 (9) (2021) 6505–6517.
[71] D. Wang, et al., A hybrid framework for improving uncertainty quantification in deep learning-based QSAR regression modeling, Journal of Cheminformatics 13 (1) (2021) 69. Available from https://doi.org/10.1186/s13321-021-00551-x2021/09/20.
[72] D.S. Esser, J.H.J. Leveau, K.M. Meyer, Modeling microbial growth and dynamics, Applied Microbiology and Biotechnology 99 (21) (2015) 8831–8846. Available from https://doi.org/10.1007/s00253-015-6877-6. 2015/11/01.
[73] R. Maheshwari, G. Joshi, D.K. Mishra, R.K. Tekade, Bionanotechnology in pharmaceutical research, Basic Fundamentals of Drug Delivery, Elsevier, (2019), pp. 449–471.
[74] V. Stone, et al., Nanomaterials versus ambient ultrafine particles: an opportunity to exchange toxicology knowledge, Environmental Health Perspectives 125 (10) (2017) 106002.
[75] D.H. Brouwer, et al., Occupational dermal exposure to nanoparticles and nano-enabled products: part 2, exploration of exposure processes and methods of assessment, International Journal of Hygiene and Environmental Health 219 (6) (2016) 503–512.
[76] D. Yang, et al., DNA materials: bridging nanotechnology and biotechnology, Accounts of Chemical Research 47 (6) (2014) 1902–1911.
[77] R. Falkner, N. Jaspers, Regulating nanotechnologies: risk, uncertainty and the global governance gap, Global Environmental Politics 12 (1) (2012) 30–55.
[78] G.V. Lowry, K.B. Gregory, S.C. Apte, J.R. Lead, Transformations of Nanomaterials in the Environment, ACS Publications, (2012).
[79] D. Vallero, Nano-Cerium Fuel Additives Phase Distribution, U.S. Environmental Protection Agency, National Exposure Research Laboratory, Research Triangle Park, North Carolina, (2012). J. Weinstein (Ed.).

[80] L.S. Birnbaum, P. Jung, From endocrine disruptors to nanomaterials: advancing our understanding of environmental health to protect public health, Health Affairs 30 (5) (2011) 814–822.
[81] H. Becker, F. Herzberg, A. Schulte, M. Kolossa-Gehring, The carcinogenic potential of nanomaterials, their release from products and options for regulating them, International Journal of Hygiene and Environmental Health 214 (3) (2011) 231–238.
[82] J.R. Lead, Manufactured nanoparticles in the environment, Environmental Chemistry 7 (1) (2010) 1–2.
[83] K.D. Grieger, S.F. Hansen, A. Baun, The known unknowns of nanomaterials: describing and characterizing uncertainty within environmental, health and safety risks, Nanotoxicology 3 (3) (2009) 222–233.
[84] M. Baalousha, J.R. Lead, Overview of Nanoscience in the Environment, Environmental and Human Health Impacts of Nanotechnology, Wiley-Blackwell Publishing Ltd, Hoboken, NJ, (2009), pp. 1–25.
[85] F.F. Larese, et al., Human skin penetration of silver nanoparticles through intact and damaged skin, Toxicology 255 (1-2) (2009) 33–37.
[86] S.J. Klaine, et al., Nanomaterials in the environment: behavior, fate, bioavailability, and effects, Environmental Toxicology and Chemistry 27 (9) (2008) 1825–1851.
[87] National Institute for Occupational Safety and Health, Approaches to safe nanotechnology: an information exchange with NIOSH. Available from http://www.cdc.gov/niosh/topics/nanotech/safenano/, 2007 (accessed 02.08.07).
[88] Nanotechnology White Paper, 2007.
[89] D.A. Vallero, Beyond responsible conduct in research: new pedagogies to address macroethics of nanobiotechnologies, Journal of Long-Term Effects of Medical Implants 17 (1) (2007).
[90] H.-L. Lien, D.W. Elliott, Y.-P. Sun, W.-X. Zhang, Recent progress in zero-valent iron nanoparticles for groundwater remediation, Journal of Environmental Engineering and Management 16 (6) (2006) 371.
[91] D.B. Warheit, Nanoparticles: health impacts? Materials Today 7 (2) (2004) 32–35.
[92] US Environmental Protection Agency, PFAS action plan: program update February 2020. Office of Air Quality Planning & Standards. Available from https://www.epa.gov/pfas/pfas-action-plan-program-update-february-2020, 2020 (accessed 31.07.20).
[93] US Environmental Protection Agency, Basic information on PFAS. Office of Air Quality Planning & Standards. Available from https://www.epa.gov/pfas/basic-information-pfas, 2020 (accessed 31.07.20).
[94] C. Biamis, K. O'Driscoll, G. Hardiman, Microplastic toxicity: a review of the role of Marine Sentinel Species in assessing the Environmental and Public Health impacts, Case Studies in Chemical and Environmental Engineering (2020) 100073.
[95] M.J. Carvan, T.P. Dalton, G.W. Stuart, D.W. Nebert, Transgenic zebrafish as sentinels for aquatic pollution, Annals of the New York Academy of Sciences 919 (1) (2000) 133–147.
[96] M.C. Fossi, C. Panti, Sentinel species of marine ecosystems, in: Oxford Research Encyclopedia of Environmental Science, 2017.
[97] J. Zhu, et al., Cetaceans and microplastics: first report of microplastic ingestion by a coastal delphinid, Sousa chinensis, Science of the Total Environment 659 (2019) 649–654.
[98] Y. Liu, et al., Uptake and depuration kinetics of microplastics with different polymer types and particle sizes in Japanese medaka (Oryzias latipes), Ecotoxicology and Environmental Safety 212 (2021) 112007.
[99] X. Qiu, et al., Quantifying the vector effects of polyethylene microplastics on the accumulation of anthracene to Japanese medaka (*Oryzias latipes*), Aquatic Toxicology 228 (2020) 105643.
[100] G.T. Ankley, et al., Adverse outcome pathways: a conceptual framework to support ecotoxicology research and risk assessment, Environmental Toxicology and Chemistry 29 (3) (2010) 730–741.
[101] A. Lillicrap, et al., Alternative approaches to vertebrate ecotoxicity tests in the 21st century: a review of developments over the last 2 decades and current status, Environmental toxicology and chemistry 35 (11) (2016) 2637–2646.
[102] A.L. Karmaus, D.L. Filer, M.T. Martin, K.A. Houck, Evaluation of food-relevant chemicals in the ToxCast high-throughput screening program, Food and Chemical Toxicology 92 (2016) 188–196.
[103] U.S. Environmental Protection Agency. Chemistry Dashboard. Available from https://www.epa.gov/chemical-research/chemistry-dashboard (accessed October 12, 2020, 2020).
[104] A.J. Williams, et al., The CompTox Chemistry Dashboard: a community data resource for environmental chemistry, Journal of cheminformatics 9 (1) (2017) 61.

[105] M.D. Wood, K. Plourde, S. Larkin, P.P. Egeghy, A.J. Williams, V. Zemba, et al., Advances on a decision analytic approach to exposure-based chemical prioritization, Risk Analysis 40 (1) (2020) 83–96.
[106] D.A. Vallero, Environmental Systems Science: Theory and Practical Applications, first ed., Elsevier Academic Press, (2021).
[107] P.T. Anastas, R.L. Lankey, Life cycle assessment and green chemistry: the yin and yang of industrial ecology, Green Chemistry 2 (6) (2000) 289–295.
[108] U.S. Environmental Protection Agency, Benzo(a)pyrene: 50-32-8 | DTXSID2020139. Available from https://comptox.epa.gov/dashboard/dsstoxdb/results?search=DTXSID2020139#adme, 2020 (accessed 12.10.20).
[109] D. Hattis, M. Lynch, Interspecies Differences and Human Inter-Individual Variability in Tissue-Level Pharmacokinetic Parameters Task 3 Final Report—Description and Analyses of the Data Bases, Report by Clark University to the US Environmental Protection Agency under Purchase Order no. EP05W002147. Available from http://www2.clarku.edu/faculty/dhattis/, , 2006 (accessed 12.19), p. 06.
[110] D. Hattis, A. Russ, Role of dosimetric scaling and species extrapolation in evaluating risks across life stages. II. Pharmacokinetic dosimetric considerations in old age. Report to US Environmental Protection Agency. RFQ no, DC-03-00009. Washington, DC: National Center for Environmental Assessment, 2003.
[111] E.J. Calabrese, The linear no-threshold (LNT) dose response model: a comprehensive assessment of its historical and scientific foundations, Chemico-Biological Interactions 301 (2019) 6–25.
[112] EPA/100/R-14/002F, Guidance for applying quantitative data to develop data-derived extrapolation factors for interspecies and intraspecies extrapolation. [Online] Available from https://www.epa.gov/sites/default/files/2015-01/documents/ddef-final.pdf, 2014.
[113] U.S. Environmental Protection Agency, Rapid chemical exposure and dose research. Available from https://www.epa.gov/chemical-research/rapid-chemical-exposure-and-dose-research, 2017 (accessed 28.03.17).
[114] U. EPA, Benchmark Dose Technical Guidance, US Environmental Protection Agency, (2012).
[115] J. Smith, Aflatoxins, Handbook of Plant and Fungal Toxicants, CRC Press, (2020), pp. 269–285.
[116] EPA, Exposure factors handbook - soil and dust ingestion. [Online] Available from https://www.epa.gov/expobox/exposure-factors-handbook-chapter-5, 2017.
[117] D.A. Vallero, S.S. Isukapalli, V.G. Zartarian, T.R. Mccurdy, T.E. Mckone, P. Georgopoulos, et al., Modeling and predicting pesticide exposures, in: R. Krieger (Ed.), Hayes Handbook of Pesticide Toxicology, Elsevier Science, New York, NY, (2010), pp. 995–1020 ch. 44.
[118] T. McCurdy, Modeling the dose profile in human exposure assessments: ozone as an example, Reviews in Toxicology 1 (1997) 3–23.
[119] National Research Council, Exposure Science in the 21st Century: A Vision and a Strategy, The National Academies Press, Washington, DC, (2012), p. 196 (in English).
[120] D. Vallero, Environmental Contaminants: Assessment and Control, Academic Press, (2010).
[121] E.J. Calabrese, L.A. Baldwin, The frequency of U-shaped dose responses in the toxicological literature, Toxicological Sciences 62 (2) (2001) 330–338.
[122] D.A. Vallero, Translating diverse environmental data into reliable information: how to coordinate evidence from different sources, Academic Press, (2017).
[123] M. Kamber, S. Flückiger-Isler, G. Engelhardt, R. Jaeckh, E. Zeiger, Comparison of the Ames II and traditional Ames test responses with respect to mutagenicity, strain specificities, need for metabolism and correlation with rodent carcinogenicity, Mutagenesis (2009) p. 17.
[124] S. Flückiger-Isler, et al., Assessment of the performance of the Ames II™ assay: a collaborative study with 19 coded compounds, Mutation Research/Genetic Toxicology and Environmental Mutagenesis 558 (1) (2004) 181–197.
[125] B.N. Ames, J. McCann, E. Yamasaki, Methods for detecting carcinogens and mutagens with the Salmonella/mammalian-microsome mutagenicity test, Mutation Research/Environmental Mutagenesis and Related Subjects 31 (6) (1975) 347–363.
[126] J. McCann, E. Choi, E. Yamasaki, B.N. Ames, Detection of carcinogens as mutagens in the Salmonella/microsome test: assay of 300 chemicals, Proceedings of the National Academy of Sciences 72 (12) (1975) 5135–5139.
[127] P.A. Schulte, et al., Occupational safety and health, green chemistry, and sustainability: a review of areas of convergence, Environmental Health 12 (1) (2013) 1.

[128] R. Golden, J. Bus, E. Calabrese, An examination of the linear no-threshold hypothesis of cancer risk assessment: introduction to a series of reviews documenting the lack of biological plausibility of LNT, Chemico-Biological Interactions 301 (2019) 2–5.
[129] D. Costantini, B. Borremans, The linear no-threshold model is less realistic than threshold or hormesis-based models: an evolutionary perspective, Chemico-Biological Interactions 301 (2019) 26–33.
[130] B.A. Ulsh, A critical evaluation of the NCRP COMMENTARY 27 endorsement of the linear no-threshold model of radiation effects, Environmental Research 167 (2018) 472–487.
[131] M. Doss, The conclusion of the BEIR VII report endorsing the linear no-threshold model is no longer valid due to advancement of knowledge, Journal of Nuclear Medicine 59 (11) (2018) 1777 1777.
[132] W. Weber, P. Zanzonico, The controversial linear no-threshold model, Journal of Nuclear Medicine 58 (1) (2017) 7–8.
[133] E.J. Calabrese, J.C. Hanekamp, D.Y. Shamoun, The EPA cancer risk assessment default model proposal: moving away from the LNT, Dose-Response 16 (3) (2018) 1559325818789840.
[134] U.S. Environmental Protection Agency, Cancer risk calculations. Available from http://www.epa.gov/oswer/riskassessment/sghandbook/riskcalcs.htm (accessed 01.05.15).
[135] E.E. McConnell, H.A. Solleveld, J.A. Swenberg, G.A. Boorman, Guidelines for combining neoplasms for evaluation of rodent carcinogenesis studies, Journal of the National Cancer Institute 76 (2) (1986) 283–289.
[136] I. A. f. R. o. Cancer, IARC monographs on the evaluation of carcinogenic risk of chemicals to man, IARC monographs on the evaluation of carcinogenic risk of chemicals to man., 1, 1972.
[137] EPA/630/P-03/001F, Guidelines for Carcinogen Risk Assessment, 2005.
[138] National Research Council, Science and Decisions: Advancing Risk Assessment, The National Academies Press, Washington, DC, (2009), p. 424 (in English).
[139] K.M. Seong, et al., Is the linear no-threshold dose-response paradigm still necessary for the assessment of health effects of low dose radiation? Journal of Korean Medical Science 31 (2016) S10–S23.
[140] R. Judson, et al., In vitro and modelling approaches to risk assessment from the US Environmental Protection Agency ToxCast programme, Basic & Clinical Pharmacology & Toxicology 115 (1) (2014) 69–76.
[141] D.J. Dix, K.A. Houck, M.T. Martin, A.M. Richard, R.W. Setzer, R.J. Kavlock, The ToxCast program for prioritizing toxicity testing of environmental chemicals, Toxicological Sciences 95 (1) (2007) 5–12.
[142] A.M. Evans, G.E. Rice, J.M. Wright, L.K. Teuschler, Exploratory cumulative risk assessment (CRA) approaches using secondary data, Human and Ecological Risk Assessment: An International Journal 20 (3) (2014) 704–723.
[143] P. Whaley, et al., Improving the quality of toxicology and environmental health systematic reviews: what journal editors can do, ALTEX-Alternatives to Animal Experimentation 38 (3) (2021) 513–522.
[144] C.A. Weitekamp, L.J. Phillips, L.M. Carlson, N.M. DeLuca, E.A.C. Hubal, G.M. Lehmann, A state-of-the-science review of polychlorinated biphenyl exposures at background levels: relative contributions of exposure routes, Science of the Total Environment (2021) 145912.
[145] E.A.C. Hubal, et al., Advancing systematic-review methodology in exposure science for environmental health decision making, Journal of Exposure Science & Environmental Epidemiology 30 (6) (2020) 906–916.
[146] E.A. Cohen Hubal, D.M. Reif, R. Slover, A. Mullikin, J.C. Little, Children's environmental health: a systems approach for anticipating impacts from chemicals, International Journal of Environmental Research and Public Health 17 (22) (2020) 8337.
[147] J. Mitchell, et al., A decision analytic approach to exposure-based chemical prioritization, PLoS One 8 (8) (2013) e70911. Available from https://doi.org/10.1371/journal.pone.0070911.
[148] J. Mitchell, et al., Comparison of modeling approaches to prioritize chemicals based on estimates of exposure and exposure potential, Science of the Total Environment 458 (2013) 555–567.
[149] J.F. Wambaugh, et al., High-throughput models for exposure-based chemical prioritization in the ExpoCast project, Environmental Science & Technology 47 (15) (2013) 8479–8488.
[150] G. Roman, P. Isnard, J.-M. Jouany, Critical analysis of methods for assessment of predicted no-effect concentration, Ecotoxicology and Environmental Safety 43 (2) (1999) 117–125.
[151] X. Jin, J. Zha, Y. Xu, J.P. Giesy, K.L. Richardson, Z. Wang, Derivation of predicted no effect concentrations (PNEC) for 2, 4, 6-trichlorophenol based on Chinese resident species, Chemosphere 86 (1) (2012) 17–23.
[152] B. Benson, A. Mills, B. Wood, A review of the reference dose and reference concentration processes, in: Risk Assessment Forum EPA/630/P-02 F, 2002, 2.

[153] R. Landsiedel, et al., Testing metal-oxide nanomaterials for human safety, Advanced Materials 22 (24) (2010) 2601–2627.
[154] I.M. Cozzarelli, J.R. McKelvie, A.L. Baehr, 11.12 - Volatile Hydrocarbons And Fuel Oxygenates, in: H.D. Holland, K.K. Turekian (Eds.), Treatise on Geochemistry, second ed. Elsevier, Oxford, (2014), pp. 439–480.
[155] N. Gao, L. Yang, X. Lu, Z. Duan, L. Zhu, J. Feng, A review of interactions of microplastics and typical pollutants from toxicokinetics and toxicodynamics perspective, Journal of Hazardous Materials 432 (2022) 128736. Available from https://doi.org/10.1016/j.jhazmat.2022.128736. 2022/06/15/.
[156] D.A. Vallero, Mapping Exposure onto Nanoscale Toxicity Measures, in: J.R. Lead, S.H. Doak, M.J.D. Clift (Eds.), Nanotoxicology in Humans and the Environment, Springer International Publishing, Cham, (2021), pp. 141–191.
[157] M.R. Wiesner, et al., Decreasing uncertainties in assessing environmental exposure, risk, and ecological implications of nanomaterials†‡, Environmental Science & Technology 43 (17) (2009) 6458–6462.
[158] O. Lindqvist, et al., Mercury in the Swedish environment—Recent research on causes, consequences and corrective methods, Water, Air, and Soil Pollution 55 (1) (1991) xi–261. Available from https://doi.org/10.1007/BF005424291991/01/01.
[159] S. Yan, S.B. Subramanian, R. Tyagi, R.Y. Surampalli, T.C. Zhang, Emerging contaminants of environmental concern: Source, transport, fate, and treatment, Practice Periodical of Hazardous, Toxic, and Radioactive Waste Management 14 (1) (2010) 2–20.
[160] S.-D. Li, L. Huang, Pharmacokinetics and biodistribution of nanoparticles, Molecular Pharmaceutics 5 (4) (2008) 496–504.
[161] S.G. Elci, et al., Surface charge controls the suborgan biodistributions of gold nanoparticles, ACS Nano 10 (5) (2016) 5536–5542.
[162] Toxicological Profile for DDT, DDE, and DDD, 2002.
[163] National Research Council, Risk Assessment in the Federal Government: Managing the Process, National Academy Press, Washington, DC, (1983).
[164] H. Liu, P. Gramatica, QSAR study of selective ligands for the thyroid hormone receptor β, Bioorganic & Medicinal Chemistry 15 (15) (2007) 5251–5261.
[165] S. Vilar, G. Cozza, S. Moro, Medicinal chemistry and the molecular operating environment (MOE): application of QSAR and molecular docking to drug discovery, Current Topics in Medicinal Chemistry 8 (18) (2008) 1555–1572.
[166] A. Lagunin, A. Zakharov, D. Filimonov, V. Poroikov, QSAR modelling of rat acute toxicity on the basis of PASS prediction, Molecular Informatics 30 (2-3) (2011) 241–250.
[167] T. Puzyn, et al., Using nano-QSAR to predict the cytotoxicity of metal oxide nanoparticles, Nature Nanotechnology 6 (3) (2011) 175.
[168] K. Roy, I. Mitra, Electrotopological state atom (E-state) index in drug design, QSAR, property prediction and toxicity assessment, Current Computer-Aided Drug Design 8 (2) (2012) 135–158.
[169] U. Sahlin, Uncertainty in QSAR predictions, Alternatives to Laboratory Animals: ATLA 41 (1) (2013) 111–125.
[170] U. Sahlin, N. Jeliazkova, T. Öberg, Applicability domain dependent predictive uncertainty in QSAR regressions, Molecular Informatics 33 (1) (2014) 26–35.
[171] T. Kimura, Y. Miyashita, K. Funatsu, S.-I. Sasaki, Quantitative structure– activity relationships of the synthetic substrates for elastase enzyme using nonlinear partial least squares regression, Journal of Chemical Information and Computer Sciences 36 (2) (1996) 185–189.
[172] ER-TG-03-1 7/00, Preparing information on risks, costs and benefits for applications under the Hazardous Substances and New Organisms Act 1996, 2000.
[173] Chemical Computing Group, Molecular Operating Environment: Chemoinformatics and Structure Based Tools for High Throughput Screening, ed: Montreal Canada, 2013.
[174] J.-W. Lu, N.-B. Chang, L. Liao, Environmental informatics for solid and hazardous waste management: advances, challenges, and perspectives, Critical Reviews in Environmental Science and Technology 43 (15) (2013) 1557–1656.
[175] P. Ertl, Cheminformatics and Its Role in the Modern Drug Discovery Process, University of Strasbourg. Novartis, nd Web, 18, 2014.

[176] U.S. Environmental Protection Agency, Integrated risk information system (IRIS). Available from http://www.epa.gov/iris/ (accessed 02.05.15).

[177] D.A. Vallero, Exposure space: integrating exposure data and modeling with toxicity information, in: OpenTox USA 2015, Baltimore, MD, 2015.

[178] A.L. Washington, The interoperability of US federal government information: interoperability, in: Big Data: Concepts, Methodologies, Tools, and Applications: Concepts, Methodologies, Tools, and Applications: Information Resources Management Association Ed. Hershey, Pennsylvania: IGI Global, 2016, pp. 210–228.

[179] D. Vallero, Translating Diverse Environmental Data Into Reliable Information: How to Coordinate Evidence from Different Sources, Academic Press, (2017).

[180] Elements of a Quality Assurance Project Plan (QAPP) for collecting, identifying and evaluating existing scientific data/information. [Online] Available from https://www.epa.gov/osa/elements-quality-assurance-project-plan-qapp-collecting-identifying-and-evaluating-existing, 2017.

[181] A. Takhom, S. Usanavasin, T. Supnithi, M. Ikeda, Collaborative ontology development approach for multidisciplinary knowledge: a scenario-based knowledge construction system in life cycle assessment, IEICE TRANSACTIONS on Information and Systems 101 (4) (2018) 892–900.

[182] T.I. Oprea, O. Taboureau, C.G. Bologa, Of possible cheminformatics futures, Journal of Computer-Aided Molecular Design 26 (1) (2012) 107–112.

[183] E. Schymanski, D. Aurich, Cheminformatics and the Exposome in Health and Disease, 2021.

[184] E. Schymanski, Environmental Detective Work: Environmental Cheminformatics and the Exposome, 2022.

[185] J. Krier, et al., Discovering pesticides and their TPs in Luxembourg waters using open cheminformatics approaches, Environment International 158 (2022) 106885.

[186] A. Mikolajczyk, N. Sizochenko, E. Mulkiewicz, A. Malankowska, B. Rasulev, T. Puzyn, A chemoinformatics approach for the characterization of hybrid nanomaterials: safer and efficient design perspective, Nanoscale 11 (24) (2019) 11808–11818.

[187] P. Willett, The Literature of Chemoinformatics: 1978–2018, MDPI, (2020), p. 5576 vol. 21.

[188] F. Lunghini, Comprehensive Data Analysis and Predictive Chemoinformatics Models for REACH Related Physicochemical and (Eco) Toxicity Properties, Université de Strasbourg, (2020).

[189] L. Hu, J.L. Au, M.G. Wientjes, Computational modeling to predict effect of treatment schedule on drug delivery to prostate in humans, Clinical Cancer Research 13 (4) (2007) 1278–1287.

[190] C.A. Lipinski, F. Lombardo, B.W. Dominy, P.J. Feeney, Experimental and computational approaches to estimate solubility and permeability in drug discovery and development settings, Advanced Drug Delivery Reviews 64 (2012) 4–17.

[191] P.P. Egeghy, et al., Computational exposure science: an emerging discipline to support 21st-century risk assessment, Environmental Health Perspectives (Online) 124 (6) (2016) 697.

[192] N. Brandon, K. Dionisio, K. Isaacs, D. Kapraun, W. Setzer, R. Tornero-Velez, A novel framework for characterizing exposure-related behaviors using agent-based models embedded with needs-based artificial intelligence (CSSSA2016), in: The Computational Social Science Society of the Americas, Santa Fe, NM, November 17–20, 2016, 2016. [Online]. Available from https://computationalsocialscience.org/csssa-2016-program/. [Online]; https://computationalsocialscience.org/csssa-2016-program/

[193] R.S. Thomas, et al., The next generation blueprint of computational toxicology at the US Environmental Protection Agency, Toxicological Sciences (2019).

[194] B. Pecoraro, M. Tutone, E. Hoffman, V. Hutter, A.M. Almerico, M. Traynor, Predicting skin permeability by means of computational approaches: reliability and caveats in pharmaceutical studies, Journal of Chemical Information and Modeling 59 (5) (2019) 1759–1771.

[195] N. R. Council, Science and decisions: advancing risk assessment, 2009.

[196] P. Babuji, S. Thirumalaisamy, K. Duraisamy, G. Periyasamy, Human health risks due to exposure to water pollution: a review, Water 15 (14) (2023) 2532.

[197] K. Rahman, S. Barua, H. Imran, Assessment of water quality and apportionment of pollution sources of an urban lake using multivariate statistical analysis, Cleaner Engineering and Technology 5 (2021) 100309.

[198] Y. Liu, J. Zhang, Y. Zhao, The risk assessment of river water pollution based on a modified non-linear model, Water 10 (4) (2018) 362.

[199] D. Garrick, J.W. Hall, Water security and society: risks, metrics, and pathways, Annual Review of Environment and Resources 39 (2014) 611–639.

[200] W.W. Piegorsch, H. Xiong, R.N. Bhattacharya, L. Lin, Nonparametric estimation of benchmark doses in environmental risk assessment, Environmetrics (London, Ont.) 23 (8) (2012) 717–728.
[201] G. Hanrahan, Environmental Chemometrics: Principles and Modern Applications, CRC Press, (2008).
[202] R. Vandoorne, P.J. Gräbe, Stochastic modelling for the maintenance of life cycle cost of rails using Monte Carlo simulation, Proceedings of the Institution of Mechanical Engineers, Part F: Journal of Rail and Rapid Transit 232 (4) (2018) 1240–1251.
[203] P. O'Connor, A. Kleyner, Practical Reliability Engineering, John Wiley & Sons, (2012).
[204] D.A. Vallero, Chapter 11 - Economics and feasibility assessments, in: D.A. Vallero (Ed.), Environmental Systems Science, Elsevier, (2021), pp. 509–542.
[205] I.o W.P. Control, Water Pollution Control, Institute of Water Pollution Control, (1967).
[206] J.D. Solomon, Communicating Reliability, Risk and Resiliency to Decision Makers, Duke University, (2017 October 12).
[207] U.M. Shamsi, Water Distribution Network Reliability, University of Pittsburgh, (1988).
[208] D.A. Vallero, Excerpted and Revised from: Fundamentals of Air Pollution, fifth ed., Elsevier Academic Press, Waltham, MA, (2014), p. 999 pages cm.
[209] E. Hollnagel, D.D. Woods, N. Leveson, Resilience Engineering: Concepts and Precepts, Ashgate Publishing, Ltd, (2007).
[210] J.D. Solomon, D.A. Vallero. From Our Partners – Communicating Risk and Resiliency: Special Considerations for Rare Events. Center for Infrastructure Protection & Homeland Security, George Mason University. Available from https://cip.gmu.edu/2016/06/01/partners-communicating-risk-resiliency-special-considerations-rare-events/ (accessed 01.05.17).
[211] M. Lee, D. Basu, An integrated approach for resilience and sustainability in geotechnical engineering, Indian Geotechnical Journal 48 (2018) 207–234.
[212] H.R. Heinimann, A generic framework for resilience assessment, an edited collection of authored pieces comparing, contrasting, and integrating risk and resilience with an emphasis on ways to measure resilience, 2016, p. 90.
[213] M. Dardonville, C. Bockstaller, O. Therond, Review of quantitative evaluations of the resilience, vulnerability, robustness and adaptive capacity of temperate agricultural systems, Journal of Cleaner Production 286 (2021) 125456.
[214] D.A. Vallero, Chapter 8 - Effect of waste on ecosystems, in: T.M. Letcher, D.A. Vallero (Eds.), Waste, second ed., Academic Press, (2019), pp. 171–198.
[215] R.H. Day, S.T. Allen, J. Brenner, K. Goodin, D. Faber-Langendoen, K.W. Ames, Ecological resilience indicators for mangrove ecosystems, Ecological Resilience Indicators for Five Northern Gulf of Mexico Ecosystems, NatureServe, Arlington, Virginia, USA, (2018).
[216] S.T. Allen, et al., Ecological Resilience Indicators for Salt Marsh Ecosystems, Ecological Resilience Indicators for Five Northern Gulf of Mexico Ecosystems, NatureServe, Arlington, (2018), p. 53.
[217] J. Fiksel, I. Goodman, A. Hecht, Resilience: navigating toward a sustainable future, Journal of Solutions 5 (5) (2014) 38–47.
[218] C. Reynolds, Resilience in aquatic ecosystems–hysteresis, homeostasis, and health, Aquatic Ecosystem Health & Management 5 (1) (2002) 3–17.
[219] K.S. Crump, The linearized multistage model and the future of quantitative risk assessment, Human & Experimental Toxicology 15 (10) (1996) 787–798.
[220] E. U. Health & Consumer Protection Directorate General, Training for the Health Sector. EU. Available from https://ec.europa.eu/health/ph_projects/2003/action3/docs/2003_3_09_a23_en.pdf (accessed 14.03.23).
[221] M.L. Card, et al., History of EPI Suite[trade mark sign] and future perspectives on chemical property estimation in US Toxic Substances Control Act new chemical risk assessments, Environmental Science: Processes & Impacts 19 (3) (2017) 203–212. Available from https://doi.org/10.1039/C7EM00064B10.1039/C7EM00064B.
[222] 69, Guidance Document on the Validation Of (Quantitative) Structure-Activity Relationship [(Q)Sar] Models, 2007.
[223] M.F. Dupont, et al., Chemometrics for environmental monitoring: a review, Analytical Methods 12 (38) (2020) 4597–4620.
[224] J. Chapman, et al., Combining chemometrics and sensors: toward new applications in monitoring and environmental analysis, Chemical Reviews 120 (13) (2020) 6048–6069.

[225] A. Sutcliffe, G. Rugg, A taxonomy of error types for failure analysis and risk assessment, International Journal of Human-Computer Interaction 10 (4) (1998) 381–405.
[226] M.D. Adams, P.S. Kanaroglou, A criticality index for air pollution monitors, Atmospheric Pollution Research 7 (3) (2016) 482–487. Available from https://doi.org/10.1016/j.apr.2015.11.0042016/05/01/.
[227] P. Boffetta, J.K. McLaughlin, C. La Vecchia, R.E. Tarone, L. Lipworth, W.J. Blot, False-positive results in cancer epidemiology: a plea for epistemological modesty, Journal of the National Cancer Institute 100 (14) (2008) 988–995.
[228] J.L. Furman, K. Jensen, F. Murray, Governing knowledge in the scientific community: exploring the role of retractions in biomedicine, Research Policy 41 (2) (2012) 276–290.
[229] D. Psaltopoulos, A.J. Wade, D. Skuras, M. Kernan, E. Tyllianakis, M. Erlandsson, False positive and false negative errors in the design and implementation of agri-environmental policies: a case study on water quality and agricultural nutrients, Science of The Total Environment 575 (2017) 1087–1099.
[230] D.A. Vallero, Wastewater, Waste, Elsevier, (2019), pp. 259–290.
[231] R. Harder, S. Heimersson, M. Svanstrom, G.M. Peters, Including pathogen risk in life cycle assessment of wastewater management. 1. Estimating the burden of disease associated with pathogens, Environmental Science & Technology 48 (16) (2014) 9438–9445.
[232] C.P. Gerba, I.L. Pepper, Wastewater treatment and biosolids reuse, Environmental Microbiology, second ed., Elsevier Inc., USA, (2009), pp. 503–530.
[233] B. Jiménez, D. Mara, R. Carr, F. Brissaud, Wastewater treatment for pathogen removal and nutrient conservation: suitable systems for use in developing countries, Wastewater Irrigation and Health, Routledge, (2009), pp. 175–196.
[234] EPA No. 832-B-05-001, Handbook for managing onsite and clustered (decentralized) wastewater treatment systems: an introduction to management tools and information for implementing EPA's management guidelines, 2005.
[235] O. Jolliet, A.S. Ernstoff, S.A. Csiszar, P. Fantke, Defining Product Intake Fraction to Quantify and Compare Exposure to Consumer Products, ACS Publications, (2015).
[236] O. Jolliet, A. Ernstoff, S. Csiszar, P. Fantke, Health impacts of consumer exposure during product use: near field exposure applied to risk assessment and LCA, in: International Society of Exposure Science 24th Annual Meeting, 2014, pp. 178–179.
[237] P. Fantke, A.S. Ernstoff, L. Huang, S.A. Csiszar, O. Jolliet, Coupled near-field and far-field exposure assessment framework for chemicals in consumer products, Environment International 94 (2016) 508–518.
[238] M.G. Evich, et al., Per-and polyfluoroalkyl substances in the environment, Science (New York, N.Y.) 375 (6580) (2022) eabg9065.
[239] J.C. Quezada Davalos, M.A. Michaud, L.E. Lowe, E.N. Hanson, J.E. Owens, Per- and polyfluoroalkyl substances (PFASs) in the fountain creek watershed, Colorado Springs, CO, USA: a yearlong investigation of PFAS levels in water, soils, and sediments, ACS ES&T Water 3 (1) (2023) 96–105. Available from https://doi.org/10.1021/acsestwater.2c00440. 2023/01/13.
[240] J.A. Cotruvo, S.B. Goldhaber, A.J. Cohen, EPA's Unprecedented Interim Drinking Water Health Advisories for PFOA and PFOS, Groundwater, (2023).
[241] U.S. Environmental Protection Agency, Drinking water contaminant candidate list 5-draft: a proposed rule by the environmental protection agency on 07/19/2021, Federal Register, FR Doc. 2021-15121, Filed 7-16-21; 8:45 am, 2021. [Online]. Available from https://www.federalregister.gov/documents/2021/07/19/2021-15121/drinking-water-contaminant-candidate-list-5-draft.
[242] M. Bailey, et al., Risk Assessment Guidance for Superfund: Volume I: Human Health Evaluation Manual (Part F, Supplemental Guidance for Inhalation Risk Assessment), Environmental Protection Agency, (2009).
[243] R. Lewis, S. Gordon, Sampling for Organic Chemicals in Air, Environmental Protection Agency, Research Triangle Park, NC (United States), (1994).
[244] W.F. Bleam, Soil and environmental chemistry, Academic Press, (2016).
[245] C. Stephan, D. Mount, D. Hansen, J. Gentile, G. Chapman, W. Brungs, Guidelines for deriving numerical water quality criteria for the protection of aquatic organisms and their uses, USEPA PB85-227049, 1985.
[246] Agency for Toxic Substances and Disease Registry, Calculating hazard quotients and cancer risk estimates. ATSDR. Available from https://www.atsdr.cdc.gov/pha-guidance/conducting_scientific_evaluations/epcs_and_exposure_calculations/hazardquotients_cancerrisk.html#CommonNonCancerGuidelines.

[247] J.M. Neff, S.A. Stout, D.G. Gunster, Ecological risk assessment of polycyclic aromatic hydrocarbons in sediments: identifying sources and ecological hazard, Integrated Environmental Assessment and Management: An International Journal 1 (1) (2005) 22–33.

[248] L.D. Duke, M. Taggart, Uncertainty factors in screening ecological risk assessments, Environmental Toxicology and Chemistry: An International Journal 19 (6) (2000) 1668–1680.

[249] U.S. Environmental Protection Agency, Guidelines for ecological risk assessment, U.S. EPA, (2024). Available from https://www.epa.gov/sites/default/files/2014-11/documents/eco_risk_assessment1998.pdf. (accessed 24.01.24).

[250] E.A. Cohen Hubal, et al., Exposure science and the US EPA national center for computational toxicology, Journal of Exposure Science & Environmental Epidemiology 20 (3) (2010) 231–236.

[251] Y. Li, et al., Ecological risk assessment of environmental stress and bioactive chemicals to riverine fish populations: an individual-based model of smallmouth bass *Micropterus dolomieu*☆, Ecological Modelling 438 (2020) 109322.

[252] U.S. Environmental Protection Agency, Technical overview of ecological risk assessment: risk characterization. Available from https://www.epa.gov/pesticide-science-and-assessing-pesticide-risks/technical-overview-ecological-risk-assessment-risk (accessed 05.09.18).

[253] ChemSafetyPRO, How to Calculate Hazard Quotient (HQ) and Risk Quotient (RQ). Available from https://www.chemsafetypro.com/Topics/CRA/How_to_Calculate_Hazard_Quotients_(HQ)_and_Risk_Quotients_(RQ).html (accessed 05.09.18).

[254] Y. Zhang, Z. Sun, J.S. Ji, X. Li, J. Yang, Measuring perceived green volume for quantifying urban green exposure, Urban Forestry & Urban Greening 116 (2026) 129231. Available from https://doi.org/10.1016/j.ufug.2025.1292312026/02/01/.

[255] P. Grahn, U.K. Stigsdotter, The relation between perceived sensory dimensions of urban green space and stress restoration, Landscape and Urban Planning 94 (3-4) (2010) 264–275.

[256] J. Barton, J. Pretty, What is the best dose of nature and green exercise for improving mental health? A multi-study analysis, Environmental Science & Technology 44 (10) (2010) 3947–3955.

[257] S. Kaplan, The restorative benefits of nature: toward an integrative framework, Journal of Environmental Psychology 15 (3) (1995) 169–182.

CHAPTER

11

Engineering education

OUTLINE

> If we consider the customer to be potential employers, it is easier to see where universities must focus attention. Employers value professional certifications – such as Chartered Financial Analyst or Professional Engineer certification.... Universities should focus more on integrating industry standards of competence into their curriculum.
>
> Dean Emeritus Charles Crespy ***Central Michigan University (2026) [1]***

11.1 Introduction

With the emergence of new technologies and engineering challenges, the way engineers are trained has never been more important. This chapter addresses two aspects of engineering education. First, what does the future engineer need to know after the undergraduate experience? Second, how does the engineering pedagogy need to change to achieve this? Crespy's quote adds a bit more on the need for credentialing the next generation of engineers [1]:

> While a four year degree now demonstrates merely acceptable attendance, a certification demonstrates skills, acumen and judgment that portend value in the workplace. Additionally, employers value work experience – internships directly related to the student's career path. In many cases, such experiences are prerequisites for professional certifications.

Technical and Ethical Optimization: Challenges and Opportunities for the Modern Engineer and Scientist
DOI: https://doi.org/10.1016/B978-0-443-44925-3.00007-7

That is a tall order for a four-year program. This is such little time to prepare the next generation of professional engineers. Of course, this credentialing path has milestones, the first being the successful completion of the fundamentals of engineering (FE) exam, upon which the student becomes a candidate for the designation of engineer-in-training (EIT) or engineering intern (EI), ultimately leading to full licensure under the principles and practice of engineering (PE) exam [2–7]. Thus, the student-candidate-professional credentialing path provides academic, experiential, and professional development that promises the engineer has demonstrated and will continue to possess the competence and the ability to safeguard the public's safety, health, and welfare [3]. Encouraging the future engineer to gain this level of competence requires both technical inculcation and an introduction to standards of ethical awareness, written and unwritten rules of professional decision-making, and behavior fitting of the professional [8].

11.1.1 Ethics across the curriculum

Engineers are agents of change. They endeavor to hold paramount the health, safety, and welfare of all the public, which requires both technical competence and professionalism, as well as an appreciation for objectivity and rigor codified in the scientific method. The credo of any professional is *credat emptor*, that is, let the client trust [6,9,10]. Professional engineers are charged with responsibilities to protect the public as they fulfill their duties. The Greeks called this *ethike aretai*, or "skill of character" [11]. It is not enough to be excellent in technical competence [12]. Such competence must be coupled with trust gained from ethical practice.

Undergraduate engineering schools offer two major ways of introducing ethics. Some offer a separate course that covers the ethical theory and PE. Others offer ethics across the curriculum (EAC), which incorporates ethics into several engineering courses beginning when the student arrives on campus, adding ethical case studies and topics throughout the curriculum, and culminating in a senior design course with direct and indirect ethical content, for example, optimizing the best design of a highway with ethical constraints like fairness to neighborhoods or an embedded counterexample like a design that is the most efficient but morally impermissible.

The EAC approach can offer two modules in the introductory engineering course: academic integrity and professional ethics[1]. The academic integrity module centers on the student's obligation to abide by standards of conduct, such as avoiding plagiarism, showing responsibility in study habits, and being a good citizen [6]. The professional engineering module will introduce the students to case studies that bring to life all the tenets of the engineering codes of ethics [3].

As mentioned, instead of offering a single ethics course, EAC embeds ethical modules directly into several engineering courses. To ensure that EAC covers the relevant ethical content, the university must provide training and resources to help instructors who have strong expertise in the course's subject matter also gain knowledge of ethical matters. This can include guest lectures and one-on-one discussions with faculty from departments outside the engineering school, and outside experts from the field. This includes making ethical theories

[1] Indeed, this was an innovation at Duke University's Pratt School of Engineering, when my colleague, Tod A. Laursen, asked me to teach these two modules in his Introduction to Engineering first-year course some decades ago.

applicable to real-world engineering, including the societal impact of technology and the demands of professionalism. Ultimately, the student gains moral literacy, that is, critical reasoning about ethical challenges to prepare them for a professional engineering career.

The preparation for an engineering career, or any career for that matter, is not one size fits all. Every course has an array of student interests and rationales for enrolling. Most join the class because they need the course to graduate, whether as a core part of their major, or because it is on a short list of allowed electives. Other students are intrinsically interested in the topic. Some are there because the time was convenient, or a friend would also be enrolling.

Whatever the motivation, the class is a step in the journey from academic learning to a lifelong career. If the instructor fails to convey useful information and skill sets during the brief encounter with the student, at best, it has been a waste of the student's time. At worst, the failed instruction may have diverted them from their rightful path. Learning at a university is a social contract between the student and instructor. To achieve enlightenment, the student may become aware of gaps in knowledge, skills, and abilities that must be closed to move through the academic training onto a professional career.

Balancing core academic requirements, such as those dictated for engineering programs by accrediting institutions, can be a challenge. Most engineering faculty can see the inherent unevenness of an engineering degree's rigor and demands compared to those of most other majors. Yet, all receive a similar "bachelors" designation. Faculty can remind the student that the difference will pay off in the future. The skills attained in any engineering major will continue to be more marketable and fungible than those of most academic endeavors. Certainly, many engineering majors go on to become highly successful professionals and researchers in the engineering disciplines, but a large number become physicians, lawyers, and highly sought-after modelers and experts in the financial world. Admittedly, saying this to an undergraduate engineering major who must spend hours on homework and studies, while fellow students in their dorms are not, provides little comfort. Indeed, an engineering student must be prepared to ignore the so-called "fear of missing out" (FOMO), that is, the fear of missing out. To their credit, one of the qualities of most engineering students is the ability to delay gratification.

An academic engineering path is challenging and rewarding. So, what are the reasons that one would pursue an engineering degree in the first place? In the author's case, it was initially the result of the high school counselor's take on college entry exam scores and suggesting consideration of an engineering degree. He was the first in his family to enter a four-year college and had no context for what the counselor was suggesting. However, he was impressed that she thought he was smart, so he followed her suggestion. Upon arrival at the college for freshman[2] initiation, he met with an academic advisor who asked him which engineering academic path he wished to pursue. She went through the litany of majors, whereupon the author stopped her at electronic engineering. His love of music and being an avid consumer of vinyl long-playing records (LPs) made that an obvious choice. What could possibly go wrong? Well, the coursework was a long way from the author's 18-year-old expectations, consisting exclusively of math and physics courses, with all electrical subject matter delayed for the future.

Growing weary of the journey, the author still had little idea of what engineering was all about, yet he was expected to invest all this time and energy in an ephemeral career goal. As it

[2] That is what it was called then. Now, the preference is "first-year student."

happened, Southern Illinois University (SIU) required all students to take general studies courses. Yes, even engineering majors had to take a modicum of courses in the life sciences, social sciences, and the humanities. This introduced the author to a range of exciting topics in philosophy, psychology, history, and biology. The environmental movement was in its early stages, so there was often a teachable moment when earth systems were described and analyzed in the spatial sciences courses at SIU. This was the foray into the earth sciences, which offered something that the engineering courses did not. They included content that was immediately understandable and useful to an academically developing young man, so he decided not to wait two more years to gain the psychic and intellectual investment. The author eventually returned to engineering, but only after years of working with many gifted engineers who somehow were able to delay gratification to land jobs that they loved. In this case, we were all engaged in cleaning up the environment, one wastewater treatment plant at a time. They reviewed and approved plans and specifications for wastewater treatment facilities, while the author wrote environmental impact statements on their projects. This provided a real-world context for engineering.

Some years ago, Duke's Engineering Dean, Kristina Johnson, advocated for two important goals for engineering education. She wished to invert the curriculum and to introduce engineering students to entrepreneurship early in their academic experience. Inverting the curriculum meant first introducing the student to real-world experiences *before* immersing them in the mathematical and physical fundamentals. This can be likened to woodworking. If you want to teach someone how to work with wood, it is not ideal to engage them first in the "theory of the saw." Rather, first show them how to use the tool, rather than teaching the statics, dynamics, laws of motion, thermodynamics, metallurgy, optics, and material science of a rip saw, then a cross-cutting saw, then a table saw, then a band saw, etc. Working with a seasoned woodworker in the shop would lead to mastery. Analogously, doing the work of the engineer under the mentorship of faculty and outside experts would give the student engineering context.

The inverted curriculum has a long history. As evidence, the term "university" is derived from *universitat* which means "guild," in this case, a guild for the clergy. Mentorship at the university began as part of the Cathedral and was conceived as a means for the formation and education of the clergy. The clerical guild was not simply for theological and religious instruction but also embodied an extensive liberal arts and science education. For example, the Catholic priest Copernicus, who established the heliocentric view of the solar system, was a product of these early universities [13].

Entrepreneurial engineering introduces the student to the marketplace. This provides important context, since engineers often do all the hard work and provide the intellectual firepower for a design, a new product, or an improved process, only to turn the project over to a manager who takes the engineer's innovations to the next step. Even if the engineer wishes not to be involved in the marketing, the engineer needs to know where his or her brainchild is going and, if not, only to hear years down the road that the idea was realized with little or no recognition of the engineer's work in the early, difficult stages.

11.1.2 The driver's education analogy

The transition from college to the real world can be compared to that of driver's education. Society does not ask an untrained, inexperienced driver to merge onto an interstate highway at high speed in heavy traffic. The future licensed driver must first gain some literary context, that is, a book on the rules of the road. They learn the signs, distances, speeds, etc. This information is very abstract. The hopeful driver can pass a test on the rules but remains far from ready to merge onto an interstate highway. There needs to be experience under supervision. Likewise, the engineering student gets ample book learning but needs to apply this learning quickly in real-world scenarios since the slope of the memory extinction curve is very steep [14]. As the student's confidence and experience grow, so do the knowledge base, judgment, motor memory, and intuition.

Thus, academics and professionalism lie on a continuum. They are not discrete endeavors. In fact, training and technology transfer intertwine with career advancement throughout professional work life. Education and practice are mutually beneficial. They are both essential to the first canon of all engineering disciplines. They are the only way to "hold paramount the safety, health, and welfare of the public" [3].

11.2 The engineering curriculum

Every engineering student needs to have a working knowledge of foundational mathematics and the physical sciences. Biomedical and environmental engineering students also need a strong grasp of the life sciences. Engineers must also understand human factors in the application of engineering knowledge. These have been and continue to be core capabilities of the engineering graduate. However, the engineering curriculum content needs to address topics that were not covered even a decade ago.

The Accreditation Board for Engineering and Technology (ABET) provides guidance on meeting professional engineering quality standards. Established in 1932, ABET has accredited more than 4800 programs at 950 institutions in 42 countries [15]. ABET requires that eight criteria be met for an institution's accreditation [16]:

Criterion 1 requires the evaluation of each student's performance by enforcing policies for accepting both new and transfer students, awarding appropriate academic credit for courses taken at other institutions, and awarding appropriate academic credit for work in lieu of courses taken at the institution. The program must have and enforce procedures to ensure and document that students who graduate meet all graduation requirements.

Criterion 2 requires that the institution publish its program educational objectives that are consistent with the mission of the institution, the needs of the program's various constituencies, and these criteria. The objectives must be documented, systematically utilized, and have an effective process, involving program constituencies, for the periodic review of these program educational objectives that ensures consistency with the institutional mission, the program's constituents' needs, and these criteria.

Criterion 3 states that, to ensure proper preparation for the professional PE, the institution must document student outcomes that support the program's educational objectives. In addition to any additional outcomes that may be articulated by the program. Each student must have:

1. an ability to identify, formulate, and solve complex engineering problems by applying principles of engineering, science, and mathematics.

2. an ability to apply engineering design to produce solutions that meet specified needs with consideration of public health, safety, and welfare, as well as global, cultural, social, environmental, and economic factors.
3. an ability to communicate effectively with a range of audiences.
4. an ability to recognize ethical and professional responsibilities in engineering situations and make informed judgments, which must consider the impact of engineering solutions in global, economic, environmental, and societal contexts.
5. an ability to function effectively on a team whose members collectively provide leadership, create a collaborative environment, establish goals, plan tasks, and meet objectives.
6. an ability to develop and conduct appropriate experimentation, analyze and interpret data, and use engineering judgment to draw conclusions.
7. an ability to acquire and apply new knowledge as needed, using appropriate learning strategies.

Criterion 4 requires that the institution must regularly assess the extent to which student outcomes are being met.

Criterion 5 requires that the institution's curriculum specify adequate content for each required subject area to meet standards needed to enter the PE. The curriculum must include a minimum of 30 semester hours of college-level mathematics and basic sciences, with an appropriate amount of experimental work, a minimum of 45 semester hours of engineering topical coursework, a broad education component to complement the technical content of the curriculum, and design experience that incorporates engineering standards and constraints that is based on the comprehensive knowledge gained from the courses.

Criterion 6 requires that the faculty size and competencies are sufficient to accommodate effective student-faculty interaction.

Criterion 7 requires that the institution has sufficient facilities, including space and equipment in classrooms and laboratories, as well as computing and information infrastructure.

Criterion 8 calls for institutional support to ensure that all the other criteria continue to be met. This includes financial, administrative, and technical staff support.

Ethics is mentioned six times in the criteria. It is mentioned under Criterion 3 for both undergraduate and master's student outcomes, as the ability to recognize ethical and professional responsibilities in engineering situations and make informed judgments, which must consider the impact of engineering solutions on global, economic, environmental, and societal impacts. Codes of ethics are also delineated under the civil engineering, construction engineering, and cybersecurity engineering curricula.

Interestingly, ethics is not mentioned in the required curriculum for biomedical engineering.

Following are a few content areas and how they may be incorporated into the engineering pedagogy.

11.2.1 Computational content

In several ways, undergraduate engineering students may be ahead of the instructors in their grasp of most computational subject matter. For one thing, they have much less to unlearn than those who primarily worked with analogs. Before entering college, many students have already employed machine learning and artificial intelligence in decision-making, pattern recognition, problem solving, and query design. Chances are good that they have written training and testing algorithms, scraped databases, and deployed a large language model (LLM). Recall from Chapter 1 that the training set for an LLM includes text and coding from which the LLM learns to understand information. From this, the model will generate and predict terms and a human-like language that will be used for pattern recognition, context construction, and relationships within the accessed internet data. The testing dataset is written to determine the performance and the ability of the model to generalize the accessed data, especially those not previously accessed by the LLM. That is, the model's training set allows builds, whereas the testing data determine the LLM's ability to learn, and the performance of LLM is determined by its ability to extend the building of scenarios not previously seen by the model [17].

A particular skill set that will enhance the success of the graduating engineer is additive manufacturing and geometric dimensioning and tolerancing (GD&T) [18]. Therefore, students and newly minted engineers are likely to offer ways to address hyperscaling issues like training infrastructure, cloud economics, software orientation, and inference delivery. This and other computational tools will be needed to support drawings and three-dimensional models to define form, orientation, and location of items to define allowable and precise geometry needed in manufacturing and reliability engineering. GD&T will be increasingly needed by companies to develop and manufacture products. GD&T involves complicated concepts that must be learned through repetition and practice. For example, the Georgia Institute of Technology deploys active learning interventions to enable students to learn basic manual inspection methods and to communicate through detailed drawings [19]. This approach provides a foundational, practical understanding of GD&T as the students incorporate the knowledge gained into a part drawing and inspection plan. The GD&T intervention is enhancing the students' knowledge base of the topic while increasing their confidence in their ability to perform identification and inspection tasks that will be used in the immediate and future workforce.

Quantum computing is another emergent area that needs to receive at least a modicum of attention in the engineering curriculum. Certainly, the quantum mechanics is taught as a stand-alone course at Northwestern University, Rice University, the University of Washington, the Colorado School of Mines, and elsewhere. However, every engineering student should have a working knowledge of the future of quantum computing, given its likely impact on society, including its ethical implications. The technical topics can be blended with information theory, computer science, and computer engineering. These can include qubits [20], entanglement [21–23], Shor's and Grover's algorithms [24,25], and quantum gates [26].

Robotics will also be part of every engineering discipline [27]. For example, construction engineering will likely accelerate its use [28–32]. This will include prefabrication, clash detection, and prediction in building information modeling (BIM) by identifying spatial encounters in a virtual, three-dimensional space before the onset of construction. Robotics will also integrate big data and historical project data with AI and machine learning design constraint algorithms to forecast conflicts based on the spatial relationships of the many building components. Students can learn to apply robotics for the automatic identification of critical clashes to enhance efficiency and safety [33]. For example, biomedical students will learn how to improve surgeries by combining robotics with diagnostic tools [34].

The robotics curriculum must include ethical and economic content, such as human job replacement, security concerns [35], and loss of privacy and confidentiality [36,37].

Certainly, ethics and morality content should be included in the computational coursework. Perhaps, modules could be offered on the ethics of computational advances. One interesting topic is the ethics of digital phenotyping to identify observable physical or behavioral traits with the goal of better understanding of disease states. The collection of data from wearable sensors and mobile phones is a way to acquire insight into people's behaviors and habits in real time and in a more accurate way than using self-report questionnaires. The benefits and drawbacks, such as disempowerment and loss of privacy, could be discussed.

11.2.2 Environmental content

Future civil and environmental engineers will need knowledge about human health and the environment. However, all engineering disciplines should have a working knowledge of the potential impacts of their projects and designs on the environment. This knowledge should range from the molecular to the planetary scale. For example, at the cellular scale, mechanical and chemical engineering in the future will support biomedical and tissue engineering to develop better drug delivery. Such work is the inverse of environmental toxicology, where efforts are made to find ways to prevent a substance, a pollutant, from reaching target cells and tissues [38].

At the other extreme, engineers will likely be called upon to work on planet-scale projects, that is, geoengineering [39]. Engineers will be drawn into debates about the need and risks of such large-scale projects. As mentioned in Chapters 1 and 12, proponents of geoengineering usually admit that these measures can be associated with significant risks to the environment and public health, but they believe that they are worth pursuing, given the failure of means to mitigate climate change problems [40]. Opponents, especially some environmental groups, are against any type of geoengineering [41,42]. Engineers will be among the arbitrators of the extent to which such visionary and risky ventures will be pursued [43–45].

11.2.3 Biotechnological content

Future engineering disciplines will need an understanding of biotechnology. Environmental engineers already have a hefty amount of coursework in areas like wastewater and hazardous waste engineering. However, all engineers will need to deal directly and indirectly with living things as a means of understanding and managing the opportunities and risks when applying living organisms to societal needs. Biotechnological advances will present societal challenges. Many engineering and science disciplines will be involved, requiring a better appreciation for and ability to predict the implications and possible drawbacks of biotechnological developments. Key among them will be biotechnological advances at smaller scales that approach the molecular level.

11.2.4 Nanotechnology

Engineering works at continually decreasing scales. Nanotechnologies will increasingly aim to take advantage of the special physical and chemical properties of substances as they approach the molecular scale, including electromagnetics, materials science, and physiology. However, the large uncertainties of operating at these minuscule scales mean that benefits must be carefully weighed against the societal, individual costs, and risks [46–51].

11.3 Tyranny of the syllabus

The teachable moment is one of the most valuable, yet ignored, pedagogies in engineering education. Events that occur during a semester can provide real-world examples of many scientific, engineering, and social concepts. For example, September 11, 2002 attack and its aftermath provided several teachable moments [52]. To deploy teachable moments, educators must be open to the needs of their students and be sufficiently flexible to teaching opportunities as they are presented.

However, more than most other curricula, engineering coursework is expected to provide a structured program of learning. At times, this rigor invites a "tyranny of the syllabus," wherein the instruction is so firmly dictated at the outset of the semester that the instructor is unwilling to incorporate important learning opportunities as they occur. The extent to which an instructor is bound to the syllabus for a course can be considered from an ethical decision-making perspective. Decisions are made under various ethical constructs, especially teleological, deontological, and relational [53,54]. Engineering faculty decisions mainly concerned with outcomes are teleological (*telos* is the Greek word for "far"). The teleological instructor asks, "What do I want from this?" Instructors who employ the deontological (*deon* is the Greek word for "obligation") model ask, "What is my duty?" The relational instructors are also concerned about achieving results and living up to duties, so they ask, "What is going on around me and how does it relate to experience?"

Teleology was described by Aristotle (384–322 BC) and by St. Thomas Aquinas (c. 1270), who held that "every active cause acts for an end; otherwise, from its activity, one effect would not result rather than another, except by chance." The teleological educator follows the syllabus to guide the course to a particular set of endpoints, such as preparation for a certification examination or gaining a command of a particular field in a discipline. Niebuhr's [55] analogy for this method was that humans are "builders," where they first decide what is needed and then prepare the means to attain that goal. This is particularly appropriate for engineering educators. The teleological approach asserts the reasons for learning the subject matter. Thus, it is often central to the engineering curriculum, such as when ensuring that students have command of a differential equations course that is a prerequisite for future mathematical and engineering concepts. In such instances, the amount of leeway from the structure defined in the syllabus is limited. The major weakness of the teleological method is the so-called "ends justifying the means." For example, an educator may be tempted to completely ignore current events related to a course's subject matter in order to "cover everything" laid out in the syllabus.

The deontological educator defines the course content and approach based on pedagogical and professional duties, such as those articulated in professional codes of ethics. The educator is bound to a set of standards, which may be intrinsic, that is, the instructor's own moral and ethical framework, or extrinsic, for example, those defined by the certifying body.

Socrates (469–399 BC) defined morality as how one ought to live. The deontological decision framework was more fully articulated later by Immanuel Kant (1785). This method requires that ethical decisions be made based on what is "right" rather than on the consequences of the decision. Ethics is seen through the prism of the "categorical imperative," wherein one's duty is defined by whether it should be generally prescribed from one instance to all instances. This imperative has recently been updated as "Six O'clock News" ethics, which states that a decision is right or wrong based on how it would be viewed if everyone evaluated it on the news each night [56].

The relational approach uses whatever construct is appropriate, so the educator applies elements of both teleology and deontology. This approach is neither all-purpose nor formulaic. It does not predict certainty from any model. It is complex, "messy," and is even perceived to be "unscientific." Relational teaching requires first that the educator be open to new ideas and new paradigms. In these senses, the teachable moment is relational. It is always related to the human experience. It requires individual reflection and an attention to experience, that is, experiential learning. Optimizing the teachable moment requires reflection at all scales, from the inner-personal to the neighborhood, to the community, to the world. This is what engineers are called to do. We are asked to consider how our work affects not only the individuals paying the engineer's advice, but how we can "build a better world" [57].

Besides the rigidity of the syllabus, individual educators may lack the confidence to address areas not directly within their specific disciplines. Certainly, by definition, teachable moments appear unexpectedly. However, there are a few actions that educators may take to prepare for them. The instructor may encourage open-ended discussions periodically, perhaps once or twice a week at the beginning of class, perhaps an "engineering in the news" session. The syllabus can explicitly state that it is the responsibility of the student to prepare these sessions by tracking current events and bringing them to discussion each week. This would encourage ongoing student awareness. The sessions could explore several small teachable moments, and as the larger ones come to the fore, the instructor will have gained the experience of dealing with open-ended subject matter.

A second course of action to prepare for teachable moments is to ask students to keep journals. This not only encourages the students to be on the lookout for relevant news items but also helps them learn to write about these important events. Engineering students often have excellent analytical skills, but their communication and writing skills may fall short of those of the humanities majors. After all, the first language of the engineer is often likely to be mathematics [58], more so than English or Spanish.

A third means of improving the use of teachable moments is for the faculty in a department or even the school to reach an understanding amongst themselves prior to each semester on how to deal with unforeseen events and how to use these events in the curriculum. An emergency response analogy may be useful. For example, if a natural disaster or other event were to occur, the instructors would be expected to change longer-term priorities to those needed for the immediacy of the event. The response protocol would be defined and practiced in advance, so in the unlikely occurrence, lives would be saved. Analogously, periodic practices in engineering events in the news would be good practice for incorporating events into class sessions. This may be a standing agenda item at the first department faculty meeting of each semester. If the Deans and Chairs openly support and encourage incorporating teachable moments into coursework, instructors may feel less inclined to ignore an event when it occurs.

The fourth means of using the teachable moments is to have the event be the core of a course. For example, the World Trade Center disaster could have been used to teach just about any topic in structural, environmental, mechanical engineering, transportation planning, chemistry, physics, material science, and professional ethics. At this writing, several techno-ethical concepts are providing daily teachable moments, such as the use of drones in in the Middle East conflict and the tradeoffs involved in attempting to open the Strait of Hormuz. These are not only the purview of policy faculty, but call of the expertise of creative thinking of engineering schools, such as ways to expand land-based bypass infrastructure, harden energy transport corridors, and accelerate regional grid integration. Such relational teaching is important during world-changing events, allowing the diligent educator to tie the course subject matter to unprecedented events and even to everyday occurrences. In fact, engineering capstone design courses must include numerous disciplines. For example, a structural design senior capstone addressing failures of spacecraft, buildings, bridges, rail cars, etc., must discuss the physical principles underlying the collapse as part of the root cause and failure analyses [52,59–64]. However, the capstone must be more comprehensive and include modules on environmental science and engineering (e.g., the asbestos controversies, air and water measurement approaches) and risk assessment (e.g., how engineers must deal with risk perception, how to communicate risks, and how to address the uncertainties of risk assessment).

Bottom line: students must be able to connect their future roles in engineering practice to ethics and societal needs. The teachable moment is a necessary part of the pedagogy to make that happen.

11.4 Conclusion

This chapter describes the core capabilities needed to train future engineers. It also offers insights on the expected changes in the future curriculum in light of emerging technologies and changing societal expectations of engineers.

References

[1] C. Crespy, Letter, Wall Street Journal 287 (30) (2026).

[2] S. Efe, M. Shokouhian, I. Akinkugbe, Professional Engineer Licensure Preparatory Course: An Evaluation and Analysis of CIAMTIS MSU Outreach Activities (2024).

[3] National Society of Professional Engineers, NSPE code of ethics for engineers, (2026) Available from https://www.nspe.org/career-growth/ethics/code-ethics, (accessed 05.02.26).

[4] J.S. Gonnella, et al., Professionalism: empathy, interprofessional collaboration, lifelong learning, Fifty Years of Findings from the Jefferson Longitudinal Study of Medical Education, Springer, (2022), pp. 179–221.

[5] National Society of Professional Engineers, NSPE code of ethics for engineers, (2016) Available from https://www.nspe.org/resources/ethics/code-ethics, (accessed 31.08.18).

[6] D.E. Wueste, Linking academic integrity and ethics across the curriculum: groundwork for sustainability in practical and professional ethics, Ethics Across the Curriculum-Pedagogical Perspectives, Springer, (2018), pp. 303–326.
[7] M. Kam, Engineering licensing and professional practice, in: 2011 International Workshop on Institutional and Programme Accreditation: Connections and Opportunities, 2011, pp. 1–9. IEEE.
[8] N. Pickus, J.A. Reuben, Debating Moral Education: Rethinking the Role of the Modern University, Duke University Press, (2010).
[9] E.C. Hughes, Professions, Daedalus (1963) 655–668.
[10] D.A. Vallero, Biomedical ethics for engineers, The Biomedical Engineering Series (2007).
[11] C. Cardwell, Hornbook Ethics, Hackett Publishing, (2015).
[12] O. Gingerich, Copernicus: A Very Short Introduction, Oxford University Press, Oxford, England, UK, (2016).
[13] D.A. Vallero, C.K. Gunsch, Applications and implications of emerging biotechnologies in environmental engineering, Journal of Environmental Engineering 146 (6) (2020) 03120005.
[14] R. Richardson, D.C. Riccio, Memory processes, ACTH, and extinction phenomena, in: A.R. Pratkanis, S.J. Breckler, A.G. Greenwald (Eds.), Attitude Structure and Function, Psychology Press, New York, NY, (1991).
[15] Accreditation Board for Engineering and Technology (ABET), Criteria for Accrediting Engineering Programs, 2025 – 2026. Available from https://www.abet.org/accreditation/accreditation-criteria/criteria-for-accrediting-engineering-programs-2025-2026/, (accessed 04.30.26).
[16] ABET, Criteria for Accrediting Engineering Programs, 2025–2026, ABET Baltimore, (2025).
[17] C. Paduraru, A. Staicu, A. Stefanescu, LLM-based methods for the creation of unit tests in game development, Procedia Computer Science 246 (2024) 2459–2468.
[18] R. Furferi, Integrating geometric dimensioning and tolerancing with additive manufacturing: a perspective, Applied Sciences 15 (6) (2025) 3398 [Online]. Available https://www.mdpi.com/2076-3417/15/6/3398.
[19] M. Paige, A. Schauer, Z. Klesmith, A. Davis, K. Fu, The impact of hands-on geometric dimensioning and tolerancing intervention activities on students in engineering design, Advances in Engineering Education 11 (2) (2023) 66–101.
[20] D.C. Youvan, Quantum intelligence-driven nano-cyborgs: integrating advanced qubit materials for next-generation bio-computing and medical applications, 2025.
[21] F. Peng, X. Chen, A quantum entanglement-based optimization method for complex expensive engineering problems, Applied Soft Computing 174 (2025) 113019.
[22] P. Malik, The Emperor's Quantum Computer Illustrated: An Alternative Light-Centered Interpretation of Quanta, Superposition, Entanglement and the Computing that Arises from it, Illustrated with Art, Possibilities Publishing, (2024).
[23] A. Khrennikov, Roots of quantum computing supremacy: superposition, entanglement, or complementarity? The European Physical Journal Special Topics 230 (4) (2021) 1053–1057.
[24] F.A.A. Moccia, et al., Complexity and the transition to post-quantum security: cryptographic challenges regarding Shor's and Grover's algorithms, Procedia Computer Science 265 (2025) 674–680.
[25] M.H. Jebur, S. Khaleel, Quantum cryptanalysis: evaluating the impact of Shor's and Grover's algorithms on modern encryption standards, CyberSystem Journal 2 (2) (2025) 41–55.
[26] S. Mummadi, B. Rudra, Optimisation of quantum circuits using cost effective quantum gates, International Journal of Computational Science and Engineering 28 (2) (2025) 246–253.
[27] C.-J. Liang, T.-H. Le, Y. Ham, B.R.K. Mantha, M.H. Cheng, J.J. Lin, Ethics of artificial intelligence and robotics in the architecture, engineering, and construction industry, Automation in Construction 162 (2024) 105369 https://doi.org/10.1016/j.autcon.2024.105369.
[28] F. Bademosi, R.R. Issa, Factors influencing adoption and integration of construction robotics and automation technology in the US, Journal of Construction Engineering and Management 147 (8) (2021) 4021075.
[29] M. Pan, Y. Yang, Z. Zheng, W. Pan, Artificial intelligence and robotics for prefabricated and modular construction: a systematic literature review, Journal of Construction Engineering and Management 148 (9) (2022) 3122004.
[30] S. Parascho, Construction robotics: from automation to collaboration, Annual Review of Control, Robotics, and Autonomous Systems 6 (1) (2023) 183–204.

[31] P. Pradhananga, M. ElZomor, G. Santi Kasabdji, Identifying the challenges to adopting robotics in the US construction industry, Journal of Construction Engineering and Management 147 (5) (2021) 5021003.
[32] S. Zhao, et al., Application and development of autonomous robots in concrete construction: challenges and opportunities, Drones 6 (12) (2022) 424.
[33] P.P. Borase, Revolutionizing construction: the synergy of VDC/BIM, laser scanning and dusty robotics, 2024 Innovation in the Construction Industry (2024).
[34] S. Ness, T.R. Xuan, O.O. Oguntibeju, Influence of AI: robotics in healthcare, Asian Journal of Research in Computer Science 17 (5) (2024) 222–237.
[35] Q. Chen, B. García de Soto, B.T. Adey, Construction automation: research areas, industry concerns and suggestions for advancement, Automation in Construction 94 (2018) 22–38 https://doi.org/10.1016/j.autcon.2018.05.028.
[36] K.E. Emam, L. Arbuckle, Anonymizing Health Data: Case Studies and Methods to Get You Started, O'Reilly Media, Inc, (2013).
[37] C. Elendu, et al., Ethical implications of AI and robotics in healthcare: a review, Medicine 102 (50) (2023) e36671.
[38] T. McCurdy, Modeling the dose profile in human exposure assessments: ozone as an example, Reviews in Toxicology 1 (1997) 3–23.
[39] J. Shepherd, Geoengineering the Climate: Science, Governance and Uncertainty, Royal Society, (2009).
[40] A. Robock, Benefits and risks of stratospheric solar radiation management for climate intervention (geoengineering), The Bridge 50 (1) (2020) 59–67.
[41] A. Corner, N. Pidgeon, Geoengineering the climate: the social and ethical implications, Environment: Science and Policy for Sustainable Development 52 (1) (2010) 24–37.
[42] Vasquez, Solar Geoengineering Could Wreak Havoc on the Planet vasquez, Sierra, the Magazine of the Sierra Club, no. September 21, 2023. [Online]. Available from https://www.sierraclub.org/sierra/2023-3-fall/stress-test/solar-geoengineering-could-wreak-havoc-planet.
[43] D.A. Vallero, D.B. Resnik, Geoengineering: enhancing cloud albedo, Access Science (2013).
[44] A.A. Tsonis, A. Kirwan Jr, , Geoengineering from the standpoint of uncertainty and related risks: science or science fiction? Stochastic Environmental Research and Risk Assessment 37 (9) (2023) 3325–3327.
[45] D.B. Resnik, D.A. Vallero, Geoengineering: an idea whose time has come? Journal of Earth Science & Climatic Change (2011).
[46] D.A. Vallero, Mapping exposure onto nanoscale toxicity measures, in: J.R. Lead, S.H. Doak, M.J.D. Clift (Eds.), Nanotoxicology in Humans and the Environment, Springer International Publishing, Cham, (2021), pp. 141–191.
[47] S. Arnaldi, A. Muratorio, Nanotechnology, uncertainty and regulation, A Guest Editorial, Springer, (2013).
[48] National Nanotechnology Initiative, National Nanotechnology Initiative Strategic Plan, National Science and Technology Council, (2011).
[49] J.R. Lead, E. Smith, Environmental and Human Health Impacts of Nanotechnology, John Wiley & Sons, (2009).
[50] K.A. Dunphy Guzmán, M.R. Taylor, J.F. Banfield, Environmental Risks of Nanotechnology: National Nanotechnology Initiative Funding, 2000– 2004, ACS Publications, (2006).
[51] M. Nasrollahzadeh, S.M. Sajadi, Risks of nanotechnology to human life, Interface Science and Technologyvol, Elsevier, (2019), pp. 323–336 vol. 28.
[52] D.A. Vallero, Teachable moments and the tyranny of the syllabus: September 11 case, Journal of Professional Issues in Engineering Education and Practice 129 (2) (2003) 100–105.
[53] R.M. Gula, Reason Informed by Faith: Foundations of Catholic Morality, Paulist Press, (1989).
[54] A.M.R. Cristiano Cordeiro Cruz, J. Gao, Engineering ethics education through a critical view: some philosophical foundations, The Routledge International Handbook of Engineering Ethics Education, Routledge, Oxfordshire, England, UK, (2025).
[55] H.R. Niebuhr, The Responsible Self: An Essay in Christian Moral Philosophy, Westminster John Knox Press, Louisville, Kentucky, (1999).
[56] E. Kiss, D.A. Vallero (Ed.), Responsible Conduct of Research Training Course" The 6 O'Clock News Analogy, Duke University, Durham, North Caroliina, (2009)ed.
[57] Walter L. Meyer, Professionalism in construction, Engineering Issues: Journal of Professional Activities 99 (2) (1973) 205–212. Available from https://doi.org/10.1061/EGISBD.000014.

[58] R.L. Baber, The Language of Mathematics: Utilizing Math in Practice, John Wiley & Sons, Hoboken, New Jersey, (2011).
[59] J. Bailey, The case history of a failure, Architectural Forum 123 (5) (1965) 22–25.
[60] H. Jones, The Challenger and space shuttle failures were due to bureaucratic deviance, 2025 IEEE Technology and Engineering Management Society Conference-Global (TEMSCON Global), IEEE, (2025), pp. 1–6.
[61] H. Petroski, Success Through Failure: The Paradox of Design, Princeton University Press, (2013).
[62] S.K. Pfatteicher, Learning from failure: terrorism and ethics in engineering education, IEEE Technology and Society Magazine 21 (2) (2002) 8–12.
[63] V. Srinivasan, G. Halada, J. Quinn, What makes a failure into an "engineering disaster"? Learning from Engineering Disasters, Stony Brook University, Department of Materials Science and Chemical Engineeirng, Stony Brook, New York, (2018).
[64] D.A. Vallero, T.M. Letcher, Engineering risks and failures: lessons learned from environmental disasters, Leadership and Management in Engineering 12 (4) (2012) 199–209.

CHAPTER

12

Planet-scale technologies

OUTLINE

12.1 Introduction

A key question in the ongoing climate change debate is, "can anything that can be done to ameliorate the increase in carbon being released to the atmosphere"? At the basic level, this is an engineering question since it calls for a solution to a problem. An engineer follows a few common and logical steps to reach a solution. First, the problem must be defined in a rational, often quantitative way based on sound information. The technologist and engineer's forte is not to debate a policy, but to arrive at a way to achieve some aspect of that policy's objectives by applying sound science and experiential data in a systematic way [1–4]. To begin, the technologist must collect, validate, and verify all relevant and reliable data to find the best solution and to determine the extent to which the problem has been solved.

There are vast data sets available, but with widely varying quality. Thus, the data must be validated and verified. Data validation is a process used to determine whether data meet format, type, and range rules, for example, valid data structures, whereas data verification confirms data accuracy by comparisons to a trusted source. Validation is concerned with how usable the data are, that is, the right data for the project at hand, and verification is concerned with determining the extent to which the valid data are right for the situation at hand, that is, data verification is a means for ensuring the data are telling the truth.

Validation and verification provide a measure of data quality. Based on a few select data sets, in the past 20 years, some engineers and scientists have suggested consideration of

Technical and Ethical Optimization: Challenges and Opportunities for the Modern Engineer and Scientist
DOI: https://doi.org/10.1016/B978-0-443-44925-3.00018-1

planetary-scale interventions, so-called geoengineering, to mitigate global climate change and global warming [5–8]. However, based on the majority of vetted data to date, any planetary-scale project is sufficiently uncertain in approach and outcome to support most planetary-scale technologies [9].

Geoengineering entails deliberate steps to change at least one physical or chemical climatic variable on a global scale. Most advocates of geoengineering acknowledge that such large-scale measures can be associated with substantial risks over time but argue that they should be pursued as long as other actions do not provide effective solutions. All forms of geoengineering are resisted by others, given the potential scale and irreversibility of risks and uncertainty. Between the two extremes are those who argue that geoengineering research should continue at smaller, pilot scales and that large-scale projects should await a better understanding of potential impacts [10]. Geoengineering is a unique engineering approach in both temporal and spatial terms. That is, it involves a deliberate effort to affect the climate at a rather rapid pace and on a global scale.

The climate change debate within the scientific community has mainly centered on the relative contribution of the observed and projected mean global temperature increase that has resulted and will result from human activities and the reliability of models and data that have been used to calculate these factors. Presently, environmental and climate policies are made using either evidence-based risk assessments or precaution-based risk assessments [11]. The two types of risk assessments differ in the manner in determining scientific uncertainty and the threshold of proof required to take action. Evidence-based risk assessment requires proven data, scientific facts, and established causal links. Precaution-based assessment adheres to the precautionary principle, where an action is prohibited even if scientific evidence is not yet conclusive if taking an action could lead to irreversible and severe adverse outcomes [12]. For example, if a chemical compound has not yet been shown by reliable studies to cause health problems, the evidence may allow for the compound to enter the marketplace. However, if there is some similarity to another compound with evidence of a causal link to the disease, the precautionary approach would prohibit the entry of the new compound to the marketplace. This becomes complicated for large-scale engineering projects where there is some evidence that the project could have beneficial results, but there is much uncertainty of other effects. Conversely, the precautionary approach supports waiting until ample evidence becomes available.

Both types of risk assessments require information on both the hazard and the likelihood of exposure to that hazard. When this information is unreliable, the risk assessment will either underestimate or overestimate the risk. A planet-scale problem like climate change may affect large numbers of people and large geographic areas (see Fig. 12.1). It is also likely to be irreversible, which calls for a large amount of precaution before any action is taken.

Planetary-scale changes to the atmosphere would likely induce numerous adverse consequences for global human health and large environmental changes that include:

- Increases in size and frequency of flooding in coastal areas [54].
- Increased frequency and strength of storms, resulting in greater incidents of flooding, property damage, and loss of life [14].
- Flooding in regions at higher altitudes, due to increased precipitation, and droughts in subtropical regions, due to decreased precipitation [15].

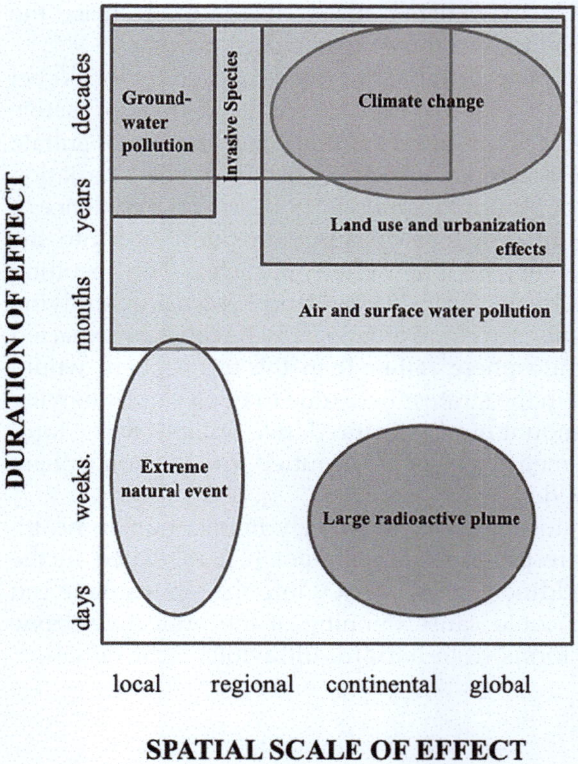

FIGURE 12.1 Temporal and spatial dependencies of environmental conditions. Immediate, short-term stress can result from a spill or an emergency situation. At the other extreme, global climate change can result from chronic releases of greenhouse gases with large-scale impacts in direct proportion to changes in global climate, such as temperature increases in the troposphere and oceans, shifting biomes, sea level rise, and migratory patterns). [7,13]. *Data from D.A. Vallero, D.B. Resnik, Geoengineering: enhancing cloud albedo, Access Sci. (2013); D.A. Vallero, Environmental Systems Science: Theory and Practical Applications, first ed., Elsevier Academic Press, 2021.*

- Increased famine and malnutrition in drought-vulnerable regions [55].
- Threats to the availability of clean drinking water due to increased drought frequencies [55].
- Increased forest fire areas because of drought and heat [54].
- Increased summer incidence and prevalence of cardiovascular and other diseases due to heatstroke and heat incapacitation [54].
- Increased prevalence of water-borne and vector-borne diseases during and after flood events [16].
- Greater frequency and intensity of algal blooms [56].

12.2 Active technologies

Compared to passive approaches, active technologies to address climate change carry more risk and uncertainty, especially in terms of future consequences. Of these, two basic types of geoengineering schemes have been most often offered as partial solutions to climate change: solar radiation management and GHG removal from the atmosphere [8], which are summarized in Table 12.1. Solar radiation management's major aim is to decrease the amount of GHGs in the atmosphere by impeding incoming solar radiation.

Several solar radiation schemes have been proposed, but the one proposed by the Nobel Prize-winning, atmospheric geochemist Paul Crutzen in 2006 is illustrative of active geoengineering [17]. Crutzen based his approach on increasing the concentration of sulfate particles in the atmosphere to increase cloud albedo, so that a portion of solar radiation would be reflected into space. The eruption of Mount Pinatubo in 1991 provides evidence of this phenomenon. The volcano emitted 10 million tons of sulfur dioxide (SO_2) into the stratosphere, which, when exposed to oxygen, became sulfate particles. The resulting enrichment of cloud albedo cooled the Earth by 0.5°C [18,19]. By logical extension, spraying gas-phase SO_2 into the stratosphere would enhance cloud albedo, which would be expected to cool the planet. Spraying SO_2 in the stratosphere rather than the troposphere would increase the time that sulfate particles reside in the stratosphere (one to two years), allowing for greater increases in cloud albedo. The amount of SO_2 required, two million metric tons, would be only about 3.6% of the 55 million metric tons of SO_2 emitted into the atmosphere each year from the burning of fossil fuels [19].

Unfortunately, spraying SO_2 into the stratosphere may adversely impact human health and the environment, given that SO_2 is a respiratory air pollutant [20] regulated in the United States under the Clean Air Act. In addition, spraying SO_2 into the stratosphere can also interfere with ozone chemistry, which could cause thinning of the protective ozone layer [21]. The ozone layer protects all life forms from harmful ultraviolet light radiation, which causes skin cancer.

TABLE 12.1 Geoengineering classifications.

Solar radiation management
• Spraying SO_2 into the stratosphere enhances cloud albedo
• Spraying engineering nanoparticles into the stratosphere to enhance cloud albedo
• Placing reflective mirrors, discs, or particles in Earth's orbit
• Painting roofs and other structures with reflective materials
• Placing solar reflectors in the desert
Carbon dioxide removal
• Using physical and chemical processes to remove and store CO_2
• Enhancing CO_2 removal and storage by terrestrial plants
• Burial of biomass
• Reforestation and afforestation (creating new forests)
• Fertilizing the oceans to stimulate CO_2 removal by plankton

Data from: D.B. Resnik, D.A. Vallero, Geoengineering: An idea whose time has come?, Journal of earth science & climatic change (2011).

Another potential problem of this method is that SO_2 is converted to sulfuric acid in the atmosphere and is a key component of acid deposition [22,23]. Many carbon-based molecules exist in equilibrium with one another. The quality of the earth's water is affected by the equilibrium chemistry of carbonates, bicarbonates, organic compounds, carbonic acid, and CO_2. Thus, on a global scale, uncontaminated rain's mean pH would already be about 5.6, owing to its dissolution of CO_2. As the water droplets fall through the air, the CO_2 in the atmosphere becomes dissolved in the water, creating an equilibrium condition:

$$CO_2(\text{gas in air}) + CO_2(\text{dissolved in water}) \tag{12.1}$$

The CO_2 in the water reacts to produce hydrogen ions, as:

$$CO_2 + H_2O \leftrightarrow H_2CO_3 \leftrightarrow H^+ + HCO_3^- \tag{12.2}$$

$$HCO_3^- \leftrightarrow H^+ + CO_3^{(2-)} \tag{12.3}$$

Sulfur is released into the atmosphere as either reduced forms, for example, hydrogen sulfide, or oxidized forms. Both forms include air pollutants. Hydrogen sulfide is oxidized to sulfur dioxide in a three-step process. The hydroxyl radical initiates the transformation from hydrogen sulfide to sulfur dioxide:

$$H_2S + HO^{\cdot} \rightarrow HS^{\cdot} + H_2O \tag{12.4}$$

$$HS^{\cdot} + O_2 \rightarrow HO^{\cdot} + SO^{\cdot} \tag{12.5}$$

$$SO^{\cdot} + O_2 \rightarrow SO_2 + O^{\cdot} \tag{12.6}$$

The atmospheric reactions of SO_2 are complex and proceed through three different pathways to the sulfate ion (SO_4^{2-}). Sulfur dioxide and the hydroxyl radical may react to form HSO_3 radical, with a cascade reaction involving another hydroxyl radical to form water and SO_3 or H_2SO_4. Sulfur dioxide's solubility allows it to dissolve in water droplets and to react with O_2 to form SO_4^{2-}. The third pathway to sulfate formation occurs when SO_2 reacts with hydrogen peroxide (H_2O_2) to form sulfuric acid:

$$HO^{\cdot} + SO_2 \rightarrow HOSO_2^{\bullet} \tag{12.7}$$

$$HOSO_2^{\bullet} + HO^{\cdot} \rightarrow H_2SO_4 \tag{12.8}$$

$$SO_{2(\text{aqueous})} + [O]_{(\text{aqueous})} \rightarrow H_2SO_4 \tag{12.9}$$

$$SO_2 + H_2O_2 \rightarrow H_2SO_4 \tag{12.10}$$

Another drawback of sulfate sprays is that SO_2 conversion to acids would exacerbate ocean acidity already occurring when greater amounts of atmospheric CO_2 dissolve in the water. Indeed, oceans have increased in acidity by about 30% since 1750, due to increases in anthropogenic carbon in the atmosphere [15]. This increase in the acidity decreases the availability of calcium carbonate and threatens the organisms that form shells from this compound, for example, corals and some types of plankton. Losses of these species can cascade the effects on other marine species, since many organisms seek shelter and feed on shellfish in coral reefs [23,24]. Greater cloud albedo to stabilize global temperatures could change precipitation patterns, tropical storm activity, temperature distribution, and wind [21]. Much of the radiation reflected by sulfate particles strikes the Earth as diffuse light, which increases the whiteness of the daytime sky and can diminish the efficiency of plant photosynthesis and solar power [21,23]. Miscalculating and spraying too much SO_2 into the stratosphere can trigger excessive cooling and potentially induce an early ice age [23].

Another active geoengineering proposal by D.W. Keith makes use of disc-shaped, engineered nanoparticles composed of layers of aluminum oxide, metallic aluminum, and barium titanate, to be released into the stratosphere [21]. The particles would take advantage of photophoretic and electromagnetic forces to levitate above the stratosphere. Photophoretic forces occur when there is a temperature differential between an aerosol and the surrounding gas. A chief advantage of these nanoparticles is that they would remain in the atmosphere much longer than SO_2. Their specific reflective properties could also be controlled through engineering and design, so that they would produce correct amount of diffuse light. Also, since they would be above the stratosphere, they would not interfere with ozone chemistry [21]. Further, the nanoparticles would not produce acid rain.

Keith's proposal also has disadvantages. Uncertainty is high, given that the proposal has not been tested. The eruption of Mt. Pinatubo provides evidence that in theory, Crutzen's idea would work, but it is not known whether the nanoparticles envisioned by Keith would behave as predicted. Like Crutzen's proposal, stabilizing global temperatures at higher CO_2 concentrations would not address ocean acidification. It could affect precipitation patterns, temperature distribution, tropical storms, and winds. The environmental and health impacts of spraying nanoparticles into the stratosphere are not well understood. In addition, Keith's proposal could be excessively expensive, given the costs and state of the science of nanotechnology [25].

12.3 Carbon geoengineering

The process of storing a substance for a longer period of time than would naturally occur, that is, sequestration [26–28], is a less intrusive approach than the previously mentioned technologies. Of all projects sequestering global greenhouse gases (GHGs), carbon dioxide (CO_2) is currently the largest type of geoengineering project in terms of mass and volume of GHG being stored [25]. Sequestration is an example of engineering and technological mimicry since it is an existing geophysical process. Sequestration technologies simply accelerate the rate and enlarge the amount of CO_2 being stored anthropogenically. As both the atmosphere and biological systems have contained carbon compounds, especially CO_2 and methane (CH_4), gaseous forms have continued to reach water, soil, and rock formations as carbon sinks. That is, carbon sequestration is a part of the earth's source-sink system [29–34]. Indeed, CO_2 has been in the atmosphere for an estimated 4.5 billion years, but plant life, the major biological systems that store carbon, did not appear until much later, that is, less than 500 million years [35,36].

Like many biogeochemical processes, human activities affect carbon source-sink dynamics [37,38]. Emissions of GHGs have increased in last few hundred years from anthropocentric sources that include combustion of fossil fuels, but also from less direct activities like wetland destruction [39–41] and urban development [42].

Speeding up and tailoring concepts, processes, and mechanisms that already occur in nature is a common theme of environmental engineering. The engineering process can cascade to new applications. For example, advances in air pollution technologies have been adapted from wastewater treatment and other biotechnologies, such as biofilters that originally mimicked soil and water microbial systems [43,44]. Likewise, evolving sequestration technologies include new ways to either sequester carbon or to enhance or expedite processes that already exist.

Conservation is an example of a more "passive" approach. There are currently enormous releases of carbon that, if eliminated, would greatly reduce the loading to the troposphere. Protecting the loss of forests, woodlands, wetlands, and other ecosystems is an efficient way of preventing future problems. In fact, much of the terrestrial fluxes and sinks of carbon involve the soil. Keeping the soil in place must be part of the overall global strategy to reduce GHGs.

Assuming that fossil fuels will continue to be used in the near future, engineers have explored various technologies to capture the CO_2 produced from fuel combustion and sequester it underground [45–48]. Sequestration is a biosystems approach, since it is an ongoing process on planet Earth, with myriad interactions between biotic and abiotic factors [43,49,50].

12.3.1 Carbon soil sequestration

Terrestrial plants need soil to live. Massive amounts of carbon are stored in root zones. Geoengineering strategies must consider soil properties, for example, texture or grain size (see Table 12.2), ion exchange capacities, ionic strength, pH, microbial populations, and soil organic matter content.

TABLE 12.2 Soil texture classifications [51].

Name	Size range (mm)
Gravel	>2.0
Very coarse sand	1.0–1.999
Coarse sand	0.500–0.999
Medium sand	0.250–0.499
Fine sand	0.100–0.249
Very fine sand	0.050–0.099
Silt	0.002–0.049
Clay	<0.002

Data from: T.E. Loynachan, K.W. Brown, T.H. Cooper, M.H. Milford, Sustaining our soils and society, American Geological Institute, 1999.

Soil is a matrix comprised of various components, including organic matter and unconsolidated material. The matrix contains liquids within its interstices, usually varying amounts of water and solutes, including carbon-based compounds. Under most environmental conditions, air and water are solutions of small amounts of chemical compounds. For example, air's solutes represent small percentages of the solution at the highest (e.g., water vapor), and most other solutes represent parts per million, including 421 ppm CO_2. A soil matrix consists mostly of solids surrounded by void spaces that can contain liquids (mainly soil water) and gases (e.g., molecular nitrogen [N_2], molecular oxygen [O_2], CO_2, and CH_4). The composition of each fraction of soil varies. For example, the amount of soil air differs from that in the atmosphere and changes with depth from the surface [52]. Table 12.3 illustrates the inverse relationship between carbon dioxide and molecular oxygen. Sediment is simply underwater soil with characteristics analogous to terrestrial soils [53].

When the soil gases are lost, carbon compounds become available to be released into the atmosphere. Preventing soil and wetland loss is a passive mechanism for sequestering GHGs.

Active geoengineering can enhance existing processes. For example, in addition to conserving current levels of carbon sequestration, technologies can be adapted to increase the rates of sequestration. The scale of such technologies can range from a single localized source, such as a coal-burning electricity generation station in which stack gases are sent to an underground rock stratum, to a widespread system of CO_2 collection and injection technologies for a whole network of GHG-emitting facilities.

In addition to rock strata, active carbon sequestration can include water sinks. Indeed, oceans have removed 48% of the CO_2 released into the troposphere from burning fossil fuels and cement manufacturing [8]. However, we are reminded that the risks and uncertainties grow with time and space. A system as large as an ocean is limited in its capacity for sequestration [54]. Most CO_2 generated by human activities resides in the upper ocean layers. Carbon compounds enter and exit oceans predominantly as a function of the compound's solubility and water temperature, so that more of the compound will be

TABLE 12.3 Composition of two important gases in soil air.

Depth from surface (cm)	Silty clay		Silty clay loam		Sandy loam	
	O_2 (% volume of air)	CO_2 (% volume of air)	O_2 (% volume of air)	CO_2 (% volume of air)	O_2 (% volume of air)	CO_2 (% volume of air)
30	18.2	1.7	19.8	1.0	19.9	0.8
61	16.7	2.8	17.9	3.2	19.4	1.3
91	15.6	3.7	16.8	4.6	19.1	1.5
122	12.3	7.9	16.0	6.2	18.3	2.1
152	8.8	10.6	15.3	7.1	17.9	2.7
183	4.6	10.3	14.8	7.0	17.5	3.0

Data from: V. Evangelou, Environmental Soil and Water Chemistry: Principles and Applications (no. 631.455 E6), 1998.

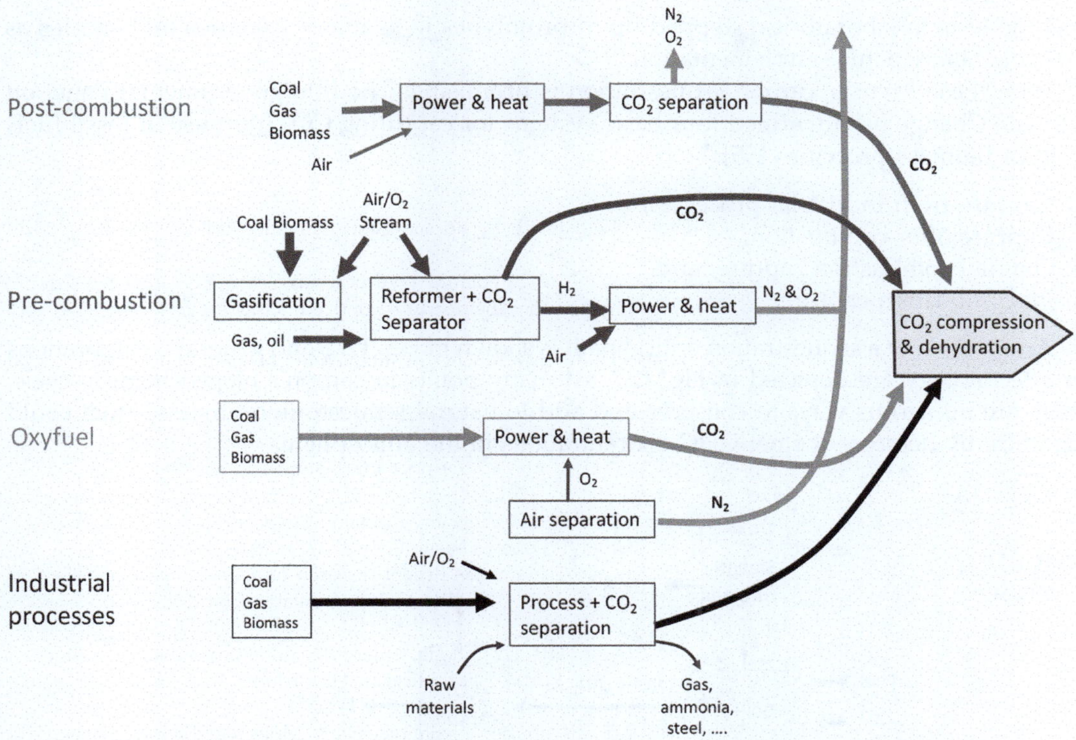

FIGURE 12.2 Schematic diagram of potential means of capturing carbon dioxide. *Data from O. Davidson, B. Metz, Special Report on Carbon Dioxide Capture and Storage, International Panel on Climate Change, Geneva, Switzerland 7 (13) (2005) 1–100.*

retained in the ocean water with decreasing temperature. About half of the anthropogenic CO_2 taken up by the ocean for the previous two centuries has stayed in the upper 10% of the ocean, given the slow mixing [30].

In addition to transporting carbon to a sink, sequestration involves holding the carbon for long time periods. Thus to keep CO_2 sequestered in the ocean, it must remain in the cooler, deeper regions. Otherwise, when CO_2 resides near the warmer surface, it is more likely to be emitted back to the atmosphere. The actual carbon mass can be enhanced by management. For example, some species of plankton are limited in growth by iron. Thus increasing the iron concentrations in certain ocean layers could considerably increase the ability of these organisms to absorb and store carbon. Even if the technological means of moving carbon to the ocean reservoirs are workable, the sequestration may lead to large-scale and virtually irreversible consequences. Evidence of ocean acidification is already a concern due to increases in atmospheric CO_2 concentrations, notwithstanding the effects of intentional sequestration.

Again, any large-scale endeavor like geoengineering must be approached with an appropriate amount of caution. The best decisions are those that account for all possible outcomes, certainly not exclusively the ones hoped for. Such an approach would likely include tests in laboratories, stepped up to prototypes in as many scenarios and species as possible before actual implementation.

With these caveats, carbon sequestration is promising. The Intergovernmental Panel on Climate Change has identified four basic systems for capturing CO_2 from use of fossil fuels and/or biomass processes [55]:

1. Capture from industrial processes;
2. Capture after combustion;
3. Oxyfuel combustion capture; and
4. Precombustion capture.

An example of a sequestration critical path is shown in Fig. 12.2, and general configurations of a technology are depicted in Fig. 12.3. At many points in carbon's biogeochemical cycle, there are numerous ways to conserve and add to natural sequestration processes that could significantly decrease the net GHG concentrations in the atmosphere.

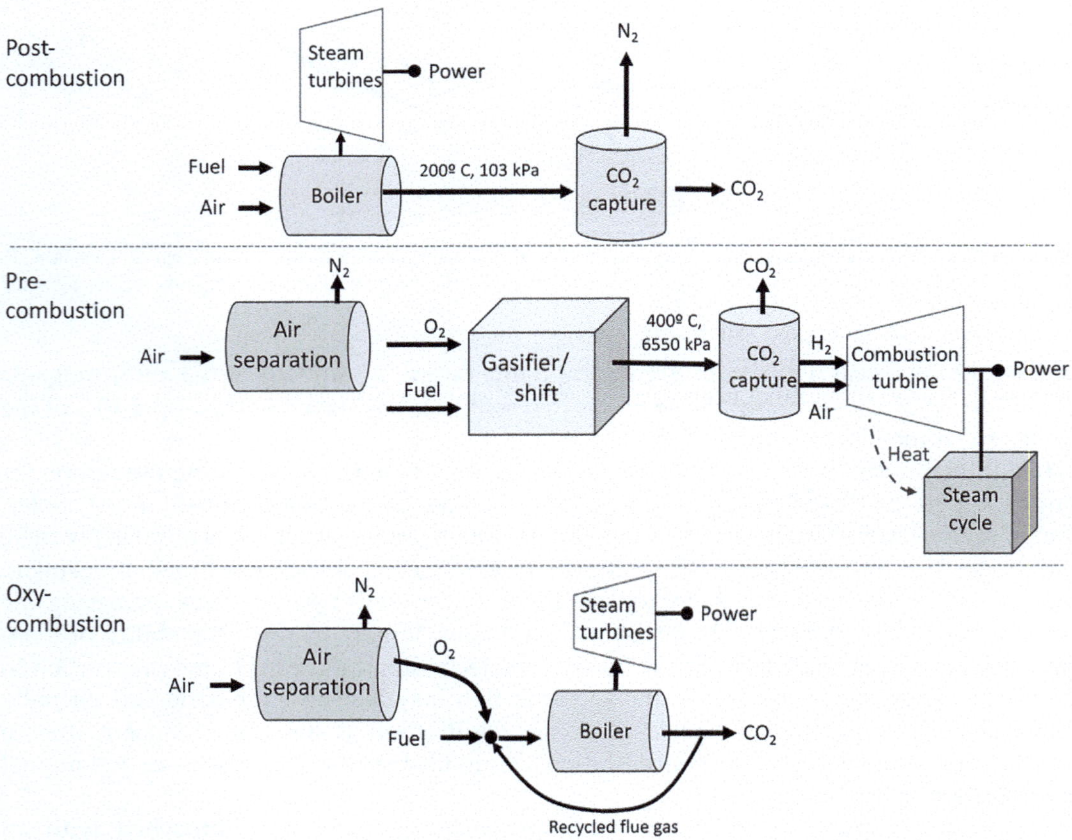

FIGURE 12.3 Block diagrams illustrating postcombustion, precombustion, and oxy-combustion systems. *Data from J.D. Figueroa, T. Fout, S. Plasynski, H. McIlvried, R.D. Srivastava, Advances in CO2 capture technology—the US Department of Energy's Carbon Sequestration Program, International Journal of Greenhouse Gas Control 2 (1) (2008) 9–20.*

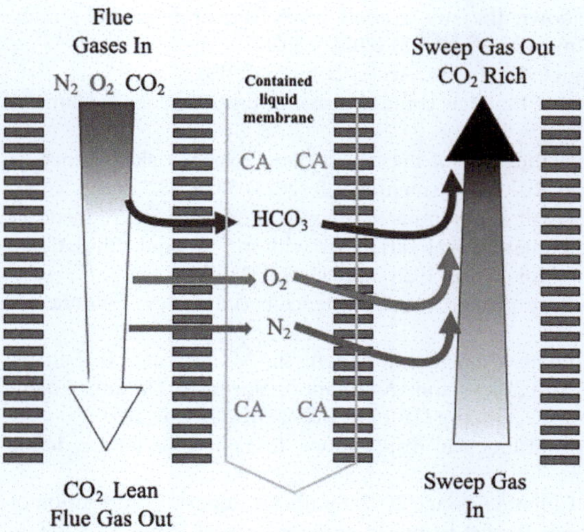

FIGURE 12.4 Schematic of carbozyme permeation process for carbon dioxide sequestration. *CLM*, Contained liquid membrane; *CA*, carbonic anhydrase. *Data from J.D. Figueroa, T. Fout, S. Plasynski, H. McIlvried, R.D. Srivastava, Advances in CO2 capture technology—the US Department of Energy's Carbon Sequestration Program, International Journal of Greenhouse Gas Control 2 (1) (2008) 9–20.*

The sequestration can take place throughout the carbon cycle. For example, an encouraging approach is to deploy the biochemistry of organisms to sequester CO_2 during respiration. Technologies can take advantage of other metabolic reactions involving CO_2 in living organisms, chiefly enzymatic processes. In one enzyme-based system [56], CO_2 is captured and released by emulating the mammalian respiratory system (see Fig. 12.4). The process stores carbonic anhydrase (CA) in a hollow fiber liquid membrane. The CA-developing membrane is employed to separate CO_2 from the flue gas. Modifying the membrane enhances the rate of diffusion of CO_2 due to the strong interactions between the permeating CO_2 molecules and the amine-functional membrane pores. Movement of O_2, N_2, and SO_2 is blocked, which increases selective sequestration of CO_2 [56].

12.4 Conclusions

This chapter has introduced the pros and cons of geoengineering to reduce the amount of GHGs in the atmosphere. The uncertainties and risks of various geoengineering methodologies are discussed, including those posed by both passive and active approaches.

References

[1] D.A. Vallero, Environmental Systems Science: Theory and Practical Applications, first ed, Elsevier, Amsterdam, NV, (2020).
[2] C. Whitbeck, Ethics in Engineering Practice and Research, Cambridge University Press, (2011).
[3] C. Whitbeck, Ethics as design: doing justice to moral problems, Hastings Center Report 26 (3) (1996) 9–16.
[4] H. Petroski, To Engineer is Human: The Role of Failure in Successful Design, St Martins Press, (1985).
[5] J. Wieding, J. Stubenrauch, F. Ekardt, human rights and precautionary principle: limits to geoengineering, SRM, and IPCC scenarios, Sustainability 12 (21) (2020) 8858.

[6] D. Visioni, D.G. MacMartin, B. Kravitz, Is turning down the sun a good proxy for stratospheric sulfate geoengineering? Journal of Geophysical Research: Atmospheres 126 (5) (2021) e2020JD033952.
[7] D.A. Vallero, D.B. Resnik, Geoengineering: enhancing cloud albedo, Access Science (2013).
[8] D.B. Resnik, D.A. Vallero, Geoengineering: an idea whose time has come? Journal of Earth Science & Climatic Change (2011).
[9] A.A. Tsonis, A. Kirwan Jr, Geoengineering from the standpoint of uncertainty and related risks: science or science fiction? Stochastic Environmental Research and Risk Assessment 37 (9) (2023) 3325–3327.
[10] B.A. Gay, L. Mandrake, K.R. Miner, C.E. Miller, Assessing the impacts of mitigation and geoengineering intervention scenarios on Earth system dynamics and climatological variability with multimodal simulations, Scientific Reports 15 (1) (2025). Available from https://doi.org/10.1038/s41598-025-91195-6.
[11] M. Monaghan, R. Pawson, K. Wicker, The precautionary principle and evidence-based policy, Evidence & Policy 8 (2) (2012) 171–191.
[12] N.A. Manson, The Precautionary Principle and Climate Change Mitigation, in: M. Lackner, B. Saijadi, W.Y. Chen (Eds.), Handbook of Climate Change Mitigation and Adaptation, Springer Nature, Cham, Switzerland, (2025), pp. 4559–4579. Available from https://doi.org/10.1007/978-3-031-84483-6_203.
[13] D.A. Vallero, Environmental Systems Science: Theory and Practical Applications, first ed., Elsevier Academic Press, (2021).
[14] Intergovernmental Panel on Climate Change (IPCC), Climate Change 2007: Synthesis Report. Contribution of Working Groups I, II and III to the Fourth Assessment Report of the Intergovernmental Panel on Climate Change [Core Writing Team, Pachauri, R.K and Reisinger, A. (eds.)]. IPCC, Geneva, Switzerland, 2007. ISBN: 978-92-9169-143-2.
[15] National Aeronautics and Space Administration, Global Climate Change, February 3, 2026. Available at: http://climate.nasa.gov/. Accessed: May 8, 2026.
[16] Interagency Working Group on Climate Change and Health, A Human Health Perspective on Climate Change, 2011. Available at: http://ehp03.niehs.nih.gov/static/climatechange.action. Accessed: August 12, 2011.
[17] O. Morton, Climate change: is this what it takes to save the world? Nature 447 (7141) (2007) 132–137.
[18] T.M. Wigley, A combined mitigation/geoengineering approach to climate stabilization, Science (New York, N.Y.) 314 (5798) (2006) 452–454.
[19] P.J. Crutzen, Albedo enhancement by stratospheric sulfur injections: a contribution to resolve a policy dilemma? Climatic Change 77 (3-4) (2006) 211.
[20] M.L. Bell, J.M. Samet, Chapter 12: Air pollution, Environmental Health: From Global to Local, Jossey-Bass (Wiley), San Francisco, California, (2010), pp. 387–416.
[21] D.W. Keith, Photophoretic levitation of engineered aerosols for geoengineering, Proceedings of the National Academy of Sciences 107 (38) (2010) 16428–16431.
[22] G.M. Lovett, et al., Effects of air pollution on ecosystems and biological diversity in the eastern United States, Annals of the New York Academy of Sciences 1162 (1) (2009) 99–135.
[23] A. Robock, Benefits and risks of stratospheric solar radiation management for climate intervention (geoengineering), The Bridge 50 (1) (2020) 59–67.
[24] V.J. Fabry, B.A. Seibel, R.A. Feely, J.C. Orr, Impacts of ocean acidification on marine fauna and ecosystem processes, ICES Journal of Marine Science 65 (3) (2008) 414–432.
[25] J.L. Jones, et al., Positioning nanotechnology to address climate change, Environment Systems and Decisions 44 (4) (2024) 1039–1053.
[26] J.T. Litynski, S.M. Klara, H.G. McIlvried, R.D. Srivastava, An overview of terrestrial sequestration of carbon dioxide: the United States Department of Energy's fossil energy R&D program, Climatic Change 74 (1) (2006) 81–95.
[27] M. Ghommem, M.R. Hajj, I.K. Puri, Influence of natural and anthropogenic carbon dioxide sequestration on global warming, Ecological Modelling 235 (2012) 1–7.
[28] J.C. Curran, S.A. Curran, Natural sequestration of carbon dioxide is in decline: climate change will accelerate, Weather 80 (3) (2025) 85–87.
[29] X. Wei, J. Yang, P. Luo, L. Lin, K. Lin, J. Guan, Assessment of the variation and influencing factors of vegetation NPP and carbon sink capacity under different natural conditions, Ecological Indicators 138 (2022) 108834.
[30] C.L. Sabine, et al., The oceanic sink for anthropogenic CO2, Science (New York, N.Y.) 305 (5682) (2004) 367–371.

[31] F. Meng, et al., Organic carbon sink dynamics and carbon sink-source balance in global lakes during the Anthropocene, Fundamental Research (2024).
[32] A. Grelle, P.-O. Hedwall, M. Strömgren, C. Håkansson, J. Bergh, From source to sink–recovery of the carbon balance in young forests, Agricultural and Forest Meteorology 330 (2023) 109290.
[33] T. DeVries, Atmospheric CO2 and sea surface temperature variability cannot explain recent decadal variability of the ocean CO2 sink, Geophysical Research Letters 49 (7) (2022) e2021GL096018.
[34] A. Cabon, et al., Cross-biome synthesis of source versus sink limits to tree growth, Science (New York, N.Y.) 376 (6594) (2022) 758–761.
[35] L.P. Knauth, M.J. Kennedy, The late Precambrian greening of the Earth, Nature 460 (7256) (2009) 728–732.
[36] A.R. Waldeck, et al., Marine sulphate captures a Paleozoic transition to a modern terrestrial weathering environment, Nature Communications 16 (1) (2025) 2087.
[37] J.G. Canadell, M.U. Kirschbaum, W.A. Kurz, M.-J. Sanz, B. Schlamadinger, Y. Yamagata, Factoring out natural and indirect human effects on terrestrial carbon sources and sinks, Environmental Science & Policy 10 (4) (2007) 370–384.
[38] J.-l Jiang, J.-g Xu, W.-j Wu, L. Zhou, D.-q Sun, Patterns and dynamics of China's human-nature carbon source-sink system, Journal of Natural Resources 29 (5) (2014) 757–768.
[39] R. Records, M. Arabi, S. Fassnacht, W. Duffy, M. Ahmadi, K. Hegewisch, Climate change and wetland loss impacts on a western river's water quality, Hydrology and Earth System Sciences 18 (11) (2014) 4509–4527.
[40] K.L. Erwin, Wetlands and global climate change: the role of wetland restoration in a changing world, Wetlands Ecology and Management 17 (1) (2009) 71–84.
[41] H. Desta, B. Lemma, A. Fetene, Aspects of climate change and its associated impacts on wetland ecosystem functions: a review, Journal of American Science vol. 8, (no. 10) (2012) 582–596.
[42] A. Humbal, N. Chaudhary, B. Pathak, Urbanization trends, climate change, and environmental sustainability, in Climate Change and Urban Environment Sustainability, Springer, (2023), pp. 151–166.
[43] D.A. Vallero, C.K. Gunsch, Applications and implications of emerging biotechnologies in environmental engineering, Journal of Environmental Engineering 146 (6) (2020) 03120005.
[44] A.W. McCumber, Y.J. Kim, O.S. Isikhuemhen, R.M. Tighe, C.K. Gunsch, The environment shapes swine lung bacterial communities, Science of the Total Environment 758 (2021) 143623.
[45] J.-Y. Yoon, M.R. Riley, Grand challenges for biological engineering, Journal of Biological Engineering 3 (1) (2009) 16.
[46] D.A. Vallero, An engineer's assessment on adapting to global warming and climate change, Living with Climate Change, Elsevier, (2024), pp. 21–43.
[47] D.A. Vallero, T.M. Letcher, Unraveling Environmental Disasters, second ed., Elsevier Academic Press, Amsterdam, The Netherlands, (2023).
[48] D.A. Vallero, Potential impacts of climate change on biogeochemical cycling, Water and Climate Change, Elsevier, (2022), pp. 41–62.
[49] D. Vallero, Environmental Biotechnology: A Biosystems Approach, Elsevier Science, (2015).
[50] J. Mitchell, et al., A decision analytic approach to exposure-based chemical prioritization, PLoS One 8 (8) (2013) e70911. Available from https://doi.org/10.1371/journal.pone.0070911.
[51] T.E. Loynachan, K.W. Brown, T.H. Cooper, M.H. Milford, Sustaining Our Soils and Society, American Geological Institute, (1999).
[52] V. Evangelou, Environmental soil and water chemistry: principles and applications (no. 631.455 E6), 1998.
[53] D.A. Vallero, Dicarboximide Fungicide Flux from an Aquic Hapludult Soil to the Lower Troposphere, Duke University, (2000).
[54] S. Solomon, et al., Intergovernmental panel on climate change fourth assessment report (AR4), Climate Change vol. 374, (2007).
[55] O. Davidson, B. Metz, Special report on carbon dioxide capture and storage, International Panel on Climate Change, Geneva, Switzerland 7 (13) (2005) 1–100.
[56] J.D. Figueroa, T. Fout, S. Plasynski, H. McIlvried, R.D. Srivastava, Advances in CO2 capture technology—the US Department of Energy's Carbon Sequestration Program, International Journal of Greenhouse Gas Control 2 (1) (2008) 9–20.

CHAPTER

13

Human factors engineering

OUTLINE

It is evident that human reliability is too often (some would argue always) the weak link in the chain of events which leads to loss of safety and catastrophic failure of even the most advanced technological systems. ***B.A. Sayers (2011) [1]***

13.1 Introduction

Human factors engineering (HFE) is a discipline concerned with the design of systems based on human capabilities, limitations, and characteristics [2]. HFE is needed to design for safe and effective human use of items, incorporating many engineering, biomedical, anatomical, physiological, psychological, and other disciplines [3]. Subdisciplines include anthropometrics, biomechanics, and sensation biology. The cognitive psychological aspects of HFE include models of human performance, memory, and attention. Indeed, the engineering science and psychology disciplines have had meaningful input in technical designs because of their shared focus on the immediate human-technology interfaces, where both disciplines are a part of any strategy to assure occupational safety and efficiency of individuals [3].

HFE focuses on user needs, user characteristics, and end-user testing of human–machine interface. It involves iterative design phases and testing, repeatedly refining the design development based on feedback from users, beginning in the early stages of the design cycle and continuing through completion. This allows for ongoing corrections to the design as

Technical and Ethical Optimization: Challenges and Opportunities for the Modern Engineer and Scientist
DOI: https://doi.org/10.1016/B978-0-443-44925-3.00005-3

flaws arise, especially before the HFE-designed product reaches the marketplace. This is essential for products before they are used in biomedical settings [4].

Emerging technologies present additional challenges, given the uncertainty of their hazards and risks [5]. Graumann [6] argued that the so-called "the world of things" was of less interest to psychologists because of psychology's obsession with individual organismic perspectives like motivation and perception, compared to the interaction of humans with their environment. Whereas psychology addressed the environmental conditions that influence behavior in a known environment, engineering creates environments that are expected to elicit certain behaviors [5].

The need for an interdisciplinary approach to the human-technology interface has grown substantially in the past few decades. Wilpert [5] put it this way: at least

> ... these installations are usually high hazard systems that are characterized by multiple actors, by enormous degrees of complexity, by tightly coupled sub-systems and, in consequence, by considerable lack of transparency with concomitant control problems. Furthermore, the analysis of all major industrial incidents and accidents in the past (e. g. Bhopal[1], Eschede[2], Challenger[3]) convincingly show that it was never the failure of a technical component or arrangement alone which "caused" the event but the complex interaction of factors that may emanate from any of the different system levels: from technical design features, individual hazardous actions, inadequate team performance, inappropriate management, ill-structured organization and even from actors external to the facility. Hence the conclusion seems self-evident: if an adequate understanding of incident causes in industrial installations requires different disciplinary competencies, the design of such installations to be safe, functional and reliable would benefit from interdisciplinary cooperation in their design.

Thus, there is a forecasting aspect of HFE. Given the rapid rate of innovations, technologies emerge without much warning. As such, HFE must consider all reasonable possibilities of use or misuse of technologies likely to reach the marketplace. This calls for proactive risk assessments that consider reasonably foreseeable failures and misuses of technologies on the horizon [9].

[1] See Chapter 16.

[2] On June 3, 1998, cars on the ICE 1 train on the Hanover–Hamburg railway near Eschede, Germany, derailed and collided with an overpass. 101 people were killed, and no fewer than 88 were injured. A single crack in one wheel caused a part of the wheel to become caught in a railroad switch which, when the train passed over it, changed the direction of the switch. This led to the train moving on two separate tracks, leading to the derailment and crash into the concrete pillars of a bridge. The bridge, in turn, collapsed, crushing two coaches. The rest of the coaches and the rear engine crashed into the wreckage [7]. After the incident, many investigations into the wheel fracture were conducted. These analysis concluded that the accident was caused by poor wheel design, which allowed a fatigue fracture to develop on the wheel rim.

[3] The Challenger space shuttle explosion in 1986 resulted in the death of seven astronauts. In this case, "the failure of the solid rocket booster O-rings to seat properly allowed hot combustion gases to leak from the side of the booster and burn through the external fuel tank. The failure of the O-ring was attributed to several factors, including faulty design of the solid rocket boosters, insufficient low-temperature testing of the O-ring material and the joints that the O-ring sealed, and a lack of proper communication between different levels of management." See: [8].

13.2 Offsetting behavior and risk homeostasis

A paramount engineering concern is to avoid design failure. Product liability is one of the public's responses to past failures and a means to induce technologists to build in safeguards to prevent future failures. One example of product liability oversight is that of the US Food and Drug Administration (FDA). Fig. 13.1 depicts a simplified model of the FDA's licensing process.

Innovations fail for two main reasons, that is, design flaws and misuse. Good engineering practice prevents the first. Good engineering practice should also prevent the second, but at what point in the future are engineers to blame for the misuse of their products? This is even more complicated for bioengineering than for many other endeavors, since biotechnologies involve living organisms, which are always more complex than abiotic systems.

The response to biotechnological failure can be dramatic and severe. People process failures both cognitively *and* motivationally [10]:

> Cognitively, injured parties see themselves as having been doing something that seemed sensible at the time, and not looking for trouble. As a result, any accident comes as a surprise. If it was to be avoided, then someone else needed to provide the missing expertise and protection. Motivationally, no one wants to feel responsible for an accident. That just adds insult to injury, as well as forfeiting the chance for emotional and financial redress. Of course, the natural targets for such blame are those who created and distributed the product or equipment involved in an accident. They could have improved the design to prevent accidents. They should have done more to ensure that the product would not fail in expected use. They could have provided better warnings and instructions in how to use the product. They could have sacrificed profits or forgone sales, rather than let users bear (what now seem to have been) unacceptable risks.

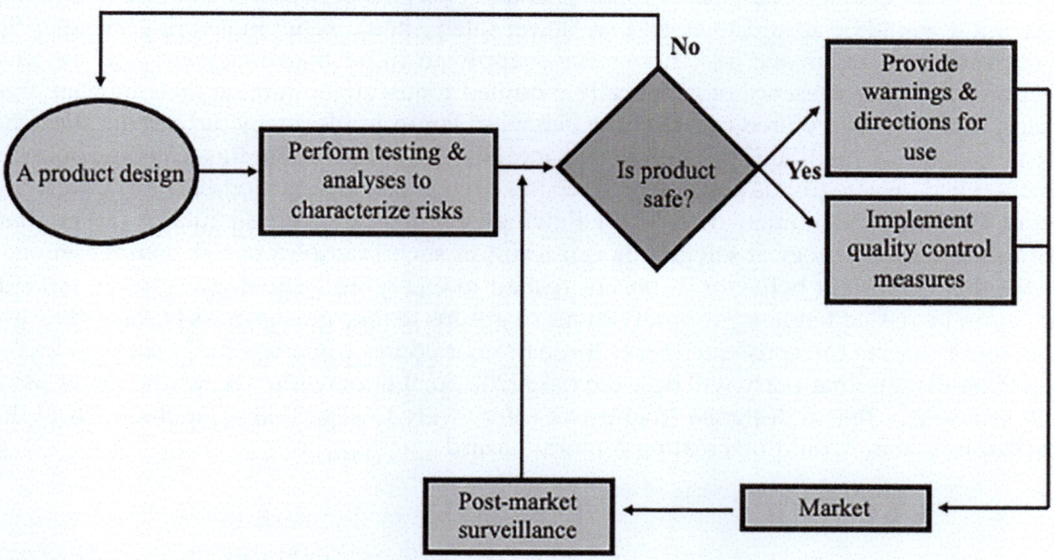

FIGURE 13.1 Licensing process of the US Food and Drug Administration [10,11]. *Data from B.M. Fischhoff, F. J., The inconvenient public, Chemtech - American Chemical Society 25 (3) (1985) 47–54; National Academy of Engineering, Product Liability and Innovation: Managing Risk in an Uncertain Environment. Washington, DC: The National Academies Press (in English), 1994, p. 216.*

Even a well-designed technology can be defeated by risk homeostasis, that is, the theory that people have a target level of risk they are willing to accept and adjust their behavior to maintain this level, notwithstanding safety improvements [12]. Basically, users defeat built-in factors of safety by asserting new ways to use and misuse the products. If such aggressiveness occurs, it may simply be because consumers want more utility or efficiency from their products. One could argue that the instructions and warnings accompanying products should be sufficient consideration, and if a user willfully ignores this information, it constitutes the user's autonomy in a rational decision made with "informed consent." Another view is that liability extends beyond labeling and reasonable use [13,14]. This is particularly problematic for complicated systems like those employed in biomedical settings and even driverless vehicles, for example, continuous defensive braking by autonomous vehicles significantly elevates the likelihood of rear-end collisions [15].

Offsetting behavior or risk compensation is the phenomenon in which actors adjust their behavior to respond to perceived changes in safety or risk. This can mean that the actor will take greater and more frequent risks when the presence of a technology, for example, air bags in a vehicle, causes them to feel safer than when the technology is not present [16]. Economists have found that policies designed to protect the public may inadvertently lead to "attenuation and even reversal of the direct policy effect on expected harm … because of offsetting behavior by potential victims as they reduce care in response to the policy" [17]. Much research has been related to transportation, especially road safety. For example, drivers with antilock brake systems were found to tailgate more closely than those without this technology. However, there is ample evidence that the arguments against safety technologies like better braking systems, air bags, and helmet wearing have been employed as a "smokescreen by some who are opposed to tough safety measures" [18]. In other words, an existing or emerging technology that improves safety must not be discouraged because of arguments that offsetting behaviors will occur. They may well, but it would be absurd to assert that "fewer safety should result in less risk taking" [18].

Offsetting behavior and risk homeostasis apply to more than machines and devices. Events such as the presence of genetically modified foodstuffs in human diets indicate that society's response is colored by offsetting behavior. For example, many did not question the use of genetically modified corn as food for animals. Is it possible that those feeding animals overly relied on safeguards and did not recognize when genetically modified corn indirectly found its way into human diets? [19] Relying on conventional safeguards rather than instituting biotechnological safeguards can result in such examples of risk compensation.

Another important behavioral concept related to emerging technology is moral hazard, that is, an increased tendency for individuals or groups to take greater risks because they are insulated from the full consequences of those risks. It occurs when one party acts recklessly, expecting that another party will bear the risks [20]. So, if opponents who would never use a new technology that is designed to improve safety were to keep that technology out of the marketplace, they would be creating a moral hazard.

13.3 Artifacts

Technologists are vulnerable to artifacts. That is, their currently acceptable designs may lead to impacts and harm in the future. Considering genetically modified corn in the human diet, offsetting behavior was not that of the consumer, but of the gatekeepers expected to keep genetically modified organisms (GMOs) segregated from regular grain. These gatekeepers may have become overly reliant on the backup and redundant systems to the extent that normal operations were not adequately monitored [19].

The engineering canon that an engineer must be competent in an area of expertise is not an excuse to disregard possible flaws that may arise down the road. In fact, just the opposite is true. That is, part of the competence requires consideration of offsetting behavior. So, what actions can lead to artifacts? The FDA and others offer some insights about how to avoid offsetting behaviors [21–23]:

1. Beware of overconfidence of computer systems—Engineers must overcome the temptation to assume that problems are usually about hardware.
2. Do not confuse reliability with safety—Software can work thousands of times without error, but even these probabilities are unacceptable when consequences are potentially widespread and long-term.
3. Include a defensive design—Redundancies as well as self-checks, troubleshooting, and error detection and correction systems are vital. A worst-case design scenario should be identified.
4. Address root causes—Causes may be misidentified, and the corrections may solve problems other than those leading to the real failures.
5. Avoid complacency—In many engineering failures, the contingency that led to the artifact was part of a routine process. Continued success does not mean due diligence can be shortcut.
6. Conduct realistic risk assessments—The failure analysis should not assume independence and miss key factors. Seemingly mundane or benign technologies can have adverse effects.

13.4 Conclusions

HFE is a vital part of any technology. The human-technology interface must be adequately designed for safety. Forecasting the use of a technology is key, with special attention to potential misuses after it reaches the marketplace. The technology may be ethically neutral, but the risk homeostasis and offsetting behaviors that it spawns are not. This is another type of the consequence that must be considered in a bioengineer's design.

References

[1] B.A. Sayers, Human Factors and Decision Making: Their Influence on Safety and Reliability, Springer, New York, NY, (2011).

[2] L. Chadwick, S. Jeffcott, A brief introduction to human factors engineering, Transfusion 53 (6) (2013) 1166–1167. Available from https://doi.org/10.1111/trf.12177.

[3] J. Gosbee, Human factors engineering and patient safety, Quality and Safety in Health Care 11 (4) (2002) 352–354.

[4] M. Wertheimer, Produktives Denken [Productive thinking]. Frankfurt, Germany: M.: Waldemar Kramer, 164. (no.date).
[5] B. Wilpert, Psychology and human factors engineering, Cognition, Technology & Work 10 (1) (2008) 15–21. Available from https://doi.org/10.1007/s10111-007-0073-0.
[6] C.F. Graumann, Psychology and the World of Things, Journal of Phenomenological Psychology, 4, 2, p. 389, Spring 1974, 2013–02–24 1974. [Online]. Available: https://login.proxy.lib.duke.edu/login?url=https://www.proquest.com/scholarly-journals/psychology-world-things/docview/1308102846/se-2?accountid=10598, https://duke.userservices.exlibrisgroup.com/openurl/01DUKE_INST/01DUKE_INST:Services?&genre=article&issn=00472662&title=Psychology+and+the+World+of+Things&volume=4&issue=2&date=Spring+1974&atitle=Psychology+and+the+World+of+Things&spage=389&sid=ProQ%3Apao&author=Graumann, https://libkey.io/libraries/229/openurl?genre=article&au=Graumann%2C+Carl+F&aulast=Graumann&issn=00472662&isbn=&title=Psychology+and+the+World+of+Things&jtitle=Journal+of+Phenomenological+Psychology&pubname=Journal+of+Phenomenological+Psychology&btitle=&atitle=Psychology+and+the+World+of+Things&volume=4&issue=2&spage=389&date=Spring+1974&doi=&sid=ProQuest.
[7] D. Rozumek, The development of fatigue cracks in metals, Experimental Mechanics of Solids (2019).
[8] H. Jones, The challenger and space shuttle failures were due to bureaucratic deviance, in: 2025 IEEE Technology and Engineering Management Society Conference-Global (TEMSCON Global), 2025: IEEE, pp. 1–6.
[9] H. Schaub, P. Badke-Schaub, Design to fail? The reasonably foreseeable failure and misuse, Proceedings of the Design Society 4 (2024) 1055–1064. Available from https://doi.org/10.1017/pds.2024.108.
[10] B.M. Fischhoff, F. J., The inconvenient public, Chemtech - American Chemical Society 25 (3) (1985) 47–54.
[11] National Academy of Engineering, Product Liability and Innovation: Managing Risk in an Uncertain Environment, The National Academies Press (in English), Washington, DC, (1994), p. 216.
[12] G.J.S. Wilde, The theory of risk homeostasis: implications for safety and health, Risk Analysis 2 (4) (1982) 209–225. Available from https://doi.org/10.1111/j.1539-6924.1982.tb01384.x.
[13] P. Slovic, Perception of risk, Science (New York, N.Y.) 236 (4799) (1987) 280–285.
[14] P. Slovic, B. Fischhoff, Targeting risks, (in En), Risk analysis 2 (4) (1982) 227–234. Available from https://doi.org/10.1111/j.1539-6924.1982.tb01385.x [Online].
[15] T. Liu, Z. Zhao, M. Yang, T. Han, Analysis of secondary risks induced by defensive braking in autonomous vehicles: a study based on stochastic distribution of drivers, Accident Analysis & Prevention 220 (2025) 108176. Available from https://doi.org/10.1016/j.aap.2025.108176.
[16] B. Dulisse, Methodological issues in testing the hypothesis of risk compensation, Accident Analysis & Prevention 29 (3) (1997) 285–292. Available from https://doi.org/10.1016/S0001-4575(96)00082-6.
[17] J.C. Hause, Offsetting behavior and the benefits of safety regulations, Economic Inquiry 44 (4) (2006) 689–698. Available from https://doi.org/10.1093/ei/cbj041.
[18] B. Pless, Risk compensation: revisited and rebutted, Safety 2 (3) (2016) 16 [Online]. Available https://www.mdpi.com/2313-576X/2/3/16.
[19] D. Vallero, Environmental Biotechnology: A Biosystems Approach, Elsevier Science, (2015).
[20] C.H. Lean, Navigating the 'moral hazard' argument in synthetic biology's application, Synthetic Biology 9 (1) (2024) ysae008. Available from https://doi.org/10.1093/synbio/ysae008.
[21] E. Hollnagel, D.D. Woods, N. Leveson, Resilience Engineering: Concepts and Precepts, Ashgate Publishing, Ltd, (2007).
[22] C.T.N. Leveson, An investigation of the Therac-25 accidents, in: K.W. Bowyer (Ed.), Ethics and Computing: Living Responsibly in a Computerized World, 26 IEEE Computer Society Press, First Published in Computer, Los Alamitos, California, (1993), pp. 18–41 1996.
[23] D.A. Vallero, Biomedical ethics for engineers, The Biomedical Engineering Series (2007).

CHAPTER

14

Professionalism

OUTLINE

14.1 Introduction

When asked to define what it means to be a professional, most will have some idea of what differentiates a professional from most other occupations. However, they will vary in whom to include. Most will include physicians, attorneys, and engineers. Others will include airline pilots, government officials, and teachers. Few would include salespeople and politicians. So, then, what distinguishes a professional from everyone else?

An artificial intelligence query asking, "What is a professional?" yielded this result:

> A professional is an individual who earns a living by engaging in a specific, often specialized, occupation, typically requiring advanced training, education, or skill. They are characterized by adhering to strict ethical codes, maintaining high standards of competence, and acting with accountability and integrity in their work.
>
> **Key aspects of a professional**
>
> **Specialized expertise:** They possess deep knowledge and skills, often validated by credentials, licenses, or academic degrees.
>
> **Ethical standards:** They adhere to a formal code of conduct and ethical guidelines, often enforced by professional associations.
>
> **Behavioral conduct:** Professionals display reliability, accountability, and a commitment to quality.
>
> **Accountability & autonomy:** They often operate with autonomy, taking responsibility for their decisions and actions within their field.

Technical and Ethical Optimization: Challenges and Opportunities for the Modern Engineer and Scientist
DOI: https://doi.org/10.1016/B978-0-443-44925-3.00011-9

Service orientation: Their work often goes beyond self-interest to benefit clients or the public.

Difference between professional and amateur A professional engages in an activity as a means of livelihood (e.g., a professional athlete or musician), whereas an amateur engages in it for pleasure or as a hobby. In many fields, being a professional is also associated with working under a salary rather than hourly wages.

Not bad. The emphasis on requiring a strict ethical code is particularly gratifying, given the subject matter of this text. If that were a requirement for being a professional, most scientists would not be considered professionals.

As mentioned in Chapter 4, the designation of an occupation as professional is often intuitive. Ascribing professionalism to an occupation is not that one is more valuable to society than another. Evidently, expertise is a necessary but not sufficient quality of professionalism. As the query states, the level of accountability and degree of responsibility are essential qualifications. On the other hand, some occupations require ample and unquestionable accountability and rare expertise, like an airplane mechanic or a NASCAR pit crew boss but are not widely considered to be professionals. It has something to do with a strictly defined span of control and jurisdictional responsibility, given that someone else has the authority over the plane, the racecar team, etc. All actions by the mechanic, the copilot, and everyone else working on or with the airplane, as well as decisions about the weather and other the flying conditions, and whatever it takes to transport the plane, passengers, and cargo safely to the destination, are the responsibility of the pilot. The pilot is also accountable for a large span of control. A set of norms and mores must be distributed throughout the span of control. Indeed, mistakes by every member within the professional's span of control must be monitored by the professional.

Mistakes must be discovered and corrected. Those who make the mistakes will be punished, usually severely, but if the mistake propagates a problem, the professional is the leader who takes the blame. The professional cannot pass off the blame to a corporate or other organizational problem. The professional has extraterritorial accountability. No company policy or business decision lessens the responsibilities of the professional. The adage "when in Rome do as the Romans" does not apply to the professional. More correctly, when in Rome, be professional, even if all the Romans are not. Or, when in Rome, continue to practice state licensure procedures [1,2].

14.1.1 Specialized expertise

A professional must possess deep knowledge and skills, often validated by credentials, licenses, or academic degrees. The technical professional must master the relevant subject matter. Every patient, client, project, and situation is unique. This means that the physician, engineer, lawyer, or other professionals must continue lifelong learning until retirement.

14.1.2 Ethical standards

Professionals adhere to a formal code of conduct and ethical guidelines, often enforced by professional associations. In the United States, physicians, engineers, and some other professions are predominantly regulated by licenses issued by a state. So, for example, a physician will likely need to obtain multiple state licenses to work with a telemedicine company, since physicians are only legally permitted to practice in states where they are licensed. Most state medical boards permit physicians outside the United States to see patients in their state through telemedicine, but each state's medical board has unique requirements [3].

Usually, an out-of-state and out-of-country physician is not allowed to treat a patient located within the physician's jurisdiction without a license issued by that state, although some offer special, limited, or telehealth-specific registrations for out-of-state providers. However, even these still require approval from the state medical board. Practicing without a license can lead to cease-and-desist orders, fines, public disciplinary action, and potential criminal charges, which can jeopardize future licensing in the United States.

Licensure for the professional engineer (PE) is similar to that of the board-certified physician. In the United States, the PE is licensed by and receives the PE stamp from the individual state. The PE must practice in the state of licensure, must only seal documents that the engineer has personally prepared or supervised, must ensure that the stamp is valid in the specific jurisdiction, and must sign and date across the seal to accept professional responsibility. The seal must comply with state-specific design regulations and be secured to prevent unauthorized use. The engineer must only stamp work within his or her specialty based on licensure requirements and experience in that area of expertise. As such, the PE must not submit to pressure from a client or employer to use the stamp in any other way [4–6].

In addition to governmental licensures, professionals must also abide by the tenets of their respective professional groups, such as the American Medical Association, the National Society of Professional Engineers, the American Society of Mechanical Engineers, the American Bar Association, and the American Institute of Certified Public Accountants. These organizations grant credentials, advocate for the goals of their members, provide training, hold conferences, publish journals, and help members connect with one another. All professional organizations require adherence to codes and tenets of behavior as articulated in codes of ethics.

14.1.3 Conduct

Professionals must be trusted. As such, they must be reliable and accountable for their work and demonstrate a commitment to quality. Professional characteristics can be summarized as honesty, integrity, fairness, and competence. Professionals must act as faithful agents for their clientele and the public.

14.1.4 Accountability

To paraphrase US President Harry S. Truman, the buck stops with the professional. A professional must take responsibility for the outcomes and products of the actions, decisions, and performance of duties as articulated in the codes and tenets of the profession. Part of accountability is reliability. For example, engineering has entire subdiscipline devoted to reliability.

Reliability extends beyond scientific soundness, economics, and item availability. It must consider the tradeoffs between risks and benefits. Costs and benefits should not be based on financial and monetary costs to achieve a social need. Costs are a subset of risks, that is, environmental risks, health risks, safety risks, as well as financial risks to the public. A project can either decrease or exacerbate these risks. For example, a project may decrease economic and safety risks but impair environmental resources and public health. Or, a project may enhance the environmental, aesthetic, and/or public health conditions for one community, but harm the environment of another community, even within the same governmental jurisdiction [7–9].

The PE must ensure that projects provide an adequate margin of safety. For example, if the expected operational life of boiler is 25 years and it fails and needs to be replaced after 30 years, this has met the margin of safety. However, if it fails after 15 years, new costs of repair may still be acceptable if failure rate and margin of safety had been fully disclosed, and the reason for the failure adequately explained. If it fails after 10 years, this may very well constitute an unacceptable risk. Extenuating circumstances may affect reliability, for example, if the water acidity and corrosiveness increase above the original design specifications beyond the control of the engineer, such as the city deciding to change the water source. Even in this situation, however, the design assumptions should have been stated clearly that the operational life of the boiler depends on expected knowledge of the engineer. In this case, incoming water was a critical factor in the operational life of the boiler.

As discussed in Chapter 16, reliability engineering considers the expected or actual reliability of a process, system, or piece of equipment to identify the actions needed to reduce failures, and once a failure occurs, how to manage the expected effects of that failure. Thus, reliability is the mirror image of failure. Risk is the probability of failure, so risk and reliability are two sides of the same coin.

14.1.5 Autonomy

Professionals are completely responsible for their work. They take responsibility for their decisions and actions within their field. Although they are accountable to the institutions and the jurisdictions where they practice, they are autonomous agents who ensure that the work is acceptable and ethical.

Unfortunately, not all cultures value moral autonomy. In the worst case, some nations discourage the autonomy in the workplace. It therefore cannot be assumed that in a global climate, the question of autonomy can serve as an uncontested universal foundational assumption for building an engineering ethics [10]. Thus, the professional can expect some pushback when doing the morally acceptable thing. This requires fortitude and the recognition that it may sometimes be necessary to decline work in such environments.

14.1.6 Service orientation

Their work often goes beyond self-interest to benefit clients or the public. The professional has an anticipatory mindset. The client does not understand the specificity of the project because the professional is to be trusted to meet the client's unstated needs for any project in the professional's area of expertise.

Professional ethics differ from that of a purely profit-oriented or willingness-to-pay perspective. Their credo is not *caveat emptor*, "let the buyer beware," but instead *credat emptor*, that is, "let the client trust" [11–13]. PEs, for example, must strive to hold paramount the health, safety, and welfare of all the public. This duty requires both technical competence and morality. Environmental engineers are specifically charged with responsibilities to protect the public and ecosystems. As mentioned, when failures occur, the professionals are accountable. When a manufacturing, transportation, or other process works well, the professional can rightly take pride in its success. The professional is responsible for the successful project. That is why they studied a rigorous curriculum and are highly trained in their disciplines. Each professional has individual responsibilities in a well-running system. Conversely, when things go wrong, that person is responsible and must account for every step, from the largest and seemingly most significant to those we perceive to be the most minuscule, in the system that was in place. Professional responsibility cannot be divorced from accountability. The Greeks called this *ethike aretai*, or "skill of character" [14]. It is not enough to be excellent in technical competence. Such competence must be coupled with trust gained from ethical practice [15].

14.2 Active engineering

An important tenet of ethics and engineering communications is that we be clear in what we say and mean. The first step in bioethical analysis is to reach an understanding of the facts of the matter. So, then, what is meant when we say that engineering is an "active" process? The first definition in the dictionary is "characterized by action rather than by contemplation or speculation" [16]. This connotation is interesting and valuable from several perspectives. Obviously, the noun "action" drives the adjective "active." But among the definitions of the noun that best fit the adjective's definition, one seems to stand out, that is, "the accomplishment of a thing usually over a period of time, in stages, or with the possibility of repetition" and "the bringing about of an alteration by force or through a natural agency."

Given that alteration means that the something has changed, the technologist hopes and expects the change to be better than what existed before. The innovation moves from the existing condition to a desired condition. However, if this does not happen, it is a type of failure. Failure in itself is not unethical. Only some failures are rooted in ethical breaches.

Another interesting aspect of the first dictionary definition of "active" is the contrast between an action *and* contemplation or speculation. Contemplation should precede any action. Prudence dictates that "you should look before you leap." The proper sequence of ethical or any decision-making is "ready, aim, shoot"; although many of us frequently "ready, shoot, aim." In other words, action-oriented people can have a natural proclivity to act, even before much or any thinking. This can lead to addressing symptoms of problems but not solving the problems themselves. The definition is not a value judgment but simply recognizes that the two steps, contemplation and action, are unique and sequential.

The other contrast between action and speculation, however, is value laden. There are times, deservedly or not, when decision-makers appear to suffer from "analysis paralysis" [17]. Decisions are always made with some degree of uncertainty. The key dilemma for the technologist is to know when enough information about the risks and benefits of a design has been ascertained. Data and information sufficiency is a function of the severity of the outcome, that is, costs of being wrong, and the loss of a benefit, that is, opportunity risk. In some cases, a "50/50," flip of the coin decision-making approach is sufficient, such as whether a bridge between Chapel Hill and Durham should be painted Carolina or Duke blue (although such decisions carry great local import in the North Carolina Piedmont). Conversely, a decision as to whether to set a drinking health standard at 10 ppb or 15 ppb for a pollutant can account for a margin of waterborne diseases and even allow for increased mortality if the higher concentration is applied. Disagreements regarding the bridge color may lead to some unhappy fans, but speculation in a safety or health can translate into increased morbidity and mortality. Unethical aspects of the latter would include dishonesty and obfuscation, but not honest disagreement about the 5-ppb difference.

Of course, engineers do things, and what we do should give us great pride [9]. Engineers are, by nature, a thoughtful and outwardly directed lot. They are sympathetic to the needs of others, beginning with their clients. In fact, all designers in general are optimistic and future-oriented. They foresee outcomes long before they take on physical reality. Thus, engineers are highly suited to take the long view. Some faith traditions may characterize engineers as having much "faith." The engineering codes of ethics call for them to be "faithful agents" [1]. While this is a statement about the trust that the clients place in the engineer, it is also a statement about engineers' confidence and faith in themselves as professionals. This is the engineering perspective. We "expect" the bridge, or building, or water supply to become real, to become "substance."[1] Even those engineers who are called in to fix things, such as biomedical engineers finding ways to ameliorate human suffering, environmental engineers who remediate hazardous waste sites or failure analysts who look for ways to prevent future disasters, always expect things to be different.

Engineers and technologists are willing to stand behind our work (e.g., we build feedback into the operation and maintenance [O&M] process, we continuously check progress on biomedical devices, install monitoring wells to ensure waste cleanup is going as planned, and we work closely with inspectors to incorporate lessons learned from failure analyses). Hence, in addition to being faithful, we adapt. In fact, adapting is a big part of design. The dynamism of biomedicine has been fertile ground for engineering for many centuries. A classic example was the 1592 visit of Galileo to one of the first medical institutions in Padua, Italy [18]. Galileo, acting as an engineer, took the opportunity to lecture the future medical practitioners and researchers on mathematical principles, including his own theories, as well as their applications; notably the pendulum, thermoscope, and telescope. Such applications led to an enhanced understanding of physiology, such as studies by Galileo's student, Sanctorius, of human body temperature and pulse rates. One student at Padua during Galileo's visit was William Harvey, renowned for characterizing the circulatory system, which he based on the mechanical and motion laws expounded by Galileo. Such interplay between medicine and engineering, while often subtle and indirect, has continued to the more formal collaborations we see today. But the dynamic design

[1] Christian Scripture has defined faith as "the substance of things hoped for, the evidence of things not seen" (Hebrews 11:1). This is an extension of the Judaic outlook for the "promised land" (a metaphor to the engineer's expectation to see the designs reach the build phase, to the medical researcher's search for cures of obdurate maladies, or to the city planner's long-range plan, envisioning green spaces and copious public amenities). This optimistic view lends a temporal dimension to faithfulness. For example, the concept of sustainability in terms of public health and environmental quality is a requirement to take (or to avoid) actions that consider their impact on future generations. This means that engineers must be faithful agents to distant and future people. An action that is most expedient for the present may not be the best if it has severe effects in the long run.

processes of the contemporary engineers make for complicated ethical challenges, and none more so than those related to biomedicine.

In biomedicine, it is truly and literally a matter of life and death. Advancing the state of biomedical science is an ethical imperative. Consequently, the typical precautions that may hold for other areas of engineering need special consideration when addressing biomedical challenges. For example, the "no action" alternative is seldom satisfying to biomedical practitioners. Not looking for a better device or system that can improve the quality of a patient's life is seldom acceptable. Risk assessors refer to this as "opportunity risk." That is, if we simply seek the refuge of no added risk, we may lose an opportunity to really improve things. For example, nanotechnologies (using systems that are only a few hundred molecules in size) are fraught with risks, such as the chance that working at this level may cause changes to self-replication and other cellular signals, which could cause unknown damage. However, if we do not seek uses of nanotechnologies, such as the application of highly efficient, biologically inspired processes for drug delivery, we may lose the opportunity to make the drugs more efficacious. For example, if we can use biologically inspired nanomanufacturing to synthesize and deliver tumor-reducing drugs, it may be possible to treat cancers that have heretofore been untreatable. So, we must be bold in applying nascent sciences and simultaneously take great care to ensure that we are not opening some awful "Pandora's box."[2] Finding this balance of acceptable and reasonable risk goes beyond science, engineering, and technology and enters the realms of ethics.

The advancement of science and its applications must follow a type of ethical index. Indices have weighted values for each of the important variables. Some more sophisticated indices have operators that can shut the index down, such as a water quality index that automatically goes to zero if certain levels of dissolved oxygen are not available, even if all the other values are fine. Thus, engineers need some type of working index for biomedical ethics. They must be able to tell when something is going awry. This is difficult, since most of us are well trained in the physical sciences and mathematics, but few in the intricacies of biology associated with healthcare. The exception, of course, is the biomedical engineer, but even their perspective on biomedicine varies from others in the healthcare profession.

14.3 Empathy

Attitudes are complex mental processes that influence how a person processes information and motivates behavior [19]. They have been explored in depth in the psychological

[2] The Pandora's Box is an interesting and useful ethical metaphor that conveys the potential for unintended, negative outcomes from a seemingly innocuous or beneficial act. The term comes from the Greek myth about a box left with Epimetheus and Pandora for safekeeping by Mercury. Epimetheus warned Pandora not to open the box, but eventually, upon hearing voices asking to be freed, Pandora's curiosity got the best of her. She opened the lid, and out came diseases, vices, and other ills to humanity. Hence, the myth has been used as a warning not to unadvisedly or prematurely rush into the unknown (e.g., viral research, neurotechnologies, nanotechnologies, and new drug therapies). However, an often-forgotten part of the story is that Pandora opened the box a second time, and "hope" was released. Biomedicine and engineering are professional manifestations of this hope.

literature, where "attitude" has been defined as a psychological tendency expressed by evaluating a particular entity with some degree of favor or disfavor [20]. An empathetic attitude is a key aspect of professionalism [20–25]. The concept of empathy is central to moral judgment about just treatment of patients, subjects, and society. Justice is the virtue that enables us to give others what is due to them as our fellow human beings. This means we must not only avoid hurting others through our actions, but we ought to safeguard the rights of others in what we do and what we leave undone.

The categorical imperative is emblematic of empathy. Kant uses this maxim to underpin duty ethics or "deontology" with empathetic scrutiny. However, empathy is not the exclusive domain of duty ethics. In teleological ethics, empathy is one of the palliative approaches to deal with the problem of "ends justifying the means." Other philosophers also incorporate the empathic viewpoint into their frameworks. In fact, Mill's utilitarianism axiom of "greatest good for the greatest number of people" is moderated by his "harm principle" which, at its heart, is empathetic [26,27]. That is, even though an act can be good for the majority, it may still be unethical if it causes undue harm to even one person. Empathy also comes into play in contractarianism, as articulated by Thomas Hobbes in social contract theory [28,29]. For example, Rawls [30–33] has moderated the social contract with the "veil of ignorance" as a way to consider the perspective of the weakest, one might say "most disenfranchised," members of society. Finally, the rationalist frameworks incorporate empathy into all ethical decisions when they ask the guiding question of "what is going on here?" In other words, what benefit or harm, based on reason, can I expect from actions brought about by the decision I am about to make? One calculus of this harm or benefit is to be empathetic to all others, particularly the weakest members of society, those with little or no "voice."

The word "empathy" has an interesting beginning. It comes from the German word *einfühlung*, which means the ability to project oneself into a work of art, like a painting [34]. Psychologists at the beginning of the 1900s searched for a word that meant the projection of oneself into another person and chose the German word, translated into English as empathy [35]. The concept itself was known, such as the Native Americans' admonition to walk in another's moccasins, but it needed a construction. The early meaning of empathy was thus the ability to project oneself into another person, to imitate the emotions of that person by physical actions. For example, watching someone prick a finger would result in a visible winching on the part of the observer because the observer would emulate how this feels.

From that notion of empathy, it was natural to move to more cognitive role-taking, the imagining the other person's thoughts and motives. From here, empathy began to be thought of as the response a person has to another's situation. Psychologists and educators, especially Jean Piaget,[3] began to believe that empathy develops throughout childhood, beginning with the child's first notion of others who might be suffering personal stress. The child's growing cognitive sense eventually allows him or her to experience the stress of others. Because people are social animals, this understanding of the stress in others, according to the psychologists, eventually leads to true compassion for others.

Empathy depends on one's ability to identify someone else's emotions as well as on the capacity to put oneself in the other person's place [36].

[3] Piaget, J. (1965) *The Moral Judgment of the Child*. New York: The Free Press.

14.4 Conscience

A person's conscience is the inner voice that orders knowledge of the rightness or wrongness of an action, albeit in the past, present, or future. Conscience directs future acts. It may indict or excuse past actions. According to St. Thomas Aquinas, "synderesis" is the person's inclination to shape understanding. Synderesis turns human nature toward good and away from evil. It stirs a person to become aware of the principles of morality as applied to actions. Synderesis concerns knowing the principles that apply to actions, whereas a person's conscience applies knowledge to a specific act. Thus, synderesis informs the conscience through the application of synderesis, which is always ordered to moral truth. Indeed, Aquinas argued that synderesis is a habit [37], much like Aristotle's argument that excellence, really moral character, is a habit [38].

So then, what is the professional's right of conscience in deciding which types of care and projects to perform? For example, in the United States, healthcare professionals and institutions that refuse on the basis of moral, ethical, or religious beliefs to perform, assist in, or fund abortions are protected legally [39,40]. Such concerns are also present in other aspects of professional life, including matters like euthanasia and assisted suicide. Indeed, the concept of "conscientious objection" began with those opposed to warfare [41]. Consider this thought experiment [13]:

> Betty is a licensed electrical engineer who has been developing new designs for optical systems used to miniaturize various products. In fact, she has become so good designing these systems, she was "hired away" from a small company two years ago by a large conglomerate that owns, among many ventures, a medical equipment company. Up to six months ago, her optical systems were used in diagnostic and telemetric devices, but since then she has been spending most of her time adapting them to miniaturize and to improve the efficiency of implanted pumps and regulators. Last month, however, she was asked to start work on adapting an optical system to miniaturize a wired transducer that is presently used in a triggering mechanism for a secret system that is being funded by a defense contract. Rumor has it that the mechanism is designed to improve the efficiency of a landmine.
>
> Betty is generally opposed to armed conflict but is especially against landmines. What should she do?

Every profession consists of people from various walks of life, with numerous perspectives, and a widely ranging set of beliefs. There is much commonality in thinking among people practicing in the same discipline, but there are also substantial differences in perspectives. Some of this divergence lies in what a professional believes constitutes right and wrong behavior. In fact, each professional discipline allows a certain degree of latitude on the types of work that an individual *must* perform. A professional practitioner may refuse to perform certain tasks, represent certain clients, and take certain ideological positions that conflict with the individual's conscience. This is known as the right of professional conscience, which is [42]:

> ...the right of an employee to refuse to partake in unethical conduct when forced to do so by an employer. This may occur in work or non-work situations and may not necessarily involve breaking the law. Conscientious refusal may be done by either simply not participating in the activity that one sees as immoral, or it may be done with the hope of making a public protest that will draw attention to the situation that one believes is wrong.

The idea of conscience-driven morality is complex. For example, you may completely agree with Betty about landmines. However, what if you found out that the work she is doing would prevent landmines from remaining active after their military use? For example, what if the mines could be remotely destroyed using Betty's device? Is it morally permissible to argue that the mines are going to be made anyway, so why not make them safer? Or is it Betty's "duty" to quit the project, no matter the potential value, simply because the whole warfare scenario is immoral?

To understand the right of conscience, both terms are important: *right* and *conscience*. Let us first consider the second term. Theologians spend a lot of time thinking about the conscience, so that might be a good place to start to find a definition. For example, Richard M. Gula [43] explains that conscience is:

> ... often used but little understood. Trying to explain conscience is like trying to nail Jell-O to the wall, just when you think you have it pinned down, part of it begins to slip away.... We all know we have a conscience, yet our experiences of conscience are ambiguous. We struggle with conscience when facing great decisions of life, such as the choice of a career, or of conscientious objection to war, or whether to pay or taxes which support defense projects. Yet we even feel the pangs of conscience over petty matters, such as jaywalking....

Gula considers the quandary that although we hear that "conscience enjoys inviolable freedom," society and its institutions have codified formal and informal rules "so absolute in character we wonder whether conscience matters at all" [43]. This uncertainty is at the heart of understanding when an engineer or other professional may invoke the right of conscience.

Distinguishing conscience from other human qualities may help us understand this right, or in engineering parlance, we can find the solution by difference. Conscience is not the same as personality or even the Freudian concept of superego. Sigmund Freud [44] hypothesized in his structural theory that the human personality consisted of three parts: id; ego; and superego. The id is the most primitive, seeking maximum pleasure and oriented toward "primitive desires" (e.g., sexual, rage, and hunger). The superego is the normative part of the personality, which polices behavior and enforces standards of morality. The ego attempts to reconcile urges of the id with the norms of the superego. This balance gives rise to the sense of self.

The superego operates to allow the person to live in society. That is, it commands an act to gain approval, whereas the conscience is owned by the person. Thus, conscience is oriented toward a value, but the superego is oriented toward an authority. Conscience attends to a future large pattern of a desired personality archetype, whereas superego attends to past individual acts in themselves (strongly influenced by trying to avoid previously punished acts). Since conscience informs behavior based on moral principles, a "good" conscience is a valuable commodity in discerning ethical choices. As such, the right of conscience of individual engineers must be respected to some extent in mediating ethical issues.

With the distinction between conscience and superego, let us modify the earlier thought experiment:

The use of landmines in military operations is opposed by many. Even many of those who see the usefulness of the technology in certain applications agree that mines left in the ground pose a standing threat to innocent people. Jan is a mechanical engineer developing a trigger mechanism for landmines that is 25% more reliable than existing mechanisms. Jan is opposed to the use of landmines but works for the defense contractor because she needs the job. She also believes that if good people like her refuse to work in the industry, only those who support landmine use will work there and that will translate into little internal opposition. As evidence, Jan has convinced her employers that they need to work on a trigger mechanism that biodegrades within one year after deployment, so that no orphan mines would be active a year after military operations cease. ***Is Jan behaving morally?***

The answer is informed by Jan's conscience, so it can only be answered by Jan. Does she, in fact, want to change the organization from within, or is this merely a rationalization for her source of income? Few of us work in organizations that allow complete autonomy. Institutional and organizational demands are overlain onto the engineer's professional responsibilities. The problem arises when personal professional integrity is sacrificed for the organization, for example, witness the Shuttle *Challenger* disaster when certain engineers put the company and agency desires above their own engineering calculations [45].

Another aspect of this thought experience is that of mitigation and risk tradeoffs. Jan's work could improve the technology, making it safer in the long run. Rather than being dangerous for decades, modifying the mine's peril would be limited if she is successful in the biodegradable design. Some might argue that the best and most ethical approach is never again to manufacture any landmines. While this is true and is arguably the positive paradigm [46] (i.e., the most ethical course of action), it is also highly unlikely to occur soon. Jan is proposing a pragmatic improvement to the design, albeit not the textbook paradigm. Sometimes ethical improvement is incremental, just as technological advancement. Revolution can be great, but evolution is usually the norm.

Often, defenders of the right of conscience are swayed by the issue at hand. What if, for example, Betty and Jan were mechanical engineers working on devices that could be used to improve abortion or assisted suicide?

How about another thought experiment:

The late term abortion procedure, dilation and extraction (D&X), commonly known as partial birth abortion, is a highly controversial medical procedure. Many people strongly oppose this procedure due to the aggressive and active means of terminating the life of the unborn child. By 1999, the legislatures of 29 states had banned the procedure. In 2000, the U.S. Supreme Court in a 5-4 ruling in the case of Stenberg *v.* Carhart decided that the State of Nebraska's ban was unconstitutional [47]. The grounds for the rejection were two-fold:

- The law did not clearly distinguish between partial-birth abortion (D&X) and the most used late-term abortion procedure (dilation and evacuation, "D&E"); and,

- The law did not provide an exception for the health of the mother.

Although this procedure has not been banned, many doctors refuse to provide it. Bill is employed by a biomedical device company that develops nanomachinery (i.e. artificial molecular machines created through molecular manufacturing). One of these angstrom-scale devices is being adapted to deliver drugs to tumors. Bill has been modifying the design to deliver psychotropic drugs *in utero* to treat brain disorders in the third trimester. He was recently asked by his company's vice president (VP) for research and development if the nanomachinery could also deliver a neurotoxin (i.e., a substance that destroys or harms nerve cells) to an unborn child's brain just before extraction. The VP implied that this could be a way to prevent the legal problems of partial birth abortions, since the child would not be alive when the extraction occurs. Bill can see no structural difference in delivering either chemical and even suggests that this might be an untapped market.

Sitting in the same meeting is another engineer, Mary, who returned to work last year after two years away from the company. She had gone through a very difficult pregnancy, and her daughter was born six weeks prematurely. The experience has reinforced her belief that life begins at conception and that all human life is sacred. In fact, she was one of the advocates for the *in utero* research because she foresaw the importance of neural fetal nanoscale devices. She now feels betrayed by her company. When she raises her concerns, other meeting participants reassure her that their company is merely a device company and that it is entirely up to the attending physician to decide what is medically and morally appropriate. In fact, one of the engineers quotes the canon from the engineering code of ethics:

"Engineers shall perform services only in areas of their competence."

And he reminds Mary that engineers are beholding to the client (loyal and "faithful agents").

Mary feels outnumbered, so she acquiesces.

What should Mary have done? Was the discussion at the meeting ethical? Is it correct that the response is not the engineer's concern? What does this say about an engineer's right of conscience? Many of us would be offended by the "bait and switch" from delivering an ethical drug to delivering a toxin.

The argument that engineers must be competent is not an invitation to ignore important aspects of a design, including possible misuses. For example, a biomedical engineer working on team design may notice some mechanical or structural miscalculations. Although the engineer's "area of competence" is biomedical, she still has the responsibility to point out problems (or opportunities) if they are known.[4]

Other canons of the codes of engineering ethics support Mary's position, including the admonition that the safety, health, and welfare of the public are paramount, and that we avoid deceptive acts. The arguments involve immoral means, for example, developing technologies to be used to facilitate harmful outcomes, or to support convenient outcomes, for example, increasing company profits. By extension, one type of deception is lying to oneself, including deluding ourselves about possible misuse of technologies. Mary provides a case study of the need to consider unlikely but possible downsides and malicious uses of beneficial technologies.

Once they are available, innovations sought for one purpose are frequently usable for others. The same technological capacity to influence and control bodily processes for medical ends may lead (wittingly or unwittingly) to non-therapeutic and malicious uses [48–57].

It suffices to say that, sometimes, professional codes are merely guides that provide general moral direction.

Sometimes institutions and organizations militate against right of conscience. For example, many engineering failures occurred because of institutional mechanisms that stifled dissent and buried "bad news." The techniques used by organizations to control their

[4] A video produced by Duke University's Program in Science, Technology, and Human Values presents a case of a foundation engineer who knew that the soil type and mechanics were unacceptable to support the foundation. However, the engineer said nothing because it "was not his job." As a result, a wall failed, killing and injuring several people. The case is complicated because the engineer was trained and practiced in a former East Bloc communist system, where people who did not mind their own business could be severely punished. This does not excuse the myopia, but it does explain it to some extent.

membership have been logged by psychologist Irving Janis, who coined the term *groupthink* to describe the collective set of biases held by and perpetuated by a group [58].

The video *Incident at Morales* [59] provides a very useful, hypothetical composite of how groupthink can affect an engineer and compete with professional conscience. In the video, Fred Martinez is a young licensed PE who has recently joined the professional staff at Phaust Chemicals, a US division of a French multinational corporation. Fred is asked to design a system to reformulate a paint-stripping product to meet new regulations. A realistic constraint, however, is that a competing firm has a similar product. In the process, Fred identifies numerous environmental and technical problems that threaten the health and safety of workers and nearby residents. Although Fred tries to be true to his professional training and his own moral compass, design and construction compromises led to failures.

Several lessons can be found in this story. One is that the changes and shortcutting can be quite gradual. Another is that institutions can have priorities that are different and opposed to professional responsibilities. The engineer may have a strong core set of beliefs, but these can be stretched away from the center. The elements of groupthink include [58]:

- Illusion of invulnerability of group
- "We feeling" very strong (i.e., outsiders are the enemy)
- Rationalization (mainly to shift responsibilities)
- Illusion of morality
- Self-censorship ("don't rock the boat")
- Illusion of unanimity (silence means consent)
- Direct pressure (against those who disagree)
- Mind-guarding (or gatekeeping)

To help understand the power of groupthink and the role it can play in bioethics, let us consider another thought experiment:

Public health and social sciences researchers, from various universities, government agencies and companies have been collaborating on interdisciplinary project that is trying to characterize the factors that lead to terrorism, especially in hopes of finding ways to prevent it. The United Nations is the major funding source for the research.

Bob, the only biomedical engineer on the research team, is charged with aiding the others in finding "low-tech" devices to extract and store fluid samples and to assist behavioral scientists in data management and informatics. Sid, one of Bob's colleagues, had an opportunity to visit Pakistan. While there, Sid interviewed young people in a terrorist training camp, using a questionnaire developed by the investigators. Sid also decided to retain a local health worker to draw blood and take urine samples from the subjects to compare hormone levels with age-matched controls in the United States. Before Sid left, however, he neglected to ask the young people for their consent to participate in the research project. Sid had previously exclusively relied on anonymous databases from which he extracted data, and he had not had experience in getting consent.

Sid asked Bob to look at the protocol for taking and handling the samples. Bob gave Sid some technical advice that greatly improved the local health worker's extraction process and would double the shelf-lives of the samples to allow for lab analyses of better resolution and sensitivity upon the team's return to the U.S. Bob, however, had enough experience to know that Sid should have gotten advance consent to obtain fluid samples, but didn't think it was any of his business since this was an ancillary study separate from the larger study and Bob's role was clearly to provide engineering, not policy, advice.

After Sid returned home, he discussed his findings with the eight co-principal investigators at his university. Most of his colleagues wanted to publish *post haste*, but a few thought that the results should be given to the UN in confidence, because of the timeliness and nature of what was found. Disagreements began to arise about the appropriate journal where the data could be published, most wanting the journal that best fit their own respective academic discipline. When word got out that there was a new emphasis on endocrine differences, one key collaborator from another university became angry about being excluded.

In addition, major arguments and disagreements centered around who should be first author, who should be co-authors, and who should simply receive an acknowledgement. The grant project period has two more years remaining.

This thought experiment illustrates that having a professional perspective not shared by the group can be difficult to manage. However, the professional must not be swayed from doing what is right. The problem of authorships and acknowledgments is an ongoing challenge in professional and scientific journals. In this case, care must be taken to avoid honorary authorships and any undeserved acknowledgments and authorships, such as gift authorships, pressured authorships, and ghost authorships, that is, denying recognition to a worthy author [60,61].

14.5 Conclusions

This chapter focused on the roles of the professional. Professions share many things in common, including specialized expertise, ethical standards, expectations of appropriate behavioral conduct, accountability, autonomy, and service orientation. Thought experiments were offered to address some of the challenges facing the professionals, including dual-use possibilities, rights of conscience, and empathy.

References

[1] National Society of Professional Engineers, NSPE code of ethics for engineers. https://www.nspe.org/career-growth/ethics/code-ethics (accessed 05.02.26).

[2] R.E. Bucknam, When in Rome...(Case 1021), Journal of Professional Issues in Engineering Education and Practice 127 (4) (2001) 144–147.

[3] J. Ivanova, et al., Regulation and compliance in telemedicine, Journal of Medical Internet Research 27 (2025) e53558.

[4] S. Efe, M. Shokouhian, I. Akinkugbe, Professional engineer licensure preparatory course: an evaluation and analysis of CIAMTIS MSU outreach activities, 2024.

[5] M.K. Swenty, B.J. Swenty, A study of EAC-ABET civil engineering accreditation curriculum requirements and exemption provisions of state licensure laws and rules, in: 2023 ASEE Annual Conference & Exposition, 2023.

[6] M. Kam, Engineering licensing and professional practice, 2011 International Workshop on Institutional and Programme Accreditation: Connections and Opportunities, IEEE, (2011), pp. 1–9.

[7] R.D. Bullard, Unequal Protection: Environmental Justice and Communities of Color, Sierra Club Books, (1994).

[8] B.S. Matulis, The economic valuation of nature: a question of justice? Ecological Economics 104 (2014) 155–157. Available from https://doi.org/10.1016/j.ecolecon.2014.04.010no. 0.

[9] D.A. Vallero, P.A. Vesilind, Socially Responsible Engineering: Justice in Risk Management, John Wiley & Sons, Hoboken, New Jersey, (2007).

[10] H.C. Luegenbiehl, Ethical autonomy and engineering in a cross-cultural context, Techné: Research in Philosophy and Technology 8 (1) (2004) 57–78.

[11] E.C. Hughes, Professions, Daedalus (1963) 655–668.

[12] D.E. Wueste, Linking academic integrity and ethics across the curriculum: groundwork for sustainability in practical and professional ethics, Ethics Across the Curriculum—Pedagogical Perspectives, Springer, (2018), pp. 303–326.
[13] D.A. Vallero, Biomedical ethics for engineers, The Biomedical Engineering Series (2007).
[14] C. Cardwell, Hornbook Ethics, Hackett Publishing, (2015).
[15] National Society of Professional Engineers, NSPE code of ethics for engineers. https://www.nspe.org/resources/ethics/code-ethics, 2016 (accessed 31.08.18).
[16] M.W. Dictionary, Merriam-Webster.com, ed, 2026.
[17] E. Petrova, Avoiding traps in group decision-making process, Review of General Management 40 (2) (2024).
[18] J.D. Enderle, S.M. Blanchard, J.D. Bronzino, Introduction to Biomedical Engineering (Academic Press Series in Biomedical Engineering), Academic Press, San Diego, (2000).
[19] I. Ajzen, Attitude structure and behavior, in: S.F.B.A.R. Pratkanis, A.G. Greenwald (Eds.), Attitude Structure and Function, Psychology Press, New York, NY, (1989).
[20] A.H. Eagly, S. Chaiken, The Psychology of Attitudes, Harcourt Brace Jovanovich College Publishers, (1993).
[21] L. Wang, L. Xie, M. Jia, How can empathy be achieved?—A comparative study between the christian "golden rule" and the buddhist "five precepts and ten virtues" in China, Religions 16 (10) (2025) 1229 [Online] https://www.mdpi.com/2077-1444/16/10/1229.
[22] L.G.D. Barrio, C. Rodríguez-Díez, A. Gea, L. Arbea, J. Pereira, N. Díez, Impact of a longitudinal course on medical professionalism on the empathy of medical students, Patient Education and Counseling 119 (2024) 108042.
[23] M. Xue, et al., Narrative medicine as a teaching strategy for nursing students to developing professionalism, empathy and humanistic caring ability: a randomized controlled trial, BMC Medical Education 23 (1) (2023) 38.
[24] J.S. Gonnella, et al., Professionalism: empathy, interprofessional collaboration, lifelong learning, Fifty Years of Findings from the Jefferson Longitudinal Study of Medical Education, Springer, (2022), pp. 179–221.
[25] J. Veloski, M. Hojat, Measuring specific elements of professionalism: empathy, teamwork, and lifelong learning, Measuring Medical Professionalism (2006) 1.
[26] P. Foot, Utilitarianism and the virtues, Mind: A Quarterly Review of Psychology and Philosophy 94 (374) (1985) 196–209.
[27] J.S. Mill, Utilitarianism, Parker, Son and Bourn, London, (1863).
[28] S. Shapin, S. Schaffer, Leviathan and the Air-Pump: Hobbes, Boyle, and the Experimental Life (New in Paper), Princeton University Press, (2011).
[29] T. Hobbes, Leviathan: Or, the Matter, form, and Power of a Commonwealth Ecclesiastical and Civil, G. Routledge and Sons, (1894).
[30] J.V. Platz, The veil of ignorance in Rawlsian theory, 2017.
[31] H. Runcheva, John Rawls: justice as fairness behind the veil of ignorance, Iustinianus Primus Law Review. 4 (2013) 1.
[32] J. Rawls, Justice as rational choice behind a veil of ignorance, Alternative Political Perspectives, second ed., Wadsworth Publishing Company, (1996).
[33] J. Rawls, A Theory of Justice, Belknap Press of Harvard University Press, Cambridge, MA, (1971).
[34] F. Iannelli, Robert Vischer and the Aesthetic Foundation of Einfühlung, Forgotten Streams in the History of 19th-Century German Psychology: Volume 2: Late Idealist, Cultural, and Phenomenological Psychologies, Springer, (2025), pp. 145–165.
[35] R. Curtis, R.G. Elliott, An introduction to Einfühlung, Art in Translation 6 (4) (2014) 353–376.
[36] C.G. Morris, Psychology: An Introduction, Prentice-Hall, Inc, (1996).
[37] Summary of results for the 2002 national-scale assessment. [Online] Available from: http://www.epa.gov/nata2002/risksum.html, 2013 (accessed 26.09.13).
[38] M. Konch, R.K. Panda, Aristotle on habit and moral character formation, International Journal of Ethics Education 4 (1) (2019) 31–41.
[39] R. Hopkirk, L. Sergi, J. Sturges, Religious exemptions, Georgetown Journal of Gender and the Law 25 (2023) 941.
[40] S.L. Betz, Private right of inaction: the decades-long struggle to enforce the Church Amendments, Arizona State Law Journal 56 (2024) 465.

[41] C. Harrison, Conscientious objectors and the war, The Routledge History of the First World War, Routledge, (2024), pp. 613–623.
[42] C. Vee, C. Skitmore, Professional ethics in the construction industry, Engineering, Construction and Architectural Management 10 (2) (2003) 117–127.
[43] R.M. Gula, Reason Informed by Faith: Foundations of Catholic Morality, Paulist Press, (1989).
[44] S. Freud, The Origin and Development of Psychoanalysis, University of Illinois Press, (1910) (no. 2).
[45] H. Jones, The challenger and space shuttle failures were due to bureaucratic deviance, in: 2025 IEEE Technology and Engineering Management Society Conference-Global (TEMSCON Global), 2025, pp. 1–6, IEEE.
[46] C.B. Fleddermann, Engineering Ethics, Prentice Hall, Upper Saddle River, NJ, (1999).
[47] B.A. Bostrom, Stenberg v. Carhart, Issues in Law & Medicine 16 (2000) 179.
[48] K. Eskandar, Artificial intelligence and synthetic biology: biosecurity risks, dual-use concerns, and governance pathways, AI and Ethics 6 (1) (2026) 66.
[49] M. Marsili, Emerging and disruptive technologies: strategic implications and ethical challenges of dual-use innovations, in: Strategic Leadership Journal. Challenges for geopolitics and organizational development, 1: Centro Alti Studi Difesa/Scuola Superiore Universitaria (CASD/SSU), 2025, pp. 57–71.
[50] M. Hukic, E. Hukić, Ethical and biosafety challenges in modern microbiology: navigating the risks of emerging technologies, Spectrum of Dual-Use Technologies: Unforeseen Risks Versus Returns, Springer, (2025), pp. 163–182.
[51] A. Vaseashta, With emerging security threats from dual-use technologies, in: Developments and Advances in Defense and Security: Proceedings of MICRADS 2024, 423, p. 125, 2025.
[52] A. Vaseashta, Existential risks associated with dual-use technologies, in: Proceedings of the Stanford Existential Risks Conference 2023, 2023, pp. 156–170. Stanford University.
[53] L.-A. Kaffee, A. Arora, Z. Talat, I. Augenstein, Thorny roses: investigating the dual use dilemma in natural language processing, arXiv preprint arXiv:2304.08315, 2023.
[54] M. Ienca, E. Vayena, Dual use in the 21st century: emerging risks and global governance, Swiss Medical Weekly 148 (2018) w14688.
[55] R. Danzig, Innovation, Dual Use, and Security: Managing the Risks of Emerging Biological and Chemical Technologies, MIT Press, (2012).
[56] F. Kuhlau, S. Eriksson, K. Evers, A.T. Höglund, Taking due care: moral obligations in dual use research, Bioethics 22 (9) (2008) 477–487.
[57] J.B. Tucker, Dilemmas of a dual-use technology: toxins in medicine and warfare, Politics and the Life Sciences 13 (1) (1994) 51–62.
[58] I.L. Janis, Groupthink and group dynamics: a social psychological analysis of defective policy decisions, Policy Studies Journal 2 (1) (1973) 19.
[59] M.C. Loui, et al., Incident at morales: an engineering ethics video, IEEE, (2003), pp. S1H–S11 33rd Annual Frontiers in Education. FIE 2003.32003.
[60] V.R. Tandon, A. Mahajan, S. Sharma, S.K. Gupta, Unethical publication practices, JK Science 8 (3) (2006) 123–124.
[61] V. Gureev, I. Lakizo, N. Mazov, Unethical authorship in scientific publications (a review of the problem), Scientific and Technical Information Processing 46 (4) (2019) 219–232.

CHAPTER

15

Technological wisdom

OUTLINE

15.1 Introduction

There are numerous ways to identify, characterize, and analyze success and failure of a technology, including its scientific soundness, effectiveness, and ethical content. The key to any analytical approach is to describe accurately the events, people, and decisions that were made, strive for objectivity, and be fair and consistent in assessing and ascribing moral intent and consequences. Such an analysis is particularly challenging for emerging technologies given the paucity of reliable information to build an acceptable knowledgebase. At best, the assessment must use experiences with similar existing technologies as "analogs" of success or failure. However, some new technologies are so novel that such a knowledgebase is almost impossible to build.

Ethics shape how technical information is shared and applied. As mentioned, the professional requirements and codes of ethics for engineers and physicians ascribe to deontological ethics theory. Of the seven fundamental canons of ethics for engineers, two are of particular interest to technical presentations. The first and foremost is that engineers must hold paramount the safety, health, and welfare of the public and shall strive to comply with the principles of sustainable development in the performance of their professional

Technical and Ethical Optimization: Challenges and Opportunities for the Modern Engineer and Scientist
DOI: https://doi.org/10.1016/B978-0-443-44925-3.00002-8

duties. A second is that engineers shall issue public statements only in an objective and truthful manner [1,2].

15.1.1 The wisdom cascade

Analyzing a pending technology not only requires knowing how it works but also having the wisdom in deciding when conditions warrant changes. Intuition has always been an asset for technologists, and its value is increasing. The term "intuition" is used in several ways. One of the things that set technical types apart is their ability to access and use information. There are two ways of looking at data to derive information and, one hopes, to gain knowledge. These are deductive and inductive reasoning. When we "deduce," we use a general principle or fact to give us information about a more specific situation. This is the nature of scientific inquiry. We use general theories, laws, and experiential information to provide accurate information about the problem or the situation we are addressing. A classic example in environmental engineering is deducing from a cause to the effect. Low dissolved oxygen levels in a stream will not support certain fish species, so we reason that the fish kill is the result of low O_2. This demonstrates a product of deductive reasoning, that is, "synthesis."

Technologists also engage in inductive reasoning or "analysis." When they induce, they move from the specific to the general and from the effect to the cause. For example, experiments are conducted in microcosms that allow us to understand certain, well-defined, and well-controlled aspects of a system. Technologists induce knowledge from these observations, so larger principles can be gained beyond the specific study. The peril of induction is that any conclusion must be limited. For example, an experiment may show a direct relationship between an independent and dependent variable, but one does not know just how far to extend the relationship beyond the controlled environment of the laboratory. The experiment may show that increasing X results in growth of Y, but what happens in the presence of A, B, C, and Z? Technologists must recognize this uncertainty and must be arbiters of what information is useful and give enough thought to what will happen in real-world settings.

So, technologists help compile a body of information and knowledge from deductive and inductive reasoning. They must rigorously apply scientific theory (deduction) and extend specific laboratory and field results (induction). Over time and through experiences, their aptitude increases. Observing the decision-making of a seasoned technologist might well lead to the conclusion that the technologist is using a lot of "intuition." For example, engineers learn about how their designs and plans will work in two ways: [3]

1. Their formal and continuing education, that is, what others tell them; and
2. What they have experienced personally.

The engineer learns both subject matter, that is, "content," and processes, that is, "rules." Each engineer has learned a modicum of scientific and practical content that grows with experience. Facts and information about matter and energy, and the relationships between them, are the content of engineering. Rules are the sets of instructions that each engineer has written (literally and figuratively) over time on how to do things [4].

The accumulation of content and rules over one's academic experience and practice leads to intuition. Thus, intuition can be explained as the lack of awareness of why or how professional judgments have come to be. Kenneth Hammond [5], a psychologist who has investigated intuitive processes, says that intuition is, in fact, "a cognitive process that somehow produces an answer, solution, or idea without the use of a conscious, logically defensible step-by-step process." So, intuition is an example of something that is known to occur, and probably quite frequently, but it is not deliberative, nor can it be explained explicitly after it occurs. Arguably, it is at least in part a collective memory of the many deductive and inductive lessons learned (content), using a system to pull these together, sort out differences, synthesize, analyze, and come to conclusions (rules).[1] The more one practices, the more content that is gathered, and the more refined and tested the rules become.

Certainly, "engineering is a profoundly creative process" [6]. Engineers and technologists must design within constraints and tolerances called for by the problem at hand. For example, for environmental engineers, this is a balance between natural and artificial systems. This balance depends on data from many sources. Good data make for reliable information. Reliable information adds to scientific and societal knowledge. Knowledge, with time and experience, leads to wisdom. Effective technologists follow this "wisdom cascade" (see Fig. 15.1).

15.2 Effective communication

Technologists must clearly present scientifically sound information. Presentation is not the product but a means toward the end of obtaining feedback, approval, and acceptance of a technology. This is less about "selling" the technology than it is about receiving a critical review of potential problems. This goal requires that stakeholders and decision makers provide early input and that all affected parties understand the options. It helps to have participation from diverse backgrounds and perspectives. Arguably more important, the options and process must be sufficiently clear to a wider group of stakeholders, especially those most directly or indirectly affected by the ultimate decisions. This may appear to be an impossible task, but the inventor and technologist must be able to communicate effectively to ensure the optimal choices are made regarding designs, to ensure that these technically sound designs are accepted by all appropriate levels of governance and stakeholders, and to convey sufficient information to the technical personnel so that the designs are implemented, operated, and maintained satisfactorily [7].

Technical communication can be seen as a critical path, where the engineer sends a message and the audience receives it (see Fig. 15.2). The means of communication can be either perceptual or interpretative [8]. Perceptual communications are directed toward one's

[1] My wife and I have some disagreements. She maintains that intuition is not learned but is something inherent to each person. We share the same faith tradition, so I certainly agree with her that the conscience is indeed a source of what appears to be intuition. God instills knowledge in each person. We only disagree on the way he makes this happen. By the way, I am mismatched in this debate, since she is steeped in Jesuit training.

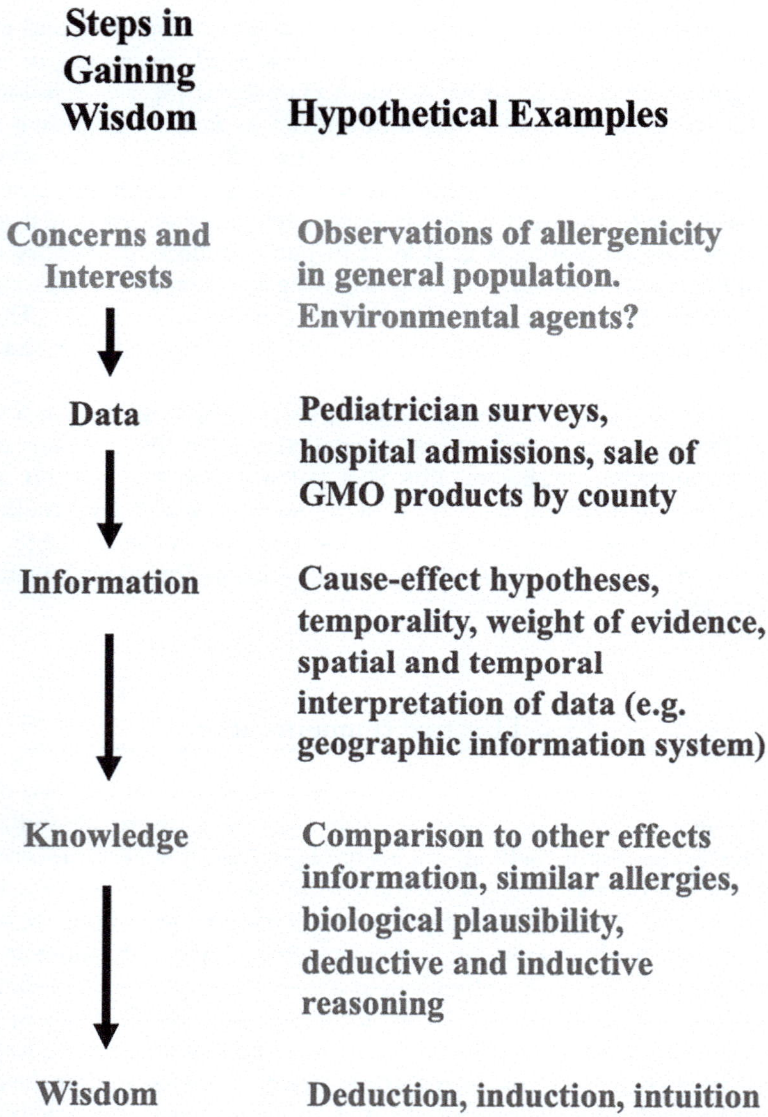

FIGURE 15.1 Value-added chain from data to knowledge and, with experience, professional wisdom. *Data from D.A. Vallero, Environmental contaminants: assessment and control, Elsevier Academic Press, Amsterdam; Boston, 2004, 801. pp. xxxix.*

senses. Human perceptual communications are similar to those of other animals. We react to sensory information, for example, reading body language or assigning meaning to gestures, like holding out one's hands with palms out, meaning "stop" or smile conveying approval.

Whatever technology is to be introduced, regulatory and corporate governance will often apply a collaborative approach to consider the views of multiple pertinent stakeholders [12]. Although technical types may spend most of the workday with fellow technologists, they

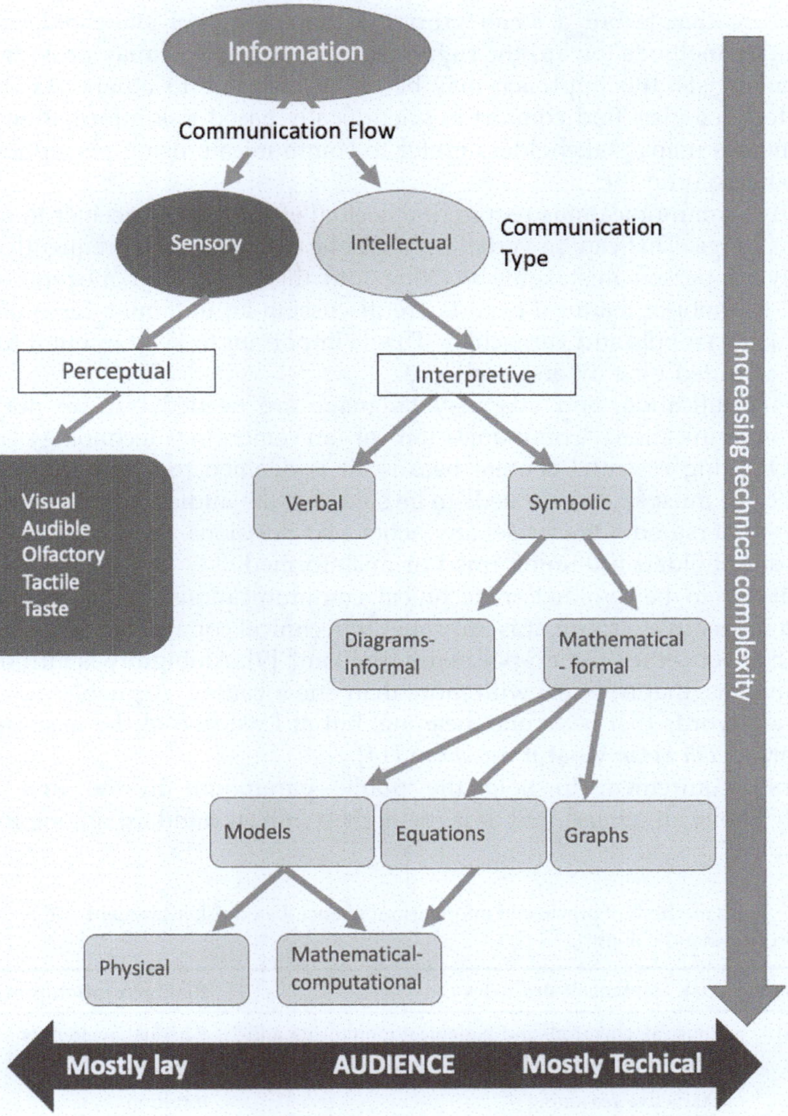

FIGURE 15.2 Methods of communicating technical information. All humans use perceptual communication, such as observing body language of a colleague in a meeting or smelling odors released by a bioreactor. The right side of the figure is the domain of technical communication. This type of information can overwhelm the nontechnical audience or even a technical audience with different expertise, who may be unfamiliar with perceptive cues or may not understand the symbolic, interpretive language being used by a bioengineer and others in the risk communication process. Thus, the communication method is usually different for a scientific briefing than for an update at a public meeting or a briefing for a neighborhood group potentially affected by a risk management decision. *Data from: [7,9–11].*

must be aware that is but a small group of the full set of stakeholders. Whereas the communication methods toward the right-hand side of Fig. 15.1 may provide clarity for the fellow technologists, this approach may be undecipherable to others. As shown in Table 15.1, most technologists find comfort in scientifically based risk information in the middle column, whereas many stakeholders prefer to communicate using perception processes in the right-hand column [9].

Interpretive communications require intellectual effort by the receiver to understand the sender's meanings. This can be verbal or symbolic. Technologists frequently communicate symbolically. A person mistakenly walking into the seminar or meeting will likely not understand the subject matter if experts are discussing an unfamiliar area of science using unrecognizable symbols and vernacular. This is important to keep in mind for any briefing or meeting attended by a diverse audience.

Clarity, simplification, and everyday language are needed but are not sufficient for effective communication. Communication of an emerging technology must also be complete. Omitting essential content because it is deemed too technical is unacceptable. The technologist must say what needs to be said, *and* the audience must hear and see what needs to be understood. Otherwise, any actions or decisions made will have failed to the extent that stakeholders are uninformed or misinformed.

Ambiguity is to be avoided in technical communications. While ambiguity may be encouraged in creative writing, it is improper in technical communications. This is what the author has coined as the "techno-polysemy problem" [9]. Ambiguity is introduced with the use of homonyms, that is, terms with more than one meaning. Arguably more problematic, polysemes are words that have one meaning, but at least two of the meanings overlap. A word or phrase *will mean* what it *can mean* [14].

Engineers' requirement to hold the public paramount in the first canon means "definitively above all others" [15]. It is the highest ethical standard among the engineering

TABLE 15.1 Two methods of processing information: science-based risk assessment (right-hand column) and risk perception (left-hand column).

Analytical phase	Risk assessment processes	Risk perception processes
Identifying risk	Physical, chemical, and biological monitoring and measuring the event	Personal awareness
	Deductive reasoning	Intuition
	Statistical inference	
Estimating risk	Magnitude, frequency, and duration calculations	Personal experience
	Cost estimation and damage assessment	Intangible losses and nonmonetized valuation
	Economic costs	
Evaluating risk	Cost/benefit analysis	Personality factors
	Community policy analysis	Individual action

Data from: D.A. Vallero, 33. Waste Governance, in: T.M.a V. Letcher, A. Daniel (Eds.), Waste: A handbook for management, Elsevier, 2019, pp. 651–671; K. Smith, Environmental hazards: assessing risk and reducing disaster, Routledge, London, UK, 1992 [9,13].

disciplines. This means that the order of priority for the public communication of technical information is first and foremost for the public good, followed by the interests of their clients and employers, and finally the engineer's personal good. The same cannot be said for all professions, including the general categories of researchers, academics, and scientists. Several experts and organizations have called for a universal code of ethics from different fields, most recently biomedical engineering and related research [16].

Ethics can be taught and reinforced by an organization. Ethical knowledge results from a three-step process: (1) awareness; (2) decision-making; and (3) behavior.

Often, the best that a university or research institution can do is to make the student aware of potential ethical dilemmas and to provide a framework for ethical decision-making. This is one of the goals of responsible conduct of research (RCR) programs at most graduate schools [17]. Similarly, research involving human subjects must meet the approval of an institutional review board (IRB). The intent of the RCR and IRB is to ensure best practices and to prevent ethical breaches. The RCR introduces the researcher to ethical norms and expectations, and the IRB requires that human subjects be protected prior to subjects' recruitment, followed by an IRB review and approval, with modifications if necessary, each year thereafter [18].

Emerging technological uncertainties begin with the realization that there is a lack or paucity of templates and cases from which to draw when making decisions about what to do. Genetic engineering essentially narrowed the changes of the microbe to one trait that needed to be "knocked" in or out. So, arguments and concerns centered around that single change. However, with synthetic biology, a whole new creature is built, so the uncertainty propagates with each change. Indeed, many decisions on whether it is right or wrong to build an organism to perform a service to humankind do not enjoy useful precedent, that is, the decision-making processes are data- and knowledge-poor. The simpler the organism at the outset, the more novel the modified organism will be. Thus, uncertainty increases moving from the natural progenitor to the adapted species to the genetically modified organism (GMO) and, finally, to the synthetic organism. Thus, the potential downstream harm is most uncertain moving from editing a "chassis" organism to manufacture a biologically synthesized one: [19]

> ... gene editing technology may allow for precise and effective artificial mutation production (by effecting exact deletions of very small parts of a genetic sequence, thereby reliably and precisely boosting or shutting off very specific cellular functions) that is more effective and accurate than traditional genetic modification technologies while leaving no trace behind of such interference, thereby being indistinguishable from the way in which genetic changes in organisms and cells are being produced artificially using traditional techniques or spontaneously in nature. It also can boost the precision and effectiveness of adding entirely new components into a genome, regardless of where these come from – other organisms of the same species, other species, or synthetically produced genetic material.

This characterization of precision and accuracy refers to the target organism, but not to the ethical outcomes. Researchers are often challenged with intellectual "tunnel vision," which is supported by strong, societal rationales. Society needs better waste treatment technologies. Thus, gene editing's anticipated value to society rests in what ethicists refer to as a "teleological" argument, that is, it is good because the outcome is good. However, the precision and accuracy of predicting other outcomes, for example, side effects and unintended consequences, are woefully data- and knowledge-poor.

Technological advances discussed in this book vary from relatively well-studied adaptations of existing technologies to the most complex and uncertain gene-editing biotechnologies and barely understood aspects of the implications of artificial intelligence. Ethical awareness is arguably easier to enhance for the more accepted and established technologies than for the data-poor, highly uncertain processes [7].

15.3 Case analysis for ethical optimization

Technologists can learn much from previous successes and failures. The analysis of information under the wisdom cascade sets the stage for optimizing a planned or existing technology. The best optimization approaches are those in which relevant and reliable data are objectively analyzed. The ethical analysis of an existing or planned technology consists of three basic steps and a few substeps [20].

15.3.1 Step 1—factual description

Key characters and events pertinent to the case are identified. This description includes narrative, tables, figures, maps, organization charts, critical path diagrams, and photographs that are needed to place the case in ethical context. As in any case analysis, the first step is descriptive. Who is involved, and what is the role of each person in the case? What happened, and where did each key step that led to the ultimate result occur? In what manner did the steps occur? Were they sequential, contemporary, or combinations of these? Describing an ethical issue makes use of photos, maps, profiles, and any other descriptive materials that will help show what happened and/or what could happen. During the descriptive stage, we are not yet ready to discuss the ethics or to ascribe blame or credit. What is required in this step is an honest retelling of all the events related to the failure.

15.3.2 Step 2—logical arguments and syllogisms

Based on the findings in Step 1, the validity of the decisions or lack thereof is analyzed. The syllogism includes a factual premise, a connecting fact-value premise, and an evaluative premise to reach an evaluative conclusion. Many moral and scientific arguments fail because of weaknesses in any of these components of the syllogism. Depending on the case, numerous arguments must be evaluated.

15.3.3 Step 3—ethical problem-solving analysis

Upon sufficiently identifying and interpreting the facts and ethical problems of the case, the issues must be classified as factual, conceptual, or moral.

From the descriptions in Step 1, the depth of each type of event and decision points in the critical path can be assessed. Factual issues are those that are known. This can sometimes be apparent simply by reading about the events, but in certain cases, the facts may not be completely clear. For example, you and I may agree on the "fact" that carbon dioxide is a radiant gas, but we may disagree about the extent to which the buildup of CO_2 in the troposphere will lead to increased global warming. This may mean that we agree on first principles of science and even the data being used, but we may disagree on the relative weightings in indices and models. This leads to a need to ascribe causality. The difficulty in reaching a consensus on the importance and relevance of fact, not the fact itself, increases with the number of different points of view. It is even more challenging for ethical points of view than for scientific perspectives.

Step 3a—identify possible causal relationships: To begin evaluating whether a causal link is valid, oftentimes the best that science can do in this regard is to provide enough weight-of-evidence between a cause and an effect. The medical research and epidemiological communities use a number of criteria to determine the strength of an argument for causality, but the first well-articulated criteria were Hill's Causal Criteria [21], including weight-of-evidence, coherence, biological plausibility, and several other criteria. Depending on the case, some of Hill's criteria are more important than others.

Conceptual issues involve different ways that the meaning may be understood. For example, what you and I consider to be "pollution" or "good lab practices" may vary, although the scientific community strives to bring consistency and, hopefully, consensus to such definitions.

Many engineers and scientists believe it is the job of technical societies and other collectives to try to eliminate factual and conceptual disagreements. Most of us agree on first principles, for example, fundamental physical concepts like the definitions of matter and energy, but unanimity wanes as the concepts drift from first principles. For example, the late John Ahearne, former President of Sigma Xi, the Scientific Research Society, recently told an audience of engineers that they should not be disagreeing about the facts.[2] The progress of research and knowledge eventually helps to resolve factual issues. The elicitation of experts aids in resolving conceptual and perception differences [22,23]. But since complete agreements are not generally possible even for the factual and conceptual aspects of a case, the moral or ethical issues are further complicated.

Step 3b—apply decision force-field analysis: Whether scientists and engineers want to believe it, few, if any, decisions can be made exclusively from physical scientific principles. We can visualize these decision-forcing functions within a force field, where a magnet is placed in each sector at point equidistant from the center of the diagram (see Fig. 15.3). The force field assumes that the decision being considered will be pulled in the direction of the strongest magnetic force. The stronger the magnet, the more likely the decision that will actually be made will be pulled in that direction. For example, lawyers may proceed in one direction, while engineers may proceed in another, and the business clients in another, all applying different forces that "deform" the decision.

A decision that is almost exclusively influenced by strong science will appear something like force field in Fig. 15.4. However, if other factors, such as a plant closing or the possibility of new jobs being attracted at the expense of environmental, geological, or other scientific

[2] The author heard Ahearnes' comments at the National Academy of Engineering workshop on emerging technologies and ethics, held in Washington, DC, in November 2003.

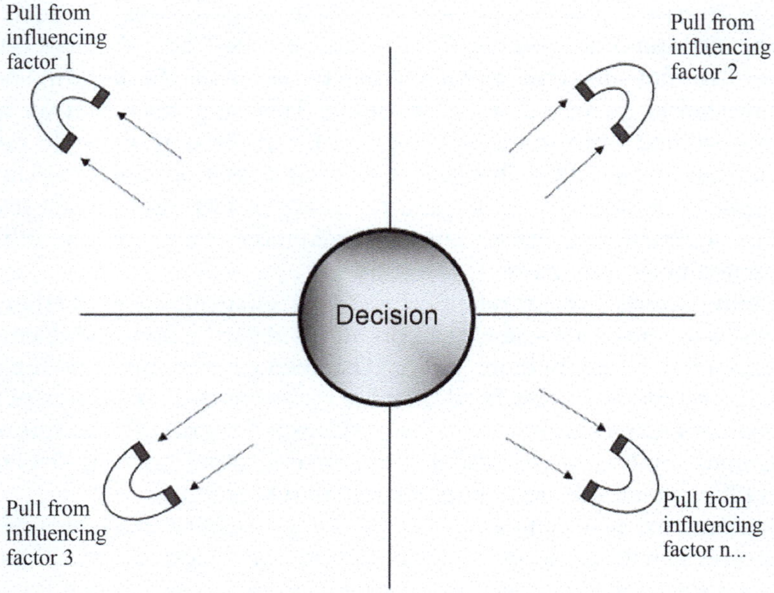

FIGURE 15.3 Decision force field.

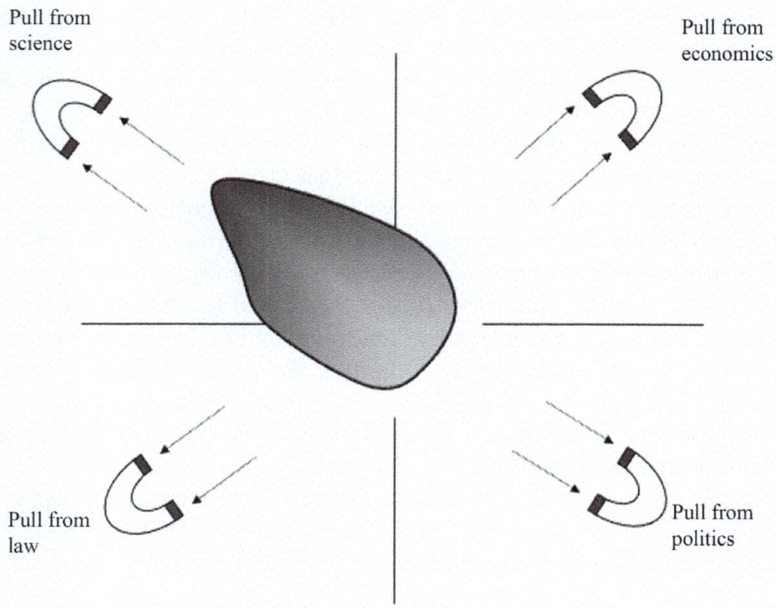

FIGURE 15.4 Decision force field for a scientifically based decision, although other factors (legal, financial, and political) are influencing the outcome.

influences, arise, the decision would migrate toward these stronger influences, as shown in Fig. 15.5.

Thus, engineers by their call are risk managers. All engineering decisions are made under risk and uncertainty, which is why factors of safety are a part of every engineering recommendation. The engineering risk management process is informed by the quantitative results of the risk assessment process. The shape and size of the resulting decision force field diagram give an idea of what the principal driving factors are that lead to decisions. A force field diagram can be a useful, albeit subjective, tool to assess each.

Step 3b—net goodness (NG) analysis: This is a subjective analysis of whether a decision will be moral or less than moral. It puts the case into perspective by looking at each factor driving a decision from three perspectives: (1) how good or bad would the consequence be; (2) how important is decision; and (3) how likely is it that the consequence would occur. These factors are then summed to give the overall net goodness of the decision:

$$NG = \Sigma(\text{goodness of each consequence}) \times (\text{importance}) \times (\text{likelihood})$$

Thus this can be valuable in decisions that have not yet been made, as well as in determining what decisions "should" have been made in a case. For example, these analyses sometimes use ordinal scales, such as 0 through 3, where 0 is nonexistence (e.g., zero likelihood or zero importance), and one, two, and three are low, medium, and high, respectively. Thus, there may be many small consequences that are near zero in importance, and since NG is a product, the overall net goodness of the decision is driven almost entirely by one or a few important and likely consequences.

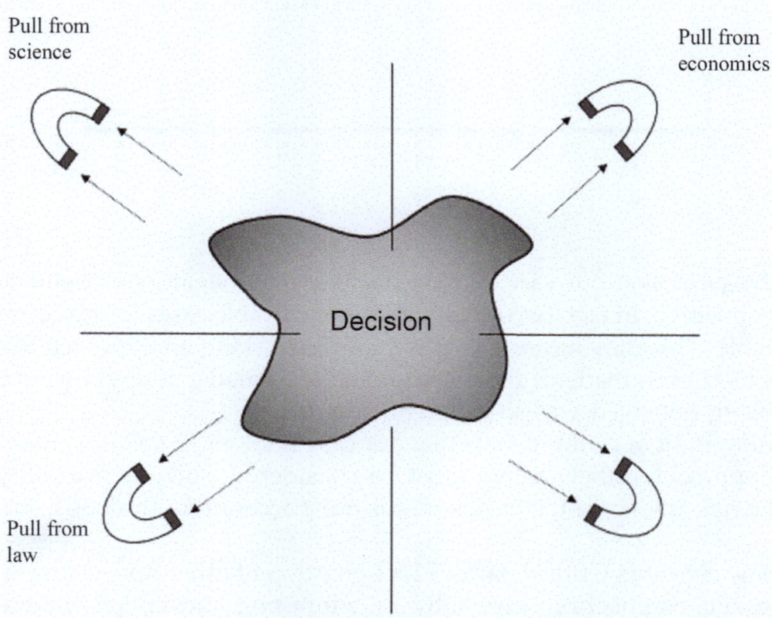

FIGURE 15.5 Decision force field for complex decision with multiple influences that have about equal influence on the outcome.

There are two cautions in using this approach. First, although it appears to be quantitative, the approach is quite subjective. Second, as we have seen many times in cases involving health and safety, even a very unlikely negative consequence is unacceptable.

Step 3c—line drawing: Graphical techniques like line drawing, flow charting, and event trees are very valuable in assessing a case. Line drawing is most useful when there is little disagreement on what the moral principles are, but when there is no consensus about how to apply them. The approach calls for a need to compare several well-understood cases for which there is general agreement about right and wrong and to show the relative location of the case being analyzed. Two of the cases are extreme cases of right and wrong, respectively. That is the positive paradigm (PP) is very close to being unambiguously moral and the negative paradigm (NP) unambiguously immoral:

Negative paradigm	Our case	Positive paradigm
Negative feature 1	X	Positive feature 1
Negative feature 2	X	Positive feature 2
Negative feature 3	X	Positive feature 4
Negative feature *n*	X	Positive feature *n*

Next, our case (T) is placed on a scale showing the PP and the NP, as well as other cases that are generally agreed to be less positive than PP but more positive than NP. This illustrates the relative position of our case T:

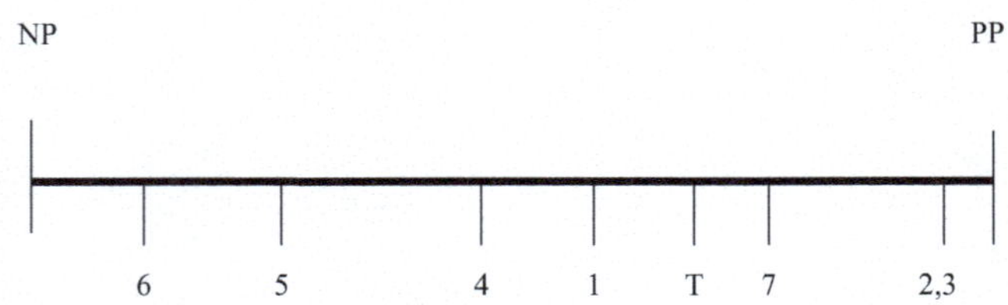

This gives the sense that our case is more positive than negative, but still short of being unambiguously positive. In fact, two other actual, comparable cases (2 and 3) are much more morally acceptable. This may indicate that we consider taking an approach like these if the decision has not yet been made. If the decision has been made, you will want to determine why the case being reviewed was so different from these.

Although being right of center means that our case is closer to the most moral than to the most immoral approach, other factors must be considered, such as feasibility and public acceptance. Like risk assessment, ethical analysis must account for tradeoffs, such as security versus liberty.

"Step 3d—flow charting: Critical paths, PERT charts, and other flow charts are commonly used in design and engineering, especially in computing and circuit design." ("How to Analyze Ethical Cases - Carleton College") They are also useful in ethical analysis if sequences and contingencies are involved in reaching a decision, or if a series of events and ethical and factual decisions lead to the consequence of interest. Thus, each consequence and the decisions that were made along the way can be seen and analyzed individually and collectively. For example, a flow chart for the Bhopal incident deals with only one of the

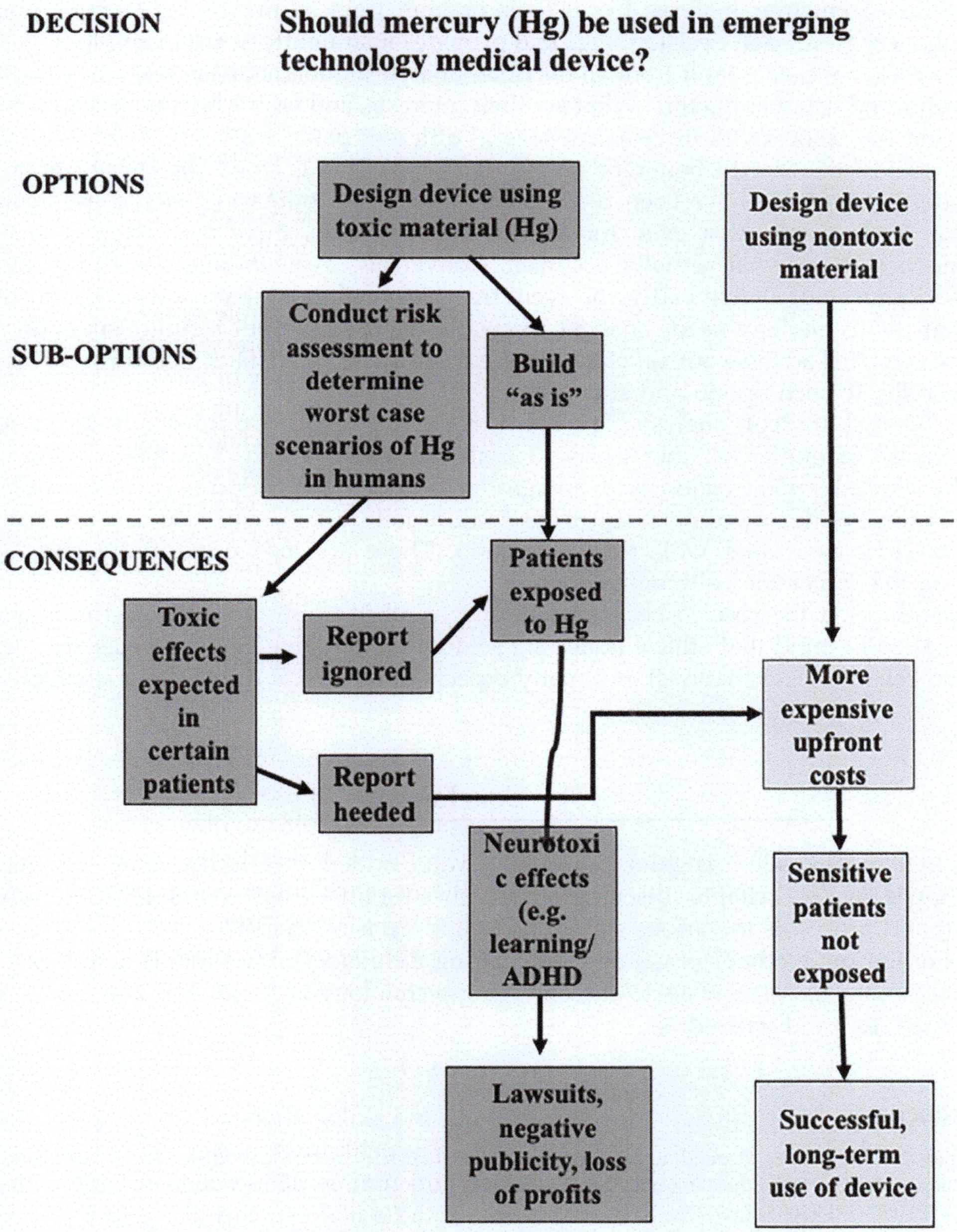

FIGURE 15.6 Event tree on whether to use mercury in a medical device. Note: ADHD is attention deficit hyperactivity disorder.

decisions involved in the incident, that is, where to site the plant [24]. But other charts need to be developed for safety training, the need for fail-safe measures, and proper operation and maintenance. Thus, a "master flow chart" can be developed for all the decisions and subconsequences that ultimately led to the disaster.

Step 3e—event tree analysis: Event trees or fault trees allow you to examine possible consequences from each decision. Fig. 15.6 provides a straightforward example.

The event tree can be built from all the other analytical tools, starting with the timeline of key events and list of key actors. What are their interests, and why were the decisions made? The event tree allows you to visualize several different paths that could have been taken, which could have led to better or worse decisions. This is done for every option and suboption that should have been considered in the case, comparing each consequence. It may be, for example, that even for a disastrous outcome, there may have been worse consequences than what actually occurred. Conversely, even though something did not necessarily turn out all that badly, the event tree could point out that we were just fortunate and that the technology needs changes to ensure that our present fortuitousness does not become an actual adverse outcome. In fact, the more quantitative fault tree approach applies a probability to each option and suboption.

Step 3f—risk tradeoff analysis: The event tree above depicts an unusually dichotomous decision, for example, whether to use a toxic substance. Often, however, decisions are fraught with advantages and disadvantages, no matter what is decided. For example, not using Hg in the device could result in premature failure that may lead to complications, such as risks associated with re-implantation. There are also opportunity risks when choosing the "no action" alternative.

Depending on the case to be analyzed, any or all of these tools are useful. However, always keep in mind that ethical issues are seldom quantitative. Some aspects are (such as risk and reliability calculations), but many aspects of an ethical decision are qualitative.

15.4 Conclusions

Technological wisdom is gained through a value-added approach to data, information, and knowledge. This chapter discusses approaches for effectively communicating technical content and processes to various stakeholders and decision makers.

One of the key methods of gaining and sharing technological wisdom is to analyze cases that are similar in some ways to the emerging technology at hand. The ways to conduct these analyses are discussed.

References

[1] Sarah K.A. Pfatteicher, Depending on character: ASCE shapes its first code of ethics, Journal of Professional Issues in Engineering Education and Practice 129 (1) (2003) 21–31. Available from https://doi.org/10.1061/(ASCE)1052-3928(2003)129:1(21).

[2] American Society of Civil Engineers. ASCE Code of Ethics, 2020. https://www.asce.org/career-growth/ethics/code-of-ethics. Accessed on May 7, 2026.

[3] D.A. Vallero, Environmental Contaminants: Assessment and Control, Elsevier Academic Press, Amsterdam, Boston, (2004), p. 801 pp. xxxix.
[4] R.M. Hogarth, Educating Intuition, University of Chicago Press, (2001).
[5] K.R. Hammond, Human Judgment and Social Policy: Irreducible Uncertainty, Inevitable Error, Unavoidable Injustice, Oxford University Press, (2000).
[6] National Academy of Engineering, The Engineer of 2020: Visions of Engineering in the New Century, National Academies Press, Washington, DC, (2004).
[7] D.A. Vallero, C.K. Gunsch, Applications and implications of emerging biotechnologies in environmental engineering, Journal of Environmental Engineering 146 (6) (2020) 03120005.
[8] C. Mitcham, R.S. Duval, Engineering Ethics (2000).
[9] D.A. Vallero, 33. Waste governance, in: T.M.a V. Letcher, A. Daniel (Eds.), Waste: A Handbook for Management, Elsevier, (2019), pp. 651–671.
[10] M. Myers, A. Kaposi, A First Systems Book: Technology and Management, Imperial College Press, (2004).
[11] T.R. Green, Cognitive dimensions of notations, People and Computers V (1989) 443–460.
[12] I. Linkov, et al., Comparative, collaborative, and integrative risk governance for emerging technologies, Environment Systems and Decisions (2018). Available from https://doi.org/10.1007/s10669-018-9686-5. May 04.
[13] K. Smith, Environmental Hazards: Assessing Risk and Reducing Disaster, Routledge, London, UK, (1992).
[14] Forgotten author, "echnical writing course, D. A. Vallero, Ed., Instructor said that "a word will mean what it can mean" and "a phrase will mean what it can mean". ed. Kansas City, Missouri: U.S. Environmental Protection Agency, Region VII, circa 1978.
[15] P.A. Vesilind, A.S. Gunn, Hold Paramount: The Engineer's Responsibility to Society, Cengage Learning, (2010).
[16] Vallero, D.A., 2007. *Biomedical ethics for engineers*. Amsterdam: Elsevier/Academic Press. ISBN-13: 978-0750682275.
[17] National Science Foundation, Responsible conduct of research, Federal Register 74 (160) (2009) 42126–42128.
[18] M.I. Lapid, B.L. Clarke, R.S. Wright, Institutional review boards: what clinician researchers need to know, Mayo Clinic Proceedings 94 (3) (2019) 515–525.
[19] C. Munthe, Precaution and Ethics: Handling Risks, Uncertainties and Knowledge Gaps in the Regulation of New Biotechnologies, Federal Office for Buildings and Publications and Logistics (FOBL), (2017).
[20] C.E. Harris, M.S. Pritchard Jr, R.W. James, E.E. Englehardt, Engineering Ethics Concepts and Cases, Cengage Learning, Inc, (2019).
[21] A.B. Hill, The environment and disease: association or causation? Proceedings of the Royal Society of Medicine 58 (5) (1965) 295.
[22] P.A. Bottomley, J.R. Doyle, A comparison of three weight elicitation methods: good, better, and best, Omega 29 (6) (2001) 553–560.
[23] U. S. EPA, Expert elicitation white paper, Science and Technology Policy Council, Washington, DC, 2011. [Online]. Available from: http://www.epa.gov/stpc/pdfs/ee-white-paper-final.pdf.
[24] C.B. Fleddermann, Engineering Ethics, Prentice Hall, Upper Saddle River, NJ, (1999).

CHAPTER

16

Optimizing for success and avoiding failure☆

OUTLINE

Men wiser and more learned than I have discerned in history a plot, a rhythm, a predetermined pattern. These harmonies are concealed from me. I can only see one emergency following another, as wave follows upon wave, only one great fact with respect to which, since it is unique, there can be no generalizations, only one safe rule for the historian: that he should recognize in the development of human destinies the play of the contingent and the unforeseen. This is not a doctrine of cynicism and despair. The fact of progress is not a law of nature. The ground gained in one generation may be lost by the next. ***H.A.L Fisher [2].***

☆ This chapter includes adaptations and updates to Chapter 8 of [1].

Technical and Ethical Optimization: Challenges and Opportunities for the Modern Engineer and Scientist
DOI: https://doi.org/10.1016/B978-0-443-44925-3.00016-8

16.1 Introduction

A daunting ethical challenge for engineers is that change is so rapid. We live in a time when misreading the slightest nuance of a design or operation can be the difference between success and failure. Often, we are not immediately even sure whether we are succeeding or failing. In fact, we may not know whether the fruition of our ideas in many emerging technologies will be for good or ill until well after the research. The explosion in emergent technologies is like having countless chests before us; some are treasure troves, but others are Pandora's boxes.

Fisher's advice and admonition to the historian holds for the contemporary engineer. It envisages the famous counsel of George Santayana [3]:

Progress, far from consisting in change, depends on retentiveness.... Those who cannot remember the past are condemned to repeat it.

Santayana's advice is widely quoted but too often ignored. What we remember can save us from much despair, needless waste, and failure in the long run. We forget important events at our own peril. It is one thing to fail, but quite another not to refuse to learn from our failures. As engineers, we must consider the reasons and events that led to the failure in hopes that corrective actions and preventive measures are put in place to avoid their reoccurrence. This is not easy and is almost always complicated, especially for systems as elegant and complex as the human species.

Anything that changes constantly and rapidly is difficult to measure with, at best, yielding finite snapshots. As such, there can be myriad views on the extent to which a new technology is a temporary success with failures down the road. Finite monitoring of a dynamic system like an emerging technology is likely to miss flaws.

Indeed, those who are engaged in the research and the practice of new technologies may have the data and information needed to recognize the hazards and pitfalls. In fact, those directly involved in the advancement may be the worst at appraising its worth and estimating its risks. The innovators have a built-in conflict of interest, which works against their ability to serve as society's honest brokers of emerging technologies.

Surely, most innovators who have pushed harmful technologies have not been inherently "bad" people but merely single-minded in their advocacy of the benefits of their ideas. Arguably, auditing "good" people who are simply missing flaws is harder than auditing people with malintent. For example, designers, engineers, and technologists in Silicon Valley strived for the noble goal of an open and connected world by developing the infinite scroll and algorithmic news feeds to maximize user engagement. Instead, however, the technology resulted in many addictions to the feeds that stressed many users' mental health, detached them from normal human relationships, polarized their thinking, and increased their viewing of harmful content [4–8]. Aza Raskin, the infinite scroll inventor, has stated regret for his lack of foresight and for missing the potential for unintended adverse consequences of the technology [5].

One example of the moral concerns with the infinite scrolling technology is the recent show, *The Secret Lives of Mormon Wives* [9]. The show exhibits patterns of chronic social media consumption among a group of social media, TikTok influencers whose lives are in a nearly constant state of performance, often revealing what should arguably be private, personal struggles. The members are chronically online with seemingly endless content that is monetized to sell various products, ideas, and ideologies, many of which are unhealthy. As such, many social media and television fans are manipulated and influenced to buy not only products but also injurious lifestyles, including social addiction. To wit, the TV show is intensely interwoven into TikTok's infinite scrolling, as the cast and other influencers' real-time social media posts often reveal plot points prior to the airing of an episode of the TV show. In this case, the success of a show can be a failure for society as a whole.

The difference between success and failure often hinges on very subtle hints in the design process. Every failure results from a unique series of events. Human factors must always be considered in any design implementation. Often, seemingly identical situations lead to very different conclusions. In fact, the mathematics and statistics of failure analysis are some of the most complicated, relying on nonlinear approaches and chaos, making use of nontraditional statistical methods, such as Bayesian theory ("Biomedical Ethics for Engineers - ScienceDirect")[1]. This is not an excuse to acquiesce. To the contrary, the complexities require that engineering creativity and imagination can and must be used to envisage possible failures and to take steps to ensure success.

16.2 Human factors engineering

As discussed in Chapter 13, human factors engineering (HFE) addresses the interface between humans and things, especially how we use these things. At the interface, human factors studies address [10]:

- systems performance;
- problems encountered in information presentation, detection, and recognition;
- related action controls;
- workspace arrangement; and
- skills required.

Any of these factors can contribute to failure. Sometimes, the use itself is problematic, such as the repetitive actions of people using manufacturing or office equipment. From an ethical perspective, HFE is needed to determine how devices fail and to suggest ways that such failures can be prevented. A design that does not properly account for problems in usage by medical practitioners or by the patients themselves is morally unacceptable. For example, a device is unacceptable if it does not account for variability in dexterity and finger length of practitioners in the operating room. Likewise, a pump delivering medication to a patient is also unacceptable if it requires an inordinate degree of user sophistication, especially if operator error can lead to major medical complications, particularly if there is an available alternate design that is more user-friendly.

Concern with emerging technological developments is not the exclusive domain of the so-called neo-Luddites, those who oppose most innovations and new technologies that change

[1] Thomas Bayes, an English preacher and mathematician, argued that knowledge of prior events is needed to predict future events. Thus Bayes, like Santayana for political thought, advocated for the role of memory in statistics. Bayes' theorem, which was published two years after his death in 1761 in *An Essay Towards Solving a Problem in the Doctrine of Chances*, introduced the mathematical approach to predict, based on logic and history, the probability of an uncertain outcome. This is very valuable in science, that is, it allows uncertainty to be quantified.

constantly and rapidly, and view such advances as detrimental to society. Thoughtful people are simply asking for protections and commitments to precaution before entering an unbridled mode of research and application.

Incidentally, the original Luddites were 19th-century English skilled textile workers who destroyed machinery (1811–17) to protest against labor-displacing technologies, poor working conditions, and reduced wages during the Industrial Revolution [11–13]. This has some unmistakable correlation with present-day concerns with emerging technologies [14], especially robotics and AI. Named after the mythical figure Ned Ludd, they were less opposed to technology itself than to capitalist system using machines to destroy their livelihoods. They were not merely antitechnology but were skilled weavers and other artisans struggling against the introduction of machinery that they believed would lead to low-quality goods, diminish the roles of the skilled workforce, and depress wages. They operated in secret, conducting raids to enter factories and disable stocking frames, power looms, and other machinery.[2]

Some would also argue that research in many areas, including genetic engineering, systems biology, artificial intelligence (AI), and human enhancement, has already passed the point of precaution, so that any attempts to control or even moderate the direction of societal risk are akin to changing a tire on a bus as it moves down the highway. It is physically possible, but certainly not very probable. You probably will have to stop the bus to change the tire, at least for a while, but many of the AI and other emerging technology advocates want the bus to continue at breakneck speeds with a head start over the ethics vehicles. To continue to torture this metaphor, the AI ethics cops are riding bicycles, lagging farther and farther behind the AI bus. One of the first steps in catching up is knowing when to stop the bus. This will require objective measures of success and failure.

16.3 Utility as a measure of success

Quantitative types, like most engineers, have a strong affinity for objective measures of successes. Thus, we gravitate toward usefulness as a measure of success. Such utility is indeed part of any successful engineering enterprise. After all, engineers are expected to provide reasonable and useful products. In a word, we look for utility. It turns out that one of the major ethical constructs is utilitarianism. The first two dictionary definitions of utilitarianism (Latin *utilis*, useful) are relevant to the ethics of emerging technology [16]:

1. The belief that the value of a thing or an action is determined by its utility.
2. The ethical theory proposed by Jeremy Bentham and James Mill states that all action should be directed toward achieving the greatest happiness for the greatest number of people.

Since in utilitarianism the morality of an action or decision is based on outcome, it is a form of consequentialism. This ties in with the ethos of inventors and innovators. They have

[2] The Luddite movement took its name from a legendary apprentice named Ned Ludd, or "General Ludd," who was reputed to have destroyed a stocking frame in 1779. The British government treated Luddism as terrorism and sent troops to the Britain's industrial north. More troops were sent there than were fighting against Napoleon in the Iberian Peninsula [15].

the end in mind. Engineers and technologists are expected to provide products and useful results, so morality based on consequence, that is, the end not the means to that end, is attractive. Furthermore, the utilitarian tools used to determine the morality of a decision are objective and quantifiable.

One vexing aspect of utilitarianism is how to identify the "greatest good" for the "greatest number" [17]. The first challenge is identifying who decides what the greatest good should be. The second challenge is what precisely is being counted to arrive at the greatest number, for example, all people, certain subpopulations of people like children or immunocompromised individuals, all species, endangered species, etc.? The third challenge is what happens to the rest of the lesser number who are ignored when serving the greatest number?

Societies apply various values to technological decisions, but there is usually some type of balance between a public or personal good or service. Preventing health, safety, and welfare problems must involve standards be met to ensure certain activities be avoided. Thus, a decision is acceptable if it provides for a desired utility and, to the extent that it does not, it is unacceptable [18].

From an environmental sustainability and public health standpoint, the dependability of a technology rests on a triple bottom line, that is, in its ultimate effect on societal, economic, and environmental quality [19–21]. Decisions based on the triple bottom line require balancing and optimization among mutual, competing, or neutral factors. As such, optimization is a selection among alternatives that provide the most social, economic, and environmental utility [17]. This means that decisions that go too far in protecting one value at the expense of other worthwhile values are bad decisions. For example, allowing the destruction of sensitive habitat to make a monetary profit for a few is arguably an immoral decision. The soundness of the decision must be based on all three bottom lines. A very useful technology, for example, a reduced cost of an energy storage system that reduces fossil fuel dependence, may be counterbalanced by factors that make it unacceptable environmentally in other ways, for example, the potential release of toxic substances during natural resource extraction, manufacturing, or operation [22].

This does not necessarily mean banning a new technology or action completely. Indeed, optimization often provides a middle ground between complete acceptance and rejection. For example, mitigative measures can ameliorate potential environmental problems of an otherwise desirable approach. For instance, microbes can be adapted to degrade hazardous compounds that would have otherwise polluted the air [23], water [24], soil [25,26], or sediments [25]. Recent advances in genomics and data science are promising to aid in these mitigations [27] and help to prevent future problems [28,29].

16.3.1 The benefit-cost ratio

The most prevalent of the consequentialist tools is the benefit-to-cost (B/C) ratio. It is an attractive metric due to its simplicity and seeming transparency. To determine whether a project is worthwhile, one need only add up all the benefits and place them in the numerator and all the costs (or risks) and put them in the denominator. If the ratio is greater than 1, its benefits exceed its costs.

One obvious problem is that some costs and benefits are much easier to quantify than others. In developed economies, monetizable benefits are often given prominence in utilitarian calculations. Some costs and benefits, like those associated with quality of life, are nearly impossible to monetize and to quantify accurately. Further, the comparison of doing anything versus doing nothing cannot always be captured with a benefit/cost ratio. Opportunity costs and risks are associated with taking no action (e.g., loss of an opportunity to apply an emerging technology may mean delay or nonexistent treatment of diseases). Simply comparing the status quo to costs and risks associated with a new technology may be biased toward no action. But, beyond the costs (time and money) are not the only reasons for avoiding action. The availability of a new monitoring device may invite more tests or drugs that, if not managed properly, could interfere with quality of life of patients, could carry its own risks, and could add costs to the public and the community (e.g., "preventive medicine" to avoid liability), with little actual benefit to anyone. So, it is not simply a matter of benefits versus costs but often one risk being traded for another.

Often, addressing contravening risk is a matter of optimization, which is a proven analytical tool in engineering. However, the greater the number of contravening risks that are possible, the more complicated such optimization routines become. Risk tradeoff is a very common phenomenon in everyday life. For example, local governments enforce building codes to protect health and safety. Often, these added protections are associated with indirect, countervailing risks. For example, the costs of construction may increase safety risks via "income" and "stock" effects. The income effect results from pulling money away from family income to pay the higher mortgages, making it more difficult for the family to buy other items or services that would have protected them. The stock effect results when the cost of the home is increased; families have to wait to purchase a new residence, so they are left in substandard housing longer [30]. Such countervailing risks are common in healthcare decisions. Patients demanding immediate and state-of-the-art technologies could also be arguing for increased risks from income and stock effects by imposing needlessly on the healthcare system.

The engineer is frequently asked to optimize two or more conflicting variables in many situations. The engineering requirement to hold paramount the health, safety, and welfare of the public allows no room for spin. However, the public's perception of risks may be quite different from the physician's or engineer's perception. So, abating risks that are, in fact, quite low could mean unnecessarily complicated and costly measures. It may also mean choosing the less acceptable alternative, that is, one, that in the long run, may be more costly and deleterious to the public health. The risk assessment and risk perception processes differ markedly, as discussed in Chapter 10, a risk assessment relies on problem identification, data analysis, and risk characterization, including cost-benefit ratios. It provides the physical and social scientific basis for risk management decisions.

Once the decision process reaches decisions by governmental, corporate, and other stakeholders, factors other than scientific credibility emerge. What the client perceives often becomes the predominant driver. Perception relies on thought processes, including intuition, personal experiences, and personal preferences. This is not necessarily a strong skill set of many engineers but worthwhile knowing that even though an engineer's hard work in designing a product or process, with a concomitantly sound set of plans and specifications, the project may not succeed because of misperceptions or even accurate perceptions if a great plan is not what the client needs. Urban planner, Daniel Burnham, offered both encouragement and solace in this regard [31]:

Make no little plans; they have no magic to stir men's blood and probably will not themselves be realized

Everyone has a way of identifying, estimating, and evaluating risks. However, engineers tend to be more comfortable with "left brained" thinking but must try to communicate with a "right brained" audience. It can be done, so long as preconceived and conventional approaches do not interfere.

16.3.2 Say's Law

If a technology exists, it will probably be used because often supply creates demand [32–36]. The familiar concept of supply following an impetus of need, that is, demand, is often supplanted, according to Say's Law of the Markets. French economist, Jean-Baptiste Say (1767–1832), posited that when a product is produced, the demand will increase because of the aggregation of supply [37]. As evidence, 20 years ago, not many people were calling for a smartphone with the array of features seen today. However, innovators gambled that once it appeared in the marketplace, many people would eventually forget what life was like without it. However,

> Forecasting has always been perilous. Throughout history, most long-term forecasts have been wrong, often blind-sided by the arrival of a new technology like smartphones. But in the coming decades the rate and scale of change will be so great that the future will become mysterious in a new way. So much so that people talk about a coming technological singularity [38].

In the case of this quote, "singularity" has a slightly different connotation than its meaning in mathematics and physics, for example, the center of a black hole becomes so compact that its density becomes infinite [39]. Here, it simply means that rules governing the ordinary and expected no longer apply. Such is the case for some emerging technologies. Artificial intelligence, in many ways, does not follow normal rules of ethics because so little is known by those already affected and even more by those whose lives will change in the future. Those who encourage the rapid growth of the new technology are not necessarily the honest brokers needed, but often single-minded advocates who have a biased, positive view of the benefits and an almost absent view of the costs and risks. For example, singularity for technologies, or technological singularity, is what occurs when the first artificial general intelligence (AGI) appears, that is, a contrivance that can execute any intellectual task that formerly only a human could perform [38]. After singularity, the new technology is hypothetically beyond control and is irreversible. Thus, the threat of AGI is that it represents an irreversible superintelligence that "self-improves" [40]. A few years ago, this would have been in the realms of science fiction, but the likelihood is increasing that we are nearing an exponentially increasing intelligence with a "runaway reaction of cycles" involving ever-refining intelligence generations that will exceed humans' cognitive abilities, including the capability of halting the superintelligence's exponential self-improvement [41].

16.4 Failure

The reasons for failure vary widely, but can be broadly categorized as mistakes, mishaps, and misdeeds. The terms all include the prefix "mis-" that is derived from Old English, "to miss." This type of failure applies to numerous ethical failures. However, the prefix "mis-" can connote that something that is done "poorly," that is, a mistake. It may also mean that an act leads to an accident because the original expectations were overtaken by events, that is, a mishap. This is an all-too-common shortcoming of professionals, that is, not upholding the levels of technical competence called for by their field. Medical and engineering codes of ethics, for example, include tenets and principles related to competence, such as only working in one's area of competence or specialty. Finally, "mis-" can suggest that an act is immoral or ethically impermissible, that is, a misdeed. Interestingly, the theological derivation for the word "sin" (Greek: *hamartano*) means that when a person has missed the mark, that is, the goal of moral goodness and ethical that person has sinned or has behaved immorally by failing to abide by an ethical principle, such as honesty and justice. Bioethical failures have come about by all three means. The lesson from Santayana is that we must learn from all these past failures. Learning must be followed by new thinking and action, including the need to forsake what has not worked and shift toward what needs to be done.

The reliability of a technology must be matched to an adequate margin of safety. For example, if the municipal solid waste incinerator fails and needs to be replaced after 30 years, this is an acceptable margin if the expected operational life is less than 30 years. If its expected life is 20 years and it fails after 17 years, when stakeholders had been assured 99.9999% efficiency for 20 years, this can be perceived as a failure. However, even though the complaints may be justified mathematically, the new costs of repair may be acceptable if the failure rate and margin of safety have been fully disclosed and the reason for the failure adequately explained, for example, an unexpected increase in the town's population above reasonable estimates during the incinerator planning stages. As the planned failure rate diverges in the wrong direction from the actual failure rate, the more likely it is that the project is indeed a failure. For example, if the incinerator fails after 10 years, but its specifications require 20 years, this is likely an unacceptable failure.

Reliability engineering considers the expected or actual reliability of a process, system, or piece of equipment to identify the actions needed to reduce failures, and once a failure occurs, how to manage the expected effects of that failure. Thus, reliability is the mirror image of failure. Risk is the probability of failure, that is, the probability that our system, process, or equipment will fail ("Waste Governance - ScienceDirect"). Therefore, risk and reliability are two sides of the same coin. The technology must have a sufficient operational life based on its design specifications. This graphical representation of engineering reliability is the so-called bathtub curve (see Fig. 16.1). The U-shape indicates that failure will more likely occur at the beginning and near the end of the life of a system, process, or equipment, known as "infant mortality."

Early in the life of a project, technology promoters and adopters seek information about the scale-up and make changes to improve the failure rate, that is, "beta testing." After this stage, the technology remains on the market with increasing reliability, followed by a period of constant reliability. After this optimal period, the failure rate again begins to increase,

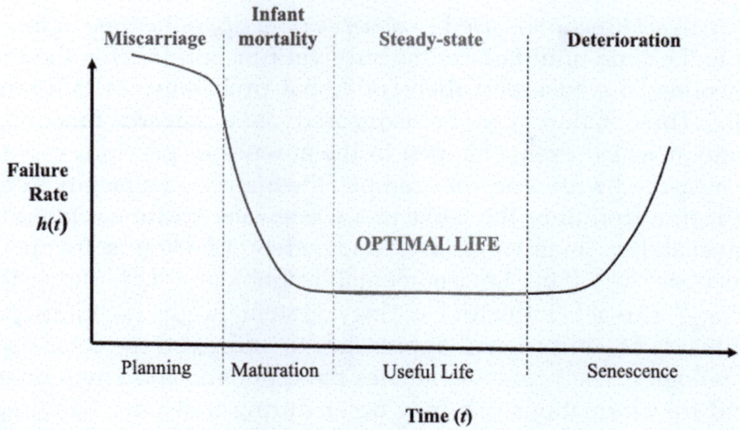

FIGURE 16.1 Prototypical reliability curve known as the bathtub distribution, with an additional stage showing failure early in a technology's planning stage, that is, miscarriage. The highest rates of failure, $h(t)$, occur during the early stages of system implementation and when the systems, processes, or equipment become obsolete or begin to deteriorate. The technologist must foresee potential problems and incorporate factors of safety accordingly, for example, closer monitoring and more extensive safety measures in the early stages of adoption and toward the end of the steady-state period when the failure rates are likely higher. *Data from D.A. Vallero, Fundamentals of air pollution, 6th ed, Academic press, Amsterdam, NV, 2025; D. Vallero, Paradigms lost: learning from environmental mistakes, mishaps and misdeeds, Butterworth-Heinemann, 2005.*

with the indication that the technology is nearing the end of its useful life or that substantial fixes and upgrades must be made to return to the ideal failure rate [42]. Indeed, failure can occur even before acquiring equipment or starting a new project. Many problems occur during the planning and idea stage. A promising energy technology may be terminated prematurely. The miscarriage stage has the highest rate of failure, $h(t)$. It occurs even before the system, process, or equipment has been made a reality. Exclusion of ideas that require a paradigm shift in decision-making or failure to get input about key scientific or social variables can create a high hazard [42–45].

Predicting the useful life of a technology requires credible data and information from which to extrapolate and interpolate from what is known to what is less known, that is, from the data-rich to the data-poor [46]. Risk management models, science-based decision-making, and forward trajectories that begin from laboratory studies can be enhanced with information from experiences of similar technologies elsewhere. Prevention of failure is best, so exhaustive literature reviews may uncover similar failures, especially if the failures have been documented by root cause analysis, decision trees, deconstruction/reconstruction, and failure analysis [47]. An important and common technique to assess the availability and reliability, $R(t)$, of a system is by calculating the mean time between failures (MTBF) [48]:

$$R(t) = \exp\left(-\frac{t}{\text{MTBF}}\right) = \exp(-\lambda t) \tag{16.1}$$

Where, t=operation time and λ=reciprocal of MTBF.

The survival analysis function is used to observe data up to the time of interest. Since our time of interest is the time until failure, the survival function [$S(t)$] is the same as $R(t)$. In reliability engineering, the hazard is the conditional probability of failure of a currently working system. Thus, failure can be expressed as a hazard function, that is, the instantaneous rate at which events happen in the absence of previous events. The hazard function can be measured with time, for example, the lifetime of a facility's components, for example, the cells in a landfill or the tanks in a wastewater treatment plant [49].

Reliability, availability, maintainability, and safety (RAMS) software packages are available to calculate $R(t)$ [50]. Environmental technologies fall under the domain of "industrial ecology" [51–53]. Industrial ecology systematically identifies potential waste streams, contaminants of concern, and approaches to address these concerns, for example, conversion technologies. The process evaluates the economic and environmental effects of using wastes and transformations that may occur during transport, handling, and processing. Depending on the various waste sources, the optimal methods will differ.

The acceptability of a technology or other engineered system is determined and can be evaluated according to the design and operation standards and criteria, that is, functionality, safety, feasibility, sustainability, etc. The reliability can be compared in the forward direction using models, life cycle analyses, cost-effectiveness studies, sensitivity analysis, and uncertainty analyses. When something goes wrong, inverse methods are applied to investigate the contributing factors to the failure, for example, root cause analysis [54].

Engineers must optimize designs for risk reduction and system reliability, but with attention to resilience [55–57]. Indeed, resilience is a key part of risk and reliability. A system that is overly sensitive to perturbations is risky and unreliable, requiring that downstream users must constantly deploy controls to render the system sufficiently safe and reliable. Conversely, a more resilient system can withstand a wider range of stressor intensity that returns to optimal conditions relatively quickly. A system that lacks sufficient resilience is dangerously sensitive and vulnerable to changing conditions. Pollutants can vary considerably. This calls for ongoing monitoring of all aspects of treatment processes. Over the operational life, environmental and operational conditions change, so that a resilient system is likely to be less expensive and in need of fewer repairs and replacements than a more inflexible system.

Resilience must be built into risk and reliability specifications. The systems must be able to return to their original form or state after being bent, compressed, or stretched, or they must have the ability to recover and return to the desired state readily following the application of some form of stress [56–58]. Of course, good practice and an eye to O&M have long been part of the engineer's tool kit, though they likely would not have readily thought of themselves as resilience engineers ("[PDF] Communicating risk and resiliency: Special considerations for …"). The specifications of a technology or item include tolerance ranges during the operational lifetime. For example, an item may perform its function for 10 years, so long as the environmental conditions, like temperature, pressure, corrosiveness, and other operational factors, remain within specified tolerances. Thus, when the engineer finds ways not only to lengthen the lifetime but also to broaden the tolerances, the design becomes more resilient. To achieve this, the item must become [59]:

1. more robust by widening the upper-to-lower performance limits;
2. more resistant to, and recoverable from, disruptions;

3. reconfigurable by incorporating adaptive properties; and,
4. more adaptable to changing conditions.

Thus, the engineer must be an eager and diligent learner, that is, attentive to changes and disruptions to anticipate future problems early or even before they occur [59,60].

Resilient systems have wide tolerances, compared to an overly sensitive system that is readily harmed when conditions fall outside of narrow tolerance ranges. For example, the hazardous waste treatment system in the marketplace must have much greater resilience than when one of its components was in nascent, highly controlled lab clean room stage. Real-world conditions vary during the operational life of the renewable.

Ecology has taught us that organisms in a resilient habitat are able to survive in broad ranges of moisture, temperature, light, and nutrient concentrations. For example, exposure to a pollutant in a biologically diverse ecosystem may lead to little harm or a short time to recover, whereas the same pollutant concentration in a nonresilient, less diverse system, for example, monoculture corn crop, could overwhelm the ecosystem [61]. The stressors on a habitat or ecosystem can be synergistic, for example, although drought conditions could be withstood, the addition of the deposited pollutant causes an ecological disaster [62]. Metaphorically, a waste technology that builds in resilience is in less need of intricate engineering controls against insult than a nonresilient system [55,63,64]. To be widely adopted, this is crucial. It is important to keep these in mind when selecting a treatment technology, such as those described in the following section.

At least two more "R-terms" can be added to this list of engineering metrics related to climate change, that is, robustness and redundancy [65]. Though related, resilience and robustness differ in that resilience is reactive, whereas robustness is proactive. The time it takes for a system to recover from a failure is its resilience. If a system is inherently strong and resists insults well, it is robust. Historically, the atmosphere has been robust and resilient, but in recent decades it has shown difficulty in recovering from the chemical loads of modern societies. Ways of improving its resilience by making it more robust are currently limited. Geoengineering and other large-scale solutions being proposed all seem to do little to improve the robustness of the atmosphere itself, but there are ways to make adaptation and response to the threats more robust [66–72]. These include better land use planning and conservation to protect sensitive habitats, improving crop management to protect the food supply, reduced greenhouse gas (GHG) emissions, and plans to address potential population migrations.

Engineering solutions must always expect failures to occur in time and space, so two more "R-terms" must be added to this discussion, that is, robustness and redundancy [65]. Improved reliability lengthens the time to failure but does not eliminate it. An acceptably redundant system acknowledges this and builds in safeguards and redundant systems to meet the failures when they occur [68,70,73–75]. Though related, resilience and robustness differ in that resilience is reactive, whereas robustness is proactive. The time for a system to recover from a failure is its resilience. If a system is inherently strong and resists insults sufficiently, it is robust. Historically, the Earth's water systems have been robust and resilient, but since the Industrial Revolution, they have increasingly shown difficulty in recovering from the pollutant loads of modern societies. Building in robustness to systems to handle increasing pollutants is much less desirable than preventing pollutants from entering the environment. Assuming increasing amounts of pollutants and then applying large-scale solutions like geoengineering would do comparatively little to improve the robustness of the hydrosphere and atmosphere, but would potentially lead to irreversible adverse consequences [66–72]. Improved land use planning, cleaner fuels, and pollution prevention strategies are preferable for sustainable and resilient ecosystems.

Redundancy is also important to engineering resilience. Backup systems for pollution control equipment during low-probability events, for example, prevent and shorten failures. Indeed, engineered systems must be prepared for failures to occur in time and space. Improved reliability lengthens the time to failure but does not eliminate it. An acceptably redundant system acknowledges this and builds in safeguards and redundant systems to meet the failures when they occur [68,70,73–75].

Engineers must continue to design and devise innovative and larger-scale solutions to the direct and indirect problems brought on by environmental issues, many of which we do not yet know are even problems. Engineers tend to be technological optimists and agents of change. [76] Binary perspectives can be obstacles to optimization. Engineers often can satisfy seemingly incongruent perspectives to protect the public *and* provide practical, sustainable, and cost-effective solutions to the problems now and those that will unfold in the future. After all, the first canon of the engineering profession is to hold paramount the safety, health, and welfare of the public. Engineers must be faithful agents to their clients [77]. Their principal client is the public, and the public expects a climate that supports healthy human lives and sustainable ecosystems.

When selecting a technology, one of the selection criteria should be considerations of what remains after the operational life has ceased, for example, will long-term disposal or special handling be needed (such as radioactive isotopes in monitoring equipment and potentially toxic materials in control equipment filters and cartridges)? Obviously, the best strategy is to have nothing to address, that is, zero waste at end of life. The design of disassembly (DfD) and design for recycling (DfR) approaches require that the entire life cycle be considered before use, especially the times near the end of useful life of the system.

Cause-and-effect diagrams identify and characterize the totality of causes or events that contribute to a specified outcome event. For example, Fig. 16.2 shows the "fishbone" diagram, which arranges the categories of all causative factors according to their importance (i.e., their share of the cause). The construction of this diagram begins with the failure event to the far right (i.e., the "head" of the fish), followed by the spine (flow of events leading to the failure). The "bones" are each of the contributing categories. This can be a very effective tool in explaining failures to clients and the public. Even better, the engineer may construct the diagrams in "real time" at a design meeting. This will help to open the design for peer review and will help to gain insights, both good and bad, early in the design process [78].

The premise behind cause-and-effect diagrams like the fishbones and other failure logic graphics is that all the causes must connect through a logic gate. This does not always hold true, so another, more qualitative tool may need to be deployed, such as the Bayesian belief network (BBN). Like the fishbone, the BBN starts with the failure (see Fig. 16.3). Then, the most immediate contributing causes are linked to the failure event. The next group of factors that led to the immediate causes is then identified, followed by the remaining contributing groups. This diagram helps to identify and weight the contributing factors, allowing for investigations of how one group of factors impacts the others.

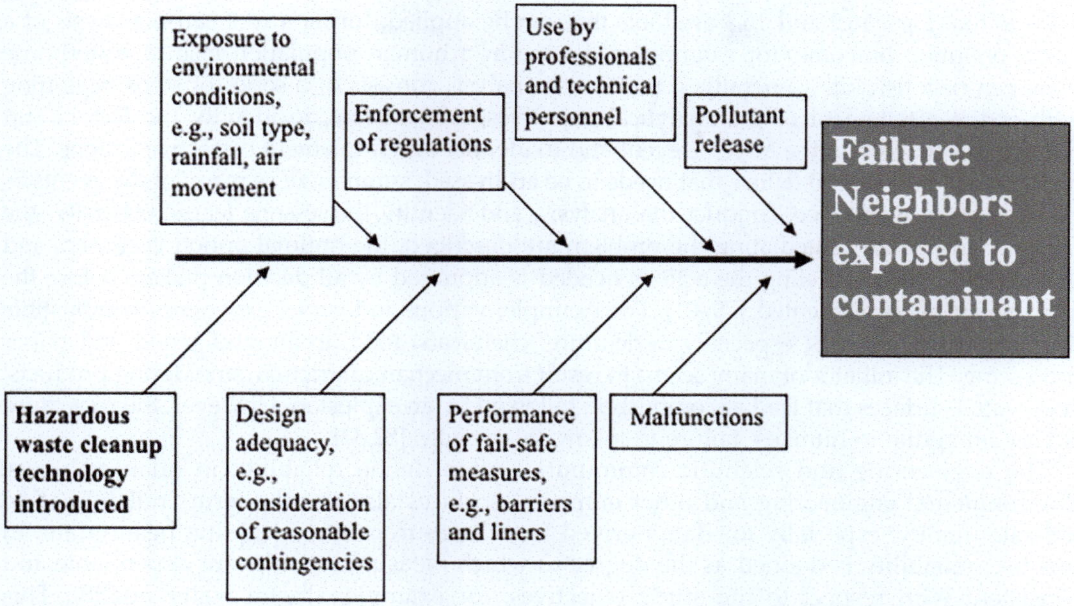

FIGURE 16.2 Fishbone reliability diagram showing contributing causes to an adverse outcome (exposure to an environmental contaminant). *Data from D.A. Vallero, P.A. Vesilind, Excerpted and adapted from, edSocially responsible engineering: Justice in risk management, Wiley Online Library, 2007.*

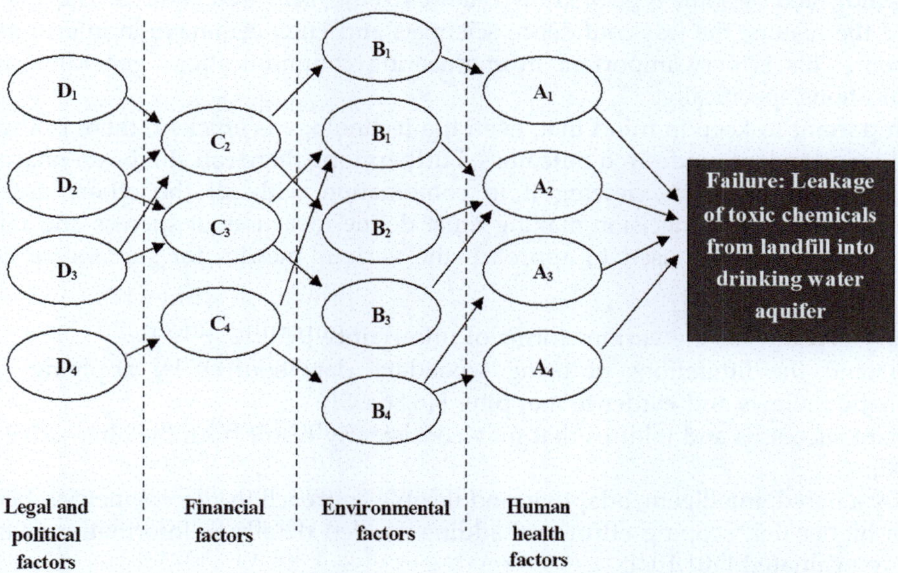

FIGURE 16.3 Bayesian belief network, with three groups of contributing causes leading to a failure. *Data from D.A. Vallero, P.A. Vesilind, Excerpted and adapted from, edSocially responsible engineering: Justice in risk management, Wiley Online Library, 2007.*

Note that Figs. 16.2 and 16.3 are tools that can be applied during a root cause analysis of a water pollution outcome, for example, a disease in a human population from a waterborne pathogen or a fish kill. Generally, a root cause analysis consists of a series of steps beginning with a direct statement of the analytical objectives, for example, to identify the factors and occurrences that led to the adverse event and to identify and implement corrective actions. The next step is to define the failure that needs to be addressed, summarize as many of the events as possible in an event tree, conduct the analysis, and identify the lessons learned. Finally, the realistic options for remediating the problem are described, the optimal option is chosen, and the schedule for completing the actions needed is approved by all decision makers before the actions can be implemented [79–81]. For example, various industries experience combustible dust explosion hazards, especially agriculture, chemicals, food, grain, and wood and paper processing. The initiator of many seems to result from mechanical friction, smoldering products, and heated surfaces that lead to combustion followed by an explosion. However, the root cause is very often due to humans' failure to follow safety rules [82,83].

The engineering and scientific communities often define reliability in nuanced ways. Environmental engineering and other empirical sciences also use the term "reliability," to indicate quality, especially for data derived from measurements, including health data. In this use, reliability is defined as the degree to which measured results are dependable and consistent with respect to the study objectives, for example, stream water quality. This specific connotation is sometimes called "test reliability" in that it indicates how consistent measured values are over time, how these values compare to other measured values, and how they differ when other tests are applied. Test reliability, like engineering reliability, is a matter of trust. As such, it is often paired with test validity, that is, just how near to the true value (as indicated by some type of known standard) the measured value is. The less reliable and valid the results, the less confidence scientists and engineers have in interpreting and using them. This is very important in engineering communications generally, and risk communications specifically.

It is important to keep in mind that, even if a technology is efficient, there is always the potential for attendant, undesired outcomes. Furthermore, there can still be an unacceptable cumulative outcome when viewing it in combination with all the other "acceptable" projects. In other words, decision-making must define measures of success and failure. A multistep process can be used to approach the various metrics for success for complex systems [84]:

1. Properly account for the various forms of uncertainty [85];
2. Understand the limitations of using secondary data sources by applying rigorous systematic reviews and evidence mapping [46,86–98];
3. Compare successes and failures that may consistently result from the alternative actions [99,100];
4. Employ a tiered, intelligent, adaptive, and flexible approach to success metrics; beginning with a higher-tier, scoping effort and adding needed details as information is continuously re-evaluated [101,102];
5. Evaluate the value of information on which decisions will be based, including assignment of quality and relevance thresholds below which the data and information should be avoided or, at a minimum, given appropriate caveats [103].

6. Validate and evaluate the models and other tools used to assess success [104]; and,
7. Consider the cumulative impacts, that is, physical, chemical, and biological [105,106], in water and wastewater decisions and assessments, and the effect any proposed actions will have on the integrity of ecosystems in time and space. A practical means of estimating this is to apply ecosystem indices and models [107].

The feasibility of a technology goes beyond sufficient assessments of risks, reliability, and the costs and benefits of a renewable energy project. A more inclusive definition of costs and benefits must be applied, that is, not simply financial and monetary costs to achieve a social need. Indeed, costs are often better defined as a subset of risks, that is, safety risks, health risks, environmental risks, health risk, safety risks, as well as financial. Indeed, the same technology may decrease economic and safety risks, while impairing environmental resources and public health. Or, a technology may enhance the environmental, aesthetic, and/or public health conditions for one community but harm the environment of another community, even within the same governmental jurisdiction [78,108,109].

Engineering failure can be categorized into five types. Whether the failure is deemed unethical is determined by the type of failure and the circumstances contributing to the failure.

16.4.1 Failure Type 1: mistakes and miscalculations

Occasionally, engineers make mistakes, and their works fail due to their own miscalculations, such as when parentheses are not closed in computer code, leading to errors in predicting pharmacokinetic behavior of a drug or a miscalculation of wind shear. Some failures occur when engineers do not correctly estimate the corrosivity that occurs during sterilization of medical devices, for example, not properly accounting for fatigue of materials resulting from high temperature and pressure of an autoclave. Such mistakes are completely avoidable if the physical sciences and mathematics, often simply arithmetic and unit conversions, are properly applied. Avoiding such errors is reflective of engineering competence and the need for thorough review by other component engineers and experts prior to finalization of plans.

16.4.2 Failure Type 2: extraordinary natural circumstances

Failure can also occur when factors of safety are exceeded due to extraordinary natural occurrences. Engineers can, with fair accuracy, predict the probability of failure due to natural forces like wind loads, and they design the structures for some maximum loading, but these natural forces can be exceeded. Engineers design for an acceptably low probability of failure—not for 100% safety and zero risk ("Engineering Risks and Failures: Lessons Learned from Environmental ..."). However, tolerances and design specifications must be defined as explicitly as possible.

Unfortunately, previous engineering and governmental decisions can increase the likelihood of failure in a natural disaster, such as poorly planned road systems, lack of available water supplies, poor forest management to reduce excess vegetation and underbrush, and inadequate evacuation planning before the California wildfires in January 2025. This is an instance where an emerging technology can decrease the likelihood of failure, such as improving wildfire detection using large language models (LLMs) and other emergent technologies [110–112].

The tolerances and factors of safety must match the consequences. For example, while a failure rate of 1% may be acceptable for the general-use light bulb, it is grossly inadequate for the light system (laser) used in medical systems. Furthermore, the failure rate of devices may spike up dramatically during an extreme natural event (e.g., power surges during storms). Equipment failure is just one of the factors that lead to medical failure, so resilience and redundancy must be built in for systems that can affect life and limb, but not so much for the need to change a light bulb in a home.

A study [113], for example, identified the reasons for relatively high failure rates associated with 180-degree selective laser trabeculoplasty (SLT) to treat intraocular pressure (IOP) in glaucoma patients. In this procedure, a highly focused beam of light is used to treat the eye's drainage angle, that is, the point in the eye where the iris and sclera meet. The researchers suggested that the poor results can be attributed to:

1. The patients studied had advanced glaucoma and were on an average of two to three medications.
2. The baseline IOP was lower than that reported in other studies, and success is less likely with a lower baseline IOP.
3. 75% of eyes were on a prostaglandin, which increases uveoscleral outflow and may result in hypoperfusion of the trabecular meshwork, thus decreasing the possible action of SLT.
4. SLT is thought to upregulate metalloproteinases similar to prostaglandins, and the prior use of prostaglandins may decrease this action of SLT and reduce its efficacy.

Note that none of the postulated reasons for failure were associated with the equipment itself, but primarily with the physiological and anatomical features of the patient. In fact, devices *should* be one of the smallest contributors to risk since they are in the span of control of the operator. At the very least, the failure rate should be characterized as best as possible by the designers and clinical engineers. Conditional probabilities of failure should be known to optimize the application of any equipment, which is relied upon for healthcare, environmental protection, and safety. That way, backup systems can be established in the event of extreme natural events, like earthquakes and tornadoes. If appropriate contingency planning and design considerations are factored into operations, the engineer's device may still fail, but the failure would be considered reasonable under the extreme circumstances. ("Engineering Risks and Failures: Lessons Learned from Environmental …")

16.4.3 Failure Type 3: critical path

No engineer can predict all the possible failure modes of every structure or other engineered device. ("Engineering Risks and Failures: Lessons Learned from Environmental …") Unforeseen situations can occur. A classical case is the Holy Cross College football team hepatitis outbreak in 1969 [114]. Circumstances, which could not have been foreseen, occurred that allowed water contaminated with hepatitis virus to enter a drinking water system. A water pipe connected the college football field with the town, passing through a golf course. Children had opened a water spigot on the golf course, splashed around in the pool they created, and apparently discharged hepatitis virus into the water. A low pressure was created in the pipe when a house caught on

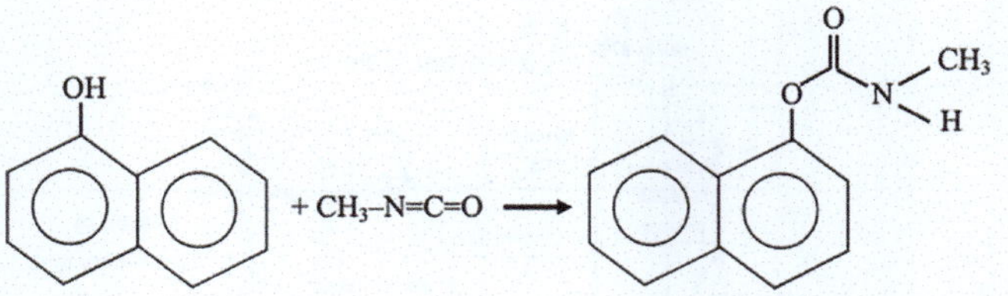

FIGURE 16.4 Chemical reaction producing methyl isocyanate at the Union Carbide plant in Bhopal, India.

fire and water was pumped out of the water pipes. This low pressure pulled the hepatitis-contaminated water into the water pipe. The next morning, the Holy Cross football team drank water from the contaminated water line, and many came down with hepatitis. The case is memorable because it was so highly unlikely—a combination of circumstances that were impossible to predict. Nevertheless, the job of engineers is to do just that, to try to predict the unpredictable and thereby to protect the health, safety, and welfare of the public [115,116].

The Holy Cross case demonstrates how engineers can fail but may not be blamed for the failure. If the public or their peers agree that the synergies, antagonisms, and conditional probabilities of the outcome could not reasonably be predicted, the engineer is likely to be forgiven. However, if a reasonable person deems that a competent engineer *should* have predicted the outcome, the engineer is, to that extent, accountable ("Engineering Risks and Failures: Lessons Learned from Environmental ...").

16.4.3.1 The Bhopal disaster

The tragic industrial accident at Bhopal, India, is one such engineering failure [117–123].[3] Perhaps the biggest air pollution disaster of all time occurred in Bhopal, India, in 1984 when a toxic cloud drifted over the city from the Union Carbide pesticide plant. This gas leak managed to kill 20,000 people and permanently injure an enormous population of 120,000. We often talk about a failure that results from not applying the science correctly, for example, the Type 1 mathematical error and an incorrect extrapolation of a physical principle. Another type of failure results from misjudgments of human systems. Bhopal had both.

Although the Union Carbide Company was headquartered in the United States, as of 1984, it operated in 38 different countries. It was quite large (35th largest US company) and was involved in numerous types of manufacturing, most of which involved proprietary chemical processes. The pesticide manufacturing plant in Bhopal, India, had produced the insecticides, Sevin and Cararyl since 1969, using the intermediate product, methyl isocyanate (MIC) in its gas phase [125]. The MIC was produced by the reaction shown in Fig. 16.4. This process was highly cost-effective, involving only a single reaction step. The schematic of the MIC process is shown in Fig. 16.5. MIC is highly water reactive (see Table 16.1); that is, it reacts violently with water, generating a very strong exothermic reaction that produces carbon dioxide. MIC vaporizes into a highly toxic gas that, when concentrated, is highly caustic and burns biological tissues. This can lead to scalding nasal and throat passages, blinding, and loss of limbs, as well as death.

On December 3, 1984, the Bhopal plant operators became concerned that a storage tank was showing signs of overheating and began to leak. The tank contained MIC. The leak rapidly increased in size, and within one hour of the first leakage, it exploded and released

[3] This section is largely taken from: [124].

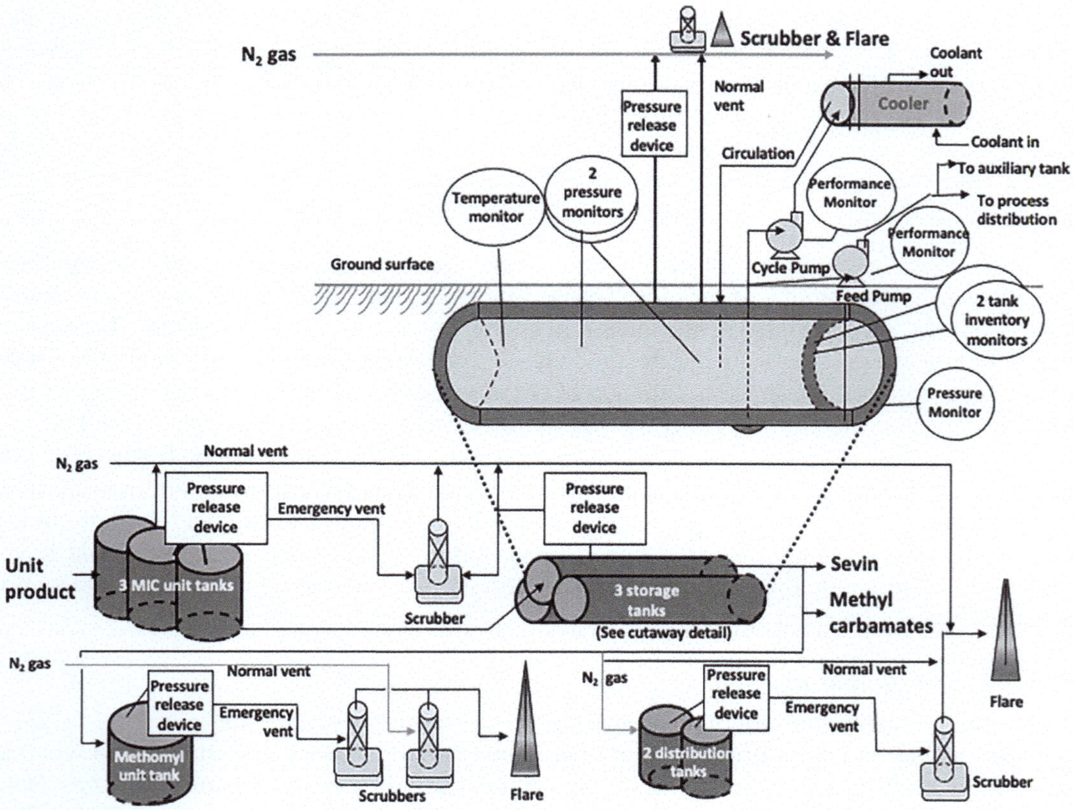

FIGURE 16.5 Schematic of methyl isocyanate processes at the Bhopal, India, plant (c. 1984). *Data from W. Worthy, Methyl isocyanate-the chemistry of a hazard, Chemical & Engineering News 63 (6) (1985) 27–33. [126].*

approximately 80,000 lbs (4×10^4 kg) of MIC into the atmosphere. The release led to arguably the worst industrial disaster on record. The human exposure to MIC was widespread, with half a million people exposed. Nearly 3000 people died within the first few days after the exposure, 10,000 were permanently disabled. 10 years after the incident, 12,000 death claims had been filed, along with 870,000 personal injury claims. However, only $90 million of the Union Carbide settlement agreement had been paid out.

Introduction of water to the MIC storage tank resulted in a highly exothermic reaction generating CO_2, which would have led to a rapid increase in pressure, which could have caused the release of 40 metric tons of MIC into the atmosphere. As of 2001, many victims did receive compensation, averaging about $600 each, although some claims are still outstanding.

The Indian government required that the plant be operated exclusively by Indian workers, so Union Carbide agreed to train them, including flying them to a sister plant in West Virginia for hands-on sessions. In addition, the company required that US engineering teams make periodic on-site inspections for safety and quality control, but these ended in 1982, when the plant decided that these costs were too high. So, instead, the US contingency was only responsible for budgetary and technical controls, but not safety. The last US

TABLE 16.1 Properties of methyl isocyanate.

Property	Experimental average	Predicted average	Experimental median	Predicted median	Experimental range	Predicted range	Unit
Log octanol-water coefficient	–	0.5	–	0.8	–	–0.332 to 0.881	–
Melting point	–59.0 (5)	–53.5	–45.0	–44.9	–80.0 to –45.0	–77.6 to –38.0	°C
Boiling point	39.3 (3)	44.9	39.4	40.7	39.0 to 39.5	35.0 to 63.3	°C
Water solubility	–	2.8	–	0.7	–	0.512 to 7.00	mol L^{-1}
Flash point	–	5.6	–	5.6	–	–6.67 to 17.8	°C
Viscosity	–	0.4	–	–	–	0.440	cP
Vapor pressure	348 (1)	362.1	–	344.9	348	210 to 531	mm Hg
Density	–	0.9	–	0.9	–	0.881 to 0.936	g cm^{-3}
Surface tension	–	28.3	–	–	–	28.3	dyn cm^{-1}
Index of refraction	–	1.4	–	–	–	1.39	–
Molar refractivity	–	15.5	–	–	–	15.5	cm^{-3}
Polarizability	–	6.1	–	–	–	6.14	$Å^{-3}$
Molar volume	–	64.8	–	–	–	64.8	cm^{-3}
Log octanol-air coefficient	–	1.0	–	–	–	0.967	–
Henry's Law	–	0.0	–	–	–	5.62e-5	atm-m^{-3} $mole^{-1}$

Data U.S. Environmental Protection Agency. "Methyl isocyanate 624-83-9 | DTXSID1023786." https://comptox.epa.gov/dashboard/dsstoxdb/results?search=DTXSID1023786#properties (accessed October 14, 2020, 2020). [127].

inspection in 1982 warned of many hazards, including a number that have since been implicated in contributing to the leak and release.

From 1982 to 1984, safety measures at the plant declined, attributed to high employee turnover, improper and inadequate training of new employees, and low technical savvy in the local workforce. On-the-job experiences were often substituted for reading and understanding safety manuals. In fact, workers would complain of typical acute symptoms of pesticide exposure, such as shortness of breath, chest pains, headaches, and vomiting, yet they would typically refuse to wear protective clothing and equipment. The refusal in part stemmed from the lack of air conditioning in this subtropical climate, where masks and gloves can be uncomfortable. For the most part, compared to regulated companies in the United States and the West, less stringent Indian safety and other standards were applied at the plant after 1982, and the ineffective enforcement of even those standards was among the contributing factors [83,118,119,123,128–130].

The Bhopal case is an archetype of a "confluence of events" leading to a disaster. The failures uncovered in the root cause analysis are still quite instructive for industrial community and warnings to the need for due diligence in governmental oversight. The major findings included [83]:

- The tank that initiated the disaster was 75% full of MIC at the outset.
- A standby overflow tank for the storage tank contained a large amount of MIC at the time of the incident.
- A required refrigeration unit for the tank was shut down five months prior to the incident, leading to a three to fourfold increase in tank temperatures over expected temperatures.
- "One report stated that a disgruntled employee unscrewed a pressure gauge and inserted a hose into the opening (knowing that it would do damage, but probably not nearly the scale of what occurred)." ("Failure Analysis - an overview | ScienceDirect Topics").
- A new employee was told by a supervisor to clean out connectors to the storage tanks, so the worker closed the valves properly, but did not insert safety discs to prevent the valves from leaking. In fact, the worker knew the valves *were* leaking, but they were the responsibility of the maintenance staff. Also, the second-shift supervisor position had been eliminated.
- When the gauges started to show unsafe pressures, and even when the leaking gases started to sting mucous membranes of the workers, they found that evacuation exits were not available. There had been no emergency drills or evacuation plans.
- The primary fail-safe mechanism against leaks was a vent-gas scrubber, that is, normally, this release of MIC would have been sorbed and neutralized by sodium hydroxide ($NaOH$) in the exhaust lines, but on the day of the disaster, the scrubbers were not working. (The scrubbers were deemed unnecessary, since they had never been needed before.)
- A flare tower to burn off any escaping gas that would bypass the scrubber was not operating because a section of conduit connecting the tower to the MIC storage tank was under repair.
- Workers attempted to mediate the release by spraying water 100 feet high, but the release occurred at 120 feet.

Thus, according to the audit, many checks and balances were in place, but the cultural considerations were ignored or given low priority, such as when the plant was sited, the need to recognize the differences in land use planning and buffer zones in India compared to Western nations, or the differences in training and oversight of personnel in safety programs.

Every engineer and environmental professional needs to recognize that much of their responsibility is influenced by geopolitical realities and that we work in a global economy. This means that engineers must have respect and appreciation for how cultures differ in their expectations of environmental quality. One cannot assume that a model that works in one setting will necessarily work in another without adjusting for differing expectations. Bhopal demonstrated the consequences of ignoring these realities. Chaos theory tells us that even very small variations in conditions can lead to dramatically different outcomes, some of which are disastrous.

Dual use and bioterrorism fears can be seen as somewhat analogous to the lack of due diligence at Bhopal. For example, extra care is needed in using similar strains and species in genetically modified microbes (e.g., substantial differences and similarities in various strains of *Bacillus spp.*). The absence of a direct analogy does not preclude that even a slight change in conditions may elicit unexpected and unwanted outcomes (e.g., weapon-grade or resistant strains of bacteria and viruses).

Characterizing as many contingencies as possible and potential outcomes in the critical path is an essential part of many biohazards. The Bhopal incident provides this lesson. For example, engineers working with bioreactors and genetically modified materials must consider all possible avenues of release. They must ensure that fail-safe mechanisms are in place *and* are operational. Quality assurance officers note that testing for an unlikely but potentially devastating event is difficult. Everyone in the decision chain must participate in monitoring, identifying, and removing flaws. Even if no incidents have yet occurred (thankfully), it means that no one really knows what will happen in such an event. That is why health and safety training and prospective "stress testing" are critical parts of the engineering and technology development processes.

The fact that no incidents have yet occurred (thankfully) means that no one really knows what will happen in such an event. That is why health and safety training is a critical part of every phase of a project or process.

Smaller versions of the Bhopal incident are more likely to occur, but with more limited impacts. For example, two freight trains collided in Graniteville, South Carolina, USA, just before 3:00 a.m. on January 6, 2005, resulting in the derailment of three tanker cars carrying chlorine (Cl_2) gas and one tanker car carrying sodium hydroxide (NaOH) liquid. The highly toxic Cl_2 gas was released into the atmosphere. The wreck and gas release resulted in hundreds of injuries and eight deaths.

The Bhopal case is relevant to Type 3 failures even today. It was interesting in the latest prospectus review that one of the comments was that this case is outdated. It did occur four decades ago. However, the lessons continue to be relevant to technology applications, especially the importance of due diligence and the reminder that black swan events do occur.

16.4.4 Failure Type 4: negligence

In many ways, the first three failure types discussed have a component of negligence. Given the first canon of engineering professions, engineers must protect the public from the carelessness of the decisions and actions of their most careless members. The case of the person who attempted to open a 2-L soda bottle by turning the aluminum cap the wrong way, with a pipe wrench no less, and having the cap fly off and into her eye, is a famous example of unpredictable ignorance. "She sued for damages and won, with the jury agreeing that the design engineers should have foreseen such an occurrence." ("Engineering Risks and Failures: Lessons Learned from Environmental ..."). Engineering solution: Since then, plastic caps have interrupted threads that cannot be stripped by turning in the wrong direction.

Designs must consider misuse and abuse. For example, most municipalities hire well-qualified water and wastewater treatment plant operators. However, in the design of treatment plants, engineers are taught to design the plants so that it is easy to do the right thing and very difficult to do the wrong thing [115]. This includes:

- Color-coding and well-defined identification: piping, valves, and components are color-coded, for example, red for hot, blue for cold, and labeled to delineate flow direction, maintenance points, and chemical types.
- Contamination prevention and misuse: nonpotable water systems are strictly and clearly labeled to indicate that the water is unsafe for drinking or washing, and other steps to prevent unintentional access consumption.
- Automation: devices can feature built-in, secure mechanisms, such as item replacement indicators and self-cleaning functions.
- Adherence to safety standards: labeling must comply with US Environmental Protection Agency, Office of Safety and Health, and other regulatory requirements to ensure the safety of everyone who ventures into the plant, for example, any accessible water is correctly identified as potable or nonpotable.
- Improved equipment: including self-cleaning pumps; long-lasting cartridges, and resilient systems, for example, able to accommodate drops or freezing and lessening the need for complex maintenance.

These and other measures have been referred to as making the treatment plant "operator proof" [115]. This certainly is not limited to treatment plants and is a nearly universal expectation. Design engineers recognize that if something can be done incorrectly, sooner or later it will happen, and that they must minimize such possibilities.

Another example of negligence occurred in the early 1970s, when jet-powered airliners were replacing propeller aircraft, the fueling system at airports was not altered, and the same trucks fueled both types of craft. The nozzle fittings for both types of fuels were therefore the same. A tragic accident occurred near Atlanta, where jet fuel was mistakenly loaded into a Martin 404 propeller craft [131]. The engines failed on takeoff, resulting in fatalities. A similar accident occurred in 1974 in Botswana with a DC-4 [132] and again in 1976 near Fairbanks, Alaska, with a DC-6 [133]. Engineers should have recognized that increasing the number of signs and training was not enough to prevent such tragedies. The solution was to modify the fuel delivery systems so that it was impossible to put jet fuel into a propeller-driven airplane and vice versa. A more recent example of how this can be done is the modification of the nozzles used in gasoline stations. The orifice in the gas tank is too small now to take the nozzles used for either leaded fuel or for diesel fuel.

16.4.5 Failure Type 5: lack of imagination

Every time something fails, whether a manufactured product (such as a tire blowout) or a constructed facility (such as a bridge collapse), it is viewed as an engineering failure. The job of engineers has historically been to predict the problems that can occur and to design to minimize these events, protecting people's health, safety, and welfare by avoiding design errors, building resilient systems to withstand natural events, designing for unforeseen events, and visualizing future misuses, abuses, ignorance, and carelessness.

This objective of engineers changed on September 11, 2001, with the attack on the United States by terrorists. Today, the engineers are also required to protect the health, safety, and welfare of the public from acts of terrorism. It had yet to enter the minds of most engineers that they would have a responsibility of protecting people from those who would want to intentionally harm other people or to intentionally destroy public facilities. This was a completely new failure mode in engineering. Sarah Pfatteicher calls such failures "intentional accidents," or failures resulting from intentional actions [134,135].

Engineers found themselves in the position of having to address these intentional accidents. Military engineers, of course, have had to design against such destructive actions since the days of moats and castles, but those were all structures built explicitly to withstand attack. Civilian engineers have never had to think in these terms but are now asked to design structures for this contingency. Engineering and engineers face a new challenge—to prevent such "accidents" on civilian targets by terrorists bent harming to the public. We recommend that the response to this threat requires a two-pronged approach— *technical* and *social*.

16.4.5.1 Biotechnology, biometrics, and bioterrorism

The engineering community has responded enthusiastically to the call for better technology to fight terrorism and crime. This effort has included developing sensing devices with continuously improved resolution and sensitivity to detect explosives and weapons, enhanced communication systems to warn and predict when terrorist activity might occur, better identification systems for suspect individuals, innovative ways of identifying used explosives, and many more such advances and improvements. An issue of *Prism*, the magazine of the American Society of Engineering Education, had several articles summarizing technology that can be brought to bear in the fight against terrorism, and these developments no doubt will have a beneficial effect [136]. But the use of such technology also signals four potential problems.

First, the application of surveillance technologies will almost always diminish or adversely alter our sense of liberty. Witness, for example, the facial identification program used on attendees at Super Bowl XXXV [137]. Protests by the manufacturer and the security personnel to the contrary that such equipment only identifies those faces that are on file; many people were upset about having their pictures taken. We Americans are just not used to having people or machines knowing where we are and what we do, and we resent such an intrusion on our privacy. We believe, and rightly so, that equipment and procedures such as face identification have the potential for abuse and loss of our liberty. This conflict between security and liberty is an example of competing values in ethics. For example, security can be seen as a necessary end, but some of the means to achieve this end will detract from liberty. This points to one of the problems with utilitarian ethics. Likewise, the

United States was founded on principles espoused by rights ethicists like John Locke (1632–1704). In fact, three of these rights, life, liberty, and the pursuit of happiness, are foundational to the US Constitution. Recently, however, the words of Benjamin Franklin have taken on added currency [138]:

They that can give up essential liberty to obtain a little temporary safety deserve neither liberty nor safety.

Thus, the balance between security and liberty can be viewed as an "optimization problem" between these two societal variables.

The effect of security at airports and public places continues to be bothersome. However, most people recognize that some security, even loss of certain previously held privacy, is part of the post-9/11 world. However, the threshold of where security has eroded liberty is ill-defined.

Second, a malicious person with the financial wherewithal and knowledge can circumvent most technologies. Witness the scourge of viruses in computers. As soon as the antivirus programs are modified to remove the latest virus, unknown persons with devious minds and malicious intent can develop new viruses that get around the defenses. So, it is with antiterrorist technology. Sophisticated hackers can devise ways of getting around our present safety procedures. It is a continuing chess match of cyber warfare.

The third problem is that hackers and terrorists can use an existing technology to create havoc; the prime example being the use of fully loaded airplanes on September 11. The terrorist organizations have misused internet banking to pay organizers and combatants, accessed defense communication networks to keep in touch, and devised sophisticated explosives to kill innocent people—technologies long developed as part of a nation's national defense. The more beneficial technologies are produced, the more likely they are towreak harm by misuse [124].

The fourth problem with the development of emerging antiterrorist and surveillance technologies is that, by definition, they are emergent. When first introduced, most engineers and technical decision makers have yet to develop an aptitude for their uses and vulnerabilities. As evidence, few resources that would protect us from low-technology assaults have been developed to date. Traditionally, our research funds have been spent either to enhance our own health or to develop sophisticated weapons for countering threats from similarly technically sophisticated enemies. For example, the use-inspired and applied research budgets of the National Institutes of Health and the Department of Defense greatly surpass the budgets for basic science research efforts of the National Science Foundation. Layered defense shields, like the proposed Gold Dome the missile defense shield, are not protective against suicide bombers, for example. Compared to defense contracts, little research has been conducted on how to protect the safety of the public from terrorist threats. Unfortunately, even if such new technologies had been developed, the saboteurs would simply use them for their own purposes, thus negating the original intent.

We must conclude, therefore, that better technology cannot be the only, or maybe not even a primary, engineering response to the threat of terrorism and crime. Bad actors with funding and skill can get around our technology and sometimes can use our own technology against us.

This is not to argue against the use of antiterrorism technology but to suggest that such technology is only part of the answer. It is not possible for engineers and officials to prevent all such deeds, just as it is not possible for engineers to make anything 100% safe from other kinds of failure. In a manner of speaking, this is analogous to our reliance on pharmaceuticals to treat diseases. One of the fears of medical researchers is that we may destroy most of the microbes, but the remaining strains rapidly become resistant to even the newest antibiotics. So, physicians try not to prescribe these drugs indiscriminately and to use a holistic approach to prevent certain contagious diseases. No matter what technology we employ, the possibility always remains that terrorists can cause great harm by using the very same technology against us. We cannot be completely "inoculated" against these diseases, nor can we expect "quick technological fixes" every time. Engineering technology is therefore only partially able to protect the health, safety, and welfare of the public from intentional destruction.

16.5 Success

There are two approaches to solving engineering problems; attack the symptoms or attack the cause, and often the best solution to a problem is to change what is causing the it. For example, in hazardous waste treatment, it is almost always cheaper and safer to travel "up the pipeline" to find out where the waste is coming from and then change the process to eliminate or minimize this hazardous material. The alternative option, which is almost always more expensive, is to treat the waste once it arrives at the treatment plant.

The first step in finding solutions to engineering challenges is to define the problem—to understand all the variables and constraints, and then to optimize the solution. Using this approach to combating terrorism, for example, we must first understand why people would cause these intentional accidents. To gain such understanding, we must develop a greater knowledge of diverse cultures and societies and ask why people would want to cause harm, what their motives are, and try to understand what drives their actions. There is always a driving force behind every action, especially those that result in risk and even suicide. Understanding this driving force and taking multiple perspectives in the solution of engineering problems can be extremely helpful in preventing such events.

A common human failing is that we tend to keep doing things the way we have always done them. Maybe this is not so much a "failing" as an adaptive skill for the survival of the species. If we suffered no ill effects after eating that berry, we will eat more of that particular species. But even a similar-looking species may be toxic, so we are careful to eat only the species that did not kill us. So, professionals experience a paradox. They do not wish to expose their clients or themselves to unreasonable risks, but they must, to some extent, "push the envelope" to find better ways of doing this. Therefore, they are tempted to suppress new ways of looking at problems.

Human beings use prior knowledge and experiences when they observe and evaluate phenomena. They seldom view anything de novo, that is, as a completely new phenomenon. This makes sense, since we are constantly observing new things, it would be intellectually impossible to process all incoming information without applying many preliminary rules and mental shortcuts. The downside is what is known as confirmation bias, that is, tending to favor observations that agree with what we already think.

Objectivity is essential to any scientific endeavor, including the initiation and development of a new technology. If we only see things through previous intellectual filters, we will miss potentially important deviations from previous experiences [139]. For example, in scientific research, when an investigation's results do not confirm previously published results, the investigators have failed to replicate the original findings. This does not necessarily mean that these investigators are wrong, but they must explain the different results. However, bias is built into this reasoning, since the onus is on the those who find things other than what is supported by an existing body of literature. The researchers are thought to have failed unless they can make a strong case against the previous investigations.

The quality assurance plan is a means of avoiding bias, including confirmation bias. It lays out steps to ensure that if a discrepant finding does show up, errors in performing the experiment would be obvious.

Confirmation bias is at its worst when a hypothesis or viewpoint becomes so hallowed that scientists and engineers resist acceptable weight of evidence. People can become sufficiently convinced that their way of doing something is unreproachable, so they reject even an obviously improved approach. Therefore, the technology development study and plan should have enough metrics to ensure that even a surprising, yet correct, result is accepted and that the results can be replicated using the same plan and protocols, so that one does not fool oneself. Recall Feyman's caution from Chapter 3 [140]:

The first principle is that you must not fool yourself – and you are the easiest person to fool. So, you have to be very careful about that. After you've not fooled yourself, it's easy not to fool other scientists. You just have to be honest in a conventional way after that.

Sometimes, the facts and theories may be so overwhelmingly convincing that we must change our world view. Thomas S. Kuhn refers to this as a "paradigm shift" [45]. Scientists are often very reluctant to accept these new ways of thinking (Kuhn said such resistance can be "violent"). In fact, even if we do accept them, they are often not dramatic reversals (revolutions), but modifications of existing designs (evolutions). Some say that the bicycle was merely a mechanical re-rendering of the horse (e.g., the saddle seat, the linear and bilaterally symmetrical structure, and the handlebars); so was the automobile. Often, as the great philosophers tell us, "We may look but we do not see!" The deductive reasoning and methods about this approach *is* what stimulates mathematicians. The engineer is stimulated by the application of these methods. For example, an exciting development in mathematics, like Fermat's Last Theorem for which the proof was published in 1986. However, the theorem was published by 17th-century mathematician Pierre de Fermat, 357 years ago. In other words, when we set aside all the applications of discoveries in mathematics, development of the subject *in and of itself* becomes much less urgent, as "urgency" here is used to refer to the degree to which discoveries in a science directly impact man's life. Now, when we consider the concepts of efficiency, efficaciousness, and effectiveness within this framework, we must consider that mathematicians are examining relationships that already exist. Therefore, the only place where the concept of effectiveness becomes applicable is in the process that involves proving or explaining these already existing phenomena. Therefore, success and failure in this case become terms to describe the ability to access a system that already exists (or the lack thereof). Mathematics is the deductive study of shape, quantity, and dependence. One could write a section on each of these three concepts, contrasting their use by engineers, mathematicians, and economists. Also, whenever one talks about math, deductive reasoning is there. One could also contrast applied versus pure math (former is linked to physical phenomena). Conversely, pure math is the intrinsic study of math structures. So, the term "structure" means something different to all three disciplines, too! Thus, even seemingly simple and generally accepted terminology will vary among ontologies.

According to Kuhn, there are essentially two types of paradigm shifts: those that result from a discovery caused by encounters with anomalies, and those that result from the invention of new theories brought about by failures of existing theories to solve problems the theory defines. In the case of a paradigm shift brought about by discovery, the first step in shifting said paradigm is the discovery of the anomaly itself. This is a phenomenon that the paradigm cannot be used to treat—in this case, Pasteur's work was the anomaly that could not be treated using the paradigm. Then, the anomaly is explored. Once the paradigm has been adjusted so that the anomalous becomes the expected, it is said that the paradigm change is complete. In the case of a paradigm shift that results from the invention of new theories caused by the failure of existing theory, the first step is the failure itself (when the system in place fails, the creation of a new system is necessary). Several things can bring about this failure: observation of discrepancies between the theory and the facts, changes in the surrounding social or cultural climate, and academic criticism of the existing theory. While these problems are generally known for a long time, Kuhn has noted several times that the scientific community is highly resistant to change in paradigms. Kuhn also notes that in the early stages of a paradigm, it's easy to invent theoretical alternatives that can be placed on a given set of data. However, once the paradigm is more well established, these theoretical alternatives are strongly resisted [45].

Similarly, the process through which advances are made, and paradigms developed, may also be examined through the lens of the social sciences, particularly economics. For example, economist John Maynard Keynes wanted to fundamentally understand the dynamics of economics. At the same time, he wanted to solve the problem of economic depression. The fusion of these goals illustrates the joint desire to extend basic knowledge and reach applied goals.

For engineers, society evaluates our success and failure in accordance with the performance of a design. Does the resulting product or system "work" as designed (effectiveness)? Is it the best way to reach the end for which we strive (efficiency)? Next, we must consider whether it will likely continue to "work" (reliability), and our due diligence causes us to reflect on what adverse implications we might face (risk). Thus, the risk associated with a design is used to refer to the possibility and likelihood of undesirable and possibly harmful effects. Errors in engineering can range from those that are merely annoying, like when a concrete building develops cracks that mar it as it settles, to those that are seemingly unforgivable like the collapse of a bridge, which causes human death [141]. This can be illustrated by the Ford Pinto, a subcompact car produced by Ford Motor Company between 1971 and 1980. The car's fuel tank was poorly placed in such a way that it increased the probability of a fire from fuel spillage due to a rear collision. The ensuing adverse implications manifested in the series of injuries that resulted from the defect.

Engineers in such instances have been criticized for missing key facts, or worse yet, knowing the facts but allowing a dangerous design to go into production. In the bioethical arena, similar criticisms have been lodged against designs of devices and systems.

Thus, risk *here* takes on a new meaning for engineering, as engineers were designing and building at personal risk to themselves and their families. Should they face failure, their lives were at stake, which is clearly different from today's protocols. Consequently, it is important to consider the evolution of the meaning of the term "risk" for engineers over time.

In *To Engineer is Human*, Henry Petroski discusses success and failure in engineering as part of an ongoing process of trial and error that leads to innovation. For example, he discusses the effects of the growth of the railroads on engineering structures. As the network of railroads in the United States was quickly expanding, there were more incentives for heavier trains to travel progressively faster on more rugged terrain. But soon enough, it was obvious that the strength that was required for earlier railroads to facilitate the earlier trains was no match for the new trains. As collapses occurred, "each defective bridge resulted in demands for excess strength in the next similar bridge built, and thus the railroad bridge evolved through the compensatory process of trial and error" [141]. As the processes of engineering have become progressively more scientific in nature over the decades, engineers have had to deal with "success" and "failure" more explicitly. Thus, the real *trial and error* that has been passed down to today's engineers by their predecessors is that of "mind over matter." According to Petroski, while no one *wants* to learn by mistakes, they are an integral part of the process. As engineers perpetually strive to employ new concepts to create lighter, more cost-effective structures, each new structure can be seen as a trial of sorts [141].

16.5.1 Characterizing success and failure

Academic and professional disciplines differ in their characterizations of success and failure. For example, economists view them from two perspectives. First, how well is the distribution of goods and services and management of economic systems working? What is the best way to allocate scarce resources? When we have found the best solution to any problem of allocation or distribution, we are at Pareto optimality, the point where no one can be made better off without making someone else worse off. Therefore, "success" in this case involves attaining Pareto optimality.

The second economic perspective of success and failure is based on how an individual or organization makes economic decisions. This can follow a version of ethical egoism, hedonism, and utilitarianism. How do we optimize *our* profits or lower *our* individual costs? Here, our goal becomes to make ourselves or our individual firms as well-off as we can, given the constraints of the marketplace, including legal constraints. Since such a view is not concerned with the degree of improvement or degradation of others, Pareto optimality is not the goal. The duty-based aspects of the engineering profession are in conflict with such a system. For example, in the consumer model, the goal is to maximize the level of utility a consumer achieves while subject to the constraints of the budget constraint (essentially the total amount of money we have to spend), but most engineering codes of ethics require that the more expansive "public" be considered. In fact, engineering norms dictate that the public is *more important* than the individual and the organization.

The free market economic model is based purely on maximizing a predefined utility. In the producer model, firms optimize by maximizing profits, given the limitations they face due to the wage (the price of labor) and the rental rate (the price of capital).

This is not to say that engineers can operate without regard to economics. To the contrary, they are mandated to be faithful agents to their clients. Thus, optimization for engineers involves an economic element. Essentially, engineering involves design under constraint [142]. Engineers may strive to optimize the level of elegance of a given structure while subject to the constraints of fixed physical and mental resources, as well as the demands of the marketplace. At the same time, engineers additionally face the greatest constraint of all: safety. This constraint becomes more important than the aesthetic and economic aspects because "the loss of a single life due to structural collapse can turn the most economically promising structure into the most costly and can make the most beautiful one ugly [141]."

16.5.2 Accountability

The level and type of accountability for success and failure of a design is affected by the setting in which the engineer works. However, in every engineering office or department, there is a designated "engineer in responsible charge" whose job it is to make sure that every project is completed successfully and within budget. Biotechnological and other life science institutions are no different. This responsibility is often indicated by the fact that the engineer in responsible charge places his or her professional engineering seal on the design drawings or the final reports. By this action, the engineer is telling the world that the drawings, plans, programs, or whatever are correct, accurate, and that they will work.[4] In sealing drawings or otherwise accepting responsibility, the engineer in charge places his or her professional integrity and professional honor on the line. There is nowhere the engineer in charge can hide if something goes wrong. If something *does* go wrong, arguing that a younger engineer did it is not a plausible defense because it is the engineer in charge who is supposed to have overseen the calculations or the design.

For very large projects where the responsible engineer may not even know all the engineers working on the project, much less be able to oversee their calculations, this is clearly impossible. In a typical engineering office, the responsible engineer depends on a team of senior engineers who oversee other engineers, who oversee others, and so on down the line [78]. How can the responsible engineer at the top of the pyramid be confident that the product of collective engineering skills meets the client's requirements?

The rules governing this activity are simple. Central to this is the concept of truthfulness in engineering communication. Such technical communication up and down the organization requires an uncompromising commitment to tell the truth, no matter what the next level of engineering wants to hear.

What is valued will influence the attention to detail. For example, an engineer in the lower ranks may develop spurious data, lie about test results, or generally manipulate the

[4] In some countries, not too many years ago, the engineer in charge of building a bridge was actually required to stand under the bridge while it was tested for bearing capacity. Talk about standing behind your work, or literally under your work! From: [78].

basic design components. Such information might not be readily detected by supervisory engineers if the bogus information is beneficial to the completion of the project. If the information is not beneficial, on the other hand, everyone along the chain of engineering responsibility will give it a serious and critical look. Therefore, the inaccurate information, if it is the desired information, can steadily move up the engineering ladder because, at every level, the tendency is not to question good news. The superiors at the next level also want good news and want to know that everything is going well with the project. They do not want to know that things may have gone wrong somewhere at the basic level. Knowing this, and fearing being shot as the messenger, engineers tend to accept good news and question bad news. In short, the axiom that "good news travels up the organization – bad new travels down" holds for engineering as well [78]. And bad news will travel down very quickly (e.g., "You're fired"! or worse, "You're under arrest"!).

The only correcting mechanism in engineering exists at the very end of the project if failure occurs: the software crashes, the bridge collapses, the project is grossly overbid, or the refinery explodes. Then the search begins for what went wrong. What were the root causes of the failure? Eventually, the truth emerges, and often the problems can be traced to the initial level of engineering design, the development of data, and the interpretation of test results [78].

It is for this reason that engineers, especially early-career engineers, must be extremely careful of their work. It is one thing to make a mistake (we all do), but it is another thing to use misinformation intentionally in the design. Fabricated or spurious test results can lead to catastrophic failures because there is an absence of a failure detection mechanism in engineering until the project is completed. Without trust and truthfulness in engineering, the system will fail.

16.5.3 Value

The various definitions of success and failure require a means for determining success. We need ways to measure the level of success of a design in terms of effectiveness. For instance, assuming that pump Y has been designed to administer drug X to a patient's liver, the efficiency of drug delivery can be measured by a typical mass balance, such as the dose of X administered compared to the mass reaching the liver. Thus, calculating efficiency in this case is straightforward. Effectiveness, however, may not be. For example, even if the total efficiency is high, for example, 99.99% of the dose of X reaches the liver, it may be ineffective (nonefficacious). This could be because metabolism changes X to an oxidized metabolite X′ before it reaches the liver that is not effective in treating the level. Thus, effectiveness assumes some definition of value.

"In engineering, one might consider "value" through the idea of value engineering." ("Value Engineering - an overview I ScienceDirect Topics"). This concept was created at General Electric Co. during World War II [143–146]. As the war caused shortages of labor and materials, the company was forced to look for more accessible substitutes. Through this process, they saw that the substitutes often reduced costs or improved a product. Consequently, they turned the process into a systematic procedure called "value analysis" ("Value Engineering - an overview I ScienceDirect Topics").

"Value engineering consists of assessing the value of goods in terms of *function*." ("Value Engineering - an overview I ScienceDirect Topics"). Value, as a ratio of function to cost, can be improved in various ways. Often, value engineering is done systematically through the Job-Plan method, which involves four basic steps [147–149]:

1. information gathering;
2. alternative generation
3. evaluation; and
4. presentation.

In the information-gathering step, engineers consider the requirements for the object. Also, function analysis attempts to determine which functions or performance characteristics are important. In the next step, alternative generation, value engineers consider the possible alternative ways of meeting requirements. Next, in evaluation, the engineers assess the alternatives in terms of functionality and cost-effectiveness. Finally, in the presentation stage, the best alternative is chosen and presented to the client for the final decision.

In economics, "value" is the worth of one commodity in terms of other commodities (or currency). There are three main value theories in economics. The first, an intrinsic theory of value, is a theory, which holds that the value of an object, good, or service is contained in the item itself. These theories tend to consider the costs associated with the process of producing an item when assigning value to the item. For example, the labor theory of value, a model developed by David Ricardo, holds that the value of a good is derived from the effort of its production, reduced to the two inputs in the production frontier, labor and capital [150]. In this model, if a lamp is produced in 5 hours by 3 persons, then the lamp is worth 3 x 5=15 man-hours. On the other hand, the subjective theory of value holds that goods have no intrinsic value other than the desire of individuals to have the items. Here, value of an item becomes a function of how much an individual is willing to give up to have that item [150].

Similarly, the marginal theory of value accounts for both the scarcity and desirability of a good, holding that the utility rendered by the last unit consumed determines the total value of a good. The main difference between David Ricardo's labor theory of value (and the concept of intrinsic value) and marginal theory of value is that the former takes into account a form of value derived from utility—from satisfying human desire [151].

Furthermore, value can be attributed to the common member of society as when one values something. This generally means that the item has importance to that member for an individual reason. Therefore, person A may value her $10 Mickey Mouse watch because she derives pleasure from it, but her Rolex watch would be considered valuable because of the large number of other members of society who also believe they would derive pleasure from it, which causes a high demand that raises the price.

This is an entree to a discussion of the diamond-water paradox and the role of scarcity in economic value theory [152]. Diamonds, which have relatively little use and only aesthetic value, have an extremely high price when compared to water, which is essential to life itself. The example illustrates the importance of scarcity in economic value. Here, as the mass and volume of diamonds are far rarer on Planet Earth than is water, diamonds demand the higher price in the marketplace. However, this is situational. If one were in the middle of the desert, where water is extremely scarce, that person would almost certainly be willing to pay more for water than for diamonds. Thus, in that situation, that person would value water more than diamonds [124]. Therefore, success and failure are value-laden.

16.6 Conclusions

Any ethical and scientific optimization requires measures of success and ways to avoid failures. This chapter has introduced several success-failure concepts, including utility, benefit-cost ratios, engineering economics, human factors engineering, engineering reliability, and root cause and other case analysis procedures.

References

[1] D.A. Vallero, Environmental Systems Science: Theory and Practical Applications, first ed., Elsevier Academic Press, (2021).
[2] H. Fisher, A History of Europe (1936), I, p. vii, 1960.
[3] G. Santayana, The Life of Reason: Five Volumes in One, Echo Library, (2006).
[4] A.B. V. de Miranda, The Lawfulness of Infinite Scroll Down on Facebook, Universidade NOVA de Lisboa (Portugal), 2025.
[5] L.B. Frankel, The inventor of the infinite scroll is sorry for ruining your life. Medium. Available from: https://medium.com/pitfall/the-inventor-of-the-infinite-scroll-is-sorry-for-ruining-your-life-be953bf0ccfb (accessed 25.01.26).
[6] D. Mackay, Infinite Scrolling, Dissociation, and Boredom Spiraling as the Drivers of Habitual Social Media Use. Southern Connecticut State University, 2023.
[7] L.-M. Meinhardt, et al., Scrolling in the deep: analysing contextual influences on intervention effectiveness during infinite scrolling on social media, in: Proceedings of the 2025 CHI Conference on Human Factors in Computing Systems, 2025, pp. 1–17.
[8] J.O. Rixen, et al., The loop and reasons to break it: investigating infinite scrolling behaviour in social media applications and reasons to stop, Proceedings of the ACM on Human-Computer Interaction 7 (2023) 1–22 no. MHCI.
[9] R.J.-S. Jeff Jenkins, A. Metz, The secret lives of mormon wives, ed: Hulu, 2025.
[10] Human Factors and Ergonomics Society, HFES history. HFES. Available from: http://www.hfes.org/WEB/AboutHFES/history.html, 2006 (accessed 31.07.06).
[11] K. Sale, The achievements of'General Ludd': a brief history of the Luddites, The Ecologist 29 (5) (1999) 310–314.
[12] B. Watson, For a while, the Luddites had a smashing success, Smithsonian 24 (1) (1993) 140–153.
[13] J. Dinwiddy, Luddism and politics in the northern counties, Social History 4 (1) (1979) 33–63.
[14] J. Baggaley, The luddite revolt continues, Distance Education 31 (3) (2010) 337–343.
[15] K. Navickas, The search for 'General Ludd': the mythology of Luddism, Social History 30 (3) (2005) 281–295.
[16] H.M.C. Trade, R. Division, The American Heritage Dictionary of the English Language, Houghton Mifflin, (2026).
[17] J.S. Mill, Utilitarianism, Parker, Son and Bourn, London, (1863).
[18] D.A. Vallero, C.K. Gunsch, Applications and implications of emerging biotechnologies in environmental engineering, Journal of Environmental Engineering 146 (6) (2020) 03120005.
[19] W. Ahmed, B. Sarkar, Management of next-generation energy using a triple bottom line approach under a supply chain framework, Resources, Conservation and Recycling 150 (2019) 104431.
[20] E. Casey, S. Beaini, S. Pabi, K. Zammit, A. Amarnath, The triple bottom line for efficiency: integrating systems within water and energy networks, IEEE Power and Energy Magazine 15 (1) (2017) 34–42.
[21] S. Hendiani, E. Sharifi, M. Bagherpour, S.F. Ghannadpour, A multi-criteria sustainability assessment approach for energy systems using sustainability triple bottom line attributes and linguistic preferences, Environment, Development and Sustainability (2019) 1–35.
[22] D.A. Vallero, Fundamentals of Air Pollution, sixth ed, Academic press, Amsterdam, NV, (2025).
[23] C.K. Gunsch, K.A. Kinney, P.J. Szaniszlo, C.P. Whitman, Quantification of homogentisate-1, 2-dioxygenase expression in a fungus degrading ethylbenzene, Journal of Microbiological Methods 67 (2) (2006) 257–265.
[24] M.F. Verce, C.K. Gunsch, A.S. Danko, D.L. Freedman, Cometabolism of cis-1, 2-dichloroethene by aerobic cultures grown on vinyl chloride as the primary substrate, Environmental Science & Technology 36 (10) (2002) 2171–2177.

[25] L.M. Czaplicki, E. Cooper, P.L. Ferguson, H.M. Stapleton, R. Vilgalys, C.K. Gunsch, A new perspective on sustainable soil remediation—case study suggests novel fungal genera could facilitate in situ biodegradation of hazardous contaminants, Remediation Journal 26 (2) (2016) 59–72.

[26] K. Ikuma, R.M. Holzem, C.K. Gunsch, Impacts of organic carbon availability and recipient bacteria characteristics on the potential for TOL plasmid genetic bioaugmentation in soil slurries, Chemosphere 89 (2) (2012) 158–163.

[27] L.K. Redfern, C.M. Gardner, E. Hodzic, P.L. Ferguson, H. Hsu-Kim, C.K. Gunsch, A new framework for approaching precision bioremediation of PAH contaminated soils, Journal of Hazardous Materials (2019) 120859.

[28] W.A. Gerhard, C.K. Gunsch, Metabarcoding and machine learning analysis of environmental DNA in ballast water arriving to hub ports, Environment International 124 (2019) 312–319.

[29] W.A. Gerhard, C.K. Gunsch, Microbiome composition and implications for ballast water classification using machine learning, Science of The Total Environment 691 (2019) 810–818.

[30] J.K. Hammitt, E.S. Belsky, J.I. Levy, J.D. Graham, Residential building codes, affordability, and health protection: a risk-tradeoff approach, Risk Analysis 19 (6) (1999) 1037–1058.

[31] B. D, Stirred by Burnham, democracy champion, in: Chicago Record-Herald, ed. Chicago, Illinois, 1910.

[32] B. Büyükşirin, Health services supply and demand structure, in: International Research in Social, Human and Administrative Sciences XVII, 2024, p. 299.

[33] F. Wen, X. Yin, C. Wen, New demand economics, arXiv preprint arXiv:2510.17121, 2025.

[34] J. Pang, F. Jiao, Y. Zhang, An analysis of the impact of the digital economy on high-quality economic development in China—a study based on the effects of supply and demand, Sustainability 14 (24) (2022) 16991.

[35] L. Xiao, New Supply Side Economics, Springer Books, (2017).

[36] L. Xiao, Theoretical comparison of supply side economics and demand side economics, New Supply Side Economics: The Structural Reform on Supply Side and Sustainable Growth, Springer, (2017), pp. 1–29.

[37] J.B. Say, A Treatise on Political Economy: Or the Production, Distribution, and Consumption of Wealth, Grigg & Elliot, (1836).

[38] C. Chace, The Economic Singularity. Three Cs New York, 2016.

[39] N. Čeplak, H. Liu, A. Parnachev, S. Valach, Black hole singularity from OPE, Journal of High Energy Physics 2024 (10) (2024) 1–60.

[40] I. Frisch, Hegel and AGI: Geist as Solution to the Threat of Technological Singularity.

[41] N.S. Bostrom, Paths, dangers, strategies, Strategies, 2014.

[42] D. Vallero, Paradigms Lost: Learning from Environmental Mistakes, Mishaps and Misdeeds, Butterworth-Heinemann, (2005).

[43] A. Misuri, V. Cozzani, A paradigm shift in the assessment of Natech scenarios in chemical and process facilities, Process Safety and Environmental Protection 152 (2021) 338–351.

[44] T. Kuhn, What are Scientific Revolutions, MIT, Boston, (1981).

[45] T.S. Kuhn, D. Hawkins, The structure of scientific revolutions, American Journal of Physics 31 (7) (1963) 554–555.

[46] D.A. Vallero, Translating Diverse Environmental Data into Reliable Information: How to Coordinate Evidence from Different Sources, Academic Press, (2017).

[47] D.A. Vallero, T.M. Letcher, Engineering risks and failures: lessons learned from environmental disasters, Leadership and Management in Engineering 12 (4) (2012) 199–209.

[48] M. Eti, S. Ogaji, S. Probert, Integrating reliability, availability, maintainability and supportability with risk analysis for improved operation of the Afam thermal power-station, Applied Energy 84 (2) (2007) 202–221.

[49] T.M. Letcher, D. Vallero, Excerpt from: Waste: A Handbook for Management, scond ed. Academic Press, (2019).

[50] L. Sikos, J. Klemeš, RAMS contribution to efficient waste minimisation and management, Journal of Cleaner Production 17 (10) (2009) 932–939.

[51] T.E. Graedel, On the concept of industrial ecology, Annual Review of Energy and the Environment 21 (1) (1996) 69–98.

[52] R. Smith, D. Sengupta, S. Takkellapati, C. Lee, An industrial ecology approach to municipal solid waste management: I. Methodology, Resources, Conservation and Recycling 104 (2015) 311–316.

[53] R. Smith, D. Sengupta, S. Takkellapati, C. Lee, An industrial ecology approach to municipal solid waste management: II. Case studies for recovering energy from the organic fraction of MSW, Resources, Conservation and Recycling 104 (2015) 317–326.
[54] D.A. Vallero, Resilience and uncertainty in engineering decisions, in: Current Trends in Civil & Structural Engineering, vol. August 21, 2019, no. ISSN: 2643-6876 2019, https://doi.org/10.33552/CTCSE.2019.03.000568.
[55] T. Elmqvist, et al., Response diversity, ecosystem change, and resilience, Frontiers in Ecology and the Environment 1 (9) (2003) 488–494.
[56] E. Hollnagel, D.D. Woods, N. Leveson, Resilience Engineering: Concepts and Precepts, Ashgate Publishing, Ltd, (2007).
[57] J.D. Solomon, Communicating Reliability, Risk and Resiliency to Decision Makers.Duke University, October 12, 2017.
[58] J.D. Solomon, D.A. Vallero. From our partners – communicating risk and resiliency: special considerations for rare events. Center for Infrastructure Protection & Homeland Security, George Mason University. Available from: https://cip.gmu.edu/2016/06/01/partners-communicating-risk-resiliency-special-considerations-rare-events/ (accessed 01.05.17).
[59] H.R. Heinimann, A generic framework for resilience assessment, An edited collection of authored pieces comparing, contrasting, and integrating risk and resilience with an emphasis on ways to measure resilience, 2016, p. 90.
[60] P. Helm, Risk and resilience: strategies for security, Civil Engineering and Environmental Systems 32 (1-2) (2015) 100–118.
[61] D. Vallero, Environmental Biotechnology: A Biosystems Approach, Elsevier Science, (2015).
[62] E. Paoletti, et al., Advances of air pollution science: from forest decline to multiple-stress effects on forest ecosystem services, Environmental Pollution 158 (6) (2010) 1986–1989.
[63] J. Fiksel, I. Goodman, A. Hecht, Resilience: navigating toward a sustainable future, Journal of Solutions 5 (5) (2014) 38–47.
[64] M.Y. Reja, S.D. Brody, W.E. Highfield, G.D. Newman, Understanding the notion between resiliency and recovery through a spatial-temporal analysis of section 404 wetland alteration permits before and after hurricane Ike, World Academy of Science, Engineering and Technology International Journal of Environmental, Chemical, Ecological, Geological and Geophysical Engineering 11 (4) (2017) 348–355.
[65] B. Minsker, et al., Progress and recommendations for advancing performance-based sustainable and resilient infrastructure design, Journal of Water Resources Planning and Management 141 (12) (2015) A4015006.
[66] M. Dardonville, C. Bockstaller, O. Therond, Review of quantitative evaluations of the resilience, vulnerability, robustness and adaptive capacity of temperate agricultural systems, Journal of Cleaner Production 286 (2021) 125456.
[67] J.K. Vanos, J.W. Baldwin, O. Jay, K.L. Ebi, Simplicity lacks robustness when projecting heat-health outcomes in a changing climate, Nature Communications 11 (1) (2020) 6079.
[68] J.M. Anderies, O. Barreteau, U. Brady, Refining the robustness of social-ecological systems framework for comparative analysis of coastal system adaptation to global change, Regional Environmental Change 19 (2019) 1891–1908.
[69] J. Jäger, et al., Assessing policy robustness of climate change adaptation measures across sectors and scenarios, Climatic Change 128 (2015) 395–407.
[70] R. Wilby, S. Dessai, Robust adaptation to climate change, 2010.
[71] S. Dessai, M. Hulme, Assessing the robustness of adaptation decisions to climate change uncertainties: a case study on water resources management in the East of England, Global Environmental Change 17 (1) (2007) 59–72.
[72] J.A. Dunne, R.J. Williams, N.D. Martinez, Network structure and biodiversity loss in food webs: robustness increases with connectance, Ecology Letters 5 (4) (2002) 558–567.
[73] F.R. Siegel, Adaptations of Coastal Cities to Global Warming, Sea Level Rise, Climate Change and Endemic Hazards, Springer, (2019).
[74] C. Mellin, et al., Safeguarding nutrients from coral reefs under climate change, Nature Ecology & Evolution 6 (12) (2022) 1808–1817.
[75] A. Sánchez-Arcilla, M. García-León, V. Gracia, R. Devoy, A. Stanica, J. Gault, Managing coastal environments under climate change: pathways to adaptation, Science of the Total Environment 572 (2016) 1336–1352.
[76] C. Whitbeck, Ethics as design: doing justice to moral problems, Hastings Center Report 26 (3) (1996) 9–16.

[77] National Society of Professional Engineers, NSPE code of ethics for engineers. Available from: https://www.nspe.org/resources/ethics/code-ethics, 2016 (accessed 31.08.18).

[78] D.A. Vallero, P.A. Vesilind, Excerpted and adapted from, ed, Socially Responsible Engineering: Justice in Risk Management, Wiley Online Library, (2007).

[79] D.A. Vallero, Chapter 8 - Decision support tools, in: D.A. Vallero (Ed.), Environmental Systems Science, Elsevier, (2021), pp. 293–357.

[80] M.A. Latino, R.J. Latino, K.C. Latino, Root Cause Analysis: Improving Performance for Bottom-Line Results, CRC press, (2019).

[81] B. Andersen, T. Fagerhaug, Root Cause Analysis, Quality Press, Birmingham, Alabama, (2006).

[82] Occupational Safety and Health Administration, Combustible dust: an explosion hazard, in: Safety and Health Bulletin, U.S. Department of Labor, Washington, DC, SHIB 07-31-2005 [Online]. Available from: https://www.osha.gov/combustible-dust , 2023 (Accessed 15.02.23).

[83] D.A. Vallero, T.M. Letcher, Unraveling Environmental Disasters, second ed., Elsevier Academic Press, Amsterdam, The Netherlands, (2023).

[84] R.A. Hill, C. Sendashonga, General principles for risk assessment of living modified organisms: lessons from chemical risk assessment, Environmental Biosafety Research 2 (2) (2003) 81–88.

[85] ER-TG-03-1 7/00, Preparing information on risks, costs and benefits for applications under the Hazardous Substances and New Organisms Act 1996, 2000.

[86] S.I. Watson, R.J. Lilford, Integrating multiple sources of evidence: a Bayesian perspective, 2016.

[87] A. Gatherer, Evidence, values, and 'right versus right'dilemmas in public health practice, Journal of Public Health Policy 32 (4) (2011) 504–507.

[88] D.L. Weed, Weight of evidence: a review of concept and methods, Risk Analysis 25 (6) (2005) 1545–1557.

[89] N.M. DeLuca, M. Angrish, A. Wilkins, K. Thayer, E.A.C. Hubal, Human exposure pathways to poly-and perfluoroalkyl substances (PFAS) from indoor media: a systematic review protocol, Environment International 146 (2021) 106308.

[90] P. Whaley, et al., Improving the quality of toxicology and environmental health systematic reviews: what journal editors can do, ALTEX-Alternatives to Animal Experimentation 38 (3) (2021) 513–522.

[91] EPA Document# EPA-D-20-031, Draft Systematic Review Protocol Supporting TSCA Risk Evaluations for Chemical Substances Version 1.0: A Generic TSCA Systematic Review Protocol with Chemical-Specific Methodologies, 2021. [Online] Available from: https://www.epa.gov/system/files/documents/2021-12/draft-systematic-review-protocol-supporting-tsca-risk-evaluations-for-chemical-substances_0.pdf.

[92] Z. Noorimotlagh, S.A. Mirzaee, N. Jaafarzadeh, M. Maleki, G. Kalvandi, C. Karami, A systematic review of emerging human coronavirus (SARS-CoV-2) outbreak: focus on disinfection methods, environmental survival, and control and prevention strategies, Environmental Science and Pollution Research 28 (1) (2021) 1–15.

[93] K. Dack, M. Fell, C.M. Taylor, A. Havdahl, S.J. Lewis, Mercury and prenatal growth: a systematic review, International Journal of Environmental Research and Public Health 18 (13) (2021) 7140.

[94] S. Ohlwein, R. Kappeler, M.K. Joss, N. Künzli, B. Hoffmann, Health effects of ultrafine particles: a systematic literature review update of epidemiological evidence, International Journal of Public Health 64 (4) (2019) 547–559.

[95] C. Ewbank, S. Gupta, B.T. Stewart, A.L. Kushner, A. Charles, A systematic review of oil tanker truck disasters: identifying prevention targets, Burns: Journal of the International Society for Burn Injuries 45 (4) (2019) 905–913.

[96] P. de Almeida Rodrigues, R.G. Ferrari, L.N. Dos Santos, C.A.C. Junior, Mercury in aquatic fauna contamination: a systematic review on its dynamics and potential health risks, Journal of Environmental Sciences 84 (2019) 205–218.

[97] Z. Bi, Finite Element Analysis Applications: A Systematic and Practical Approach, Academic Press, (2017).

[98] F. Pianosi, et al., Sensitivity analysis of environmental models: a systematic review with practical workflow, Environmental Modelling & Software 79 (2016) 214–232.

[99] B. Hong, et al., Model-based tolerance intervals derived from cumulative historical composition data: application for substantial equivalence assessment of a genetically modified crop, Journal of Agricultural and Food Chemistry 62 (40) (2014) 9916–9926.

[100] P. Macdonald, Developing workable regulatory frameworks for the environmental release of transgenic plants, Collection of Biosafety Reviews (2012).

[101] ER-TG-02-1 1/100, Assessment of effects of hazardous substances and new organisms on human health, 2000.

[102] R.A. Hill, From science to decision-making: the applicability of Bayesian methods to risk assessment, 1996.
[103] V.F. Reyna, F. Farley, Risk and rationality in adolescent decision making: implications for theory, practice, and public policy, Psychological Science in the Public Interest 7 (1) (2006) 1–44 [Online]. Available http://www.jstor.org/stable/40062377.
[104] B. Breckling, et al., Risk indication of genetically modified organisms (GMO): modelling environmental exposure and dispersal across different scales: Oilseed rape in Northern Germany as an integrated case study, Ecological Indicators 11 (4) (2011) 936–941.
[105] J.G. Lundgren, J.J. Duan, RNAi-based insecticidal crops: potential effects on nontarget species, Bioscience 63 (8) (2013) 657–665.
[106] C.V. Rider, et al., Cumulative risk: toxicity and interactions of physical and chemical stressors, Toxicological Sciences 137 (1) (2014) 3–11.
[107] Y. Li, et al., Quantifying urban ecological governance: a suite of indices characterizes the ecological planning implications of rapid coastal urbanization, Ecological Indicators 72 (2017) 225–233.
[108] R.D. Bullard, Unequal Protection: Environmental Justice and Communities of Color, Sierra Club Books, (1994).
[109] B.S. Matulis, The economic valuation of nature: a question of justice? Ecological Economics 104 (0) (2014) 155–157. Available from https://doi.org/10.1016/j.ecolecon.2014.04.0108//.
[110] H. Gao, C. Xu, W. Wang, Z. Xiao, X. He, Prevention and defense technologies for natural disasters: a review, Evidence in Earth Science 1 (01) (2025) 17–32.
[111] L.S. Hanwacker, The role of artificial intelligence in disaster recovery, Journal of Business Continuity & Emergency Planning 18 (2) (2025) 167–179.
[112] J. Periasamy, K.S. Reddy, P.R. Salve, S. Ushasukhanya, T.N. Malleswari, AI-driven disaster forecasting by integrating smart technology, Edible Electronics for Smart Technology Solutions, IGI Global, (2025), pp. 383–414.
[113] J. Song, et al., High failure rate associated with 180 selective laser trabeculoplasty, Journal of Glaucoma 14 (5) (2005) 400–408.
[114] L.J. Morse, J.A. Bryan, J.P. Hurley, J.F. Murphy, T.F. O'Brien, W.E. Wacker, The Holy Cross College football team hepatitis outbreak, JAMA: the Journal of the American Medical Association 219 (6) (1972) 706–708.
[115] P.A. Vesilind, Engineering and the threat of terrorism, Journal of Professional Issues in Engineering Education and Practice 129 (2) (2003) 70–74.
[116] P. Vesilind, Opinion-engineering and terrorism, IEEE Technology and Society Magazine 21 (2) (2002) 3–4.
[117] B.J. Hanley, What caused the bhopal gas tragedy? The philosophical importance of causal and pragmatic details, Philosophy of Science 88 (4) (2021) 616–637.
[118] B.S. Martins, Revisiting the Bhopal disaster: times of violence and latitudes of memory, The Pluriverse of Human Rights, Routledge, (2021), pp. 85–105.
[119] O. Basha, J. Alajmy, T. Newaz, Bhopal gas Tragedy: A Safety Case Study, 2020.
[120] E. P. A. U.S, Region 10, The Bhopal tragedy - 25 years ago, Chemical Emergency Prevention & Planning Newsletter, vol. January-February 2010, p. 1, 2010. [Online]. Available from: https://nepis.epa.gov/Exe/tiff2png.cgi/P100XCBP.PNG?-r+85+-g+15+-h+14,9,9,40,0,5,40,6,7,44,12,3,45,27,5,54,22,3,56,0,3,57,0,3,64,24,3,113,14,5,113,20,7,114,1,3,114,22,5+D%3A%5CZYFILES%5CINDEX%20DATA%5C06THRU10%5CTIFF%5C00001902%5CP100XCBP.TIF.
[121] T. Chouhan, I. Jaising, C.A. Alvares, Bhopal, the Inside Story: Carbide Workers Speak Out on the World's Worst Industrial Disaster, Apex Pr, (1994).
[122] N. Meshkati, Human factors in large-scale technological systems' accidents: three Mile Island, Bhopal, Chernobyl, Industrial Crisis Quarterly 5 (2) (1991) 133–154.
[123] B. Bowonder, An analysis of the Bhopal accident, Project Appraisal 2 (3) (1987) 157–168.
[124] D.A. Vallero, Biomedical ethics for engineers, The Biomedical Engineering Series, 2007.
[125] M. Martin, R. Schinzinger, Engineering Ethics, McGraw Hill, Boston, (1996). ed.
[126] W. Worthy, Methyl isocyanate-the chemistry of a hazard, Chemical & Engineering News 63 (6) (1985) 27–33.
[127] U.S. Environmental Protection Agency, Methyl Isocyanate 624-83-9 I DTXSID1023786. AVailable from: https://comptox.epa.gov/dashboard/dsstoxdb/results?search=DTXSID1023786#properties, 2020 (accessed 14.10.20).
[128] D. Lapierre, Five Past Midnight in Bhopal: The Epic Story of the World's Deadliest Industrial Disaster, Hachette+ ORM, (2025).

[129] B. Bowonder, J.X. Kasperson, R.E. Kasperson, Industrial risk management in India since Bhopal, Social Contours of Risk, Routledge, (2022), pp. 144–163.
[130] B. Bowonder, J.X. Kasperson, R.E. Kasperson, Avoiding future Bhopals, Social contours of risk, Routledge, (2022), pp. 87–110.
[131] Aviation Safety Network, Martin 4-0-4. >>>>Aviation Safety Network, 2026 #5534<<<< (accessed 23.01.26).
[132] Aviation Safety Network, Douglas DC-4. Available from: https://aviation-safety.net/wikibase/329923 (accessed 23.01.26).
[133] Aviation Safety Network, Douglas DC-6BF. Available from: https://aviation-safety.net/wikibase/329700 (accessed 23.01.26).
[134] S.K. Pfatteicher, "A new era": the limits of engineering expertise in a post-9/11 world, 2007 IEEE International Symposium on Technology and Society, IEEE, (2007), pp. 1–4.
[135] S.K. Pfatteicher, Learning from failure: terrorism and ethics in engineering education, IEEE Technology and Society Magazine 21 (2) (2002) 8–12.
[136] Front Matter, ASEE prism 11 (6) (2002) [Online] http://www.jstor.org/stable/24160491.
[137] J.D. Jr Woodward, Super Bowl surveillance: facing up to biometrics, 2001.
[138] B. Franklin, Historical review of Pennsylvania, London: R. Griffiths, Paternoster Row, 1759, p. 289.
[139] 125C3Confirmation Bias, in: S.E. Gorman, J.M. Gorman (Eds.), Denying to the Grave: Why We Ignore the Science That Will Save Us, Revised and Updated Edition, Oxford University Press, (2021), Available from https://doi.org/10.1093/oso/9780197547458.003.0004p. 0. [Online]. Available.
[140] R. Feynman, Cargo cult science, Engineering & Science 37 (7) (1974) 10–13.
[141] H. Petroski, To Engineer is Human: The Role of Failure in Successful Design, St Martins Press, (1985).
[142] National Academy of Engineering, The Engineer of 2020: Visions of Engineering in the New Century, National Academies Press, Washington, DC, (2004).
[143] J. Ochieng'Wao, Drivers and critical success factors of value engineering in the construction industry, Proceedings of Associated Schools of Con 6 (2025) 601–610.
[144] S.M. Brondoni, Value analysis, value engineering and value management in global competition, Symphonya (1) (2024) 10–21.
[145] A. Sharma, R. Belokar, Achieving success through value engineering: a case study, Proceedings of the World Congress on Engineering and Computer Science 2 (2012) 24–26.
[146] D. Younker, Value Engineering: Analysis and Methodology, CRC Press, (2003).
[147] Q. Liang, M.-Y. Leung, Z. Zhou, Apply a fuzzy method to perform objective allocation for buildings cost in value engineering job plan process, International Symposium on Advancement of Construction Management and Real Estate, Springer, (2019), pp. 917–933.
[148] G. Kumar, H. Kumaraswamy, Optimization of product by value engineering job plan technique, IUP Journal of Mechanical Engineering 11 (4) (2018).
[149] J. Mandelbaum, "Value engineering handbook," 2006.
[150] D. Ricardo, On The Principles of Political Economic and Taxation-David Ricardo, Lebooks Editora, (2024).
[151] P. Lagasse, Value in economics, The Columbia Encyclopedia, Columbia University Press, New York, NY, (2000).
[152] M.V. White, Doctoring Adam Smith: the fable of the diamonds and water paradox, History of Political Economy 34 (4) (2002) 659–683.

CHAPTER

17

Research ethics

OUTLINE

Just the Human Genome Project alone is the Full Employment Act for bioethicists. ***Arthur L. Caplan [1]***

17.1 Introduction

Some technologies emerge from garages and basements, but most from the work of researchers in institutions, companies, and government laboratories. The very nature of research invites ethical issues. Research is systematic. It is creative. It is the discovery of new knowledge by interpreting existing data and information to find something no other human has found before. It challenges and attempts to revise established principles, theories, and even the laws that have been on the books for centuries. As such, it invites shortcuts and even dishonesty.

Research progress invites a certainty that what the researcher is doing is best it can be. It tempts the researcher to ignore warning signs. It lends stubbornness to the argument with oneself that all that work is not sunk costs if things do not pan out. The researcher may tell

Technical and Ethical Optimization: Challenges and Opportunities for the Modern Engineer and Scientist
DOI: https://doi.org/10.1016/B978-0-443-44925-3.00010-7

oneself that so much is at stake, no one needs to know about a little bit of misconduct. What are the odds anyone will find out?

Unfortunately, research occurs in a highly competitive environment where publications and funding are a zero-sum game. If one's research is not published, someone else will receive the funds and, sometimes, the fame. Not so long ago, universities and research institutions fostered a "publish or perish" environment. That still exists, but it has taken on new importance in recent decades not only the need to publish but also to seek grants. As such, a vulnerable researcher may be encouraged by themselves or by others to discount the risks of misconduct in the hopes of financial gain, reputation, and career advancement.

Sometimes, ethics takes a back seat to ideology or a firm commitment to an idea or strongly held belief. For example, it asks much of a researcher to submit a grant proposal to an institute whose mission is different from that expressed in the proposal. If a climate scientist wants to receive funds from an agency that has touted one and only one view on the causes of climate change, it seems unlikely that any research that does not follow the dogma would be funded. Even worse, it could encourage less skepticism of research by ideological scientists conducting peer reviews if the paper they are reviewing has findings and methods similar to their own expectations, that is, a type of confirmation bias.

The researcher can have his or her own confirmation biases about how the science *should* proceed. If so, the researcher is likely to design the study to elicit the expected results by selectively citing literature that supports preexisting biases, ignoring studies that do not support those biases, omitting outliers and different explanations of findings, and missing or underemphasizing negative or null findings.

There are many examples of cooking, pruning, and trimming data to render it publishable, for example, to improve correlations and other statistical metrics to meet publication standards.[1] There are three types of missing data that can be imputed [2]: missing completely at random (MCAR); missing at random (MAR); and nonrandom missing. He emphasizes that imputation to address missing data is certainly not cheating and is obligatory if missing data were to lead to errors in generalization, that is, improper deduction [2–4].

Even Nobel laureates and associated researchers have faced serious accusations of research misconduct, ranging from scientific fraud to ethical controversies. Some of these cases resulted in the retraction of journal articles, and though criminal and other charges were rescinded, their reputations were irreversibly damaged.

17.1.1 Engineering research

Beyond a unique and shared vocabulary, different professions use selective problem-solving tools. As research becomes available, practitioners have different ways of applying it. For example, mathematicians and engineers share many commonalities, such as the application of mathematical reasoning to solve problems. The author has had interesting discussions with Duke mathematician John Trangenstein about the major differences between the two professions. After all, both groups speak the same language, mathematics. Trangenstein is a professor of both Mathematics and Civil and Environmental Engineering,

[1] See: [5].

so he is directly aware of the differences. One principal difference seems to be the way practitioners of the two disciplines see problems and the methods they use to solve them.

The first difference is between the practitioner and the inquirer. Differences in vernacular can point to dissimilarities in the way various groups go about solving problems. Practitioners frequently perceive mathematics as a means to an end, that is, a tool to help solve the problem, whereas the goal of mathematics is to explain shape, quantity, and dependence. This perspective characterizes mathematicians as not really "creating" anything or solving any problem (unless it is applied mathematics, the branch of mathematics that applies math to other domains).

Fermat's Last Theorem, for which the proof was published in 1986, provides an example of inquiry. However, the theorem itself was published by 17th-century mathematician Pierre de Fermat, 357 years ago. In other words, when we set aside all the applications of discoveries in mathematics, development of the subject *in and of itself* becomes much less urgent, as "urgency" here is used to refer to the degree to which discoveries in a science directly impact human life.[2]

The concepts of efficiency, efficaciousness, and effectiveness also distinguish the practitioner within this framework. Often, mathematics requires the examination of relationships that already exist. Therefore, the concept of effectiveness for mathematicians involves proving or explaining these existing phenomena. Success and failure become terms to describe the ability or inability to access a system that already exists. For engineers and design professionals, the terms apply somewhat differently. Designing something to improve a process or to solve a problem requires that the terms efficiency and effectiveness refer to whether the design is conducive to the purpose for which it was created and whether the design function performs a prescribed task, respectively. Mathematics draws from deductive reasoning to observe patterns and relationships that already exist.

Engineering, on the other hand, is not an exclusively deductive science, as engineers also base knowledge on experience and observation. They first generate rules based on observations (i.e., the laws of nature: chemistry, physics, biology, etc.) of the world around them, the way things work. Once engineers have this understanding, they may apply it by using the rule to create something, such as a new technology designed to reach some end. History is filled with such examples, ranging from keeping food cold to delivering nutrition to comatose patients, to finding better means of delivering drugs to specific tumor sites. According to the National Academy of Engineering, "Technology is the outcome of engineering; it is rare that science translates directly to technology, just as it is not true that engineering is just applied science" [6]. Innovations in engineering occur when a need or opportunity arises (hence the adage, "Necessity is the mother of invention."). Therefore, first, engineers may develop an understanding of the thermodynamics behind a phase change heat pump and then apply this knowledge when society experiences a *need* to keep food cold. Likewise, engineering researchers must

[2] I realize that this could elicit letters from mathematicians. This is not meant as a criticism, but as a distinction between the focus of the practice-oriented researcher versus that of the inquiring researcher. Both foci are necessary. However, engineers are seldom permitted the luxury of waiting 350 years for a "proof" and often must be empirical and adaptive. As soon as a new paradigm is proven, or at least evidence shows it is superior to existing approaches, the engineer must be open-minded to accept the change.

adapt and change their understanding of electromagnetic forces at the nanoscale when developing new drug delivery nanotechnologies.

17.2 Responsible conduct of human research

Arguably, the worst kind of unethical research is disrespect of others and all the harm that brings to an individual research subject and society at large. All human beings deserve respect. Those who are subjects of research must provide informed consent. Whereas philosophers for many centuries have extolled the virtues of respect, it really was not until 1979, with the publication of *the Belmont Report: Ethical Principles and Guidelines for the Protection of Human Subjects of Research* [7], that the inviolable principles were articulated on how to treat people as research subjects. Basically, this consists of three requirements: (1) respect for persons; (2) beneficence; and (3) justice. The Belmont Report attempts to summarize the basic ethical principles identified by the Commission during its deliberations. It is the outgrowth of an intensive four-day period of discussions that were held in February 1976 at the Smithsonian Institution's Belmont Conference Center, supplemented by the monthly deliberations of the Commission that were held over a period of nearly four years. It is a statement of basic ethical principles and guidelines that should assist in resolving the ethical problems that surround the conduct of research with human subjects. By publishing the Report in the Federal Register and providing reprints upon request, the Secretary intends that it may be made readily available to scientists, members of Institutional Review Boards, and federal employees.

The needed changes outlined in the *Belmont Report* resulted from the abuses of Nazi science and the US Public Health Service (US PHS)-sponsored Tuskegee Syphilis Trials.

The so-called Tuskegee Experiment represented the ethical nadir of 20th-century biomedical research. The experiment occurred between 1932 and 1972, with the US PHS's investigation of the effects of syphilis, which is caused by the spirochete bacterium *Treponema pallidum*. With a disturbing disregard for justice, beneficence, and respect for the human person, US PHS researchers deliberately denied full knowledge and treatment for the incapacitating sexually transmitted disease to 399 African-American men infected with *T. pallidum*. The researchers took advantage of a trusted African-American medical staff's reputation, ultimately leading to the deaths of 128 men and allowing 59 spouses to contract the disease or their children to be born with syphilis.

There are several lessons that must be learned from these atrocities. First, biotechnologies in every sector, industrial, medical, agricultural, and environmental, vary in the level of trust. The trust exists between researchers and the public, but also within the scientific community. For example, scientists holding a more precautionary perspective on the advancement of science are more circumspect about potential risks from emerging technologies. Those scientists who see biotechnologies as merely extensions of a much larger knowledge base (a logical extension of centuries of genetics, hybridization, husbandry, botany, and microbiology) may prefer to address each specific nuance as a slight permutation rather than a comprehensive threat. Hence, where some may see a genetically modified (GM) strain of grain as "Frankencorn," others see an almost routine variation of *Zea mays*.

Another important lesson from the Tuskegee Experiment is how to place science within a social context. Often, this presents a paradox to researchers. Most have not been trained in sociological matters, let alone geopolitical ones. Their expertise lies in an often-esoteric area of research. Technical researchers often know much about the science, but too little about the social implications of that science (including whether to continue an investigation after evidence of ethical and justice issues—often, even when such information becomes available, the scientist is not trained to recognize it).

Some parallels with emerging technologies have been drawn, as when Rebecca Bratspies of the City University of New York Law School expresses concern for research and use of genetically modified organisms (GMOs), specifically the soil bacterium, *Bacillus thuringiensis* (Bt) [8]:

> Examining the degree to which environmental concerns have or have not been incorporated into the registration requirements of Bt crops, it becomes clear that the regulatory process suffers from many ills. In approving these genetically engineered crops for market, the United States Department of Agriculture (USDA) and Environmental Protection Agency (EPA) repeatedly disregarded significant but unresolved scientific questions about these GM crops. Seed companies agreed to environmentally protective measures in their crop registrations, but assumed no responsibility for implementing those measures ... No regulatory framework existed (or, for that matter, exists) to monitor and enforce these registration restrictions ... These serious regulatory deficiencies call into question the soundness of the entire biotechnology regulatory process, a question ultimately much broader than any particular GMO.

Research scientists and technologists work within an uncertain regulatory environment. Since the distribution of goods and services is so uneven, scientists may be tempted to "regress to the mean" and to satisfy some part of society by catering to a population within some level of variance around a measure of central tendency for that population (e.g., persons falling within ±2 standard deviations of an arithmetic mean). That way, "most" people would be satisfied with the investigation. There are entire philosophical schools of thought surrounding the ethical treatment of populations. The most moral approach in science is to protect public health and the environment by ensuring that *all* persons are adequately protected.

These travesties led to consensus among the scientific community on the need to regulate research more diligently and to codify regulations to ensure that researchers abide by these principles. The other principle, nonmaleficence, follows ethical precepts required in many ethical frameworks, including harm principles, empathy, and consideration of special populations, such as the infirm and children.

17.3 Major research issues

Technical work, especially engineering practice and research, is deeply committed to and involved in the advancing technologies that will benefit humankind. However, this commitment and involvement call for deliberate and serious consideration of actual and potential ethical issues. The President's Council on Bioethics has summarized the dichotomy between the promise and ethical challenges [9]:

> For roughly half a century, and at an ever-accelerating pace, biomedical science has been gaining wondrous new knowledge of the workings of living beings, from small to great. Increasingly, it is providing precise and sophisticated knowledge of the workings also of the human body and mind. Such knowledge of how things work often leads to new technological powers to control or alter these workings, powers generally sought in order to treat human disease and relieve suffering. But, once available, powers sought for one purpose are frequently usable for others. The same technological capacity to influence and control bodily processes for medical ends may lead (wittingly or unwittingly) to non-therapeutic uses, including "enhancements" of normal life processes or even alterations in "human nature." Moreover, as a result of anticipated knowledge of genetics and developmental biology, these transforming powers may soon be able to transmit such alterations to future generations.

So, let us consider some of these technological areas that are important to engineers and their attendant ethical concerns.

17.3.1 Genetically modified organisms and synthetic biology

> While engineering is a rapidly evolving field that adapts to new knowledge, new technology, and the needs of society, it also draws on distinct roots that go back to the origins of civilization. Maintaining a linkage of the past with the future is fundamental to the rational and fact-based approaches that engineers use in identifying and confronting the most difficult issues. ***National Academy of Engineering* (2004) [10]**
>
> Like any powerful technology, DNA synthesis has the potential to be purposefully misapplied. Misuse of DNA-synthesis technology could give rise to both known and unforeseeable threats to our biological safety and security. ***Hans Bügl, International Consortium for Polynucleotide Synthesis* (2007) [11]**
>
> There are known knowns. These are things we know that we know. There are known unknowns. That is to say, there are things that we know we don't know. But there are also unknown unknowns. There are things we don't know we don't know. ***Donald Rumsfeld, US Secretary of Defense* (2002) [8,12,13]**

These three quotations illustrate the range of thinking about the ethics of most emerging technologies, particularly those that affect life. The engineering perspective of the National Academy is that of a technological optimist, relying on reason and facts. At the same time, the quotes remind us to have the humility to respect the past, to understand culture and society, and to avoid the hubris and confirmation biases tempting the scientist and engineer. The second quote serves as a warning that the hubris and myopia of the researcher can bring rewards, but arguably even greater risk. Rumsfeld's famous quote is a catch-all for the possible risks associated with decisions under uncertainty. The theme of all three is that there is a lot that even the most well-educated technologist and researcher do not know.

Most of the history of genetic manipulation is not emergent. Indeed, genetic modification of organisms is a very old endeavor. Humans have changed the characteristics of numerous plants and animals through selective breeding techniques, beginning with attempts to encourage offspring from organisms with favorable traits, such as size, color, texture, and taste. Growing up in southern Illinois, I had a very good idea of what an ear of corn looks like. Like many moments in my undergraduate experience, this "Eureka moment" arrived in a course on plants and civilization. The instructor stated that the pre-Columbian people of Mexico domesticated corn (i.e., maize) about 9000 years ago from teosinte, a wild grass, by selective breeding. The original "ear" of teosinte had but a few kernels and was one or two centimeters (cm) long, compared to the 20 cm or longer ears of today. This amazed me. What patience!

In recent decades, the process, however, has become supercharged with the onset of direct genetic manipulations of DNA and RNA. Like several bioethical issues, genetic manipulation is associated with the slippery slope fear.[3] The slippery slope occurs when allowing an act makes it "impossible to hold the line and prevent extension to less justifiable situation" [14]. The goal of genetic modification is to delete specific phenotypic characteristics from or introduce new characteristics to an organism's progeny. Prior to the late 20th century, this was done externally but is now increasingly accomplished internally within the cell's genetic material. The resulting progeny is known as a transgenic organism. Transgenesis occurs when DNA from another organism is introduced using artificial gene transfer techniques.

Such techniques are allowing researchers to understand the interactions of certain genes more completely. The major ethical issues involved in GMOs often center on animal welfare, risks to human health, and the environment [15]. For example, what if a new creature is so different in kind that it has such a competitive advantage (i.e., now effective predators) or an ability to self-replicate that it would pose risks to public health and welfare, in violation of the engineer's first ethical canon? And how many animals' lives are worth an important discovery? Are we decreasing the genetic diversity of our wildlife or destroying the habitats of other animals?

Such concerns come from within and outside the scientific community and have at least a basis in utilitarianism, that is, disagreement about the utility for the greater good versus risks to the public. Others oppose the research based on religious and moral concerns, arguing that the researchers and the biotechnological companies are immorally attempting to "play God" by creating entirely new beings and unnaturally altering the genetic makeup of progeny. The Church of Scotland put it this way [16]:

Synthetic biology brings together the two disciplines of biology and engineering and is essentially about the redesign and reassembly of biological systems; in other words, redesigning life. It is about modifying present life forms or creating new life forms. The biologist wants to understand living systems better, and the engineer wants to create new things. While synthetic biology may be seen as a further development of "genetic engineering," which has given us GM crops, human growth hormone, and human insulin, the key difference is the application of techniques used in engineering design and development to biology.

Furthermore, some or perhaps most of the currently available GMOs can in part be supported due to the double effect principle, which considers an act for some beneficial purpose to be morally permissible in circumstances where a harmful effect is unintended, but merely a side effect of the act [17]. In this case, some believe that manufacturing resulting from GMO research could lead to more profitable or productive plants and animals. Second, scientists hope that this experimentation could aid humans by developing treatments for deadly diseases or methods to assist in the creation of tissues and organs. Thus the GMO in these

[3] This discussion draws upon the ideas and writings of Duke University student Zach Abrams, who conducted undergraduate research on GMOs in the author's Ethics in Professions course at Duke.

instances is intended to be beneficial. The double effect also requires that the good effect of the act must be sufficiently desirable to compensate for the bad effect and that no other less harmful acts can produce the same benefits. If these criteria are met, GMOs would be morally permissible. However, the debate would continue about the sufficiency of the good effect of GMOs compared to no GMOs and that there are other ways to produce similar goods and services, for example, natural, unmanipulated plants and animals.

Another argument against many genetic engineering, and especially synthetic biology, endeavors is the so-called "precautionary principle," which states that an act must not be taken if the consequences are severe and irreversible [18–25].

17.3.1.1 Transgenic species research

An organism that has been genetically modified it becomes a transgenic species. There are at least two ethical considerations regarding transgenic species [15]. First, what impact might research and marketing of these species have on society, such as threats to health? Second, what ethical considerations are needed on behalf of the creatures themselves? We will address the first issue next in discussing GM food. The second is predominantly about animals. For example, is it ethical to modify a monkey so that its fur glows in the dark? [15] It seems not to fit the requirements of the double effects principle.

A key question regarding transgenic species is about who benefits. For example, Harvard University modified the genetic structure of a mouse, known as the "oncomouse," to produce a human cancer in the animal [1]. From an anthropocentric viewpoint, this can be justified as long as it helps advance cancer research, but it does not stand against a biocentric view, or even a deontological view if one considers the protection of lower sentient species to be obligatory.

The European Union has taken several recent actions, many precautionary, to address GMOs [26]. These actions can be treated as a bioethical case. In 1999, a four-year ban on all GM crops was the result of several studies indicating that GM food could be harmful to humans. One study of the effects of GM potatoes on the digestive tract of rats found that there were large differences between the intestines of rats fed GM potatoes and those fed unmodified potatoes [27,28]. Other studies have indicated that introducing a new gene to a plant could create new allergens or cause reactions in susceptible people [29]. For example, a gene from peanuts transferred to soybeans could cause those allergic to peanuts to have reactions when they ate soybeans [15].

Based on these studies, the European Union announced a ban on all GM crops. At the end of 2002, the ban was lifted, but producers were now required to label all GM goods with a special DNA sequence that identified the origin of the crops. This made it easier for regulators to spot contaminated crops and feed. Later, in June 2003, the European Parliament agreed to a UN bio-safety protocol regulating the international trade of GM food. This protocol allows countries to ban the importation of GM organisms. However, there must be credible evidence regarding the dangers of the product. This protocol, therefore, states that goods developed from new technologies should be regulated based on the precautionary principle [15].

Several ethical considerations surround GMOs. The safety and welfare of consumers appear to be the biggest concerns. The European nations have expressed worry that their inhabitants would experience sicknesses similar to bovine spongiform encephalopathy, commonly known as mad cow disease. On the other hand, some policymakers recognize that GM crops could lead to future discoveries in other fields. This begs the question of whether it is possible that certain technologies could be first tested under highly controlled conditions in the laboratory, then in microcosms, and, if found to be sufficiently safe, on crops before animals. Another question is how likely it is that methods for altering plant genetics can be tested for transferability to humans, animals, and wild forms of plants and microbes, that is, gene flow [30].

Opportunity costs, such precautions slowing possible improvements in health, need to be considered. Enhanced nutritional value, disease prevention, and other health-related positive outcomes should also be considered. Another possible negative outcome is that the effects of the crops on the environment should be investigated. An ecological question is the extent to which new plant species will interrupt predator-prey balances or become opportunistic, nuisance species [15].

A flow chart (Fig. 17.1) can be used to demonstrate the decision-making process in this case. At the beginning, Europe would like to ban the GM food. So, they approach their first question: Is there evidence to support such an action? The countries must then determine if there is sufficient evidence to support such an action. If they do have the evidence, Europe must next ensure that the studies meet all scientific requirements (i.e., they are replicable,

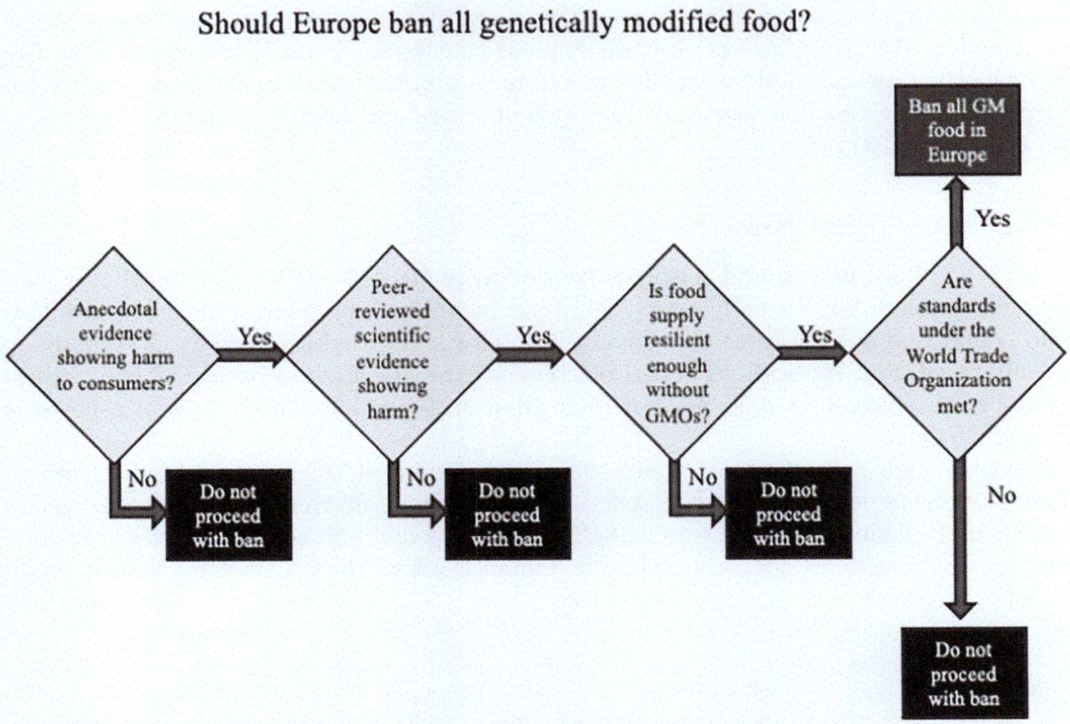

FIGURE 17.1 Decision flow chart for genetically modified (*GM*) food in Europe. *Data from D.A. Vallero, Biomedical Ethics for Engineers, The Biomedical Engineering Series (2007).*

etc.). If the studies are not sufficient, the ban cannot be established until further research has been conducted [15].

If all the scientific evidence is credible, the countries then encounter the feasibility issues. Can the countries be sustained without GM foods? Most of the food produced has some form of genetic alteration, so such a ban would be dramatic.

This question also entails a political aspect. Moreover, with such a ban, it is undeniable that many of Europe's trading partners will be infuriated. So, will the nations be able to handle the political backlash of such a regulation?

By answering yes to the above questions, Europe can move on to the final question—would this regulation be permitted under World Trade Organization guidelines? If the answer is yes, they would decide to ban the food. If the answer is no, they have no choice but to allow the food to be sold in their nations [15].

The initial decision posed was whether Europe should ban GM foods. Europe had two options: to ban or not to ban the food. If they chose to ban, they had two suboptions. First, they could attempt to obtain studies and research to support the decision. Second, they could avoid the research step and proceed to ban GM food. This demonstrates that GMOs have macroethical and geopolitical ramifications, so no matter how comfortable the GMO researcher may feel, there will likely be opposition to the research [15].

The most powerful and controversial type of genetic modification is synthetic biology, that is, an interdisciplinary approach by biologists and engineers to manipulate deoxyribonucleic acid (DNA) to design and redesign genes, biological pathways, and entirely new organisms [31–34]. There are certainly many beneficial applications of synthetic biology, like promising therapeutics and sustainable agricultural products, such as fertilizers. However, when conducting research with such high potential risk and large uncertainties, optimization must be conservative.

17.3.2 Neurological research

Ethicists, philosophers, and scientists continuously struggle with defining what constitutes personhood, but virtually every definition includes the human mind. So, the mind-brain connection encompasses elements of both ethics and neuroscience. Nanotechnology and other emergent applications of neurotechnology can affect who a person is. The medical definition of neuroethics is somewhat pedestrian and circular, that is, ethical aspects of neuromedicine.

Arguably, neuroethics encompasses more than merely a subset of biomedical ethics. Manipulations of neural tissue affect who we are, including our free will. This has intrigued ethicists for millennia. Some neuroscientists broaden the definition of neuroethics to include aspects of responsibility, personhood, and human agency. These issues are already upon us [35]:

> "How will the rapid growth of human brain/machine interfacing – a combination of neuroscience and informatics (cyborgery) – change how we live and think? ....These are not esoteric or science fiction; we aren't talking about some science- fiction prospects of human cloning, but prospects and problems that will become increasingly sharply present for us and our children in the next ten to twenty years."

Every neuroethical issue is entangled with the concept of "self." Indeed, self is more than an aggregation of neurons, or even their synaptic functions. The challenge is how to characterize the mind and self, while merging the scientific method with ethics. But this is nothing new. Rene Descartes tried but died before completing work on this conundrum [36]. The effort continued with the rationalists, who held a view closest to Descartes' [37,38]. Next, David Hume and the empiricists tried, and the search continues with the modern utilitarian and deontological viewpoints [39]. All these views fall short in dealing with the self, so they also fail in providing an ethical framework for neuroethics. For example, Hume's logical positivism contended that all knowledge is based on logical inference from simple "protocol sentences" grounded in observable facts [40]. However, this argument is self-defeating since even this postulation cannot be observed. Perhaps, this argues that mind and brain are separate but interrelated concepts.

Thus neuroethics must reconcile two perspectives [41]: (1) the doctrine of psychophysical parallelism, which holds that a mental state is always accompanied by neural activity, and (2) the concept of self, which holds that the self or person is more than mental functions, but also includes temperament or motivation, which are affected by physiological systems other than neural (e.g., endocrine) and, some would argue, nonphysiological factors (e.g., the soul). The first perspective is the most widely held by the neuroscience community, but the second is the one that is embraced by most people, and which is at the heart of many faith traditions. This difference is fodder for ethical conflict between the professional (e.g., the treating physician or hospital staff) and the patient. It also explains in part the intense conflict in coma and other cases, such as those where a person is in a permanent vegetative state (PVS), but whose loved ones "know" that that person is still living. For example, Roman Catholic tradition holds that "the administration of water and food, even when provided by artificial means, always represents a natural means of preserving life, not a medical act" [42]. However, nutrition and hydration may be discontinued if all medical interventions are deemed futile or if these actions impose disproportionate and excessive burden on the patient [43].

17.3.3 Animal testing

Respect for animals has increasingly been integrated into the Western ethos, including rethinking the majority utilitarian perspectives within the medical and biological research communities. Animals have played a key role in biomedical research and technological development. Animal research highlights some of the problems with the popularly held utilitarian ethical framework, especially how this model influences compartmentalization and objectification.

Compartmentalization in thinking is not only a division of labor, but it is often a survival mechanism for many researchers who engage in uncharted areas of science and have to live with the ethical decisions they make along the way. Sometimes, they are not even aware that they are making an ethical decision, such as the unwillingness to consider downsides of their research. Those who prefer to be blissfully ignorant about flaws and possible misuses of their research are still parties to those misdeeds. This was best demonstrated in the author's discussion with author and veterinarian, Richard Orzeck, who candidly shared the following case:

> During my second tour of duty through Cornell's pathology department, I was working late one weekday afternoon in the hospital's postmortem room performing an autopsy on a young Labrador who

had died unexpectedly from no obvious cause, when one of my friends from the junior class walked into the room leading a dapple-gray, quarter horse mare. The pathology professor and I had just finished a long and detailed study of the poor dog's lungs and heart when he decided that we both needed a short break. Having nothing else to do while he was out smoking his cigarette, I went over to chitchat with my underclassman colleague.

As she stood there holding the silent and unusually well-behaved horse by its lead, we talked about all the exciting things that vet students talk about when they're able to find the time: how our classes were going, how my rotations were going, and how demanding all the professors were. After a couple of minutes of this small talk, I casually asked her what she was doing with the horse here in this room normally set aside for studying the dead.

Her eager and straightforward answer surprised me. She said that the animal was a donation to the college by an owner who, for whatever reason, no longer wanted it. She said that she had been working with a research professor after classes on a project focused on degenerative joint diseases in racing and performance horses, and that the animal was part of their study.

Reaching over to stroke the horse's neck, I looked into eyes of my future colleague and smiled as I congratulated her on being asked to be part of this research project. As a student research assistant, even though the job mostly involves doing all the "dirty work" such as mucking out the stalls and feeding and caring for the research subjects, you do get to interact on a higher level with the doctors and professors in charge. And it can all be pretty exciting stuff, especially if a scientific or medical breakthrough is discovered. Still curious as to why she was here in the postmortem room with this obviously healthy horse, I asked her again why they were here.

A few of seconds of awkward silence ensued as I waited for her to answer. Reaching up with her free hand to pet the horse on its muzzle, she finally told me that she was waiting for the professor to arrive. After he arrived, they would *euthanize* the animal in order to harvest its healthy joint cartilage. These tissues were needed as a *positive control* in their research. And then she said no more. She seemed quite excited by the whole thing, and I remember, just briefly, being a little surprised at her enthusiasm.

I congratulated her again and not giving it any more thought at the time, returned to continue my autopsy of a Labrador retriever. Several minutes later I saw that the research professor had indeed arrived, and my friend and he carried out their grim task (I won't go into the details) to advance the "noble cause of medical science." My surprise at her enthusiasm was because I knew her to be—just like all my vet school colleagues—a dedicated and compassionate person who loved all animals, especially horses. She was subjecting herself to a rigorous, unrelenting, and expensive eight years of college and medical school to fulfill her life's dream of saving the lives and improving the welfare of her animal patients. But here she was, ready to end the life of a perfectly healthy horse and eager about the opportunity to do so.

Orzeck's case indicates the extent to which scientists can rationalize behaviors and decisions within a "research" context that we would not otherwise do. This is truly an example of the ends justifying the means. Orzeck continues:

I can't say for sure, but by the words that my fellow student used allowed her to justify and to absolve herself of what she was doing. "She and the researcher were going to euthanize the animal in order to harvest its healthy joint cartilage so that they could use the tissue as a positive control." Animal? Euthanize? Harvest? Positive control? To make what she was doing more bearable, she transformed this living, breathing horse into an object of exploitation through the use of these very specific words:

Animal: The creature was no longer a horse; it was now an animal object. It was easier to accomplish what she had to do to an animal or research subject than it was to a living horse.

Euthanize: Even though I and all other veterinarians (and, quite sadly nowadays, even the human medical profession) use the term all the time, the word *euthanize* also is quite sneaky. The term has been bastardized and twisted and applied to many situations less ennobling. It's nicer to say, "euthanize the research subject" than it is to say, "kill the innocent horse."

Harvest: The term brings to mind comforting images of strong and hardy farmers gathering up sheaves of wheat and corn, or perhaps little red-cheeked children helping their dear grandpa pick apples from well-trimmed trees in his ancient New England apple orchard. But the word also has recently been adopted by various groups of people to rationalize some despicable behavior. Credit and banking agencies "harvest" and "mine" the data that they extract and collect from their clients; various government agencies "harvest" mountains of covertly obtained information on its citizens for their so-called common good; and, most recently, scientists have sought to have the right to "harvest" stem cells from aborted babies to advance the causes of medical research. In summary, it was more pleasant to say, "harvest the tissue" than it was to say, "We're exploiting this helpless horse to steal its healthy joint cartilage for our research."

Positive control: The use of positive controls is an honorable and long-used scientific technique that allows researchers to make accurate comparisons of information they are trying to obtain for whatever study they are trying to make. The best example I can think of involves human medical research and a study in which my father was asked to participate. As a frequent hospital patient suffering from heart disease, he had been asked to participate in a study of a new drug, which, if it worked, would improve the strength of his heart. On the permission sheet, he was offered free care during the length of the study, but he would not be told whether he would receive the new drug or a placebo. (A placebo is something that looks like a drug but is actually an inactive substance.) The placebo group would then be referred to as the positive controls. Their purpose was to act as a baseline, or starting point, against which any effects of the new drug could be compared.

The problems with using positive controls are many. The individuals, whether human or animal, don't receive any potential benefits of the scientific study that's being performed. In the worst case, the positive controls could be deprived of a life-giving procedure. But the biggest problem regarding medical science and the use of positive controls is that by using the term *positive control*, researchers can make objects out of these patients. And making an object out of living and breathing people now makes them easier to use and take advantage of, all in the name of science. It is easier to justify and to appear heroic to sacrifice a positive control than it is to put to death an innocent horse.

I hasten to add that before we pass judgment on these otherwise compassionate and loving persons, we must be reminded that nearly all of us are guilty of objectification. For example, when clients bring their pets to my practice for euthanasia, I always ask them what they would like me to do with their remains; probably because the word *remains* is easier to deal with and just sounds nicer than *dead pet's body*.

Orzeck's case is yet another example of how terminology is not simply a neutral conveyance of information but is often steeped in ideology and perspective. Medical and engineering professionals, as Feynman reminded us, must be diligent and vigilant in using the correct terms to communicate. *Junk science* is fraught with the loose usage of language. In fact, strategic redefinitions and omission or selective use of actual data are common fallacies in junk science.

17.3.3.1 Is the research worth it?

Debates about animal research elucidate several ethical issues. These debates have taken place for centuries among thoughtful people. As evidence, René Descartes (1596–1650), the French philosopher and one of the founders of modern science and mathematics, saw animals as mindless machines [44]. This can mean that treating animals any way we please should not raise any ethical concerns whatsoever. Descartes' view was challenged by contemporaries, notably by Pierre Gassendi (1592–1655), a Roman Catholic priest who, in his own right, helped to shape modern scientific thought. Gassendi argued that the moral focus should be on animals' ability to perceive and to suffer, rather than Descartes' conclusion that they lack the ability to reason and speak. Further, Gassendi argued that since humans *can* reason, we have a duty to be humane and to exercise compassion toward animals rather than to dominate and exploit them.

As illustrated in the debate between Descartes and Gassendi, the difference between humans and animals is an important distinction. To most biologists, the difference is merely a continuum, as indicated by the development of the nervous system and other physiological metrics. These physiological complexities translate into sensory differences, which differentiate the species' sentience (especially self-awareness), which is one of the variables that distinguish "humanness" [15]. In fact, sentient-centered ethics falls between anthropocentric (human-centered) and biocentric. The latter holds that all living creatures have inherent worth and intrinsic value (see Fig. 8.1 in Chapter 8). Indeed, a common theme of this book, along with a systematic approach and the need for professional trust, is that of empathy. It is noteworthy that Albert Schweitzer's reverence for life has been characterized as a biocentric and a "bioempathetic" viewpoint [45]. That is, it calls us to appreciate in a personal sense the perspective of other creatures. With the advances in understanding of neurological processes, for example, we must assume that the more highly developed animals (and possibly even many of the less advanced) experience pain. That said, the bioethical view would cause us to want to do whatever we can to prevent or at least reduce suffering in these other species and to avoid perceiving them merely for their instrumental value.

The ethical problem ensues when the utility of animal-derived knowledge is presented as a dichotomy. This can cause the utilitarian view to come to the fore, forcing a choice between preventing and curing some human malady versus harming animals. This viewpoint is inherently weak, however, and such an argument is invalid. This illogical argument is referred to as the fallacy of *non sequitur* (Latin: "does not follow") since the outcome (cure) does not really depend on the condition (animal suffering). This particular *non sequitur* argument is known as "denying the antecedent." The non sequitur follows this logic:

1. If A, then B. (Animal research leads to cures, so people benefit from animal research.)
2. Not A. (Not allowing animal research would end animal suffering.)
3. Therefore, not B. (Not allowing animal suffering would prevent cures.)

Not all animal research leads to animal suffering, or at least the suffering varies substantially according to the type of research.

Another type of *nonsequitur* is affirming the consequent:

1. If A is true, then B is true (If animals suffer in research, cures are developed).
2. B is stated to be true (Animal research has resulted in cures).
3. Therefore, A must be true (Therefore, animals must suffer).

Here, the argument is made based on a false dichotomy that denies the possibility that animal research can be conducted in ways that are humane. It also denies that even if the animal research were necessary in the past (or at least beneficial to the advancement of medical knowledge), it does not mean that alternative approaches can now displace such research, such as replacing in vivo research with in vitro and in silico methods.

Moral concern grows with respect to what we value. For example, we are less tolerant of harm to our own pet than to a "generic" lab animal. Also, concern usually increases as an animal is considered more "human-like." This value can be seen even in people who believe that animals differ from humans both in degree (i.e., a continuum of self-awareness and cognitive processes) *and* in kind (i.e., inherently different, especially since many believe humans have a soul and animals do not). In both instances, people value certain animals more than others. In the first case, the species (i.e., the cat or dog) may have more value to the pet owner than species other than that of the pet; however, this value can be transferred to other animals. The second type of valuation increases with the complexity of the species. Greater concern may result because people ascribe human characteristics to nonhuman species (known as *anthropomorphism*). A recent example was the well-deserved criticism of the National Institute of Allergy and Infectious Diseases (NIAID) research on beagles that seemed to be quite cruel and arguably did not add much to the science [46].

Primates are particularly important indicators of moral value, since they share many more "human" traits than most other animals. From a biomedical perspective, the fact that nonhuman primates have similar anatomical and physiological features makes them ripe for research. From a bioethical perspective, their similarities argue against much of the research, especially that which is invasive, painful, and unpleasant. The case of psychological researcher Edward Taub and his George Washington University student Alex Pacheco, who later became cofounder of the advocacy group, People for the Ethical Treatment of Animals (PETA), dramatically illustrates the bioethical challenge.[4]

From 1981 to 1991, the National Institutes of Health provided $1.2 million to the Institute of Behavioral Research in Silver Spring, Maryland, for Taub's attempt to regenerate severed nerves in 17 Rhesus Macaque (*Macaca mulatta*), popularly known as the Rhesus Monkey. Sensory nerves were cut in the monkeys' limbs (**deafferentation**), then stimuli such as electric shock, physical restraint, and food or sight deprivation were applied to compel animals to regain use of their crippled limbs. Taub's rationale for this research was to aid stroke victims and the mentally handicapped [15].

In 1982, Pacheco, who took a summer job at the institute, visited the laboratory and took photographs of the starved, uncared-for animals. He reported the lab to the state police, who then raided the lab, removed the monkeys from the lab, and the case went to court. The monkeys were euthanized at the end of the case. Taub was convicted of six counts of animal cruelty [15].

The Taub case illustrates some of the ethical aspects of animal research: definition of human life versus animal life, the application of precautionary principles, dignity in any research methodology, the question of whether animals have rights, and whether useful results truly justify research. On this last count, Taub's research appears to advance the state of neurological science, but is it worth it?

Animal research is of three types: pure research, applied research, and testing. Pure research strives to advance the understanding of phenomena, structures, processes, and systems for the sole benefit of medical and scientific knowledge. Applied research is

[4] This discussion draws on the ideas of Diana Chang, who conducted undergraduate research in an independent study course that the author facilitated at Duke University.

research specifically conducted to address a societal need. Testing studies are special types of applied research that test the effects of a procedure, device, or drug to determine efficiencies and efficacy. Thus animal data provide a utilitarian purpose. Because animal researchers hold human life to be paramount, they may view animals as a means to an end and justify this based solely on possible benefits to humans [15].

Viewed from this absolute anthropocentric perspective, the Silver Spring monkey would be a morally and ethically sound case. However, there must be other nonutilitarian aspects to the case that are immoral. Let us consider how the court may have arrived at a guilty verdict, and arguably, a pronouncement of unethical practices using the decision tree shown in Fig. 17.2. Numerous consequences of this case are positive. Whereas the animals may have been abused and lived in deplorable conditions, there were medical advances.

It appears that Pacheco made his decisions based on mainly three factors: legality (animal rights, guidelines), duty to the health/safety of animals, and politics/public opinion. He showed less regard for the potential of these studies to aid human victims of stroke and mental retardation, as Taub proposed. His position seems to have been that not only was fruitful research not morally obligatory, but it was also not morally

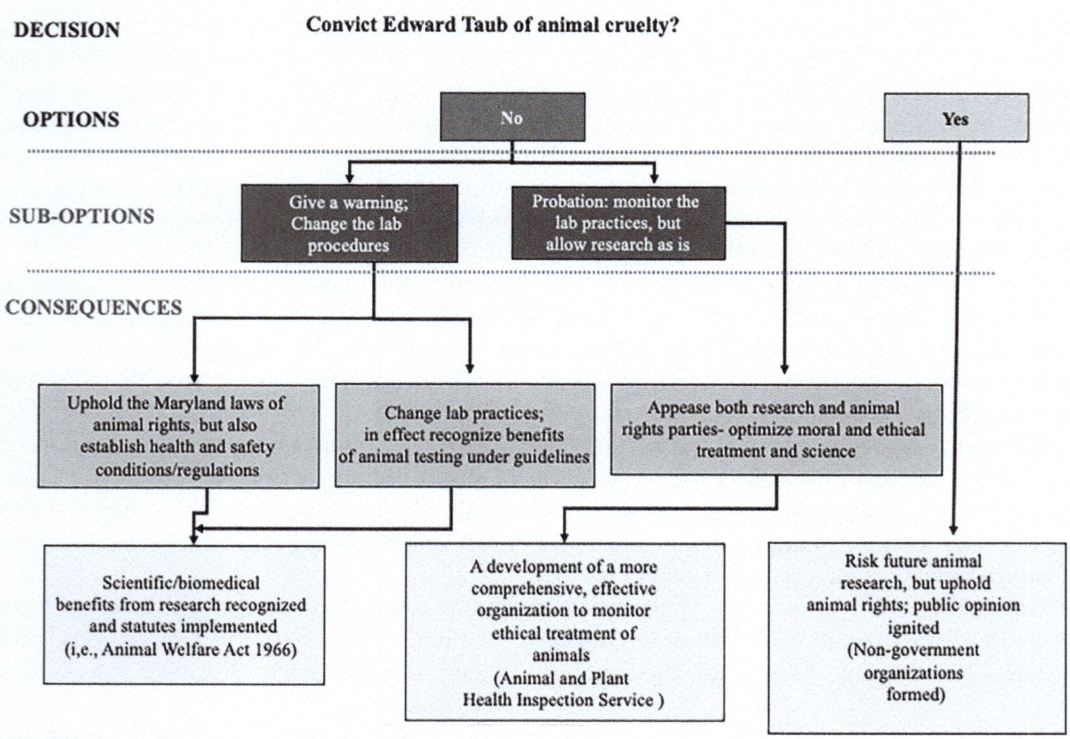

FIGURE 17.2 Event tree showing possible decision flow in Silver Spring Rhesus monkey case.

permissible. This seems to be a more deontological view than that of Taub, who was most concerned with the benefits to science/medical research, finding cures for humans, and the finance/economics of funding. In fact, many such whistleblowing cases exhibit a duty-based view. Some are indeed utilitarian, for example, if a laboratory is doing bad science that will lead to negative consequences (e.g., dangerous device or drug), but others are despite good science if immoral activity is ongoing (e.g., cruelty and harassment). In the end, the courts sided with Pacheco and convicted Taub. This logical progression can be expressed as a syllogism:

Factual premise	Animal research is a powerful tool for scientific and medical research; and Taub's monkey-based studies were intended to study nerve regeneration and recovery of crippled limbs
Fact-value premise	The animal subjects lived in deplorable conditions and were abused
Evaluative premise	Even though animal research is a powerful tool for scientific and medical research, but abuse of animal subjects in studies is morally wrong
Evaluative conclusion	Therefore, Taub's studies, while intended for research, violate animals' rights and are morally wrong

Therefore, animal research goes beyond the "whether the nonhuman nature of other animals is morally relevant, and if it is relevant, what humanity and justice permit us to do with animals" [47]. As we learn more about animal neurophysiology and psychology, it becomes harder to argue that the values of justice and humanity are not applicable to animals and do not merit ethical consideration. One does not have to be a hardcore member of the Animal Liberation Front to agree that some cases go beyond the pale. Few would disagree that the studies conducted at the University of Pennsylvania, wherein baboons were rendered brain dead from having their heads cemented to pistons that whipped the brain at accelerations up to 2000 times the force of gravity, while researchers laughed as the primates were restrained, alert, and writhing before impact [47]. However, most animal welfare cases are not so clear.

17.3.3.2 Reality check

Biomedical and biosystem engineers depend on animal studies and are more likely than many to support such research as a critical means toward a noble end. So, how can engineers and other technologists be objective and realistic about matters where they are likely to have such a bias and conflicts of interest? One way is to do what engineers do best; use the systematic approach [48]. Start by gathering the pertinent facts and identifying the uncertainties and data gaps. Ethical problem solving is fraught with ambiguities and assumptions, so it is crucial that decisions are based on the best available facts and that the decision maker is honest about what is not known.

Next, the gathered information is used to identify realistic options and alternative solutions. Conduct a sensitivity analysis, that is, compare how each alternative performs in terms of how sensitive it is to slight changes in scenarios needed to reach the decision. For example, one alternative may produce excellent outcomes (high on the utility scale), but in the process leads to injustices or disproportional harm (low on the fairness scale). The alternative of withholding the milk study results mentioned earlier, for instance, could be seen as one where one utility is optimized (i.e., security), but at the expense of another value (i.e., scientific freedom). Based on these analyses, several plans for addressing the problem can be considered.

All the reasonable plans can be compared and assessed against moral metrics, such as those embodied in moral theories like utilitarianism, Rawlsism, deontology, and rational models[5]. Key to these assessments is characterizing all the potentially affected parties and the stakes each has in the decision. Begin by asking what the risks and benefits to each party from each option are.

Finally, make a well-informed, scientifically and ethically based decision. As is the case for the ethics of a design, the engineer is not finished when the best alternative has been selected. After the decision is made, feedback must continue to be sought so that the options and alternatives are revised to ensure that the decision remains the right one. Ethical decision-making can be messy and chaotic [49,50]. Unanimous decisions are the exception. Even consensus can be difficult. Sometimes, the right decision is made in the face of a resistant majority. Thus the process is never static, and the actions may need to be adapted as new information becomes available. For example, the constraint may disappear. The choice of keeping some lead (Pb) in gasoline, even though it was known to have neurotoxic effects, eventually went away in the 1980s when alternative fuel additives became available and when engines were redesigned. Or, when a sufficient repository of data becomes available, the need for even important animal testing can be replaced by computational methods (i.e., in silico studies and informatics). Thus we need to continue to look for alternatives to animal research.

Animal welfare is sometimes categorized as the "3R's":

1. Reduction—methods that result in the use of fewer animals to obtain scientifically valid information
2. Refinement—methods to reduce stress or discomfort for the animals involved and to improve the animals' overall well-being and environment
3. Replacement—methods other than animal studies that can provide robust biomedical information and modeling

17.3.4 Cloning and stem cell research

> ... cloning represents a turning point in human history—the crossing of an important line separating sexual from asexual procreation and the first step toward genetic control over the next generation. It thus carries with it a number of troubling consequences for children, family, and society. *Leon R. Kass, Chair, The President's Council on Bioethics (2002) [9]*

In November 2001, researchers in the United States asserted that they were able to produce the first cloned human embryos, merely reaching a six-cell stage before the cells stopped division and died. In the meantime, some fertility specialists had declared a strong

[5] These classical ethical frameworks are discussed in Chapter 4.

intent to clone human beings. In response to the technical and societal uncertainties and anxieties, the US Congress began in 1998 to consider these issues, with the House of Representatives passing a strict ban on all human cloning, including the production of cloned human embryos, in July 2001. Since then, several cloning-related bills have been considered in the US Senate and several state legislatures. Numerous nations have banned human cloning, with the United Nations considering an international convention on the issue. It suffices to say that the political and societal aspects of cloning are as challenging as the technical demands.

A bit of science background is in order here. The biology of cloning begins with the ovum (egg). A human ovum consists of a single *gamete* cell, containing only 23 active chromosomes. This means that the sex has not yet been determined. Once the ovum contains a complete nucleus from any species that is activated and developing—whether that has occurred by sexual fertilization or by asexual somatic cell nuclear transfer (SCNT)—an *embryo* of that species (*Homo sapiens*, sheep, cat, dog, etc.) is produced. Cloning is a type of reproduction wherein offspring result not from the chance union of ovum and sperm (sexual reproduction); rather, it results from the deliberate replication of the genetic makeup of another single individual (asexual reproduction).

Human cloning is accomplished by introducing the nuclear material of a human *somatic cell* (donor) into an *oocyte* (egg) that has had its own nucleus removed or inactivated. This yields a product with a human genetic constitution virtually identical to that of the donor of the somatic cell. This technique is known as SCNT. Since SCNT uses human genetic material, the developing embryo is of the species *H. sapiens*.

17.3.4.1 Bioethics question: is research using human pluripotent stem cells ethical?

A *pluripotent* cell can be differentiated into more than one alternative type of mature cell, so it is able to produce all the cell types of the developing organism's body. Thus a pluripotent stem cell has the same functional capacity (i.e., pluripotency) as an embryonic stem cell, though it does not necessarily share the same origin. Stem cell research involves isolating human embryonic stem cells from embryos at the blastocyst stage or from the germinal tissue of fetuses [51]. As of this writing, such harvesting kills the donor. The embryonic stem cells have been harvested from in vitro fertilization (IVF). Human adult stem (i.e., *multipotent*) cells have been isolated from a variety of tissues. These stem cell populations can be differentiated in vitro into various cell types and are currently being extensively and intensively investigated for potential applications in regenerative medicine. Many scientists believe that embryonic and adult stem cells can lead to treatments for many human maladies. However, much of this is conjecture at this point.

The Chairman of the President's Council on Bioethics succinctly characterized the stem cell debate [52]:

> While they may well in the future prove to be of considerable scientific and therapeutic value, new human embryonic stem cell lines cannot at present be obtained without destroying human embryos. As a consequence, the worthy goals of increasing scientific knowledge and developing therapies for grave human illnesses come into conflict with the strongly held belief of many Americans that human life, from its earliest stages, deserves our protection and respect [52].

The diametric opposition lies between those who see embryos as living human beings entitled to protection versus those who consider them merely as "potential" humans that may be used as researchers see fit in an effort to advance the state of knowledge. This use includes the destruction, that is, the killing, of embryos to harvest stem cells.

One of the major ethical issues of stem cell research revolves around the first stage, especially the means of harvesting the embryonic stem cells by IVF. In particular, the argument contends that since fertilization has occurred, we are conducting research on a living human being. The IVF process itself is cavalier about the value of the individual, fertilized ova, destroying and discarding many in the process.

In an undergraduate seminar at Duke, students shared why IVF seems to "get a pass" morally, at least in comparison to the scrutiny given to stem cell and cloning research, even though these all are beginning of life (BOY) issues. One student pointed out that IVF is not new. Others noted that the technologies are more mainstream and understandable to the public. One student suggested that the fact that many people know individuals who were conceived via IVF has diminished the mystique. During the discussion, it became obvious that most of the students had previously given much thought, if any, to the morality of IVF. It simply exists as an alternative means of reproduction. Furthermore, the IVF industry is considered rather positively. The discarded embryos are just a byproduct of the process that allows people who could not ordinarily do so to have children. Thus few seem to question the morality of IVF, and the discarding of embryos is seldom seen as an immoral act of the fertility clinics. Further, those who oppose the use of embryos that would otherwise be destroyed are seen as "wasteful" and "myopic," or even as neo-Luddites who stand in the way of progress and the betterment of humankind.

As the President's Council on Bioethics puts it: "All extractions of stem cells from human embryos, cloned or not, involve the destruction of these embryos" [53]. A rather common paradox is that research may either exacerbate or ameliorate ethical issues. For example, if emerging research supports the contention that embryonic stem cells are absolutely essential to provide the treatment and cures for intractable diseases, many scientists and ethicists will push the "greater good," utilitarian viewpoint, while others will see this as malicious means (sacrificing embryos and living human beings) toward an end (treatment).

Furthermore, ethicists will be faced with the "slippery slope" dilemma. This argument sees a progression from diminished respect for embryos to a lack of respect for all unborn human beings, which leads to less respect for weaker members of society. This last step in the progression is one argument for an incremental acceptance of eugenics, the belief that information about heredity can be used to improve the human race. The dreadful truth is that the genocide and other inhumane attempts at eugenics, such as the Nazis' attempted purging of Jewish people in the 20th century, were clumsy and inefficient compared to emergent genetic manipulations available now and in the foreseeable future. The slippery slope advocates fear that the next Hitler will have subtle tools and a "rationale" supported by many to accomplish similar ends (i.e., eradicating select human subpopulations).

Conversely, scientific advances can improve the bioethical landscape. Many of us hope that the need for embryonic stem cells will soon be obviated by scientific advances that allow the same cures and treatments to be achieved using adult stem cells. Recent advances cause us to be hopefully optimistic (see discussion box: Nanog).

Nanog

In Celtic legend, Nanog (shorthand for *Tir Nan Og*) was a land of the eternally young. But the Nanog molecule is at the threshold of what *could* be the unraveling of the need for embryonic stem cells to advance medical science. Scientists have argued that embryonic stem cells are unique in their ability to develop into any type of cell in the body, meaning that they show great, yet still unfulfilled, potential to replace and to mend damaged tissue. Thus the argument goes, embryonic stem cells are essential for such medical treatments.

Researchers have become aware that Nanog may someday allow ordinary cells to make use of that wonderful attribute.

Questions:

1. What are some of the scientific obstacles that need to be overcome in Nanog research?
2. If the science supports this advance, what might be some reasons that scientists and others may still be skeptical of these findings?
3. Compare this type of skepticism to Kuhn's description of the scientific community during a paradigm shift.

17.3.5 Human enhancement

Engineering is all about enhancing our environment and ourselves. However, the tools that have emerged in recent decades are changing the nature of what we *can* do. So, let us conduct a thought experiment:

> Dave is a biomedical engineer working on an aerosolizer that will allow a steroid to be administered easily and delivered to the target organ (in this case, the lung's alveolar region) with 100% efficiency. This will be a monumental advancement in drug therapies for asthmatics where much of a particular drug does not find its way to the target alveoli and is lost to other inefficacious regions. However, in the process, since the drug is delivered with such efficiency, metabolism mimics natural corticosteroids so well that most drug testing protocols will not be able to distinguish the pharmaceutical steroids from those endogenously and naturally produced by the body.
>
> Your company forecasts a large, legal profit, but that it also predicts possible illegal human enhancement activities that may lead to a 25% increase in steroid use by at-risk populations, such as high school athletes.
>
> Is it ethical to produce such a device?

Clearly, there are pros and cons to almost any technological advancement. Often, the difference between research success and failure is measured by change; in particular, the difference between the *status quo* and the condition that is changed by the innovation. From a utilitarian view, the value that society places on each aspect of this change determines whether something is moral or immoral.

Research changes the future. Some changes are profound, such as future invoked by the designs of Copernicus on space travel, Orville and Wilbur Wright and Henry Ford on transportation, Paracelsus and Galen on medicine, and Newton and Galileo on monitoring devices (optics and remote sensing, respectively) [15]. But, as the controversial Edward O. Wilson's quote reminds us, enhancement evokes negative futures as well [54]:

> *Homo sapiens*, the first truly free species, is about to decommission natural selection, the force that made us.... Soon we must look deep within ourselves and decide what we wish to become.

So, then, is enhancement good or evil? The answer, as usual, is "it depends." Ethics operates between the realm of the clearly wrong and clearly right. This is evident by the words we use to compliment or to condemn the work of a professional. Consider the following adjectives:

- Although painful, the therapy was *worthwhile*.
- The surgeon exposed his patient to *unnecessary* pain.
- The stent's failure rate is *unacceptable*.
- The defibrillator's reliability is *tolerable*.

A common human failing is that we are frequently *xenophobic*. We fear and resist change. We tend to keep doing things the way we have always done them. Maybe this is not so much a "failing" as an adaptive skill for the survival of the species. If we suffered no ill effects after eating that fruit, we would eat more of that particular species. But even a similar-looking species may be toxic, so we are careful to eat only the species that did not kill us, as best we can differentiate. So, researchers face a paradox. We do not want to expose ourselves and others to unreasonable risks, but they must to some extent "push the envelope" to advance knowledge. This anxiety could lead to the suppression of new ways of looking at problems. Sometimes, the facts and theories may be so overwhelmingly convincing that we must change our worldview. This is consistent with Thomas S. Kuhn's concept of paradigm shift [55]. Scientists are often very reluctant to accept these new ways of thinking (Kuhn said such resistance can be "violent"). In fact, even when they have been accepted, they are often not dramatic reversals (revolutions), but mere modifications of existing designs (evolutions). To wit, the bicycle was merely a mechanical re-rendering of the horse (e.g., the saddle seat, the linear and bilaterally symmetrical structure, and the handlebars); so was the automobile.

Biomedical research has certain guiderails, especially to protect human subjects from unnecessary risks. This is principally accomplished through programs the require full disclosures of potential risks and the requirement of a subjects informed consent prior to the subject's enrollment in a study. In a sense, however, the research on a technology does not stop when the technology or product is approved by the regulatory authority. With use, certain unforeseen side effects and other surprises can occur as a larger population is exposed beyond the test groups. As such, an enhancement technology must be monitored by the practitioner to ensure that an individual's unique biome, behaviors, or other idiosyncracies do not increase the risks to that patient. Indeed, sometimes the surprises are welcomed, such as a drug approved for one disease is found to be efficacious against another, unexpected malady, such as a treatment for high blood pressure (hypertension) found to be effective in reversing hair loss [56].

17.3.6 Patenting life

Bioprospecting, the search for natural substances of medicinal value, is very divisive topic in bioethical debates. In November 1999, the US Patent and Trademark Office rescinded a patent on the plant species *Banisteriopsis caapi* held by a Californian since 1986 [57]. The plant is sacred to tribal communities living in the Amazon basin and is the source of the hallucinogen *ayahuasca*, used in their religious rituals. In addition to being a harbinger of the complications of religious and cultural respect, it presages the looming, bitter debates about the extent to which biological materials can be "owned."

In fact, humankind has been in the bioprospecting business for millennia. Like many bioethical issues, emerging technologies and research have changed both the ethical landscape and the biological landscape dramatically. Powerful interests, such as pharmaceutical companies, see natural materials (including certain genes) as lucrative ventures that need to be harnessed for profit. For example, the biotechnology firm, Diversa Inc., entered into an agreement with the National Park Service to find efficacious and beneficial microbes in the geysers and springs in Yellowstone National Park [57]. Subsequently, however, the agreement was suspended by a federal court ruling.

As controversial as patents on plant genetic material are, it pales in comparison to the bioethical debates surrounding those of animals. This is in part because patenting animals' genetic materials is linked to cloning. The larger bioethical issue is captured well by the Church of Scotland's Society, Religion and Technology Project [16]:

> Many people would also say that knowledge of a genetic sequence itself is part of the global commons and should be for all to benefit from. To patent parts of the human genome as such, even in the form of "copy genes," would be ethically unacceptable to many in Europe. In response it is argued that patenting is the legal assessment of patent claims and should not be confused with ethics. But patenting is already an ethical activity, firstly in that it expresses a certain set of ethical values of our society; it is a response to a question of justice, to prevent unfair exploitation of inventions. Secondly a clause excluding inventions "contrary to public order and decency" is part of most European patent legislation - an extreme case of something like a letter bomb would be excluded as immoral. But now we have brought cancerous mice and human genetic material in the potential frame of intellectual property, ethics has moved to a much more central position, where it sits uncomfortably with the patenting profession. They do not like the role of ethical adjudicator to be thrust upon them by society. ***Teachable Moment: Patenting Germplasm***

Consider the following statement from Keith Douglas Warner of Environmental Studies Institute at Santa Clara University:

> The privatization of germplasm formerly considered the common heritage of humankind is incompatible with notions of the common good and economic justice. The scrutiny that life industries have been receiving is well deserved, although most of this attention has been focused on the potential threats to human and ecosystem health. The economic implications of the biotechnology patent regime are less obvious because they do not impact individuals, but rather social groups. The public appears less interested in this dimension of the biotechnology revolution. Nevertheless, addressing this patent regime through the lens of the common good is a better strategy for critics of agricultural biotechnology, who will likely be more successful in slowing down the expansion of corporate control over germplasm by addressing economic issues [58].

Questions:

1. The biotechnical revolution has improved crop yields and greatly increased the world's food supply in recent centuries. What have been the "human and ecosystem health" tradeoffs associated with these benefits?
2. Is it morally preferable to engage in "slowing down the expansion of corporate control over germplasm" and other genetic materials? Why or why not?
3. Compare any opportunities lost with the risks prevented if germplasm ownership by private concerns is halted.
4. How might the experiences with germplasm relate to more recent emerging technologies, especially generative artificial intelligence and quantum computing?

17.4 The therapeutic misconception

Paul Appelbaum, a psychiatry professor at the University of Massachusetts Medical School, and his colleagues coined the term "therapeutic misconception" to refer to situations where, on the surface, a patient, client, or subject appears to be giving informed consent, but is completely convinced that they are really receiving treatment and not a placebo [59].

According to Appelbaum, despite explanations, patients recruited by their physician strongly believe that research protocols are designed to benefit them personally and directly instead of testing the effectiveness of treatment methods. This effect is not even strongly associated with understanding experiments and the scientific method, since potential subjects who understand technical concepts like placebos, randomization, and double-blind study designs are still apt to maintain their belief that the physician-researchers will give them the best treatment based on the subject's particular needs. In other words, the hospital or medical facility conveys such an atmosphere of trust that the subject expects the professional therapeutic relationship to transfer to the clinical trial. The physician is perceived to be a professional first and a researcher second.

This should serve as a warning that coercion can be direct or indirect. Overpaying someone to participate in a study is an example of direct coercion. Therapeutic misconception is an example of indirect or soft coercion.

Part of the challenge is the very ethical behavior of the physician in cultivating trust with the patient. That said, the physician-researcher must be vigilant in convincing the patient that he or she will be treated the same as any other subject in the research, which means that they have just as much likelihood of being among those who receive the placebo.

17.5 Conclusions

Research ethics has many of the same challenges and expectations as those related to the practice of engineering, medicine, and other professions. However, the work of the researcher carries more uncertainty than that of the practitioner. This chapter highlights some of the important ethical questions, obstacles, and dilemmas of contemporary research. The discussions have centered around how ethical theory relates to experiences in actual research environments.

References

[1] D.J. Haraway, Modest Witness@ Second Millennium: The FemaleMan© Meets OncoMouseTM (1993).

[2] T.D. Little, T.D. Jorgensen, K.M. Lang, E.W.G. Moore, On the joys of missing data, Journal of Pediatric Psychology 39 (2) (2014) 151–162. Available from https://doi.org/10.1093/jpepsy/jst048.

[3] J.C. Little, C.J. Weschler, W.W. Nazaroff, Z. Liu, E.A. Cohen Hubal, Rapid methods to estimate potential exposure to semivolatile organic compounds in the indoor environment, Environmental Science & Technology 46 (20) (2012) 11171–11178.

[4] T.D. Little, M. Rhemtulla, Planned missing data designs for developmental researchers, Child Development Perspectives 7 (4) (2013) 199–204.

[5] T.S.R.S. Sigma Xi, Honor in Science, Sigma Xi, Research Triangle Park, NC, (2000).

[6] National Research Council, Exposure Science in the 21st Century: A Vision and a Strategy. Washington, DC: The National Academies Press (in English), 2012, p. 196.

[7] E. Department of Health, The belmont report. Ethical principles and guidelines for the protection of human subjects of research, The Journal of the American College of Dentists 81 (3) (2014) 4.

[8] D. Rumsfeld, DoD news briefing-secretary rumsfeld and gen, Myers. February 12 (2002) (2002) 11.

[9] L.R. Kass, R. Kass, Human Cloning and Human Dignity: The Report of the President's Council on Bioethics, Public Affairs, (2002).

[10] National Academy of Engineering, The Engineer of 2020: Visions of Engineering in the New Century, National Academies Press, Washington, DC, (2004).

[11] H. Bügl, et al., DNA synthesis and biological security, Nature Biotechnology 25 (6) (2007) 627–629.

[12] D. Rumsfeld, Known and Unknown: A Memoir, Penguin, (2011).

[13] D. Rumsfeld, Knowns & unknowns, JSTOR, (2002).

[14] G. Tulloch, Euthanasia-Choice and Death, Edinburgh University Press, (2019).

[15] D.A. Vallero, Biomedical ethics for engineers, The Biomedical Engineering Series (2007).

[16] J.M. Francis, M. Macdonald, All things new: a brief history of the first forty years of the church of scotland society, religion and technology project.

[17] L. Allan, The principle of double effect, Available from: http://www.rationalrealm.com/philosophy/ethics/principle-double-effect.html, 2015.

[18] Y. Ou, S. Guo, Safety risks and ethical governance of biomedical applications of synthetic biology, Frontiers in Bioengineering and Biotechnology 11 (2023) 1292029.

[19] Y. Wang, A comparative study on the governance approaches for biotechnological risks, Frontiers in Bioengineering and Biotechnology 13 (2025) 1647204.

[20] M. Hukic, E. Hukić, Ethical and biosafety challenges in modern microbiology: navigating the risks of emerging technologies, Spectrum of Dual-Use Technologies: Unforeseen Risks Versus Returns, Springer, (2025), pp. 163–182.

[21] C.I.G. Arruti, The precautionary principle and the protection of biological diversity, American Journal of Applied Scientific Research 8 (1) (2022) 1–10.

[22] E. Persson, What are the core ideas behind the Precautionary Principle? Science of the Total Environment 557 (2016) 134–141.

[23] N.N. Taleb, R. Read, R. Douady, J. Norman, Y. Bar-Yam, The precautionary principle (with application to the genetic modification of organisms), arXiv preprint arXiv:1410.5787, 2014.

[24] M. Monaghan, R. Pawson, K. Wicker, The precautionary principle and evidence-based policy, Evidence & Policy 8 (2) (2012) 171–191.

[25] C.G. Turvey, E.M. Mojduszka, C.E. Pray, The precautionary principle, the law of unintended consequences, and biotechnology, in: Presented at the Agricultural Biotechnology: Ten Years After, Ravello, Italy, (2005).

[26] P. Rozas, E.I. Kessi-Pérez, C. Martínez, Genetically modified organisms: adapting regulatory frameworks for evolving genome editing technologies; Biological Research 55 (2022).

[27] S.W. Ewen, A. Pusztai, Effect of diets containing genetically modified potatoes expressing Galanthus nivalis lectin on rat small intestine, Alternatives Journal 27 (3) (2001) 14–14.
[28] S.W. Ewen, A. Pusztai, Effect of diets containing genetically modified potatoes expressing Galanthus nivalis lectin on rat small intestine, The Lancet 354 (9187) (1999) 1353–1354.
[29] M.A.U. Özgür A., Food allergen risks in genetically modified foods: current approaches, Food Safety (2024).
[30] P.G. Byrne, A.J. Silla, A.J. Kouba, H. Heatwole, Genetic management of threatened amphibians: using artificial fertilisation technologies to facilitate genetic rescue and assisted gene flow, Reproductive Technologies and Biobanking for the Conservation of Amphibians, CRC Press, Boca Raton, Florida, (2022).
[31] M. Sheets, J. Atkinson, Intro to Synthetic Biology (2023).
[32] S. Hirschi, T.R. Ward, W.P. Meier, D.J. Müller, D. Fotiadis, Synthetic biology: bottom-up assembly of molecular systems, Chemical Reviews 122 (21) (2022) 16294–16328.
[33] J. Li, H. Zhao, L. Zheng, W. An, Advances in synthetic biology and biosafety governance (in English), Frontiers in Bioengineering and Biotechnology, Review 9 (2021). Available from https://doi.org/10.3389/fbioe.2021.598087.
[34] B.D. Trump, C.L. Cummings, S. Galaitsi, J. Kuzma, I. Linkov, Synthetic biology: perspectives on risk analysis, governance, communication, and ELSI, Synthetic Biology 2020: Frontiers in Risk Analysis and Governance, Springer, (2020), pp. 1–18.
[35] D. Rees, S. Rose, The New Brain Sciences: Perils and Prospects, Cambridge University Press, (2004).
[36] S. Priest, Descartes, Kant, and self-consciousness, The Philosophical Quarterly (1950-) 31 (125) (1981) 348–351.
[37] D.J. Stump, Rationalism in Science (2005).
[38] S.G. Crowell, Rationalism in history, Diacritics 33 (1) (2003) 3–22.
[39] K. Sharlamanov, M. Kostovska, The Modern Construction and Postmodern Deconstruction of Self, Springer, (2024).
[40] S.I. Ogaba, An epistemic reflection on David Hume's notion of the self, Nigerian Journal of Philosophical Studies 4 (1) (2025).
[41] J.J. Giordano, B. Gordijn, Scientific and Philosophical Perspectives in Neuroethics, Cambridge University Press, (2010).
[42] P.J. Paul II, Care for patients in a "permanent" vegetative state, Origins 33 (43) (2004) 737–740.
[43] R. Hamel, The CDF statement on artificial nutrition and hydration: what should we make of it? Health Care Ethics USA: A Publication of the Center for Health Care Ethics 15 (4) (2007) 5–7.
[44] R. Descartes, The Philosophical Writings of Descartes, Cambridge University Press, Cambridge, UK, (1985).
[45] S. A., Out of My Life and Thought: An Autobiography, Johns Hopkins University Press, Baltimore, Maryland, (2009).
[46] G. Kessler, Unpacking the story of Fauci and painful experiments involving dogs, in: The Washington Post, ed. Washinton, DC, 2024.
[47] T.L. Beauchamp, F.B. Orlans, R. Dresser, D.B. Morton, J.P. Gluck, The Human Use of Animals: Case Studies in Ethical Choice, second ed., Oxford University Press, Oxford, England, UK, (2008).
[48] T.F. Budinger, M.D. Budinger, Ethics of Emerging Technologies: Scientific Facts and Moral Challenges, Wiley, Hoboken, New Jersey, (2006). ISBN:9780471692126.
[49] J.A.S. Cliffe C., The messiness of ethics in education, Journal of Academic Ethics 20 (1) (2022) 101–117.
[50] T. Miller, Messy ethics, Ethics in the Field: Contemporary Challenges 7 (2013) 140.
[51] Monitoring Stem Cell Research, Appendix A, 2004.
[52] L.R. Kass, Letter of Transmittal to President George W. Bush, Alternative Sources of Human Pluripotent Stem Cells, a White Paper of the President's Council on Bioethics, Government Printing Office, Washington, DC, (2002).
[53] President's Council on Bioethics, Executive Summary, Human Cloning and Human Dignity: An Ethical Inquiry, Government Printing Office, Washington, DC, (2002), p. xxvii.
[54] E.O. Wilson, Consilience: The Unity of Knowledge, The Vintage Press, Visalia, California, (1999).
[55] T.S. Kuhn, D. Hawkins, The structure of scientific revolutions, American Journal of Physics 31 (7) (1963) 554–555.
[56] L.A. O'Higgins, K. Eustace, A.M. Tobin, Fortuitous discoveries in dermatology. Journal of the European Academy of Dermatology & Venereology, 37 (2) (2023).

[57] H. Doremus, Contracts for bioprospecting: the Yellowstone national Park experience, Microbial Diversity and Bioprospecting (2003) 450–457.
[58] K.D. Warner, Are life patents ethical? Conflict between Catholic social teaching and agricultural biotechnology's patent regime, Journal of Agricultural and Environmental Ethics 14 (3) (2002) 301–319.
[59] L.H.R.P.S. Appelbaum, C.W. Lidz, P. Benson, W. Winslade, False hopes and best data: consent to research and the therapeutic misconception, Hasting Center Report 17 (2) (1987) 20–24.

CHAPTER

18

A resilient and sustainable future

OUTLINE

18.1 Introduction

The various topics covered in this book elicit the need for a combination of perspectives to address the opportunities and challenges of emerging technologies. This calls for a blend of numerous scientific and ethical disciplines to address impending labor disruptions, privacy threats, environmental injustice, climate change, pandemics, and the myriad implications that may result from the applications of emerging technologies. This and future generations of technologists will likely never work in a solely reductionist, single-discipline "silo," which had been the default work atmosphere for many scientific and engineering disciplines not so long ago. Instead, they will need to draw upon knowledge from numerous experts in various disciplines using systems thinking [1].

Science translation is a way of "harnessing knowledge from basic sciences to produce new drugs, devices, and treatment options ...." [2]. Especially for technical practitioners and researchers, translating information and lessons learned from various sources and applying them to the project at hand is essentially what we must always do.

Technical and Ethical Optimization: Challenges and Opportunities for the Modern Engineer and Scientist
DOI: https://doi.org/10.1016/B978-0-443-44925-3.00017-X

However, whenever this is done, we must be clear about what terms and concepts mean and how they may be misinterpreted or misapplied. Various scientific and engineering disciplines have ontologies and lexicons that give nuanced connotations to the same terms. Some ontologies use terms with identical or very similar meanings, such as those between physics and structural engineering. Other ontologies are more nuanced, like chemistry and geology, whereas others use the same term to mean completely different concepts, such as the use of the term "particle" by theoretical physicists, civil engineers, and atmospheric scientists. Indeed, even subdisciplines within the atmospheric sciences differ, as air pollution models see the particle as a point in a plume-in-grid model, while air pollution measurement scientists see a particle as synonymous with an aerosol.

18.2 Ecological resilience

Two terms that have migrated among engineering and environmental disciplines are hysteresis and resilience. Ecologists envision one type of hysteresis as a landscape, where the state of an ecological community is envisioned as a ball on the surface of the landscape (see Fig. 18.1) [3]. The ball's position on the landscape represents the present state of the ecological community, such as the species abundance of populations in the community. Lacking external interference, a single force, that is, gravity, forces the ball to roll downhill. Furthermore, alternative stable states can be represented as two basins, with the ball residing in one. To visualize a change in ecological state, the ball must exit one basin and join the other, either by the movement of the ball or by a change in the landscape (see Fig. 18.1A and B, respectively). Perturbation to the system's variables is needed to move the ball; modifying the landscape requires changing the parameters that govern interactions within the system.

For biological ontologies, hysteresis can be rate-dependent, that is, the shape of the hysteresis remains constant regardless of the rate of change in the variables [5]. For example, there is a lag between the time when a pollutant stops entering an ecosystem and the time when its concentration in that ecosystem decreases to a defined desired state [6]. A pesticide banned in 2016 could have a desired state represented by the time when that pesticide's maximum environmental concentration decreases by 99%, say, in 2026. Thus, temporal environmental hysteresis occurred for 10 years, when the desired state was reached. The hysteresis in this case was not necessarily a return to a previous state, but rather a transition from an existing state to a desired state.

Another ecological connotation of hysteresis occurs when healthy diversity and productivity are not sustainable because the system cannot adapt to insults. Thus, the ecosystem lacks resilience. The resilience of any system, whether engineered or natural, must also be considered for human populations, but in different ways; for example, the ability to adapt to a toxic spill or plume. Such a situation is at least moderately clear regarding the location of the release, but not where the stress will occur and how long the stressor will persist. Following an accidental release of a gaseous contaminant in an urban setting, the time for gas concentrations to diminish to nondetectable levels can vary widely [7]. Among the factors that determine the time are the micrometeorology, the three-dimensional shape

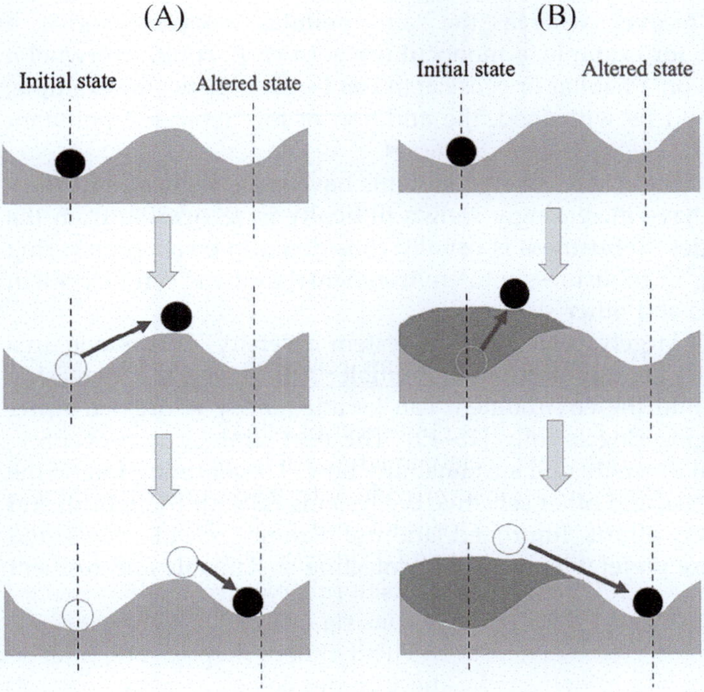

FIGURE 18.1 Heuristic representation of two possible explanations of ecological hysteresis. The change in the state of an ecosystem may occur either by (A) perturbing variables, represented by moving the ball uphill; or (B) changing the parameters that govern interactions within the ecosystem, represented by changing the landscape to move the ball upward to overcome gravity. *Data from B.E. Beisner, D.T. Haydon, K. Cuddington, Alternative stable states in ecology, Frontiers in Ecology and the Environment 1, (7) (2003) 376–382;D.A. Vallero, Applications and Calculations in Environmental Physics, AIP Publishing LLC, 2022. [3,4]*

and size of the airshed, waterbody, or aquifer, the inherent properties of the pollutant, and the type of release.

Ecological resilience depends on stressor-receptor relationships [8]. A stressor can be physical, chemical, or biological. A physical stressor can be energy, such as heat, noise, light, or even landscape disruptions. Chemical stressors can be those that cause everything from threatening endangered species habitats to cancer in human populations. Biological stress can occur when opportunistic and invasive species diminish the survival and abundance of indigenous animal and plant species, or when the genetic material of genetically modified organisms (GMOs) drifts into habitats and supplants native species. If such problems are readily reversible, the system can adapt. Irreversibility is inversely proportional to a system's resilience. Thus, resilience can be quantified as the magnitude of disturbance, that is, needed before a state shift occurs [9,10]. For example, if a hypothetical forest system loses 75% of its canopy species with an average decrease of 2.5 inches of rainfall per year, but another forest loses none of its canopy species with the same rainfall, the latter forest exhibits resilience in its function and structure when challenged with drought stress.

Resilience can also be expressed by the rate of recovery. Imagine two forests in which each loses 50% of their species abundance due to an insect infestation. The first fails to regain the diversity in five years, but the second recovers to its previous state in two years. Thus, the second forest is more resilient. Ecology has taught us that organisms in a resilient habitat are able to survive in broad ranges of moisture, temperature, light, and nutrient concentrations. For example, exposure to a pollutant in a biologically diverse ecosystem may lead to little harm or a short time to recover, whereas the same pollutant concentration in a nonresilient, less diverse system, for example, a monoculture corn crop, could overwhelm the ecosystem [11]. The stressors on a habitat or ecosystem can be synergistic; for example, although drought conditions could be withstood, the addition of the deposited pollutant causes an ecological disaster [12].

For most of its history, the Earth's environmental systems have been resilient, but since the Industrial Revolution, these have increasingly shown difficulty in recovering from the pollutant loads of modern societies. Robustness is usually considered to be an engineering and design concept, but building in robustness into environmental systems can help them handle increasing pollutant loads and other stresses.

Environmental robustness will largely come with ecosystem diversity. This is one area where technological optimists like the author must be humble. Addressing the increasing amounts of pollutants released into the environment can be adequately addressed using large-scale solutions like geoengineering is folly [13–19]. The best strategy is preventing problems and ensuring ecological diversity and sustainability now. The alarming rate of the loss of wetlands, tropical rainforests, and other sensitive ecosystems calls for thoughtful and effective conservation and recovery efforts. Improved land use planning, cleaner fuels, and pollution prevention strategies are preferable means of protecting sustainable and resilient ecosystems [20].

18.3 Engineering resilience

The connotation of resilience can be quite similar between ecology and engineering. For example, like reliability engineering's "item," a habitat that can return to its desired state in a reasonable amount of time is considered resilient. A microbial ecosystem may achieve a return to the desired state in a few hours, but a continental-scale biome can take decades or longer to recover. Both engineered systems and ecological habitats vary in their resilience and vulnerability to irreversible damage [10].

As discussed in Chapter 16, engineers must optimize designs to reduce risks and to make a system reliable and resilient [21–23]. A system that can tolerate only mild perturbations is unreliable, requiring external controls to keep it operating safely and reliably. Conversely, a more resilient system can withstand a wider range of insults at various intensities. After an insult, the system returns to optimal conditions relatively quickly. A system without sufficient resilience is perilously sensitive and susceptible to changing conditions. Over a product's or process's operational life, environmental and operational conditions fluctuate, making a resilient system less likely to need fewer repairs and replacements than a more inflexible system.

Resilience must be built into a technology's specifications. The ideal technology is able to return to its original form or state after being shocked, bent, compressed, stretched, or otherwise stressed, and yet recover to return to the desired state readily following the stress [8,22,23]. Of course, sound practice and adherence to operation and maintenance (O&M) protocols are essential parts of every engineer's toolkit, though most likely would not have readily considered themselves to be resilience engineers.

18.3.1 Specifications

The specifications of a technology or item include tolerance ranges during the operational lifetime. For example, an item may perform its function for 10 years, so long as the environmental conditions, like temperature, pressure, corrosiveness, and other operational factors, remain within specified tolerances. Thus, when the engineer finds ways not only to lengthen the lifetime but also to broaden the tolerances, the design becomes more resilient. To achieve this, the item must become [24]:

1. more robust by widening the upper-to-lower performance limits;
2. more resistant to, and recoverable from, disruptions;
3. reconfigurable by incorporating adaptive properties; and
4. more adaptable to changing conditions.

Thus, the engineer and technologist must be eager and diligent learners, that is, attentive as soon as changes and disruptions appear and be able to anticipate future problems early or even before they occur [24,25].

Resilient systems have wide tolerances. For example, the hazardous waste treatment system in the marketplace must have much greater resilience than when one of its components was in nascent, highly controlled laboratory clean room stage. Real-world conditions vary during the operational life of the any product or process outside of a tightly controlled, closed environment [7,26,27]. For example, one too many cells in an operating system can foul a healthcare technology's protocol by introducing an infectious agent and risking healthcare-associated infections [28–30].

Improved reliability lengthens the time to failure but does not eliminate it [31]. A suitably redundant system acknowledges this by establishing safeguards and redundant systems to address the failures when they occur [15,17,32–34]. Resilience and robustness differ in that resilience is reactive, whereas robustness is proactive. The time for a system to recover from a failure is its resilience. If a system is inherently strong and sufficiently resists insults, that system is robust.

Redundancy is also important for engineering resilience. Backup systems for pollution control equipment during low-probability events, for example, prevent and shorten failures. Indeed, engineered systems must be prepared for failures to occur in time and space. Improved reliability lengthens the time to failure but does not eliminate it. An acceptably redundant system acknowledges this and builds in safeguards and redundant systems to address the failures when they occur [15,17,32–34].

Engineers must continue to design and devise innovative and larger-scale solutions to the direct and indirect problems brought on by environmental issues, many of which we do not yet know are even problems. Engineers tend to be technological optimists and agents of change [35]. Binary perspectives can be obstacles to optimization. Engineers often can satisfy seemingly incongruent perspectives to protect the public *and* provide practical, sustainable, and cost-effective solutions to the problems now and those that will unfold in the future. After all, the first canon of the engineering profession is to hold paramount the safety, health, and welfare of the public. Engineers must be faithful agents to their clients [36]. Their principal client is the public, and the public expects a climate that supports healthy human lives and sustainable ecosystems.

18.4 Sustainability

The magnitude and urgency of the pollution problems in the 1950s and 1960s led to environmental actions that focused mainly on remediating the existing problems. There were numerous episodes in which people suffered from the acute problems associated with high concentrations of pollutants in their water and air. As such, very little time and effort were dedicated to preventing the pollution in the first place. Scientists and engineers enhanced their skills and creativity to find ways to remove and treat the pollutants [37]. Their success in this effort over the past half-century has been phenomenal. The needed remediation technologies, for the most part, did not exist in 1960, and if they did, they were rudimentary means of testing for and cleaning up pollution. Only the most obvious and easily measured pollutants were addressed.

Some of these early detection limits were higher than health and environmental thresholds so were little use in supporting the enforcement of environmental rules and regulations. For example, the author served as the Clean Lakes Coordinator in the Kansas City Regional Office of the US Environmental Protection Agency (EPA). One of the responsibilities of this position was to collect samples of lake water from several lakes in the State of Missouri and to analyze the samples for mercury. The freshwater aquatic life standard was 0.05 micrograms per liter ($\mu g\ L^{-1}$), and 20 $\mu g\ L^{-1}$ for domestic water supply. However, the detection limit of our laboratory at that time was 50 $\mu g\ L^{-1}$. So, if mercury were detected, we did not know if it met the standards. If mercury were not detected, we could only say that it was not detected at 50 $\mu g\ L^{-1}$, so it could have been present any concentration below that amount.

Thinking about prevention strategies and sustainability began as measurement, control technologies, and predictive and descriptive models improved. This can be traced to an evolution of engineering pollution controls. This evolution was not merely in better methods, but also in a new engineering and science ethos. The end-of-stack and end-of-pipe technologies continue to be as important today as ever, with continually improving efficiency and effectiveness. The environmental ethos now includes a systematic view of the pollutant, the environment, and the receptors of the pollutant [38–42].

Industrial, commercial, and governmental green initiatives are plentiful, including design for the environment (DfE), design for disassembly (DfD), and design for recycling (DfR) [43–45]. Systems thinking begins with concern about future generations, that is, it is not only important to achieve an acceptable condition but also that the level of acceptability be sustained. Solving an immediate problem may well be unacceptable if it leads to consequences over time. It is preferable to avoid air pollution rather than to treat it. DfD is a mindset; that is, akin to the engineering concept of reliability. Sustainable engineering must consider the time horizon before and after the useful life of a device or system.

With the mounting concerns about the need for unprecedented amounts of energy for large language model (LLM) computations, energy is a key challenge for a sustainable future [46–49]. Advancing sustainable energy production and use calls for transdisciplinary and systems thinking. The entire life cycle of any energy source is the focus of environmental research and practice worldwide. To wit, the European Union (EU) is increasingly anxious about the health of "downstream users" of products that contain potentially hazardous chemical compounds, for example, fuel additives. The EU now funds research to improve predictions of potential exposures and risks in near-field scenarios, rather than a piecemeal approach to releases during manufacturing, marketing, use, and other aspects of energy supply and demand. Similarly, the US EPA has recently begun building knowledge bases of human exposure to support life-cycle analyses' characterization factors [50].

18.4.1 Life cycle assessment

The life cycle assessment (LCA) is a commonly used tool to determine the sustainability and environmental footprint of an action, product, or process. Sound environmental management considers the entire life cycle of processes, from manufacturing to postuse (e.g., chemicals in fuels and their combustion products that reach the air, soil, and water), up to the end of life, for example, disposal, recycling, and reuse, as well as ways to prevent damage from the fuels before they are formed by choosing safer substitutes for chemicals and substituting materials and processes that lead to better environmental and public health outcomes, that is, fewer toxic substances and reduced volumes of pollutants.

The life cycle perspective is very valuable in identifying possible problems (see Fig. 18.2), even in projects that are designed to protect the environment [51–53]. The LCA is the prominent tool to identify and compare processes in a systematic way. It provides a complete annotation of product flows, critical paths, and life cycle inventories (LCIs). Building upon the benefit–cost ratio (BCR) screening approach discussed in Chapters 9 and 16, the LCA can be used to compare various technological options, including the status quo, alternative passive approaches, and more aggressive options. The comparisons are not only in terms of the principal technological objective (e.g., contamination cleanup or safer food) but also from the standpoint of various ecosystem, public health, and other societal values. In other words, the view from the past (e.g., material extraction) to the future (e.g., postproject impacts) is a critical path for each option that can be evaluated objectively.

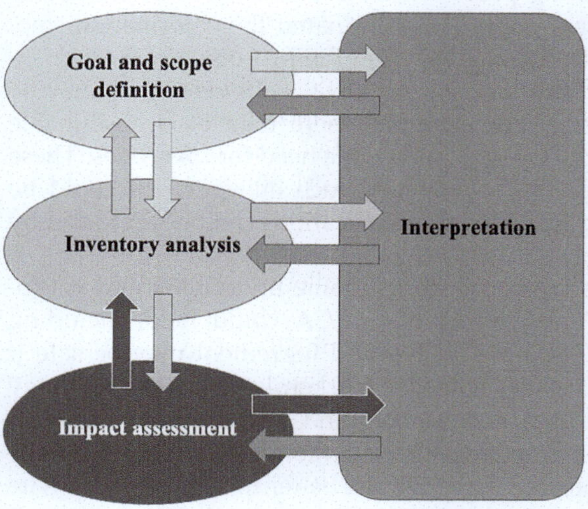

FIGURE 18.2 Life cycle assessment framework. Component 1: a specifically stated purpose of boundaries of the study (Goal and Scope Definition); Component 2: an estimate of the energy use, raw material inputs, and environmental releases associated with each stage of the life cycle (Life Cycle Inventory); Component 3: an interpretation of the results of the inventory to assess the impacts on human health and the environment (Impact Assessment); and Component 4: an evaluation of ways to reduce energy, material inputs, or environmental impacts along the life cycle (Interpretation). *Data from Systems Analysis Research: Program Brief–Life Cycle Analysis. (2009).*

Environmental systems consist of intricately interconnected components that are in balance at many levels, that is, thermodynamic, fluid dynamic, trophic, and physical-chemical-biological cycling. Introducing change, sometimes seemingly small, can have profound impacts on these systems.

This life cycle view is also the first step toward preventing problems. For example, if we consider all the possible contaminants of concern, we can compare which of these must be avoided completely, which are acceptable with appropriate safeguards and controls, and which are likely to present hazards beyond our span of control. We may also ascertain certain processes that generate none of these hazards. Obviously, this is a preferable means of reliability and prevention of problems.

The usefulness of the LCA ranges from cursory attention paid to inputs and outputs of materials and energy to multifaceted decision fields extending deeply into time and space. The latter is preferable for decisions involving large scales, such as the cumulative build-up of greenhouse gases, or those with substantially long-term implications, such as the release of genetically altered microbes into the environment. Complex LCAs are also favored over cursory models when the effects are extensive, such as externalities and artifacts resulting in geopolitical impacts. Environmental decisions may fall into any or all of these three categories. Life cycle-based waste management depends on finding alternatives to substances that are toxic, energy inefficient, and difficult to treat. This can be done using an alternatives assessment, which is a formal technique to characterize the hazards of substances based on a full range of human health and environmental information (US EPA 2012). Chemical choices made based on these assessments improve potential for unintended consequences beforehand (*a priori*). This approach, however, could advance beyond hazard to address benefits and utility, that is, look at a process that currently calls for helium (He) and find points in the life cycle where the need for He can be reduced or eliminated.

Selecting an alternative substance or process is easiest if substitutes are commercially available. For example, when pulp and paper companies wanted to reduce the generation of toxic halogenated dioxins from their bleaching process, they found that other industries were whitening without using chlorine compounds. This eliminated the chlorination step, which meant that dioxin production was also eliminated from the process.

Candidates for chemical and process substitutions are not always obvious. They may be used in industries unfamiliar to one's sector. For example, there may be a sustainable process of using microfibers in the textile industry to eliminate the need for toxic dyes. These are making use of optical properties rather than chemicals. Such an approach could be adapted to plastics, electronics, and other industries. Such innovations can be supported by clearinghouses.

Conversely, it is possible that the alternative approach has some underlying and yet-to-be-manifested deficiencies and hazards. Again, a complete LCA could help, including exploring various "what if" scenarios. For example, a chemical ingredient may be safe if used as intended. Once the product finds its way into the marketplace, however, it may encounter many unintended and incidental uses. As discussed in Chapter 10, dual use can be byproducts of a lack of due diligence. For example, extra attention is needed in synthetic biology, which can provide a solution in one area, for example, better medicines, but invite future problems because some aspects of the new organism were ignored. This can result not only from a confluence of events that lead to a surprising, adverse outcome, but also as an invitation to intentional misuse, for example, bioterrorism. The absence of a direct analogy when developing a technology does not preclude the need for precaution, as even a slight change in conditions may elicit unexpected and unwanted outcomes [54–57].

A complete life cycle takes into consideration such uses, utilizing reliable data to predict possible hazards. These include consideration of a substance's carcinogenicity, reproductive and developmental toxicity, neurotoxicity, sensitization and irritation, hormonal activity, and potential to cause ecological damage. The success of pollution prevention, waste reduction, and waste elimination is not based solely on cost–benefit economics, such as those based on the return on monetary investment that can be expected over the life of a product. Often, the policy and decision makers are presented with a list of options. Obviously, the decision makers' span of control dictates the number and diversity of options. Even if limited in options, however, sustainable management is a continuous quest for upstream improvements. Products that end up in the waste stream must also be evaluated using methods beyond a comparison of the initial investment as a fraction of the total cost of manufacture, use, and disposal.

While treating waste has been a necessary component of waste management, eliminating waste before it is generated is a preferable goal.

18.4.2 The ethanol case

The viability of ethanol as an alternative to fossil fuel is best considered from the perspective of its entire life cycle [58,59]. Accurately accounting for energy and matter losses in calculations can be challenging. From a thermodynamics standpoint, increased ethanol use could actually increase demands for fossil fuels, such as the dependence on crude oil-based products to manufacture farm chemicals derived from oil, as well as energy use of farm vehicles and equipment for planting, cultivating, harvesting, and transporting corn to markets. A comprehensive LCA is a vital tool for ascertaining the actual efficiencies.

The questions surrounding ethanol can be addressed using a three-step methodology. First, the efficiency calculations must conform to the physical laws, especially those of thermodynamics and motion. Second, the "greenness," as a metric of sustainability and effectiveness, can be characterized by life cycle analyses. "Third, the policy and geopolitical options and outcomes can be evaluated by decision force field analyses." ("Waste Technology - an overview I ScienceDirect Topics") In fact, these three approaches are sequential. The first must be satisfied before moving to the second. Also, the third depends on the first two methods. Whether politically attractive or favored by society, an alternative fuel must comport with the conservation of mass and energy. Thus, each step in the life cycle, for example, extraction of raw materials, value-added manufacturing, use, and disposal, must be considered in any benefit–cost or risk–benefit analysis. Finally, the societal benefits and risks must be part of the LCA and optimization processes. For example, even a scientifically sound, efficient, and effective fuel production process may be rejected for societal reasons, for example, religious, cultural, historical, or ethical concerns [41,60–62].

The challenge of the scientist, engineer, and technologist is to sift through the myriad data and information to ascertain whether ethanol truly presents a viable alternative fuel. Of the misrepresentations being made, some clearly violate the physical laws. Many ignore or do

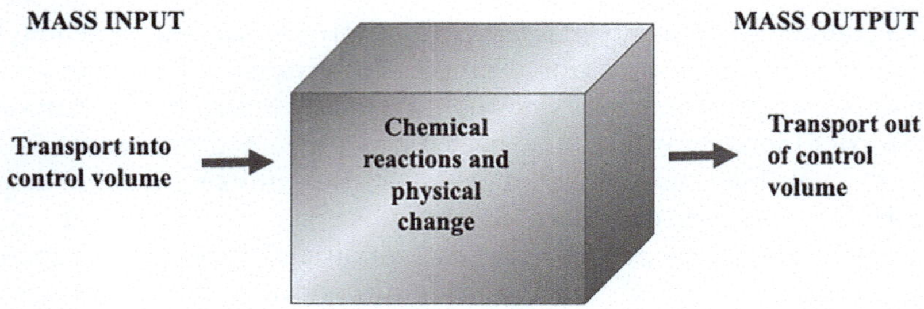

FIGURE 18.3 Control volume showing input, chemical and physical change, and output. The process applies to both mass and energy balances. *Data from D. Vallero, Environmental Biotechnology: A Biosystems Approach, Elsevier Science, 2015.*

not provide correct weights to certain factors in the life cycle. There is a continual risk of mischaracterizing the social good, service, or costs, especially for an emerging technology.

Biomass-based fuel efficiencies are evaluated in terms of net energy production, that is, based on the first and second laws of thermodynamics. The numerator of the efficiency equation includes all energy losses. However, these are dictated by the specific control volume. "This volume can be of any size, from molecular to planetary." ("Fuel Additive - an overview | ScienceDirect Topics") To analyze energy losses related to alternative fuels, every control volume of each step of the life cycle must be quantified.

The first two laws of thermodynamics drive this step. First, the conservation of mass and energy requires that every input and output be included. Energy or mass can neither be created nor destroyed, only altered in form. ("First Law of Thermodynamics - an overview - ScienceDirect") For any system, energy or mass transfer is associated with mass and energy crossing the control boundary within the control volume. If mass does not cross the boundary, but work and/or heat do, the system [63] is a "closed" system. If mass, work, and heat do not cross the boundary, the system is an isolated system. Too often, open systems are treated as closed, or closed systems include too small a control volume (Fig. 18.3). A common error is to assume that the life cycle begins at an arbitrary point conveniently selected to support a BCR. For example, if a life cycle for ethanol fuels begins with the corn arriving at the ethanol processing facility, none of the fossil fuel needs on the farm or in transportation will appear.

The second law of thermodynamics is less direct and obvious than the first. In all energy exchanges, if no energy enters or leaves the system, the potential energy of the state will always be less than that of the initial state. The tendency toward disorder, that is, entropy, requires that external energy is needed to maintain any energy balance in a control volume, such as a heat engine, waterfall, or an ethanol processing facility. Entropy is ever present. Losses must always occur in conversions from one type of energy (e.g., mechanical energy of farm equipment ultimately to chemical energy of the fuel). Thus, a series of efficiency equations must be calculated for the entire process, with losses at every step.

18.4.3 The life cycle assessment optimization

The information needed for the LCA is provided by the LCI. The LCI data are used to assess the environmental implications associated with a product, process, or service by compiling a register of relevant energy and material inputs and environmental releases. From this LCI, the potential environmental impacts are evaluated. The results aid in decision-making. Thus, the LCA process is a systematic, four-component process [64]:

1. *Goal definition and scoping*—Define and describe the product, process, or activity. Establish the context in which the assessment is to be made and identify the boundaries and environmental effects to be reviewed for the assessment.
2. *Inventory analysis*—Identify and quantify energy, water, and material usage and environmental releases (e.g., air emissions, solid waste disposal, and wastewater discharges).
3. *Impact assessment*—Assess the potential human and ecological effects of energy, water, and material usage, and the environmental releases identified in the inventory analysis.
4. *Interpretation*—Evaluate the results of the inventory analysis and impact assessment to select the preferred product, process, or service with a clear understanding of the uncertainty and the assumptions used to generate the results.

These steps track closely with the life cycle stages dictated by the laws of thermodynamics. There is both an energy and matter mass balance, since energy and matter entering the control volume are fully characterized and quantified. Thermodynamic system boundaries are drawn, and flows of energy and matter are accounted for.

This stepwise process can be used to optimize emerging technologies. First, the LCI is constructed to define the boundaries of the possible effects of a technology, for example, microbial populations, GMOs, and toxic chemical releases. If the technology is hypothetical, this can be done by analogy with a similar conventional process. Next, experts can participate in an expert panel to find the driving forces involved. This is known as "expert elicitation" [65–68]. Then, scenarios can be constructed from these driving forces to identify which factors are most important in leading to various outcomes. This last step is the sensitivity analysis. The greater the weight of the factor, the greater the change in the outcome will be. For example, if one change, for example, adding catalyst A, in the production of a chemical compound decreases the emissions of benzene from the current $100\ \mu g\ m^{-3}\ second^{-1}$ to $1\ \mu g\ m^{-3}\ second^{-1}$, the emission rate reduction would be 100-fold. However, if a different change, for example, decreasing the reactor temperature by 10 C, of the same process lowers the benzene emission rate to $0.1\ \mu g\ m^{-3}\ second^{-1}$, that is, a 1000-fold emission rate reduction, this means that the temperature change is 10 times more sensitive than adding the catalyst.

Systematic optimization should be part of a comprehensive environmental system. For example, many pollution control technologies come at the very end of manufacturing and energy production processes. However, it is preferable to prevent as much of the formation of pollutants as possible earlier in the life cycle. In fact, it may be possible to eliminate pollution even before production. For example, the extraction of crude oil produces air and water pollution, so eliminating the need for crude oil would decrease the formation of water and air pollutants early in the life cycle and obviate the need for treatment entirely.

Algal biodiesel fuel is an example of an emerging technology that can decrease the amount of fossil fuels needed. The process begins with the preparation of an inoculum by culturing an algal strain using photo-bioreactors, for example, indoor ponds, and ends with harvesting lipids that are converted to biofuel and transported for sale and use [69]. Algal-based biotechnologies can use inoculums cultured from whole organisms, which include carbohydrates, lipids, and proteins. They may also be extracted from the whole organism, for example, lipids from microalgae and carbohydrates. In one scenario, the Fischer-Tropsch catalytic process [70–73] converts a mixture of carbon monoxide and hydrogen in the presence of iron or cobalt to long-chain hydrocarbons like olefins and paraffin [74]. In this case, an emergent use of a natural, renewable resource, that is, algae, and an established synthetic gas method developed in the 1920s can be combined to provide a sustainable alternative to combusting fossil fuels and their attendant pollutants.

Biofuels certainly have advantages over fossil-fuel-derived fuels, but they are not a pollution cure-all. The LCA for the processes must not only consider the direct inputs, but also every exposure route and pathway from harvesting the algae to combusting the fuel [63]. Therefore, even a sustainable source of fuel, for example, algae, may rely on unsustainable steps (e.g., crude oil drilling and refining for materials needed to separate cell components) and toxic byproducts (e.g., metals mined for catalysts) [69].

Recent advances in integrating LCAs and risk assessment are promising, including metrics for exposure and hazard [75–78]. To improve the LCA's usefulness as a predictive tool will require enhancements of life LCIs, including more harmonized formats and wider availability to users. Recent progress has included the development of open-source software, making data more accessible to a larger audience, and improving transferability of software. Improved exposure models and data are also becoming more widely available. Many previous paper data sets have become digitized and searchable; for example, EPA's *Exposure Factors Handbook* [79,80].

Exposure models now make use of a wide array of data, including (1) product ingredients; (2) pharmacokinetic factors; (3) consumer product category-specific "exposure factor surrogates"; and (4) time/activity estimates (i.e., human factors). These disparate data streams can then be integrated within an interface such that different exposure scenarios for target groups, for example, individuals, populations, or occupational groups, yield time-use profiles that can be interchanged and quantitatively explored to prioritize tailored chemical exposure and, ultimately, dose. This allows estimates of multichemical signatures of exposure, internalized dose (uptake), remaining dose or body burden, and elimination [81]. Such models and databases should be useful for characterizing and comparing applications of synthetic biology and genetic engineering processes.

To arrive at a risk comparison, the resulting exposure data can be mapped onto toxicity data. For example, the comparisons can describe the extent to which chemical and biological substances contribute to potential toxic outcomes at each life cycle stage [82]. To answer this question, the LCA must consider how these substances move and change in the LCA to help determine which exposures are most likely to lead to various adverse effects, for example, cancer, neurotoxicity, developmental and reproductive disorders, and problems with the immune and endocrine systems in human populations, as well as ecosystem effects.

For ecological optimizations, the LCA must ascertain the extent to which potential exposures to sentinel or endangered species differ from human exposures. Thus, the LCA may account for a direct emission from a stack, an exposure to an ingredient in a product that is intentionally applied (e.g., a pesticide), or a substance that migrates to a habitat, for example, released from a stack or outfall structure and then transported to the habitat. Thereafter, the exposure in human populations or the activities of the species in an ecosystem can be modeled (e.g., predator-prey, migration, and bioaccumulation factors), including the pharmacokinetic factors within the organism that can be modeled and estimated in the LCA [75].

One means of combining biological and chemical stressors into a cumulative risk assessment is to integrate the results for the LCA with those from a quantitative microbial risk assessment (QMRA). Although QMRA has principally been a tool used for waterborne pathogens, it may be useful for bioaerosols, which can carry bacteria, viruses, protozoa, or their byproducts. The QMRA can be combined with risk assessments of chemical stressors. The combined tools may evaluate the potential for an array of risks, for example, food and water contamination after deposition. In addition to the harm caused by the bioaerosols themselves, they may also contain chemicals that induce various acute and chronic effects, for example, cancer. Pathogenic risk can be expressed as a burden of disease, for example, disability-adjusted life years (DALYs) [83]. Thus, the DALYs can be combined with risk information for the chemical stressors to depict cumulative risks, for example, as comparing combined pathogenic and chemical risks from a manufacturing process and making changes to lower those risks.

The success of an alternative technology displacing conventional manufacturing depends on the efficiency and safety with which a product can be produced and used. Complicating matters, certain materials and costs may be embedded in the technology. As mentioned, fossil fuels are used in almost every step of ethanol production and/or operation, as is the case for all biofuels.

An analysis is only as good as the availability and quality of the LCI data. Uncertainties in the inventory and in the impact assessment methodology can lead to omitted options and unforeseen impacts. The lack of agreement among elements of the impact assessment methodology, for example, internal inconsistencies, can lead to additional uncertainties and errors. Finally, many of the factors are qualitative and subjective, so differences in LCA problem formulation may, in fact, be due to differences in weights placed on factors. Such weighting can be biased according to the values of those conducting the LCA. For example, the previously mentioned QMRA expresses only the pathogenic burden of disease in the human populations, omitting numerous other biological adverse outcomes, such as the effects of a transgenic organism on the soil microbial community [84]. Thus, additional tools and data are needed for a comprehensive assessment of the life cycle in which air pollution is occurring.

18.5 Evaluating cause and effect

The underlying assumption in any environmental assessment is that "cause and effect" can be identified using credible science to connect exposure to a hazard and to estimate a negative outcome. As stated in most introductory statistics courses, association is different from causation; more properly, association is only one factor in causation and can, in some cases, be no factor at all. Often, linking a pollutant to a disease or other adverse outcome involves the proverbial "third variable" [85]. Something other than the exposure to the pollutant could be the reason for the outcome. In some cases, the presence of the third variable exacerbates the outcome. In statistics classes, we are given simple examples of such occurrences:

1. Studies show that people who wear shorts in Illinois consume more ice cream.
2. Therefore, wearing shorts induces people to consume more ice cream.

The first statement is simply a measurement. It is stated correctly as an association. However, the second statement contains a causal link that is clearly incorrect for most occurrences [14]. The facts are correct, but the logic is wrong. One or more other variables are causing both variables to appear, that is, the wearing of shorts and the consuming of ice cream. For example, if one were to plot ambient average temperature and compare it to either the wearing of shorts or the consuming of ice cream, one would see a direct relationship between the variables. That is, as ambient temperatures increase, so do short wearing and the rate of ice cream consumption.

18.5.1 Hill's criteria

The biomedical science community may help us deal with the causality challenge. The best that scientific research can usually do in this regard is to provide enough weight of evidence to support or reject a suspicion that a substance causes a disease. The medical research and epidemiological communities use several criteria to determine the strength of an argument for causality, but the first well-articulated criteria were Hill's causal criteria [15].

Of course, assigning causality when none really exists must be avoided. However, if all we can say is that the variables are associated, the public will want to know more about what may be contributing to an adverse effect, for example, learning disabilities associated with elevated blood lead levels. This was particularly problematic in early cancer research. Possible causes of cancer were being explored, and major research efforts were being directed at innumerable physical, chemical, and biological agents. So, there needed to be some manner of sorting through findings to see what might be causal and what is more likely to be spurious results. Sir Austin Bradford Hill is credited with articulating key criteria that need to be satisfied to attribute cause and effect in medical research [86].

18.5.1.1 Strength of association

Strong associations provide more certain evidence of causality than weak associations do [87]. A cause must be associated with an outcome, that is, the statistical association cannot be zero, although a cause may show up as a very low statistical association. Common metrics used to indicate association include risk ratio, odds ratio, and standardized mortality ratio.

18.5.1.2 Consistency

If an outcome occurs consistently under different studies using diverse methods of study of various populations under varying circumstances by different investigators, the causal link is stronger than when the associations occur in just a few scenarios. For example, if results are similar for an air pollution control approach studied with scenarios, simulated with models, and documented using historical information, there is greater consistency.

However, if there is no agreement among the various approaches, this could mean that the results are unique to a given set of conditions, for example, if the only scenarios were towns with populations of less than 5000, a model was parameterized only for one type of collection approach, or historical data are incomplete or erroneous.

18.5.1.3 *Specificity*

The specificity criterion holds that the cause should lead to only one outcome and that the outcome must result from only this single cause. Recall that Hill was interested in cancer and viewed causality through a biomedical perspective, that is, this criterion appears to be based in the germ theory of microbiology, where a specific strain of bacteria and viruses elicits a specific disease. This is rarely the case for environmental scenarios, where there are commonly multiple causes for multiple adverse outcomes [88,89].

18.5.1.4 *Temporality*

This criterion requires that exposure to the chemical must precede the effect. This is an inviolable criterion. In a retrospective study, the researcher must be certain that the outcome was not already present before the contributing factors were present. For example, a disease may have been present before a toxic release. However, this does not eliminate the temporal link if, for example, the exacerbation of the disease occurs, as when the incidence of existing asthma events and lung disease exacerbations spike. Thus, the previous existence of an adverse outcome does not necessarily preclude the agent in the release as a cause, but it does mean that the agent is not the sole cause of the disease (see "Specificity" above), or that the outcome is not new, but merely a worsening of an existing condition.

Temporality can be complicated for emerging technologies, especially for those where the introduction and adoption of a new technology are attenuated over time [90,91].

18.5.1.5 *Gradient*

If the force of a cause, for example, the intensity and extent, increases the amount of outcome, this strengthens the causal link. This is a common metric for risk assessment, known as the "dose-response" step in risk assessment (see Chapter 18). If the level, intensity, duration, or total amount of chemical exposure is increased, a concomitant, progressive increase should occur in the toxic effect. However, the gradient also applies to other areas, for example, if a recycling program is intensified and the amount of recycling increases concomitantly, this is a gradient [87]. So is the case when increasing the number of a species of soil bacteria increases the rate of carbon cycling in soil [92].

18.5.1.6 *Scientific plausibility*

Hill limited this to biological plausibility, but the criterion has wider application. Generally, an association needs to follow a well-defined explanation based on a known scientific system, that is, a paradigm. A paradigm, however, is mutable. Even some of the most strongly held scientific laws have had to be revised with the evolution of knowledge [93]. Outside of research, this caution has little relevance to the practice of engineering and environmental management. Engineers and other professionals involved in pollution control and management must apply best practices according to their profession. If results are implausible according to underlying scientific principles, they should not be heeded. However, simply because an approach does not follow directly from what has worked in the past, it is not rendered implausible, but merely inconvenient. Indeed, the professional, practitioner, and technologist must beware of Kaplan's law of the instrument[1], that is, if the only tool you have is a hammer, it is tempting to treat everything as if it were a nail [95]. The environmental manager's or engineer's preferred approach can be likened to a hammer, so that the entirety of air quality management is a collection of nails. There are likely to be other plausible "tools," besides one's hammer, and there are certainly myriad other problems besides the "nails." This can be particularly worrisome for environmental and climate modelers, who often must incorporate simplifying assumptions to make the math work [96]. Engineering models, for example, can be accused of assuming that a chicken is spherical, but when the designed coop fails, this may be in part due to this oversimplification. After all, the real-world coop's chickens have beaks, feathers, and claws.

18.5.1.7 Coherence

Coherence suggests that all available evidence must form a cohesive whole. Thus, the proposed causal relationship should not conflict with or contradict information from experimental, laboratory, epidemiologic, theoretical, or other knowledge sources.

18.5.1.8 Experimentation

Experimental evidence useful to air quality management can support a causal hypothesis. Laboratory experiments, tests, pilot studies, physical models, and natural experiments (e.g., a disaster response) are available. The experiment does not have to be specific but can be drawn from others' experiences, for example, an informatics experiment using "big data."

18.5.1.9 Analogy

The term analogy implies a similarity in some respects among things that are otherwise different. It is thus considered one of the weaker forms of evidence.

For environmental and public health decision-making, some of Hill's criteria are more important than others [87]. Reliability in a public health and environmental context relies heavily on strength of association, for example, to establish dose-response relationships. Coherence is also very important. Animal and human data should not be extensions of one another and should not disagree. Biological gradient is crucial, since this is the basis for dose-response, that is, the more dose, the greater the biological response.

As mentioned, temporality is crucial to all scientific research, that is, the cause must precede the effect. However, this is sometimes difficult to perceive in certain instances, such as when the exposures to suspected agents have been continuous for decades and the health data are only recently available.

[1] In fact, this concept preceded Kaplan by at least a century. The so-called " Birmingham screwdriver" was meant to humorously convey that the workers used a hammer as the single tool for all purposes. Why waste time turning a screw, for example, when you could drive it with the force of a hammer? See: [94] J. Green, *Cassell's dictionary of slang*. Sterling Publishing Company, Inc., 2005.

Linking causes to outcomes must be based on sound science and scientific judgment. Environmental engineers and managers must consider whether the selected approach will likely continue to work throughout the entire expected operational life of a product or process (reliability). Furthermore, selecting the best option requires consideration of the hazards that may arise, especially if a new and untested technology is chosen. Risk is a function of likelihood that the hazard will, in fact, be encountered, so the decision maker is tasked with predicting the adverse implications that the user and society might face. Thus, the risk associated with a selected approach refers to the possibility and likelihood of undesirable and possibly harmful effects. Errors in reliability and risk predictions can range from not foreseeing outcomes that are merely annoying (e.g., fuel efficiencies that are less than expected) to those that are devastating (e.g., the release of carcinogens from an incinerator that finds its way to a school) [97].

18.6 Time: persistence bad; sustainability good

The terms persistence and sustainability are almost synonyms and nearly antonyms. Both terms mean that something will last a long time. However, in both technical and lay usage, a problem persists, but an ideal condition must be sustained.

Time is a key aspect of pollution. Pollutants can cause harm in human populations and in ecosystems. They harm public welfare. This aspect of harm has to do with severity and spatial extent of injury. Time is certainly a part of this harm. Recall from Hill's gradient criterion that the longer the exposure, the greater the injury will be.

Indeed, many engineering and environmental equations include a time factor, beginning with the first law of thermodynamics: The rate of change of mass of an air pollutant (i.e., change in mass with respect to time) within a control volume must equal the rate of the mass transported in minus the rate of mass transported out:

$$\frac{d[A]}{dt} = -v \cdot \frac{d[A]}{dx} + \frac{d}{dx}\left(\Gamma \cdot \frac{d[A]}{dx}\right) \pm r \tag{18.1}$$

where v=fluid velocity; Γ=a rate constant specific to the environmental medium; concentration gradient of chemical A; and r=internal sinks and sources within the control volume. The r is usually considered to be a loss term, that is, the mass of the compound exiting the control volume is decreased by the amount of reactivity. Thus, even the loss term includes a time function, that is, the reaction rate (this is shown graphically in Fig. 18.4).

The hazardous aspects of a pollutant include temporal factors. For example, animal studies are part of the process to construct and interpret the dose-response curves. A chemical compound that causes cancer or noncancer outcomes in laboratory rats would be suspected of doing the same in humans. However, models must be employed to extrapolate the rat-to-human relationship. This process must compensate for the time difference (e.g., six-month exposure to an animal versus a lifetime exposure to a human), so the animal dose must usually be much higher than the dose for the human. That is, time is compensated mathematically with dose.

The basic way to express exposure to an air pollutant must account for time:

$$E = D\,t^{-1} \tag{18.2}$$

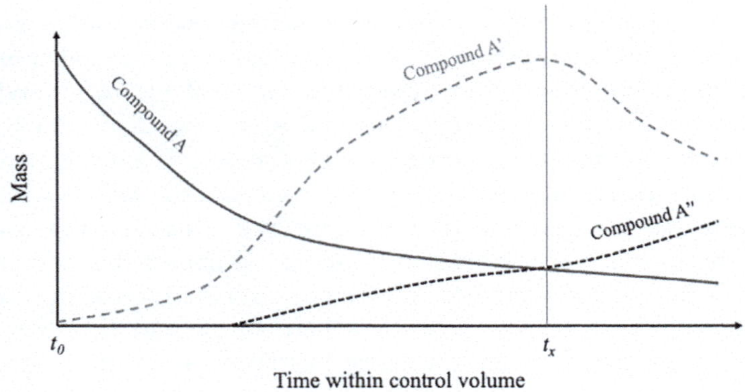

FIGURE 18.4 Hypothetical degradation of chemical compound A and its degradation products, A′ and A″. As compound A degrades in the control volume, the mass of its degradation product, compound A′, increases. However, A′ has its own degradation product, A″. The mass of A″ increases as it is produced by the degradation of A′. At *tx*, the mass of A has almost reached steady state, and the mass of A′ begins to decline at the expense of A″. This illustrates that the reaction variable in mass balance equations applies only to each compound at a specific time within the control volume. *Data from D.A. Vallero, Fundamentals of air pollution, 6th ed., Academic press, Amsterdam, NV, 2025.*

where, E=human exposure during the time period, D=mass of pollutant per body mass (dose), and t=exposure time.

The expression of dose is meaningless without factoring in time, which is demonstrated by the average daily dose (ADD), which averages exposures and doses during exposure duration periods. The ADD is ascertained by averaging an intake dose according to body weight and an averaging time:

$$\mathrm{ADD} = \frac{I}{W \times t_{\mathrm{averaging}}} \tag{18.3}$$

where, I is the intake dose; W is the body weight (kg); and $t_{\mathrm{averaging}}$ is the averaging time (days). For inhalation, I equals the concentration of the air pollutant times the inhalation rate times the exposure duration.

Thus, the temporal aspects of pollution include exposure time and duration, reaction rates, and partitioning within the environment, that is, the movement among phases and environmental compartments, such as the time over which pollutants remain in the atmosphere, water, soil, sediment, and biota. These times are usually expressed as body burdens in organisms (including humans) and residence times in environmental compartments, for which the most common metric is the half-life ($t_{1/2}$), for example, xenobiotic $t_{1/2}$ in a human body, atmospheric $t_{1/2,}$ or soil $t_{1/2}$.

Another temporal aspect of pollution is the resilience of the receptor or the reversibility of an impact. The ability of receptors to withstand insults from pollution drives decisions about how to address the pollution. For example, if the habitat of an endangered species is particularly vulnerable to deposition of acid aerosols so that the damage would be irreversible, the proper response would be to find ways to prevent the acidic deposition. However, if the habitat system can recover in a few days, the response would likely be less aggressive. This is part of the thinking behind applying "precaution" rather than "risk-based evidence" for potentially irreversible problems. If data indicate that a pollutant's damage will require years or decades to recover, then the precautionary approach would be to eliminate the source of the problem, even with lower thresholds of certainty that would be required under an evidence-based decision.

18.6.1 Persistence

Toxicity is the principal type of harm to humans and other organisms. The pollutant is the toxic agent, and the organism is the receptor of the toxicity. For harm to result, the agent must be toxic when it reaches the receptor or at least become toxic after it is taken up by the receptor. Thus, three of the most important characteristics of a pollutant are its persistence, its potential to accumulate in organisms, and its toxicity. Indeed, the most worrisome pollutants are known as PBTs, that is, persistent, bioaccumulating, toxic substances.

Persistence is related to the partitioning coefficients (see Table 18.1). As mentioned, the environmental half-life ($t_{1/2}$) is a common way to express a chemical substance's persistence, that is, the amount of time it takes to degrade one-half of the mass of a compound.

Thus, $t_{1/2}$ is a metric of persistence, that is, the larger the $t_{1/2}$, the more persistent the compound. Persistence is both an intrinsic and extrinsic property of a substance. It is dependent upon the molecular structure of the compound, such as the presence of aromatic rings, certain functional groups, isomeric structures, and especially the number and types of substitutions of hydrogen atoms with chlorine and other halogens.

Persistence is directly related to bioconcentration. Generally, the larger the $t_{1/2}$ of a substance in a compartment, the larger the bioconcentration factor (BCF). For example, the lower $t_{1/2}$ of unsubstituted aliphatic compounds often has lower BCFs, for example, acrolein's BCF is half that of the chlorinated aliphatic compound chloroform and is four

TABLE 18.1 Atmospheric persistence compared to octanol–water and Henry's law coefficient.

Compound	Half-life (days)	Log K_{ow}	Log K_H
Benzene	7.7	2.1	–0.6
Chloroform	360	1.97	–0.7
DDT	50	6.5	–2.8
Ethyl benzene	1.4	3.14	0.37
Formaldehyde	1.6	0.35	–5.0
Hexachlorobenzene	708	5.5	–3.5
Methyl chloride	470	0.94	–0.44
Methylene chloride	150	1.26	–0.9
Polychlorinated biphenyls	40	6.4	–1.8
1,1,1 Trichloroethane	718	2.47	0.77

Data from D.D. Toro and H. F., "Long-range transport and deposition: The role of Henry's law constant," in "Final report, International Council of Chemical Associations," Arlington, Virginia, 1999 [98].

orders of magnitude lower than the BCF of the chlorinated aromatic compound, 2,3,7,8-tetrachlorodibenzo-*para*-dioxin [20].

Persistence potential also depends upon the pollutant's relationship to its media. Chemical compound $t_{1/2}$ values are commonly reported for each compartment, so it is possible for a compound to be highly persistent in one environmental medium, for example, water, yet relatively reactive in another, for example, soil. The variation in persistence can be attributed to unique characteristics within each compartment, such as temperature, pH, intensity of sunlight, presence and diversity of microbes, and chemistry, for example, reactivity of the pollutant, as well as the presence of oxidizing agents, free radicals, and catalysts.

The half-lives and rate constants (k) discussed in Chapter 16 represent identically ordered decay processes and are inversely related to one another. For example, first-order decay can be expressed in terms of concentration versus time, concentration versus distance, and as abiotic (e.g., photochemical) and biotic (e.g., microbial) degradation rates. The relationships between $t_{1/2}$ and k can be stated mathematically. The first-order rates are:

$$k = \frac{0.693}{t_{1/2}} \tag{18.4}$$

and,

$$t_{1/2} = \frac{0.693}{k} \tag{18.5}$$

Thus, a half-life of two years is the same as a first-order rate constant of 0.35 year^{-1}, and a half-life of 10 years=a first-order rate constant of 0.0693 (i.e., a slower rate constant is inversely related to a longer half-life). This is an essential consideration in estimating the rate at which a contaminant plume will be attenuated and is commonly used in groundwater studies.

Note that Table 18.1 contains no metals or metalloids, but only organic compounds. Obviously, outside of nuclear fission or fusion, elements do not undergo degradation. For example, the degradation of the listed organic compounds does not involve destruction of any element. The carbon, hydrogen, oxygen, and other atoms remain intact. During degradation, bonds are broken or added, elements are substituted, and new compounds are produced. This always involves changes in oxidation states. Thus, even though a metal is an element, which will not be changed by degradation (i.e., its elemental half-life is infinite), particular metallic compounds will certainly change, just as the particular carbon compounds change in the organic degradation. Thus, even for inherently toxic compounds like mercury, $t_{1/2}$ is useful (e.g., to state the atmospheric conditions under which the methylmercury compounds degrade to elemental Hg° and simpler inorganic compounds).

The $t_{1/2}$ is also an important part of endogenous kinetics within an organism, including adsorption, distribution, metabolism, and excretion of substances. The $t_{1/2}$ of a compound is inversely proportional to that compound's body burden. A large $t_{1/2}$ increases the likelihood of accumulation over time. Thus, if substance A's $t_{1/2}$=1 day, and substance B's $t_{1/2}$=100 days, substance B would be expected to show a much larger body burden than substance A. A substance's $t_{1/2}$ also indicates which aspects of the toxicokinetics are most important. For example, <15% of inorganic Hg is absorbed by the human gastrointestinal tract [99], whereas about 95% of methylated forms of Hg are absorbed [100]. Thus, at the same intake rate, the actual uptake is more than six times larger for methylmercury than for inorganic Hg compounds.

Concentrations versus time constants, known as point decay rates (k_{point}), which are derived from a single concentration value versus time plot, can be used to estimate the length of time that a plume will persist. Bulk attenuation rates (k) are derived from concentration versus distance plots and are used to determine if the contaminant plume is expanding. Thus, they are most useful for groundwater contamination but can be applied to other environmental media [20].

Substances that remain in the environment long after their release and deposition are more likely to continue to cause problems or to be a threat to environmental quality. In the United States, for example, EPA [101] considers a compound to be persistent if it has a $t_{1/2}$ in water, soil, or sediment of greater than 60 days, and very persistent if the $t_{1/2}$ is greater than 180 days. In air, the compound is considered persistent if its $t_{1/2}$ is greater than two days. Some of the most notoriously toxic chemicals are also very persistent, such as polychlorinated biphenyls (PCBs) [102] and halogenated dioxins, especially 1,4,7,8-tetrachlorodibenzo-*para*-dioxin [103–105].

The unhealthiest combination of factors is when a compound is persistent in the environment, builds up in organic tissues, and is toxic. Such compounds are referred to as PBTs. However, each factor is a function of both the contaminating agent (i.e., the pollutant) and the conditions of the environment in which it resides.

The concept of persistence elucidates the notion of tradeoffs that are frequently needed as part of many responses to environmental insults. It also underscores the importance of sound science and reliable data. For example, the pesticide DDT [1,1,1-trichloro-2,2-bis-(4-chlorophenyl)-ethane ($C_{14}H_9Cl_5$)] is relatively insoluble in water (1.2–5.5 mg L^{-1} at 25°C) and is not very volatile (vapor pressure: 0.02×10^{-5} mmHg at 25°C) [106]. The aqueous solubility and vapor pressures alone may indicate that humans and wildlife are not likely to be exposed in the air or water. However, the compound is highly persistent in soils, with a $t_{1/2}$ of about 1.1–3.4 years, so it may still end up in drinking water in the form of suspended particles or in the air sorbed to fine particles. DDT also exhibits high BCFs, that is, approximately 50,000 for fish and 500,000 for bivalves. This indicates that once organisms become exposed, they tend to increase body burdens of DDT over their lifetimes. In the environment, the parent DDT is metabolized mainly to dichlorodiphenyldichloroethane (DDD) and dichlorodiphenyldichloroethylene (DDE).[2]

There is a noteworthy exception to the continuous increase in body burden over an organism's lifetime. During gestation, there can be in utero, cross-placental transfer of lipophilic compounds from the fat reserves of the mother to the baby. The initial peaks in both the male and the female whale represent contaminant offloading via the mother's milk. Since PCBs and pesticides are typically found in fat (i.e., they are highly lipophilic) and because milk is the means by which the baby receives the necessary fat for nutrition, the neonate receives large doses of these contaminants during the weaning stage of

[2] The two principal isomers of DDD are: *p,p'*-2,2-bis(4-chlorophenyl)-1,1-dichloroethane; and *o,p'*-1-(2-chlorophenyl)-1-(4-chlorophenyl)-2,2-dichloroethane. The principal isomer of DDE is *p,p'*-1,1'-(2,2-dichloroethenylidene)-bis[4-chlorobenzene].

development [107,108]. When rate of weaning diminishes, the uptake by the calf declines. This is a detoxification or depuration stage during which more mass of the contaminant is excreted than is absorbed. After this, for the rest of the male's life, the persistent compound body burden increases; whereas for the birthing stage of the female, there is an oscillation between contaminant offloading to the calves and accumulation periods [109].

Chemists often define persistence as an intrinsic chemical property of a compound.[3] Most environmental and biomedical engineers consider persistence to be both intrinsic and extrinsic, that is, a function of the media, energy and mass balances, and equilibria. Engineers usually want to know not only about the molecular weight, functional groups, and ionic form of the compound, but also whether it is found in the air or water, and the condition of the media. Many toxic compounds are semivolatile (i.e., at 20°C and 101 kPa atmospheric pressure, vapor pressures=10^{-5}–10^{-2} kPa), under typical environmental conditions. These low vapor pressures and low aqueous solubilities translate into low fugacities, that is, they lack a strong propensity to flee a compartment, for example, to move from the water to the air.

Thus, a multipathway perspective must be employed for pollution, with a time aspect to each pathway. After release into the environment, the deposition of air pollutants onto soil, plants, and water is followed by the uptake of the chemicals by biota, and potential exposures by organisms via contact with the contaminated soils, waters, and foods. For example, such a human health multipathway analysis goes beyond the expected inhalation pathway to include other exposure pathways [110]. Even low K_H and K_{AW} compounds, however, can be transported long distances in the atmosphere when sorbed to particles. Fine particles can behave as colloids and stay suspended for extended periods of time, explaining in part why low K_H compounds can be found in the most remote locations relative to their sources, such as in the Arctic regions. This is important, for example, when explaining to indigenous populations why they may be exposed to contaminants that are produced thousands of miles from them.

If the substrate has sufficient sorption sites, such as soil particles or carpet twill, even a substance not expected to be persistent may become tightly bound and persistent. For example, once an air pollutant is transported and deposited onto highly sorbent clays and organic matter or onto carpet and other sorbent indoor surfaces, the substance may become tightly bound and persistent. This type of substrate dependence is common in the environment and complicates $t_{1/2}$ estimates. The properties of the compound and those of the water, soil, and sediment determine the rate of sorption and other partitioning. Thus, care must be taken when considering the conditions under which published atmospheric $t_{1/2}$ values were calculated. For example, if a substance has a high affinity for particulate matter in the atmosphere, its half-life may be diminished under conditions that lead to increased particle deposition. Conversely, the half-life may be increased if the compound sorbs to ultrafine particles that remain suspended longer than the compound's gas-phase $t_{1/2}$.

[3] Daniel Chang, a computational chemist with the EPA, says that the term "inherent" is seldom an accurate depiction of a substance. Certainly, atomic mass is an inherent property, but solubility and vapor pressure, in the strictest terms, are not inherent but highly dependent on environmental conditions. Thus, modelers like Dr. Chang, who seek to parameterize based on first principles, are challenged to find those that are actually inherent to a substance.

The formation, transformation, transport, and fate of a pollutant involve both chemical persistence and environmental persistence. Henry's law, solubility, vapor pressure, and sorption, and other properties of a compound may provide *prima facie* indications that the compound is not persistent. However, in real-world scenarios, this may not be the case. Indoor environments, for example, often lack conditions conducive to degradation, including less sunlight (i.e., lower photochemical degradation rates), microbial populations (i.e., lower biotic transformation rates), and moisture regimes (i.e., lower hydrolysis rates). Thus, an indoor air pollutant, for example, a pesticide, which is not very persistent outdoors (e.g., in the garden) may become persistent indoors, for example, a reactive pesticide that is tracked into a home and becomes entrapped in carpet fibers. This is important if the pesticide label's expected half-life was calculated with the assumption that the product would be applied outdoors, but the product was used (or misused) indoors [37].

Another example of environmental versus inherent persistence is that of a steady-state condition. A repository of a source of a nonpersistent compound may exist that leads to a continuous, persistent exposure of a neighborhood population. The loading of the compound into the parcel of atmosphere is equal to the deposition and degradation of the pollutant within the parcel for a specific time period. A parcel of the air mass may be viewed as a thermodynamic control volume. Once a steady state is reached, as long as the specific contaminant input (e.g., total stack releases) equals the rate of deposition and rate of transformation, then even an inherently reactive, nonpersistent compound is environmentally persistent [37].

Fig. 18.5 depicts a source-sink scenario. Here, the source is the amount of contaminant exiting the stack, and the sinks are the deposition, storage, and transformation processes. The steady-state region exemplifies the spatiotemporal aspects of a pollutant plume.

Note that the distance from the source can be determined by the rate of movement of the plume. For example, if the contaminant in the plume is moving advectively, that is, being carried along with the air mass, at a velocity of 1 m second^{-1}, then a measurement 600 m downwind from the stack is the same as a measurement taken 600 seconds (10 minutes) after the contaminant is released. Of course, spatial measurements are constant and time-related measurements are relative, that is, 600 m is always the same distance, whereas the plume velocity is changing continuously. It is often important to average the time of flight for the contaminant since the direction of wind will also change. Thus, the control volume in Fig. 18.5 will move with the movement of the air mass. Thus, measurements may have to track with this movement, for example, a stationary monitor will miss the conditions in the control volume, but a mobile monitor can be adjusted to move with the control volume if plume velocity is known. Thus, on a day with a rapidly moving air mass, the measurement would be taken farther from the source (e.g., 1200 m at 2 m second^{-1}); and on days with a slower moving air mass, the measurement would be taken nearer the source (e.g., 300 m at 0.5 m second^{-1}).

Steady-state concentration within a control volume can be expressed as:

$$C_{SS} = \frac{S \times \tau}{V} \tag{18.6}$$

where, C_{SS}=steady state concentration in the control volume; S=source or sink rate (in steady state they are equal, so only one factor is needed); τ=residence time; and V=control volume.

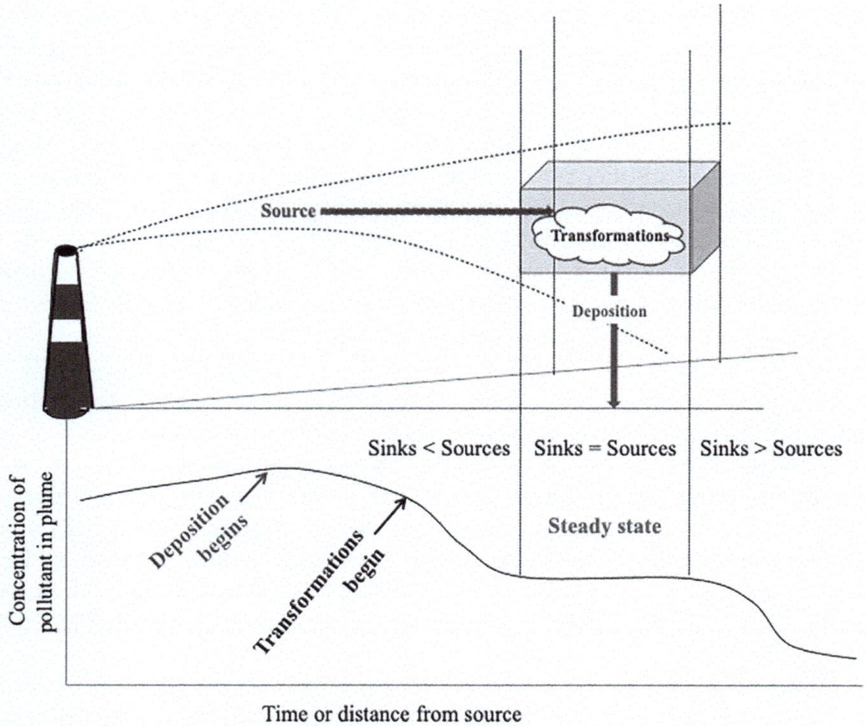

FIGURE 18.5 Steady-state period during which emissions equal rates of deposition and degradation. Sources dominate near the source, and then steady state is reached and maintained as long as the sources and sinks are equal. When deposition and transformation rates exceed source loading into the control volume, sinks begin to control, with a corresponding decline in pollutant concentration following the steady-state conditions. *Data from D.A. Vallero, Fundamentals of air pollution, 6th ed., Academic press, Amsterdam, NV, 2025.*

For example, if S=1000 μg hour^{-1} of benzene; τ=10 hours; and V=100 m^3, then C_{ss}= (1000 × 10)/100 =100 μg m^{-3} in the control volume. The concentration increases with increasing τ, that is, the longer the time in the control volume, the greater the concentration of the air pollutant. Thus, if the residence time doubles to 20 hours, the steady-state concentration also doubles to 200 μg m^{-3}. This illustrates the problem of temporal uncertainty. Overestimating the residence time can result in larger concentrations than are in fact there, but underestimating the residence time would give the incorrect estimate, for example, that a control volume such as an urban airshed is meeting an air quality standard when it is not.

The processes within the ambient plume shown in Fig. 18.5 also occur in microenvironments. There is pollutant infiltration into the microenvironment, as well as deposition to surfaces and transformation within the microenvironment. Certainly, deposition does not mean the contaminant is gone. It has merely moved to another environmental compartment (e.g., atmosphere to soil in the ambient setting; air to walls and floors in the indoor microenvironment). In some cases, new transformations will occur (e.g., soil bacteria degrading the pollutant), or merely different routes of exposure (e.g., settled particulate matter, that is, later ingested or contacted dermally). Dramatically increased environmental half-lives of certain substances will occur indoors due to the lower rate of photolysis (degradation by light energy) for indoor environments versus outdoors. Other factors increasing indoor half-lives owe to the sorptive characteristics of the surfaces like carpet twill, as well as surface characteristics, for example, if the pesticide molecule is sorbed to a soil particle, that is, embedded in the twill. For example, ionic charge can affect the sorption, for example, a negatively charged site on a particle may increase the persistence of a substance that has an abundance of negatively charged sites.

The amount and form of any substance, that is, released into a system determine its pollution potential. Total emission or other release data are not nearly as useful as speciation data. The total mass of nitrogen, for example, can be further quantified not only by the gross mass of the substance entering the atmosphere but also by the response within compartments of the atmosphere and hydrosphere. Hence, the relationship between a substance's entry and its impact begins even before release.

The first major division of loading pollutants into the environment is between point sources and nonpoint sources. This is a distinction of convenience, often used in the discussion of water, wastewater, and ecosystems. For air pollutants, the distinction is often between a stationary source and a mobile source.

After and during an emission, an air pollutant or a precursor of an air pollutant will undergo change. The change depends on physical, chemical, and biological factors. For example, a nitrogen load of an ammonium compound may either be reduced (ammonification) or oxidized (nitrification).

$$\text{Ammonification: } NH_4^+ + OH^- \Leftrightarrow NH_3 + H_2O \tag{18.7}$$

$$\text{Nitrification: } NH_4^+ + 2O_2 \rightarrow NO_3^- + 2H^+ + H_2O \tag{18.8}$$

Either of these reactions can occur abiotically, can be mediated by microbes, and/or take place within an organism. Other conditions that affect loading include temperature, the presence of aerosols on which and in which sorption can occur, humidity, pH, and ionic strength. For example, under acidic conditions, most of the ammonia nitrogen will ionize to ammonium, and under basic conditions, the nonionized ammonia concentrations will increase in proportion to the ammonium. This is important since nonionized ammonia is very toxic to aquatic fauna, whereas ionized species are nutrients needed for flora and algae. After deposition, the decomposition of organic matter makes for very complex and heterogeneous reactions:

$$\begin{aligned}\text{Organic nitrogen} + O_2 \rightarrow NH_3 \text{ nitrogen} + O_2 \rightarrow NO_2 \text{ nitrogen} + O_2 \\ \rightarrow NO_3 \text{ nitrogen} + O_2\end{aligned} \tag{18.9}$$

Even this is an oversimplification, since the oxidation and reduction depend on the types of bacteria present. For example, the O_2 required for nitrification is theoretically 4.56 mg O_2 $mg^{-2}NH_4^+$, but this is an autotrophic reaction. Thus, O_2 is being produced by the nitrifying bacteria, decreasing the amount needed. ("Nutrient Loading - an overview I ScienceDirect Topics") However, the growth rate of nitrifying microbes is far less than that of heterotrophic microbes decomposing organic wastes. Therefore, when large amounts of organic matter are being degraded, the nitrifiers' growth rate will be sharply limited by the heterotrophs, which means the rate of nitrification will commensurately be decreased [111]. Furthermore, these same responses to conditions exist for all other nutrients in the system, for example, organic phosphorus will be oxidized and oxidized phosphorus species reduced, etc. Thus, the kinetics of pollutant loading into a control volume and the transformations within that control volume are quite complex.

18.6.2 A sustainable future

All the technical and scientific information shared in this chapter calls for a systematic and sustainable view toward emerging technologies. It is difficult to place ethics ahead of innovation. Often, products reach the marketplace well before all or much of the necessary ethical thoughtfulness has occurred.

One way to get out in front with a modicum of ethics is to require sustainable methods industry-wide as technologies are being developed. For example, the pharmaceutical industry can prepare guidance and support legislation to promote sustainable processes like additive manufacturing (AM) and microfluidics (MF) [112]. AM, or three-dimensional printing, allows companies to be predictive of downstream events prior to the release of a pharmaceutical product before it enters the marketplace. This is accomplished by layering active ingredients, polymers, and other materials based on digital models. Traditional drug manufacturing has been subtractive, that is, material is removed from solid blocks or matrices of material to create pharmaceutical delivery systems or components. This can be inefficient and wasteful because much, up to 90%, of the raw material is lost in the removal process [113].

As the product is released, AM can support precision medicine with on-demand production, highly personalized dosages, and complex, multidrug tablet structures. Drugs can be customized to an individual patient's needs for exact dosage based on weight, age, metabolism, and other endogenous factors. In addition to the advantages of precision medicine over "one-size-fits-all" medications, it can also mean less waste and more targeted production. In addition to personalized medicine, AM supports rapid prototyping and faster, decentralized supply chains when compared to traditional, high-volume, subtractive manufacturing methods [113].

Microfluidic processes can support the sustainability of single-use, disposable, point-of-care diagnostic devices that are increasingly needed to meet urgent needs for testing and diagnosis in places away from sophisticated laboratory facilities. Regrettably, manufacturing and disposal of single-use devices by lateral flow assay, cartridges, cassettes, or lab-on-chip microfluidic formats can leave rather large pollution footprint [114]. This is because many point-of-care devices are manufactured using recalcitrant polymeric materials derived from fossil fuels. Incineration is often deployed to dispose of them after use to decrease infection risk, with the attendant release of volatile organic compounds and other pollutants. Many devices also contain cyanide derivatives.

Thus, there is a need to find innovative, sustainable ways to manufacture these devices; solutions are available to point-of-care test developers to start embedding sustainability at an early stage in their design and to reduce their nonrenewable plastic consumption in research and product development. Whereas many single-use, "lab-on-a-chip" (LOC) devices are traditionally made of petrochemical-based plastics, microfluidic advancements can decrease waste, lower energy consumption, and support the integration of sustainable materials like paper or biodegradable polymers [114].

Microfluidic devices operate using small fluid volumes, for example, microliter to picoliter, which lowers the need for large, expensive, and resource-intensive patient samples and reagents. By integrating mixing, separation, detection, and other analytical steps onto a single chip, these MF-manufactured devices eliminate the need for other disposable laboratory apparatus like pipettes, tubes, and plates. In addition, MF devices can reduce reagent usage by up to 95%–99% compared to traditional laboratory methods.

Computational approaches can be used to estimate and predict events before they occur, for example, what can result after a drug or other product reaches the marketplace. The computational models can virtually "tweak" molecules to see what impacts might result down the road. In some cases, the actual molecule is never synthesized but is a virtual molecule that can be run through various biochemodynamic scenarios as a first screen on possible hazards. These tools can also help with predicting hazards, exposures, and risks to humans and other organisms by improving the synthetic chemistry, for example, preventing the formulation of chiral and enantiomer compounds that are resistant to natural biodegradation (e.g., some left-hand chirals may be much more easily broken down than right-hand chirals of the same compound; also, one chiral may be toxic and the other efficacious). This may also apply to GMOs and synthetic biology. For example, subtle changes to a microbe's DNA can foster unexpected changes in ecosystem competition, which might be predicted using proteomics and other computational tools.

Green chemistry and green engineering approaches should be considered when manufacturing products. For example, there are ways to eliminate pollution at its source by minimizing or eliminating the hazards and substituting safer chemicals in feedstocks and use of sustainable reagents, solvents, and other products [115,116].

18.7 Conclusions

Discussing resilient and sustainable systems is a fitting conclusion to a book on emerging technologies. It is important for many reasons, but chief among them is that the technologist must be humble and objective about potential problems down the road of an innovation. The knowledge base from which to forecast possible flaws and adverse events is small since the world has never seen this technology, and the problems will have to unfold in the real world.

Therefore, the technology and its products must be designed to be resilient and adaptable as new findings emerge. Also, the technology and its products must be effective in providing benefits, while also being able to continue to operate with little waste and continuous improvement. That is, it must be sustainable.

References

[1] D. Vallero, Translating Diverse Environmental Data Into Reliable Information: How to Coordinate Evidence from Different Sources, Academic Press, (2017).

[2] S.H. Woolf, The meaning of translational research and why it matters, JAMA the Journal of the American Medical Association 299 (2) (2008) 211–213.
[3] B.E. Beisner, D.T. Haydon, K. Cuddington, Alternative stable states in ecology, Frontiers in Ecology and the Environment 1 (7) (2003) 376–382.
[4] D.A. Vallero, Applications and Calculations in Environmental Physics, AIP Publishing LLC, (2022).
[5] S. Gharari, S. Razavi, A review and synthesis of hysteresis in hydrology and hydrological modeling: memory, path-dependency, or missing physics? Journal of hydrology 566 (2018) 500–519.
[6] R.A. Hites, D.C. Lehman, A. Salamova, M. Venier, Temporal environmental hysteresis: a definition and implications for polybrominated diphenyl ethers, Science of the Total Environment 753 (2021) 141849.
[7] D. Vallero, S. Isukapalli, Simulating real-world exposures during emergency events: studying effects of indoor and outdoor releases in the Urban Dispersion Program in upper Manhattan, NY, Journal of Exposure Science and Environmental Epidemiology 24 (3) (2014) 279–289.
[8] J.D. Solomon, D.A. Vallero, From Our Partners–Communicating Risk and Resiliency: Special Considerations for Rare Events. Center for Infrastructure Protection & Homeland Security, George Mason University, (2017). Available from https://cip.gmu.edu/2016/06/01/partners-communicating-risk-resiliency-special-considerations-rare-events/(accessed 01.05.17).
[9] S. Carpenter, B. Walker, J.M. Anderies, N. Abel, From metaphor to measurement: resilience of what to what? Ecosystems 4 (8) (2001) 765–781.
[10] R. O'Neill, Ecosystem persistence and heterotrophic regulation, Ecology 57 (6) (1976) 1244–1253.
[11] D. Vallero, Environmental Biotechnology: A Biosystems Approach, Elsevier Science, (2015).
[12] E. Paoletti, et al., Advances of air pollution science: from forest decline to multiple-stress effects on forest ecosystem services, Environmental Pollution 158 (6) (2010) 1986–1989.
[13] M. Dardonville, C. Bockstaller, O. Therond, Review of quantitative evaluations of the resilience, vulnerability, robustness and adaptive capacity of temperate agricultural systems, Journal of Cleaner Production 286 (2021) 125456.
[14] J.K. Vanos, J.W. Baldwin, O. Jay, K.L. Ebi, Simplicity lacks robustness when projecting heat-health outcomes in a changing climate, Nature Communications 11 (1) (2020) 6079.
[15] J.M. Anderies, O. Barreteau, U. Brady, Refining the robustness of social-ecological systems framework for comparative analysis of coastal system adaptation to global change, Regional Environmental Change 19 (2019) 1891–1908.
[16] J. Jäger, et al., Assessing policy robustness of climate change adaptation measures across sectors and scenarios, Climatic Change 128 (2015) 395–407.
[17] Wilby, R., Dessai, S., 2010. Robust adaptation to climate change.
[18] S. Dessai, M. Hulme, Assessing the robustness of adaptation decisions to climate change uncertainties: a case study on water resources management in the East of England, Global environmental change 17 (1) (2007) 59–72.
[19] J.A. Dunne, R.J. Williams, N.D. Martinez, Network structure and biodiversity loss in food webs: robustness increases with connectance, Ecology Letters 5 (4) (2002) 558–567.
[20] D. Vallero, Fundamentals of Water Pollution: Quantifying Pollutant Formation, Transport, Transformation, Fate and Risks, first ed., Elsevier, Amsterdam, NV, (2024).
[21] T. Elmqvist, et al., Response diversity, ecosystem change, and resilience, Frontiers in Ecology and the Environment 1 (9) (2003) 488–494.
[22] E. Hollnagel, D.D. Woods, N. Leveson, Resilience engineering: Concepts and precepts, Ashgate Publishing, Ltd, (2007).
[23] J.D. Solomon, Communicating Reliability, Risk and Resiliency to Decision Makers, Duke University, (2017).
[24] H.R. Heinimann, A generic framework for resilience assessment, An edited collection of authored pieces comparing, contrasting, and integrating risk and resilience with an emphasis on ways to measure resilience (2016) 90.
[25] P. Helm, Risk and resilience: strategies for security, Civil Engineering and Environmental Systems 32 (1-2) (2015) 100–118.
[26] J. Zhao, X. Feng, M.-K. Tran, M. Fowler, M. Ouyang, A.F. Burke, Battery safety: fault diagnosis from laboratory to real world, Power Sources 598 (2024) 10.1016 no. 234111,.
[27] R.L. Winkler, A.H. Murphy, Experiments in the laboratory and the real world, Organizational Behavior and Human Performance 10 (2) (1973) 252–270.

[28] Rotheudt, H., et al., 2022. Distribution of microbial contaminants in operating theatres and healthcare environments. In: Proceedings of the 24th International Symposium on Contamination Control and Cleanroom Technology, the Hague, the Netherlands.

[29] I. Permana, F. Wang, Performance improvement of a biotechnology vaccine cleanroom for contamination control, Journal of Building Engineering 82 (2024) 108248.

[30] A. Čaja, M. Vantúch, N.Č. Kantová, Energy-efficient solution for cleanrooms in healthcare facilities, MATEC Web of Conferences, EDP Sciences, (2025), p. 5001 vol. 412.

[31] B. Minsker, et al., Progress and recommendations for advancing performance-based sustainable and resilient infrastructure design, Journal of Water Resources Planning and Management 141 (12) (2015) A4015006.

[32] F.R. Siegel, Adaptations of Coastal Cities to Global Warming, Sea Level Rise, Climate Change and Endemic Hazards, Springer, (2019).

[33] C. Mellin, et al., Safeguarding nutrients from coral reefs under climate change, Nature Ecology & Evolution 6 (12) (2022) 1808–1817.

[34] A. Sánchez-Arcilla, M. García-León, V. Gracia, R. Devoy, A. Stanica, J. Gault, Managing coastal environments under climate change: pathways to adaptation, Science of the Total Environment 572 (2016) 1336–1352.

[35] C. Whitbeck, Ethics as design: doing justice to moral problems, Hastings Center Report 26 (3) (1996) 9–16.

[36] National Society of Professional Engineers, NSPE code of ethics for engineers. https://www.nspe.org/resources/ethics/code-ethics, 2016 (accessed 31.08.18).

[37] D.A. Vallero, Fundamentals of Air Pollution, sixth ed., Academic Press, Amsterdam, NV, (2025).

[38] D.A. Vallero, T.M. Letcher, Society, Unraveling Environmental Disasters, second ed., Elsevier Academic Press, Amsterdam, The Netherlands, (2023). ch. 720.

[39] D.A. Vallero, Potential impacts of climate change on biogeochemical cycling, Water and Climate Change, Elsevier, (2022), pp. 41–62.

[40] D.A. Vallero, Air pollution physics, Methods and Calculations in Environmental Physics, AIP Publishing LLC Melville, New York, (2022). pp. 7-1–7-70.

[41] D.A. Vallero, Introduction: importance of systems sciences and first principles, Methods and Calculations in Environmental Physics, AIP Publishing LLC Melville, New York, (2022). pp. 1-1–1-20.

[42] D.A. Vallero, Environmental Systems Science: Theory and Practical Applications, first ed., Elsevier Academic Press, (2021).

[43] S.B. Billatos, Design methodologies for the environment, Mechanical Life Cycle Handbook, CRC Press, (2001), pp. 96–111.

[44] S. Billatos, Green Technology and Design for the Environment, CRC Press, (1997).

[45] Allada, V., 2000. Preparing engineering students to meet the ecological challenges through sustainable product design. In: Proc. 2000 International Conf on Engineering Education.

[46] Chen, X., Wang, X., Colacelli, A., Lee, M., Xie, L., 2025. Electricity demand and grid impacts of AI data centers: challenges and prospects. arXiv preprint arXiv:2509.07218.

[47] V. Veeramachaneni, Optimizing renewable energy integration in AI-driven data centers using quantum algorithms, Journal of Network Security and Data Mining 8 (1) (2025) 36–48.

[48] Ferreira, J., Callou, G., Josua, A., Maciel, P., 2018. Estimating the environmental impact of data centers. In: 2018 IEEE 17th International Symposium on Network Computing and Applications (NCA), IEEE, pp. 1–4.

[49] H. Rong, H. Zhang, S. Xiao, C. Li, C. Hu, Optimizing energy consumption for data centers, Renewable and Sustainable Energy Reviews 58 (2016) 674–691.

[50] W.W. Ingwersen, et al., A new data architecture for advancing life cycle assessment, The International Journal of Life Cycle Assessment 20 (4) (2015) 520–526.

[51] Systems Analysis Research: Program Brief – Life Cycle Analysis, 2009. [Online] Available from: http://www.epa.gov/nrmrl/std/sab/lca/lca_brief.htm.

[52] B. Hermann, C. Kroeze, W. Jawjit, Assessing environmental performance by combining life cycle assessment, multi-criteria analysis and environmental performance indicators, Journal of Cleaner Production 15 (18) (2007) 1787–1796.

[53] G. Rebitzer, et al., Life cycle assessment: part 1: framework, goal and scope definition, inventory analysis, and applications, Environment International 30 (5) (2004) 701–720.

[54] K. Eskandar, Artificial intelligence and synthetic biology: biosecurity risks, dual-use concerns, and governance pathways, AI and Ethics 6 (1) (2026) 66.

[55] Marsili, M., 2025. Emerging and disruptive technologies: strategic implications and ethical challenges of dual-use innovations. In: Strategic Leadership Journal. Challenges for geopolitics and organizational development, 1: Centro Alti Studi Difesa/Scuola Superiore Universitaria (CASD/SSU), pp. 57–71.
[56] W.H. Organization, 2021. Emerging technologies and dual-use concerns: a horizon scan for global public health.
[57] M. Ienca, E. Vayena, Dual use in the 21st century: emerging risks and global governance, Swiss Medical Weekly 148 (2018) w14688.
[58] D.A. Vallero, C. Brasier, Sustainable Design: The Science of Sustainability and Green Engineering, John Wiley & Sons, (2008).
[59] D.A. Vallero, C. Braiser, Teaching green engineering: the case of ethanol lifecycle analysis, Bulletin of Science, Technology & Society 28 (3) (2008) 236–243.
[60] D.A. Vallero, Environmental Systems Science: Theory and Practical Applications, first ed, Elsevier, Amsterdam, NV, (2020).
[61] D.A. Vallero, Chapter 11 - Economics and feasibility assessments, in: D.A. Vallero (Ed.), Environmental Systems Science, Elsevier, (2021), pp. 509–542.
[62] D.A. Vallero, Chapter 8 - Decision support tools, in: D.A. Vallero (Ed.), Environmental Systems Science, Elsevier, (2021), pp. 293–357.
[63] M. Raynolds, R. Fraser, D. Checkel, The relative mass-energy-economic (RMEE) method for system boundary selection part 1: a means to systematically and quantitatively select LCA boundaries, The International Journal of Life Cycle Assessment 5 (1) (2000) 37–46.
[64] Life Cycle Assessment: Inventory Guidelines and Principles, 1993. EPA/600/R-92/245.
[65] C.E. Beaudrie, M. Kandlikar, G. Ramachandran, Using expert judgment for risk assessment, Assessing nanoparticle risks to human health, Elsevier, (2016), pp. 91–119.
[66] M.E. Wittmann, R.M. Cooke, J.D. Rothlisberger, D.M. Lodge, Using structured expert judgment to assess invasive species prevention: Asian carp and the Mississippi—great lakes hydrologic connection, Environmental Science & Technology 48 (4) (2014) 2150–2156.
[67] U.S. EPA, Expert Elicitation White Paper, Science and Technology Policy Council, Washington, DC, (2011). [Online]. Available http://www.epa.gov/stpc/pdfs/ee-white-paper-final.pdf.
[68] J. Naimon, Using expert panels to assess risks of environmental biotechnology applications: a case study of the 1986 Frostban risk assessments, in: RM Atlas, E.P. Greenberg, AL Demain, BH Olson, G. Sayler (Eds.), Risk Assessment in Genetic Engineering: Environmental Release of Organisms. McGraw-Hill Environmental Biotechnology Series, 1991 McGraw-Hill, New York, (1991).
[69] K. Sander, G.S. Murthy, Life cycle analysis of algae biodiesel, The International Journal of Life Cycle Assessment 15 (7) (2010) 704–714.
[70] D. Katsourinis, D. Giannopoulos, M. Founti, Fischer–tropsch biofuel production from supercritical water gasification of lignocellulosic biomass: process modelling and life-cycle assessment, Processes (3) (2025) 895.
[71] M.E. Dry, The fischer–tropsch process: 1950–2000, Catalysis today 71 (3-4) (2002) 227–241.
[72] M.E. Dry, High quality diesel via the Fischer–Tropsch process–a review, Journal of Chemical Technology & Biotechnology: International Research in Process, Environmental & Clean Technology 77 (1) (2002) 43–50.
[73] A. de Klerk, Fischer–tropsch process, Kirk-Othmer Encyclopedia of Chemical Technology (2000) 1–20.
[74] F. Shi, P. Wang, Y. Duan, D. Link, B. Morreale, Recent developments in the production of liquid fuels via catalytic conversion of microalgae: experiments and simulations, RSC Advances 2 (26) (2012) 9727–9747.
[75] J.C. Bare, D.A. Vallero, Incorporating exposure science into life-cycle assessment, Access Science McGraw-Hill Education; Yearbook of Science & Technology, (2014).
[76] O. Jolliet, A. Ernstoff, S. Csiszar, P. Fantke, Health impacts of consumer exposure during product use: near field exposure applied to risk assessment and LCA, In International Society of Exposure Science 24th Annual Meeting (2014) 178–179.
[77] O. Jolliet, A.S. Ernstoff, S.A. Csiszar, P. Fantke, Defining product intake fraction to quantify and compare exposure to consumer products, ed: ACS Publications (2015).
[78] S.A. Csiszar, et al., A conceptual framework to extend life cycle assessment using near-field human exposure modeling and high-throughput tools for chemicals, Environmental Science & Technology (2016). Available from https://doi.org/10.1021/acs.est.6b02277.

[79] J. Moya, et al., Exposure Factors Handbook: 2011 ed., US Environmental Protection Agency, Washington, (2011).
[80] J. Moya, L. Phillips, J. Sanford, M. Wooton, A. Gregg, L. Schuda, A review of physiological and behavioral changes during pregnancy and lactation: potential exposure factors and data gaps, Journal of Exposure Science and Environmental Epidemiology (2014).
[81] K.K. Isaacs, et al., SHEDS-HT: an integrated probabilistic exposure model for prioritizing exposures to chemicals with near-field and dietary sources, Environmental Science & Technology 48 (21) (2014) 12750–12759.
[82] D.E. Meyer, S.C. Bailin, D. Vallero, P.P. Egeghy, S.V. Liu, E.A.C. Hubal, Enhancing life cycle chemical exposure assessment through ontology modeling, Science of the Total Environment 712 (2020) 136263.
[83] R. Harder, S. Heimersson, M. Svanstrom, G.M. Peters, Including pathogen risk in life cycle assessment of wastewater management. 1. Estimating the burden of disease associated with pathogens, Environmental Science & Technology 48 (16) (2014) 9438–9445.
[84] R.A. Souza, L.C. Babujia, A.P. Silva, M. de Fátima Guimarães, C.A. Arias, M. Hungria, Impact of the ahas transgene and of herbicides associated with the soybean crop on soil microbial communities, Transgenic Research 22 (5) (2013) 877–892.
[85] M.Z. Naser, Causality and causal inference for engineers: beyond correlation, regression, prediction and artificial intelligence, Wiley Interdisciplinary Reviews: Data Mining and Knowledge Discovery 14 (4) (2024) e1533.
[86] A.B. Hill, The environment and disease: association or causation? Proceedings of the Royal Society of Medicine 58 (5) (1965) 295.
[87] T.M. Letcher, D. Vallero, Excerpt from: Waste: A Handbook for Management, second ed., Academic Press, (2019) ed.
[88] T. Burgdorf, et al., The AOP concept: how novel technologies can support development of adverse outcome pathways, Applied In Vitro Toxicology 3 (3) (2017) 271–277.
[89] D.L. Villeneuve, et al., Adverse outcome pathway (AOP) development I: strategies and principles, Toxicological Sciences 142 (2) (2014) 312–320.
[90] M.A. Wood, Rethinking how technologies harm, The British Journal of Criminology 61 (3) (2021) 627–647.
[91] M.J.A. De Abreu, Technological indeterminacy: medium, threat, temporality, Anthropological Theory 13 (3) (2013) 267–284.
[92] T.S. Potter, et al., Applying causal reasoning to investigate multicausality in microbial systems, Ecosphere 15 (5) (2024) e4782.
[93] T.S. Kuhn, D. Hawkins, The structure of scientific revolutions, American Journal of Physics 31 (7) (1963) 554–555.
[94] J. Green, Cassell's Dictionary of Slang, Sterling Publishing Company, Inc, (2005).
[95] Maslow, A., 1966. The PSYCHOLOGY OF SCIENCE: A reconnaissance Harper & Row, New York.
[96] M. Dubova, et al., Is Ockham's razor losing its edge? New perspectives on the principle of model parsimony, Proceedings of the National Academy of Sciences 122 (5) (2025) p. e2401230121.
[97] H. Petroski, To Engineer is Human, St. Martin's Press, New York, (1985).
[98] Toro, D.D., H. F., 1999. Long-range transport and deposition: the role of Henry's law constant. In: Final Report, International Council of Chemical Associations, Arlington, Virginia.
[99] T. Rahola, T. Hattula, A. Korolainen, J. Miettinen, Elimination of free and protein-bound ionic mercury (20Hg2+) in man, Annals of Clinical Research 5 (4) (1973) 214–219.
[100] B. Aberg, L. Ekman, R. Falk, U. Greitz, G. Persson, J.O. Snihs, Metabolism of methyl mercury (203Hg) compounds in man: excretion and distribution, Archives of Environmental Health: An International Journal 19 (4) (1969) 478–484.
[101] U.S. E P Agency, Estimating persistence, bioaccumulation, and toxicity using the PBT profiler, Sustainable Futures / P2 Framework Manual 2012 EPA-748-B12-001, USEPA, Washington, DC, (2024). ch. 7.
[102] L. Montano, et al., Polychlorinated biphenyls (PCBs) in the environment: occupational and exposure events, effects on human health and fertility, Toxics 10 (7) (2022) 365.
[103] U.S. Environmental Protection Agency, 2,3,7,8-Tetrachlorodibenzo-p-dioxin - 1746-01-6 | DTXSID2021315. https://www.epa.gov/chemical-research/chemistry-dashboardhttps://comptox.epa.gov/dashboard/chemical/details/DTXSID2021315, 2021 (accessed 22.12.21).

[104] S. Kanan, F. Samara, Dioxins and furans: a review from chemical and environmental perspectives, Trends in Environmental Analytical Chemistry 17 (2018) 1–13. Available from https://doi.org/10.1016/j.teac.2017.12.001.
[105] A. Schecter, L. Birnbaum, J.J. Ryan, J.D. Constable, Dioxins: an overview, Environmental Research 101 (3) (2006) 419–428.
[106] United Nations Environmental Programme and World Health Organization, 1999. Pollution Prevention and Abatement Handbook 1998 Toward Cleaner Production. World Bank Publications.
[107] C.M. Butt, U. Berger, R. Bossi, G.T. Tomy, Levels and trends of poly- and perfluorinated compounds in the arctic environment, Science of The Total Environment 408 (15) (2010) 2936–2965. Available from https://doi.org/10.1016/j.scitotenv.2010.03.015.
[108] C. Metcalfe, B. Koenig, T. Metcalfe, G. Paterson, R. Sears, Intra-and inter-species differences in persistent organic contaminants in the blubber of blue whales and humpback whales from the Gulf of St. Lawrence, Canada, Marine Environmental Research 57 (4) (2004) 245–260.
[109] D.A. Vallero, Hazardous wastes and the environment, Hazardous Waste Management, Elsevier, (2022), pp. 3–32.
[110] U.S. Environmental Protection Agency, Chapter 9: Overview of multisource multipathway risk assessment, Technical Resource Manual. Volume 3 - Community Scale Assessment Research, USEPA, Triangle Park, North Carolina, (2006).
[111] V. Novotny, Water Quality. Diffuse Pollution and Watershed Management, John Willey & Sons, Inc, Hoboken, New Jersey, (2003).
[112] E. Weaver, C. O'Hagan, D.A. Lamprou, The sustainability of emerging technologies for use in pharmaceutical manufacturing, Expert Opinion on Drug Delivery 19 (7) (2022) 861–872. Available from https://doi.org/10.1080/17425247.2022.2093857.
[113] J.K. Watson, K.M.B. Taminger, A decision-support model for selecting additive manufacturing versus subtractive manufacturing based on energy consumption, Journal of Cleaner Production 176 (2018) 1316–1322. Available from https://doi.org/10.1016/j.jclepro.2015.12.009.
[114] A.E. Ongaro, et al., Engineering a sustainable future for point-of-care diagnostics and single-use microfluidic devices, Lab on a Chip 22 (17) (2022) 3122–3137.
[115] P.T. Anastas, R.L. Lankey, Life cycle assessment and green chemistry: the yin and yang of industrial ecology, Green Chemistry 2 (6) (2000) 289–295.
[116] P.T. Anastas, Beyond reductionist thinking in chemistry for sustainability, Trends in Chemistry 1 (2) (2019) 145–148.

APPENDIX

A

Abbreviations and symbols

The following abbreviations and symbols are used in this text and/or in cited sources. The list is not completely exhaustive, since there is much variability in the use of symbols, especially since ethics and science draw from numerous academic and professional disciplines.

$\therefore$	therefore or hence
$\because$	because or since
σ or SD	standard deviation
σ^2	variance
μg	microgram
μm	micron or micrometer
σ_n	normal shear stress (mass length^{-1} time^{-2})
$\xrightarrow{\Delta}$	energy (heat) required for reaction
$\frac{dy}{dx}$	ordinary derivative of x with respect to y
η_e	effective porosity (dimensionless)
τ	mean residence time
$\frac{\partial y}{\partial x}$	partial derivative of x with respect to y
[C]	molar concentration
∇ (nabla)	gradient of a scalar field, especially in the Navier-Stokes equations
∇ (nabla)	water table
$\cong$	approximately equal to
$\approx$	almost equal to
$\rightarrow$	kinetic reaction
$\leftrightarrow$	equilibrium reaction
°C	degree Celsius or centigrade (temperature)
°F	degree Fahrenheit (temperature)
K	Kelvin (temperature)
°lat	latitude (degrees)
°long	longitude (degrees)
°R	degree Rankine (temperature)
μE	microenvironment
μ_g	gas viscosity (mass length^{-1} time^{-1})
μ_m	arithmetic mean

μM	micromole
0	initial (subscript)
A	area, surface area (length^2)
a	acceleration (length time^2)
A	aqueous phase
A	atomic weight
A	uniform cost
ABL	atmospheric boundary layer
abs	absorbed
ABT	averaging, banking, and trading
AC	activated carbon
AC	alternating current (electricity)
ACM	asbestos-containing material
ACS	American Cancer Society
ADAF	age-dependent adjustment factor (cancer risk)
ADD	average daily dose
ADI	acceptable daily input
ADME	absorption, distribution, metabolism, and excretion (or elimination)
ADP	adenosine diphosphate
AF	absorption factor
AF	atrial fibrillation
A_{fabric}	fabric surface area (length^2)
AFO	animal feeding operation
AIRS	Aerometric Information Retrieval System (the United States)
amu	atomic mass unit
A_p	cross-sectional area of the particle (length^2) = $\frac{\pi d_p^2}{4} a_p$particle acceleration
APCD	air pollution control device
APCD	air pollution control district
APCS	air pollution control system
APG	assigned protection factor
aq	aqueous phase
AQCD	Air Quality Criteria Document (the United States)
AQI	air quality index
AQMA	air quality maintenance area
AQS	Air Quality System (part of AIRS; the United States)
ASCE	American Society of Civil Engineers
ASME	American Society of Mechanical Engineers
ASPECT	airborne spectral photometric environmental collection technology
AT or $t_{averaging}$	averaging time for exposure (time)
atm	atmosphere (unit of pressure)
ATP	adenosine triphosphate
ATSDR	Agency for Toxic Substances and Disease Registry (the United States)
ATUS	American Time Use Survey
AWMA	Air and Waste Management Association
B	background pollutant concentration (rollback model)
B(a)P	benzo(a)pyrene
BCA	benefit-cost analysis
B/C or BCR	benefit-to-cost ratio
BACT	best available control technology
BAF	bioaccumulation factor
b_{ag}	absorption of light by NO_2 gas
b_{ap}	absorption of light by particles
BCF	bioconcentration factor

BCR	benefit/cost ratio
b_{ext}	light extinction coefficient
BF	bioavailability (percent of substance absorbed per day)
BMF	biomagnification factor
BOF	basic oxygen furnace
BP	blood pressure
b_{rg}	scattering of light by gaseous molecules (Rayleigh scattering)
b_{scat}	scattering of light by particles
BTEX	benzene, toluene, ethylbenzene, and xylenes (indication of gasoline contamination)
Btu or BTU	British Thermal Unit (power) = 1060 Joules
BW	body weight (mass)
C	chemical concentration [(mass mass^{-1}), (volume volume^{-1}), or (mass volume^{-1})]
c	speed of light (3×10^8 m s^{-1})
C_a	airborne chemical concentration (mass volume^{-1} of air in breathing zone)
$\overline{C}$	time-weighted concentration
CAA	Clean Air Act (the United States)
CAAA90	Clean Air Act Amendments of 1990 (the United States)
cal	calorie
CARB	California Air Resources Board
CCR	coal combustion residuals
C_D	drag coefficient (dimensionless)
CDC	Centers for Disease Control and Prevention (the United States)
CDD	chlorinated dibenzodioxin
CDF	chlorinated dibenzofuran
CEQ	Council on Environmental Quality (the United States)
CERCLA	Comprehensive Environmental Response, Compensation and Liability Act (also known as Superfund; the United States)
CFC	chlorofluorocarbon
CFD	computational fluid dynamics
CFD	cash flow diagram
CFR	Code of Federal Regulations (the United States)
CI	confidence interval
Ci	curie (radioactivity unit)
C_{ij}	concentration of exposure by person i in microenvironment j
CNG	compressed natural gas
CNS	central nervous system
COE	US Army Corps of Engineers (also USACE)
COHb	carboxyhemoglobinemia
COPD	chronic obstructive pulmonary disease
cos	cosine
c_p	specific heat at constant pressure
cps	cycles per second (1 cps = 1 Hz)
CSF	cancer slope factor [(mass mass^{-1} time^{-1})$^{-1}$] (also SF)
c_v	specific heat at constant volume
CVD	cardiovascular disease
CWA	Clean Water Act (the United States)
CWS	Canada-wide standards
$C_xH_yO_x$	organic compound (hypothetical or representative)
D	air quality standard (rollback model)
d	day
d	diameter (length)
D	dose
D	mass of pollutant per body mass

DAS	data acquisition system
DC	direct current (electricity)
DEFRA	Department for Environment, Food, and Rural Affairs (the United Kingdom)
DfD	design for disassembly
DfE	design for the environment
DfR	design for recycling
DHHS	Department of Health and Human Services (the United States)
DHS	Department of Homeland Security (the United States)
d_i	percentage difference (precision)
DI	daily intake
dL	deciliter (volume)
DNA	deoxyribonucleic acid
DNAPL	dense (high-density) nonaqueous-phase liquid
DO	dissolved oxygen
DOE	Department of Energy (the United States)
DOI	Department of the Interior (the United States)
DOI	digital object identifier
DOS	Department of State (the United States)
DOT	Department of Transportation (the United States)
DP	degradation product
d_p	physical particle diameter (length)
$\mathbf{D_{pa}}$ **or** d_{pa}	aerodynamic diameter of a particle (length)
$\mathbf{D_{ps}}$ **or** d_{ps}	Stokes particle diameter (length)
DQO	data quality objective
DRE	destruction and removal efficiency
DRL	differential rate law
e	charge of electron (4.8×10^{-10} statcoulombs)
E	electric field strength (potential, voltage length$^{-1)}$
E	bulk modulus (fluid stress divided by strain)
E	energy required in a reaction
e	Euler's number = 2.71828183…
E	event (probability and statistics)
E	exposure
E	particle collection efficiency; inverse of penetration (filtration)
E_0	strength of electric field (volts length^{-1})
E10	Fuel mix of 90% gasoline and 10% ethanol
E85	Fuel mix of 15% gasoline and 85% ethanol
E_a	Arrhenius activation energy
EAC	Ethics across the curriculum
EC	average exposure concentration
ECA	European Chemicals Agency
ED	exposure duration (time)
EDC	endocrine-disrupting compound
EDSP	endocrine disruptor screening program
EEA	European Environment Agency
E_F	emission factor (in emission calculation)
EF	exposure frequency (time)
EF	enrichment factor (biogeochemistry)
e.g.	*exempli gratia* (Latin); for example,
EIA	Energy Information Administration (the United States)
E_{ing}	ingestion exposure (mass per time)
E_{inh}	inhalation exposure (mass per time)
EIS	Emission Inventory System (the United States)

EIS	environmental impact statement
EJ	environmental justice
EL	exposure length (time per time)
ELF	extremely low frequency
EMEP	European Monitoring and Evaluation Programme
EMR	electromagnetic radiation
EPA or US EPA	Environmental Protection Agency (the United States)
erg	energy and work unit (10^{-7} joules)
EROEI	energy return on energy investment
ERPG	Emergency Response Planning Guidelines (the United States)
ETS	environmental tobacco smoke
EUAB	equivalent uniform annual benefit
EUAC	equivalent uniform annual cost
EUAW	equivalent uniform annual worth
EVT	extreme value theory
exp(x)	exponential (e^x)
E_υ	energy at frequency υ
F	energy flux
F	force (mass length time^{-2})
F	frequency (probability and statistics)
f	fugacity
F_B	buoyancy force (mass length time^{-2})
F_C	centrifugal force (mass length time^{-2})
F_D	drag force (mass length time^{-2})
F_E	electrostatic charge (force)
FFDCA	Federal Food, Drug, and Cosmetic Act (the United States)
FHWA	Federal Highway Administration (the United States)
FIFRA	Federal Insecticide, Fungicide, and Rodenticide Act (the United States)
FOARAM	Federal Ocean Acidification Research and Monitoring Act (the United States)
FONSI	finding of no significant impact
FQPA	Food Quality Protection Act (the United States)
FR	Federal Register (the United States)
FRM	Federal Reference Method (the United States)
ft	foot (length)
g	gas or vapor phase (subscript or parenthetical)
G	Gibbs free energy
g	gram
g	gravitational force
g	gravitational acceleration (subscript)
g	growth factor (rollback model)
G	gradient maintenance cost
g_0	standard gravitational field
GAC	granulated activated carbon
gal	gallon (volume)
GC	gas chromatography
g_c	gravitational constant = $32.2\frac{\text{lb}_\text{m}\cdot\text{ft}}{\text{lb}_\text{f}\cdot\text{sec}^2}$ = 980 cm sec^{-2}
GDP	gross domestic product
GHG	global greenhouse gas
GHz	gigahertz (10^9 Hz)
GIS	geographic information system
GLP	good laboratory practice
GPS	global positioning system
grad	gradient; grad $f = \nabla f(x, y)$ in a plane and $\nabla f(x, y, z)$ in space

GWP	global warming potential
$\Delta G_f^{*\,0}$	Gibbs free energy formation at steady state
H	hazard
h or hr	hour
h	elevation
h	piezometric head
h	physical stack height
H	effective stack height
H	enthalpy
H	headspace height
H	heat [subscript (mass length^{-1} time^{-2})]
H	Henry's Law (subscript) (also K_H)
h	Planck's constant (6.62×10^{-27} erg s)
HAP	hazardous air pollutant
HC	hydrocarbon
HEDS	Human Exposure Database System (the United States)
HEPA	high-efficiency particulate air (filter)
HHM	hierarchical holographic modeling
HI	hazard index
HMIWI	hospital/medical/infectious waste incinerators
hPa	hectopascal (unit of pressure)
HPC	high-performance computing
HPLC	high-performance liquid chromatography
HPV	high production volume (chemical class)
HQ	hazard quotient
HR	hazard ratio
HSWA	Hazardous and Solid Waste Amendments (the United States)
HVAC	heating, ventilation, and air conditioning
HYSPLIT	Hybrid Single-Particle Lagrangian Integrated Trajectory (model)
Hz	hertz (cycles s^{-1})
$h\upsilon$	photon
I	in situ
I	intake dose
I	intensity
i.e.	*is est* (Latin): that is
I/M	inspection and maintenance program
IAEA	International Atomic Energy Agency
IARC	International Agency for Research on Cancer
ICRP	International Commission on Radiological Protection
IDL	instrument detection limit
IDLH	immediately dangerous to life or health
in Hg	inches of mercury (unit of pressure)
IRL	integrated rate law
IPCC	Intergovernmental Panel on Climate Change
IR	intake rate [(mass time^{-1}) or (volume per time)]
IR	infrared
IRIS	Integrated Risk Information System
IRR	internal rate of return
IUR	inhalation unit risk
J	Joule (energy unit)
J	flux
k	Boltzmann's constant (1.37×10^{-16} erg K^{-1})
K	partitioning coefficient

k	reaction rate constant
K_a	acid equilibrium constant
K_{AW}	air-water partitioning coefficient
K_b	base equilibrium constant
K_D	soil sorption partitioning coefficient
k_e	rate at which the producers use energy (via respiration)
K_F	Freundlich isotherm constant
kg	kilogram (mass)
$K_{G/A}$	partition coefficient between the gaseous and aqueous phases
K_H	Henry's law partitioning coefficient (also H)
k_{inh}	inhalation rate (volume of air per unit time)
km	kilometer (length)
K_{oa}	octanol-air partitioning coefficient
K_{oc}	organic-carbon partitioning coefficient
K_{ow}	octanol-water partitioning coefficient
k_p	rate of chemical energy storage by primary producers
kPa	kilopascal (unit of pressure)
K_{sp}	solubility product constant
K_w	water dissociation equilibrium constant
kWh	kilowatt-hour
L or l	liter
l	liquid phase (subscript or parenthetical)
LADD	lifetime average daily dose
LAER	lowest achievable emission rate
lb	pound (weight)
$\mathbf{lb}_f$	foot-pound (force)
$\mathbf{lb}_m$	international avoirdupois pound (weight)
LC	liquid chromatography
$\mathbf{LC}_{50}$	lethal concentration 50 (lethal to 50% of tested organisms)
LCA	life cycle analysis or life cycle assessment
LCI	life cycle inventory
$\mathbf{LD}_{50}$	lethal dose 50 (lethal to 50% of tested organisms)
LEV	low-emission vehicle
ln(x)	natural log of x
LNAPL	light (low-density) nonaqueous-phase liquid
LOAEC	lowest observed adverse effect concentration (also LEL or LOEL)
LOAEL	lowest observed adverse effect level
LoD	limit of detection (also known as DL, detection limit)
log	logarithm
LoQ	limit of quantitation
LOSU	level of scientific understanding
LPG	liquefied propane gas
LRTAP	long-range transboundary air pollutants
LTP	local temperature and pressure
L_v	distance at which a black object is just barely visible
m	mass
m	meter
M	mole (gram molecular weight)
MACRS	Modified Accelerated Cost Recovery System
MACT	maximum achievable control technology
max	maximum (subscript)
MCE	mixed cellulose ester
MCL	maximum contaminant level (drinking water)

Me	metal
MEI	maximum exposed individual
MFAA	Millipore cellulose ester membrane
MHz	megahertz (10^6 Hz)
MIC	methyl isocyanate
min	minimum (subscript)
min	minute (time)
MIR	maximum individual risk
mm Hg	millimeters of mercury (unit of pressure)
mm	millimeter
mM	millimole
mmBTU	10^6 BTU
MMT	methylcyclopentadienyl manganese tricarbonyl
mmt	million metric tons
MOA	mode of action
mol	mole (gram molecular weight)
MQO	model quality objective
MRL	minimum risk level
MS	mass spectroscopy
MSAT	mobile source air toxic
MSDS	Material Safety Data Sheet
MTBE	methyl tert-butyl ether
MTTR	mean time to repair
MUC	maximum use concentration
mV	millivolts (potential)
MW	megawatt
MW	molecular weight
n	number of charges (electrostatics)
N	concentration (number)
N	Newton (unit of force)
n	number (e.g., number of moles or statistical sample size)
N-100	respirator protection efficiency against particulate aerosols (99.97%), but not resistant to oil
NA	nonattainment (air quality not meeting NAAQS)
NAAQS	National Ambient Air Quality Standards (the United States)
NADP	National Atmospheric Deposition Program (the United States)
NAE	National Academy of Engineering (the United States)
NAEI	National Atmospheric Emissions Inventory (the United Kingdom)
NAMS	National Air Monitoring Station (the United States)
NAPL	nonaqueous-phase liquid
NAS	National Academy of Sciences (the United States)
NASA	National Aeronautics and Space Administration (the United States)
NATA	National Air Toxics Assessment (the United States)
ND, n.d.	nondetect
NEA	Nuclear Energy Agency
NEI	National Emissions Inventory (the United States)
NEPA	National Environmental Policy Act (the United States)
NESHAP	National Emission Standards for Hazardous Air Pollutants (the United States)
ng	nanogram
NG	net goodness
NHANES	National Health and Nutrition Examination Survey (the United States)
N_{in}	number of particles (count) entering a system (e.g., filter)
NIOSH	National Institute for Occupational Safety and Health
NIST	National Institute of Standards and Technology (the United States)

NLEV	National Low-Emission Vehicle Program (the United States)
nM	nanomole
NMHC	nonmethane hydrocarbon
NMMAPS	National Morbidity, Mortality, and Air Pollution Study (the United States)
NMOG	nonmethane organic gases
NMVOC	nonmethane volatile organic compound
NOAA	National Oceanic and Atmospheric Administration (the United States)
NOAEC	no observed adverse effect concentration
NOAEL	no observed adverse effect level (also NEL or NOEL)
NOES	National Occupational Exposure Survey (the United States)
N_{out}	number of particles (count) exiting a system (e.g., filter)
NO_x	oxides of nitrogen, specifically NO and NO_2
NO_y	oxides of nitrogen, specifically NO_x and NO_z
NO_z	$HNO_3 + HONO + N_2O_5 + HO_2NO_3 +$ PAN + organic nitrates
NP	nanoparticle
NPL	National Physical Laboratory (the United Kingdom)
NPP	net primary product (or productivity)
NPRI	National Pollutant Release Inventory (Canada)
NPV	net present value
N_R	Reynolds number (also Re)
N_{Rp}	dimensionless particle Reynolds number
NSF	National Science Foundation (the United States)
NSR	new source review
NTP	National Toxicology Program (the United States)
NTP	normal temperature and pressure
NTSIP	National Toxic Substance Incidents Program (the United States)
NVOC	nonvolatile organic compound
O	octanol (subscript)
O&M	operation and maintenance
O(1D)	O atom in an excited singlet state
OC	organic carbon (subscript)
OC/EC	organic carbon-to-elemental carbon ratio
OECD	Organisation for Economic Co-operation and Development
OP	organophosphorus (pesticide class)
OSHA	Occupational Safety and Health Administration (the United States)
OUT	mass or energy outflow from control volume; effluent (subscript)
P **or** p	pressure
p	momentum
p	partial
p	particle (subscript)
P	particle penetration; inverse of efficiency (filtration)
P	particulate matter emissions (mass $time^{-1}$)
P	present air quality (rollback model)
P	probability
P	initial cost
P^0	vapor pressure
P_1	net primary productivity (or production)
P-100	oil-proof respirator protection efficiency against particulate aerosols (99.97%)
p_a	partial pressure of gas a
Pa	pascal (unit of pressure)
PA	preliminary assessment (of a waste site)
PAD	population-adjusted dose
PAH	polycyclic aromatic hydrocarbon (sometimes polyaromatic hydrocarbon)

PAL-DS	point, area, and line model with deposition and settling
PAMS	Photochemical Assessment Monitoring Station
p_b	partial pressure of chemical substance gas in blood
PBPD	physiologically based pharmacodynamic
PBPK	physiologically based pharmacokinetic
PBT	persistent, bioaccumulative, and toxic
PCB	polychlorinated biphenyl
PCDD/F	pentachlorodibenzodioxins and furans
PCE	tetrachloroethane
pCi	picocurie (radioactivity unit)
PDF	probability density function
PEL	permissible exposure level (OSHA)
pH	negative log of H^+ or H_3O^+ concentration
PIC	product of incomplete combustion
$\mathbf{pK_a}$	negative log of the acid dissociation constant
p_L	supercooled liquid vapor pressure
PLM	polarized light microscopy
PM	particulate matter
$\mathbf{PM_{10}}$	Particulate matter, <10 μm diameter
$\mathbf{PM_{2.5}}$	Particulate matter, <2.5 μm diameter
PNS	peripheral nervous system
pOH	negative log of OH^-concentration
POM	polycyclic organic matter
POP	persistent organic pollutant
ppb	parts per billion
ppbV	parts per billion by volume
PPE	personal protective equipment
ppm	parts per million
ppmV	parts per million by volume
ppt	parts per trillion
PQAO	primary quality assurance organization
PQL	practical quantitation limit
PRTR	Pollutant Release and Transfer Register (Mexico)
PSD	Prevention of Significant Deterioration (the United States)
PSI	Pollution Standard Index (replaced by AQI)
psi	pounds per square inch (pressure)
PUF	polyurethane foam
PV	present value
PVC	polyvinyl chloride
Q	flow
q	flux
QA	quality assurance
QAPP	Quality Assurance Project Plan
QC	quality control
QSAR	quantitative structure-activity relationships
R or r	correlation coefficient
r	chemical production or degradation rate
R	percentage emission reduction (rollback model)
r	radius
R	risk
r	thermodynamic loss term
R(*t*)	reliability function
R	universal gas constant (8.315 $J\ K^{-1}\ mol^{-1}$)
R-100	oil-resistant respirator protection efficiency against particulate aerosols (99.97%)

R^2 or r^2	coefficient of determination
RACT	reasonably available control technology
RCA	replacement cost analysis
RCRA	Resource Conservation and Recovery Act (the United States)
Re	Reynolds number (also N_R)
REACH	Registration, Evaluation, Authorisation, and Restriction of Chemicals (Europe)
REL	recommended exposure limit
REL-TWA	recommended exposure limit (time-weighted average)
Re_p	dimensionless particle Reynolds number
RF	radio frequency
RfC	reference concentration
RfD	reference dose
RH	relative humidity
rms	root-mean-square
RNA	ribonucleic acid
ROD	record of decision
RR	relative risk
RT	residence time
RTP	room temperature and pressure
s or sec	second (time)
S	entropy
S	severity of risk
S	solid-phase concentration
S	solubility
S	sorption
S	source or sink
S/N	signal-to-noise ratio
S^2	variance
s^2	variance of the source
S_A	substrate concentrations in the aqueous phase
SAR	structure-activity relationships
SARA	Superfund Amendments and Reauthorization Act (the United States)
SD or σ	standard deviation
SDWA	Safe Drinking Water Act (the United States)
SE	standard error
SES	socioeconomic status
SF	slope factor
SHEDS	Stochastic Human Exposure and Dose Simulation (model)
SHS	secondhand smoke
sin	sine
SIP	State Implementation Plan (the United States)
SLAMS	State and Local Air Monitoring Station (the United States)
SMR	standard morbidity ratio
SMR	standard mortality ratio
SOM	soil organic matter
SOP	standard operating procedure
SO_x	oxides of sulfur
SS	steady state (subscript)
Stk	Stokes number
STP	standard temperature and pressure
SVOC	semivolatile organic compound
T	temperature
t	ton
T	total (subscript)

t_0	initial time
$t_{1/2}$	half-life
tan	tangent
TC	total carbon
TCDD	tetrachlorodibenzo-*para*-dioxin
TCDF	tetrachlorodibenzofuran
TCE	trichloroethane
TCLP	toxicity testing leaching procedure
TDS	total dissolved solids
TEL	tetraethyl lead
T_f	time of failure
T_{ij}	time spent in microenvironment *j* by person *i*
TL	typical lifetime
TLV	threshold limit value
TLV-C	threshold limit value (ceiling)
TLV-STEL	threshold limit value-short-term exposure limit
TNT	trinitrotoluene
TOMS	total ozone mapping spectrometer
tpy	tons per year
T-R	transformer-rectifier
TRI	Toxic Release Inventory (the United States)
TSCA	Toxic Substances Control Act (the United States)
TSP	total suspended particulate
TVA	Tennessee Valley Authority
TWA	time-weighted average
u	velocity
u	wind speed
U	epistemic uncertainty
U	internal energy
u_{rms}	root-mean-square (rms) speed (gas molecules)
UAS	unmanned aircraft system
UAV	unmanned aerial vehicle
UCR	unit cancer risk
UEL	upper explosive limit
UF_{inter}	interspecies uncertainty factor
UF_{intra}	intraspecies uncertainty factor
UF_{other}	factor indicating RfC or RfC uncertainty other than UF_{inter} or UF_{intra}
UFP	ultrafine particles
UNEP	United Nations Environment Programme
UNFCCC	United Nations Framework Convention on Climate Change
USCG	US Coast Guard
USDA	US Department of Agriculture
USF&W	US Fish and Wildlife Services
USGS	US Geological Survey
UST	underground storage tank
UV	ultraviolet
UVCB	chemical substances of unknown or variable composition, complex reaction products, and biological materials (the United States)
V	fluid velocity
V	speed
V	stochastic or aleatory uncertainty
v	vapor phase (subscript or parenthetical)
V	volume

$\overline{v}$	average linear velocity (length time^{-1})
$\overline{V}$	average speed (length time^{-1})
VC	vinyl chloride
$\mathbf{V_D}$	volume of distribution
v_g	gas velocity
v_g	geostrophic velocity
vis	visible light range
$v_{\mathbf{max}}$	frequency of maximum radiation intensity (Wien's displacement law)
VMT	vehicle miles traveled
VOC	volatile organic compound
w	work
W	body weight
W	watt (unit of power; J s^{-1})
WHO	World Health Organization
WOE	weight of evidence
x	distance (horizontal)
y	cross-sectional distance, that is, at distance *y* from *y* = 0 (subscript)
y	width
y	year
$\overline{y}$	sample mean
z	at distance *z* from *z* = 0 (subscript)
Z	atomic number
z	distance (vertical)
Z	fugacity capacity
Z	zenith angle (degrees)
α	packing density (solidity)
γ	specific weight (mass length^{-3})
Δ	change (e.g., heat)
δ	solar declination (degrees)
$\Delta G*$	standard change in Gibbs free energy
θ	potential temperature
λ	degradation term in chemical decay
λ	wavelength (length, often nm)
π	3.14159...
ρ	density (volume length^{-3})
ρ_0	standard density (1 g m^{-3})
ρ_g	gas density (mass length^{-3})
ρ_p	particle density (volume length^{-3})
σ	Stefan-Boltzmann constant
τ	tortuosity
τ_s	shear stress (mass length^{-1} time^{-2})
$\mathbf{\upsilon}$ **or** $\mathbf{\upsilon_v}$	kinematic viscosity (length^2 time^{-1}) = dynamic viscosity divided by density ($\mu\ \rho^{-1}$)
υ	specific volume (reciprocal of density, length^3 mass^{-1})
φ	angle (degrees or radians)
φ	latitude (degrees)
$\varphi°$	fraction, ratio, or coefficient
X	leaving group from organic compound
$\mathbf{X_p}$	particle diffusivity (length^2 time^{-1})
ψ	surface deposition
Ψ_D	diffusional collection parameter (dimensionless)
Ω	migration velocity (electrostatics)
Ω	angular speed of the Earth's rotation
ω	solar azimuth (degrees)

APPENDIX

B

Key equations

See the text for description of variables and coefficients.

"p" value: $pX = -\log X$

Acid equilibrium constant for acetate: $K_a = \frac{[H_3O^+][CH_3COO^-]}{[CH_3COOH]}$

Actual vapor pressure: $p^\circ_{\text{actual}} = \frac{RH \times p^\circ_{\text{saturated}}}{100}$

Advection rate: $G_i = V\ t^{-1}$

Advective and diffusive flux: $q = Cu + j$

Advective flux: $J_{\text{Advection}} = \bar{v}\eta_e [c]$

Aerodynamic diameter (aerosol): $D_{pa} = D_{ps}\sqrt{\rho_p}$

Air-water partitioning coefficient: $K_{AW} = \frac{C_A}{C_W} = \frac{Z_A f_A}{Z_W f_W}$

Airway volume (lung physiology): $V_{aw} = 1.018 \times \text{Height(cm)} - 76.2$

Annual discount rate: $i = \left(1 + \frac{r}{k}\right)^{\frac{k}{p}} - 1$

Area of a circle: $A = \pi r^2 = \pi\left(\frac{D}{2}\right)^2 = \frac{\pi D^2}{4}$

Area of a cylinder: $A = 2\,\pi rh + 2\,\pi(r)^2$

Arithmetic mean: $\bar{x} = \frac{\sum_{i=1}^{n} x_i}{n}$

Arrhenius equation: $k = A \cdot e^{\frac{E_a}{RT}}$

Aspect ratio: $\text{AR} = \frac{\text{width}}{\text{height}}$

Bernoulli equation—conservation of mass: $p_1 + ½pV_1^2 + pgh_1 = p_2 + ½pV_2^2 + pgh_2$

Bernoulli equation for pump flow: $\left(\frac{p}{\rho g} + \frac{V^2}{2g} + z\right)_1 + h_p = \left(\frac{p}{\rho g} + \frac{V^2}{2g} + z\right)_2 + h_i$

Bernoulli equation: $\left(\frac{p}{\rho g} + \frac{V^2}{2g} + z\right)_1 = \left(\frac{p}{\rho g} + \frac{V^2}{2g} + z\right)_2$

Box model mass balance: $\frac{dyCV}{dt} = QA + (\bar{u} \cdot C_{\text{in}} \cdot W \cdot H_{\text{mix}}) - (\bar{u} \cdot C_{\text{in}} \cdot W \cdot H_{\text{mix}})$

Boyle's law: $V \propto P^{-1}$

Buoyancy force: $F_B = \frac{\pi d_p^3 \rho_g g}{6}$

Cancer risk: $R = CSF \times E$

Centrifugal force: $F = \frac{MV^2}{R}$

Change of mass in control volume: $\frac{d[A]}{dt} = -v \cdot \frac{d[A]}{dx} + \frac{d}{dx}\left(D \cdot \frac{d[A]}{dx}\right) + r$

Charles's law: $V \propto T$

Chemical intake from air pathway: $I = C_A \times \frac{IR}{BW} \times \frac{ET \times EF \times ED}{AT}$

Chemical intake: $I = \frac{C \cdot CR \cdot EF \cdot ED \cdot AF}{BW \cdot AT}$

Collection efficiency: $\eta_i = 1 - e^{-X}$ and $\eta_i = 1 - e^{-\frac{v_i LW}{Q}}$

Combustion efficiency: $CE = \left[\frac{C_{CO_2}}{C_{CO_2} + C_{CO}}\right] \times 100$

Combustion reaction: $\underbrace{C_aH_bO_cN_dS_e}_{\text{fuel}} + x\underbrace{(O_2 + 3.76N_2)}_{\text{air}} \rightarrow aCO_2 + \frac{b}{2}H_2O + \left(\frac{d}{2} + 3.76x\right)N_2 + eSO_2$

Combustion with nitrogen and sulfur: $C_aH_bO_cN_dS_e + (4a + b - 2c) \rightarrow aCO_2 + \left(\frac{b}{2}\right)H_2O + \left(\frac{d}{2}\right)N_2 + eS$

Combustion with nitrogen: $4CH_3NH_2(l) + 9O_2(g) \xrightarrow{\Delta} 4CO_2(g) + 10H_2O(g) + 2N_2(g)$

Complete combustion: $a(CH)_x + bO_2 \xrightarrow{\Delta} cCO_2 + dH_2O$

Continuity equation (cylinder): $\frac{1}{r}\frac{\partial}{\partial r}(ru_r) + \frac{1}{r}\frac{\partial}{\partial \theta}u_\theta + \frac{1}{r}\frac{\partial}{\partial z}u_z = 0$

Conversion from ppm to µg m^{-3} (25°C and 760 mmHg): 1 ppm (volume) = $\frac{\left(\frac{1L}{22.4}\right) \times MW \times 10^6 \times \mu g \times g \times m^{-1}}{10^6 L \times 298K^\circ \times 10^{-3} m^3 \times L^{-1}}$

Cunningham slip factor: $C_C = 1 + \frac{6.21 \times 10^{-4} T}{d_p}$

Cylindrical bag area: $A = \pi DL$

Daily intake: $DI(mg\ kg^{-1} day^{-1}) = \frac{\text{Concentration}(mg\ volume^{-1}) \times \text{Intake}(volume\ day^{-1})}{\text{body mass}(kg)}$

Darcy's law: $Q = -KA\frac{\partial h}{\partial x}$

Decomposition reaction: $AB \rightarrow A + B$

Degree of saturation: $\mu = \frac{\omega}{\omega_{sat}}$

Density field: $\rho = \rho(x, y, z, t)$

Destruction and removal efficiency: $DRE = \left[\frac{W_{in} - W_{out}}{W_{in}}\right] \times 100$

Differential rate law: $Rate = -\frac{d[A]}{dt} = k[A]^n$

Diffusive flux: $j = C_A \cdot u' - C_B \cdot u' = u'\Delta C;\ j = -D\frac{\Delta C}{\Delta x}$

Discount factor: $P/F, i\%, n$

Discount formula: $P = F(P/F, i\%, n)$

Discount rate (spreadsheet): $i = RATE(N, A, P, F, \text{Type}, \text{guess})$

Dissociation of carbon dioxide: $CO_2 + H_2O \leftrightarrow H_2CO_3^*$

Double replacement reaction: AB + CD → AD + CB
Drag coefficient (laminar): $C_D = \frac{24}{\text{Re}_p}$
Drag coefficient (transition): $C_D = \frac{18.5}{\text{Re}_p^{0.6}}$
Drag coefficient (turbulent): $C_D = 0.44$
Drag coefficient with Cunningham slip factor: $C_D = \frac{24}{\text{Re}_p \cdot C_C}$
Drag force (laminar): $F_D = \frac{3\pi\mu_g v_p d_p}{C_C}$
Drag force (transition): $F_D = 2.30(d_p \cdot v_p)^{1.4} \cdot \mu_g^{0.6} \cdot \rho_g^{0.4}$
Drag force (turbulent): $F_D = 0.05\pi\,(d_p \cdot v_p)^2 \cdot \rho_g$
Drag force: $F_D = \frac{A_g \cdot \rho_g \cdot V_p^2 \cdot C_D}{2} = \frac{\pi d_p^2 \rho_g v_p^2 \cdot C_D}{8}$
Eco-risk quotient: $\text{RQ} = \frac{\text{Exposure}}{\text{Ecotoxicity}} = \frac{\text{EEC}}{\text{EC}_{50}}$
Enrichment factor (cadmium): $\text{EF} = \frac{\left(\frac{[Cd]}{[Al]}\right)_{\text{air}}}{\left(\frac{[Cd]}{[Al]}\right)_{\text{crust}}}$
Entropy: $q = m.\ SH.\ \Delta T$
Equivalent uniform annual worth: EUAW = EUAB – EUAC
Ethical outcome: $P = f\,(x)$
Ethical power: Five Ps = Purpose + Pride + Patience + Persistence + Perspective
Exposure concentration: $I = C_A \times \frac{IR}{BW} \times \frac{ET \times EF \times ED}{AT}$
Exposure: $E = D/t$
Exposure: $E = \int_{t=t_1}^{t=t_2} C\,(t)dt$
Extinction coefficient: $b_{\text{ext}} = b_{\text{rg}} + b_{\text{ag}} + b_{\text{scat}} + b_{\text{ap}}$
Face velocity at filter: $U_0 = \frac{Q}{A}$
Failure likelihood: $P\,\{t_1 \le T_f \le t_1\} = \int_{t_1}^{t_2} f\,(t)dt$
Fick's law: $j = -D_0.i_c;\ j = -AD\frac{\Delta C}{\Delta x};\ j = i_c R$
Filter drag: $S_f = \frac{\Delta P_{\text{total}}}{v_f} = k_1 + (k_2 \cdot c_i \cdot v_f \cdot t)$
First law of thermodynamics: $\Delta U = q + w$
Flux density for dry deposition: $F_d = v_d \cdot C_A$
Flux density: $F = vC_A$
Flux: $j = \frac{Q}{A};\ q = \frac{m}{A\,\Delta t} = \frac{m}{V} \times \frac{V}{A \cdot t} = C \cdot u$
Force on a particle in electric field: $F_E = n \cdot e \cdot E$
Freundlich sorption isotherm: $C_{\text{sorb}} = K_F \cdot C^n$
Fugacity capacity constant: $C_i = Z_i \cdot f$
Fugacity: $Z_A = \frac{1}{RT} = \frac{C_A}{f_A}$
Future worth (spreadsheet): $F = FV\,(i, N, A, P, \text{Type})$

Gaussian concentration (stable conditions): $C = Q(1/u)\{g_1/[(2\pi)^{0.5}\sigma_y]\}\{g_2/[(2\pi)^{0.5}\sigma_z]\}$

Gaussian concentration (unstable or neutral conditions): $C = Q(1/u)\{g_1/[(2\pi)^{0.5}\sigma_y]\}(1/L)$

Gaussian concentration: $C(x, y, z) = \frac{Q}{2\pi u\sigma_y\sigma_z} e^{-\frac{y^2}{2\sigma_y^2}} \left(e^{-\frac{(z+H)^2}{2\sigma_z^2}} + e^{-\frac{(z-H)^2}{2\sigma_z^2}} \right)$

Gay-Lussac's law: $P \propto T$

Generalized equilibrium constant: $K_{eq} = \frac{k_f}{k_b} = \frac{[C]^c[D]^d}{[A]^a[B]^b}$

Gravitational constant: $g_c = 32.2\frac{\text{lb}_\text{m} \cdot \text{ft}}{\text{lb}_\text{f} \cdot \text{sec}^2} = 1\frac{\text{kg} \cdot \text{m}}{\text{N} \cdot \text{s}^2}$ or 1 kg m $\text{N}^{-1}\text{s}^{-2}$

Gravitational force of a spherical particle: $F_G = \frac{\pi d_p^3 \rho_p g}{6}$

Gravitational force: $F_G = m_p \cdot g = \rho_p \cdot V_p \cdot g$

Henderson-Hasselbalch equation: $pK_a = \text{pH} + \log \frac{[\text{HA}]}{[\text{A}^-]}$

Henry's law constant (unitless): $K_{AW} = \frac{C_\text{air}}{C_\text{water}} = \frac{K_H}{R \cdot T}$

Henry's law constant: $K_H = \frac{P^\circ \times \text{MW}}{760 \times C_\text{sol}}$

Human equation: Ethics × (Beauty + Wealth + Authority) = True value

Humidity ratio: $\omega = 0.62198 \cdot \frac{p_w}{p_{at} - p_w}$

Hydrologic continuity equation: $\frac{dS}{dx} = I - O$ and $\frac{dS}{dx} = P\text{–}R\text{–}E\text{–}T\text{–}I$

Ideal gas law: $PV = nRT$

Impaction factor (dimensionless): $\Psi = \frac{C_C d_p^2 3 y_p \rho_p}{18\mu_g D_C}$

In-stream pollutant concentration: $C_r = \frac{Q_d C_d + Q_s C_s}{Q_r}$

Incompressible fluid continuity: $\nabla \cdot \boldsymbol{u} = 0$

Integrated rate law: $\ln[A] = -kt + \ln[A]_0$

Kinetic energy of gas molecule: $E_g = \frac{mu^2}{2} = \frac{3RT}{2}$

Kinetic reaction: $r = -\frac{\Delta(A)}{\Delta t}$

Lagrangian velocity field: $\vec{v} = [x(t), y(t), z(t)]$

Light interaction with atmosphere: $-dI = b_\text{ext} I \cdot dx$

Mass balance: $\frac{dM}{dt} = M_\text{in} - M_\text{out} \pm R$

Mass continuity-steady state: $\frac{dm}{dt} = \rho_1 \times A_1 \times u_1 = \rho_2 \times A_2 \times u_2 = \text{constant}$

Maximum concentration along centerline (Gaussian): $C_\text{max} = \frac{2Q}{\pi \cdot u \cdot e \cdot H^2} \cdot \frac{\sigma_z}{\sigma_y}$

Maximum use concentration: MUC = PEL × APF

Mean concentration based on mean emission rate: $\overline{C}(x, y, z) = \frac{\overline{Q}}{\pi\sigma_y\sigma_z u} \exp\left[-\frac{1}{2}\left(\frac{y^2}{\sigma_y^2} + \frac{z^2}{\sigma_z^2}\right)\right]$

Mean mass flow: $\frac{dm}{dt} = \dot{m} = \rho \times A \times u$

Mean residence time: $\tau = \frac{M}{J_\text{out}}$

Mobilization factor: $\mathrm{MF} = \frac{E_h}{E_n}$
Model uncertainty: $Y = h(U, V)$
Molecular momentum: $\Delta mu = (\Delta mu_x)_f - (\Delta mu_x)_i = -(mu_x) - (mu_x) = -2\ mu_x$
Moral importance equation: $M = B \times A$
Navier-Stokes equation: $\rho\left[\frac{\partial \boldsymbol{u}}{\partial t} + (u \cdot \nabla)u\right] = -\nabla p + \eta\nabla^2 u + \mathrm{f}$
Newton's second law of motion: $\boldsymbol{F} = \frac{d\boldsymbol{p}}{dt}$
Newton's third law of motion: $\boldsymbol{F}_{\text{reacting}} = -\boldsymbol{F}_{\text{acting}}$
Noncancer risk hazard index: $\mathrm{HI} = \Sigma_1^n\ \mathrm{HQ}$
Noncancer risk hazard quotient: $\mathrm{HQ} = \frac{\mathrm{MDD}}{\mathrm{ADI}} = \frac{\mathrm{Exposure}}{\mathrm{RfD}}$
NO_x: $NO_y = NO + NO_2$
NO_y: $NO_y = NO_2 + NO + HNO_3 + PAN + 2N_2O_5 +$
Octanol-air partitioning: $K_{OA} = \frac{K_{ow} \cdot R \cdot T}{K_H} = \frac{K_{ow}}{K_{AW}}$; $\log K_{\mathrm{OA}} = \log K_{ow} - \log K_{\mathrm{AW}}$
Partial pressure: $p_a = K_H[c]$
Particle diffusivity: $X_p = \frac{C_C kT}{3\pi\mu_g d_p}$
Particle residence time: $t_r = \frac{L}{v_g}$
Particle Reynolds number (dimensionless): $\mathrm{Re}_p = \frac{d_p v_p \rho_g}{\mu_g}$
Percent recovery: $\%R = \frac{S-U}{C} \times 100\%$ or $\%R = \frac{C_M}{C_{SRM}} \times 100\%$
pH: $\mathrm{pH} = -\log[H^+]$ or $\mathrm{pH} = -\log[H_3O^+]$
Photolytic reaction: $A + h\nu \rightarrow C + D$
Planck's law: $E = h\nu = \frac{hc}{\lambda}$
Pollutant gradient: $\frac{d[A]}{dt} = -v \cdot \frac{d[A]}{dx} + \frac{d}{dx}\left(D\frac{d[A]}{dx}\right) + r$
Pollutant transport: $N = D \cdot f = V[c]k = (V \cdot Z \cdot k) \cdot f = D_R f$
Present value: $PV = \frac{B_{y_i}}{(1+r)^{y_i - y_0}}$ and $PV = \frac{C_{y_i}}{(1+r)^{y_i - y_0}}$
Present worth (spreadsheet): $P = PV\,(i, N, A, F, \mathrm{Type})$
Pressure drop across filter: $\Delta P_{\text{total}} = \Delta P_{\text{media}} + \Delta P_{\text{dust-cake}} = = (k_1 \cdot v_f) + (k_2 \cdot c_i \cdot v_f^2 \cdot t)$
Pressure: $p = \frac{F}{A}$
Pressure: $P = \frac{F_{\text{total}}}{A} = \frac{m}{l^3}(u_{x1} + u_{x2} + \ldots u_{xn}) = \frac{mN}{V}\left(\frac{u_{x1} + u_{x2} + \ldots u_{xn}}{N}\right)$
Psychrometric constant: $\gamma = \frac{0.0016286 \cdot P}{\lambda_v}$
Radiant flux: $E_v \mathrm{d}v = c_1 v^3 [\exp(c_2 v \cdot T^{-1}) - 1]^{-1}\,\mathrm{d}v$
Raoult's law: $p_i = p_i^* x_i$
Rate law: $r = k[A]^m[B]^n$

Fuel stoichiometric combustion reaction:

$$\underbrace{C_mH_n}_{\text{fuel}} + \left(m + \frac{n}{4}\right)\underbrace{(O_2 + 3.76N_2)}_{\text{air}} \rightarrow mCO_2 + \frac{n}{2}H_2O + 3.76\left(m + \frac{n}{4}\right)N_2$$

Reference concentration: $\text{RfC} = \frac{\text{NOAEC}}{U_{\text{inter}} \times U_{\text{intra}} \times U_{\text{other}}}$

Reference dose: $\text{RfD} = \frac{\text{NOAEL}}{UF_{\text{inter}} \times UF_{\text{intra}} \times UF_{\text{iother}}}$

Relative percent difference: $\text{RPD} = \frac{(C_1 - C_2)}{C_1 + C_2)/2} \times 100\%$

Relative standard deviation: $\text{RSD} = \frac{s}{\bar{Y}} \times 100\%$

Reliability function: $R(t) = P\{t_1 \le T_f \le t_1\} = 1 - \int_0^t f(x)dx$

Reversal reaction (equilibrium) $aA + bB \leftrightarrow cC + dD$

Reynolds number: $\text{Re} = \frac{d_e v \rho}{\mu}$

Reynolds number: $\text{Re} = \frac{d_e v}{\upsilon}$

Reynolds number: $\text{Re} = \frac{\text{Inertial Forces}}{\text{Viscous Forces}}$

Risk equation: $R = p(\text{L})$

Risk function: $R = f(S, P)$

Risk-based target concentration: $\text{C} = \frac{\text{Risk} \cdot \text{BW} \cdot \text{AT}}{\text{CR} \cdot \text{EF} \cdot \text{ED} \cdot \text{AF} \cdot \text{SF}}$

Rollback model: $R = \frac{g(P) - D}{g(P) - B} \times 100$

Root mean square speed: $u_{\text{rms}} = (\bar{u}^2)^{1/2} = \left(\frac{3P}{\rho}\right)^{1/2}$

Saturation charge: $q = (3\pi\varepsilon_0 E)d_p^2$

Scavenging ratio: $S = C_p \cdot \rho \cdot C_{\text{aerosol}}{-1}$

Schrödinger equation: $-\frac{h^2}{2}\frac{d^2f}{m\,dx^2} + V(x)f = Ef$

Settling characteristic value: $K = d_p\left(\frac{g \cdot \rho_p \cdot \rho_g}{\mu_g^2}\right)^{0.33}$

Shuttleworth's evaporation: $E = \frac{mR_n + \gamma \cdot 6.43(1 + 0536 \times U_2) \cdot \delta e}{\lambda_v \cdot (m + \gamma)}$

Single replacement reaction: $A + BC \rightarrow AC + B$

Soil or sediment partitioning: $K_D = K_{OC} \cdot C_{\text{org}} = C_S(C_W)^{-1}$

Solid/liquid fuel stoichiometric combustion reaction:

$$\underbrace{C_aH_bO_cN_dS_e}_{fuel} + x\underbrace{(O_2 + 3.76N_2)}_{air} \rightarrow aCO_2 + \frac{b}{2}H_2O + \left(\frac{d}{2} + 3.76x\right)N_2 + eSO_2$$

Solubility product constant: $K_{sp} = [C]^c[D]^d$

Sorption: $S = K_D \cdot C_W$

Specific gas constant: $R_{\text{gas}} = \frac{R}{m_{\text{gas}}}$

Standard deviation: $s = \sqrt{\left[\sum_{n=1}^{n}\left[\frac{(Y_1 - \bar{Y})^2}{n-1}\right]\right]}$

Static pressure of flow: $p + ½ pV^2 + pgh = \text{constant}$

Stefan–Boltzmann law: $F = \sigma T^4$

Stokes number: $\text{Stk} = \frac{t_0 u_0}{l_0}; \text{Stk} = \frac{\rho_p d_p^2 \dot{u}}{18\mu_g d}$

Stress: $\sigma(P) = \lim_{\delta A \to 0} \frac{\delta F}{\delta A}$

Substrate partition coefficient: $K_{G/A} = \frac{S_G}{S_A^*}$

Sum of forces on a particle: $\sum F = m_p \cdot a_p = m_p \frac{dv_p}{dt}$ and $\sum F = \frac{m_p \cdot a_p}{g_c}$

Synthesis reaction: A + B → AB

Terminal velocity (laminar flow): $v_t = \frac{g \cdot C_C \cdot \rho_p \cdot d_p^2}{18\mu_g}$

Terminal velocity (transition flow): $v_t = \frac{0.153 g^{0.71} \cdot \rho_p^{0.71} \cdot d_p^{1.14}}{\mu_g^{0.43} \cdot \rho_g^{0.29}}$

Terminal velocity (turbulent flow): $v_t = 1.74 \left(\frac{g \cdot \rho_p \cdot d_p}{\cdot \rho_g} \right)^{0.5}$

Time-weighted concentration: $\overline{C} = \frac{M}{Q \times t}$

Total compartmental fugacity: $f = \frac{M_{\text{total}}}{\Sigma_i (Z_i \cdot V_i)}$

Transfer rate for gas molecules: $F_{\text{total}} = \frac{m}{l} (u_{x1} + u_{x2} + \ldots u_{xn})$

Universal gas constant: $R = \frac{0.08212 \text{L} \cdot \text{atm}}{\text{mol} \cdot \text{K}}$ or $\frac{8.314 \text{joules}}{\text{mol} \cdot \text{K}}$

Vapor pressure deficit: $\text{VPD} = p^\circ_{\text{saturated}} - p^\circ_{\text{actual}}$

Velocity field: $\vec{v} = \vec{v}(x, y, z, t)$

Velocity of flow: $u = f(x, y, z)$

Virtual temperature: $T_v = \frac{T}{1 - \left(\frac{e}{P}\right)(1 - \varepsilon)}$

Volume of a cylinder: $V = h\pi(r)^2$

Volume of a sphere: $V = \frac{4}{3}\pi r^3$

Volume of a spherical particle: $V_p = \frac{\pi d_p^3}{6}$

Volumetric flow rate: $Q = V\, t^{-1}$

Volumetric gas flow: $Q = \frac{v_g}{W \times H}$

Volumetric mass transfer rate: $K_{1a_{G/A}} \left(\frac{S_G}{K_{G/A}} - S_A \right)$

Water dissociation constant at 25°C: $K_w = [H_3O^+][OH^-] = 1.0 \times 10^{-14} \text{mol L}^{-1}$

Water dissociation constant: $K_w = \frac{[H_3O^+][OH^-]}{[HOH]^2}$

Water use efficiency: $\text{WUE} = \frac{[c_{CO_2\text{-fixed}}]}{[c_{H_2O\text{-lost}}]}[$

Wein's displacement law: $v_{\max} = 1.04 \times 10^{11} T$

Worst-case (longest path) particle residence time: $t_r = \frac{H}{v_t}$

APPENDIX

C

Glossary

These terms are the author's operational definitions. The sources are numerous, and many terms have been modified from their original definitions. However, here are the resources used to augment this glossary:

A posteriori knowledge (Latin: What comes after). Knowledge gained from experience. Modern science is founded on acquiring such knowledge. Contrast with *a prioriknowledge.*

A priori knowledge (Latin: What comes before). Knowledge independent of experience. Contrast with *a posteriori knowledge.*

ABET (Accreditation Board for Engineering and Technology, Inc.) Entity that accredits U.S. college and university programs in applied science, computing, engineering, and technology.

Abortion Premature exit of the fetus from the uterus. A medically induced abortion can be completed surgically or chemically (e.g., hormonal drugs) because the fetus is not wanted.

Acceptable engineering practice (1) Amount of engineering-related work experience, in years, to meet one of the minimal criteria to sit for the professional engineer (PE) examination. (2) Reasonably expected professional performance by an engineer demonstrating competence, especially adhering to codes for design, construction, and operation.

Accreditation Nongovernmental, peer review of an institution to ensure educational quality. Institutional accreditation evaluates overall institutional quality. One form of institutional accreditation is regional accreditation of colleges and universities. Specialized accreditation examines specific programs of study, rather than an institution as a whole. This type of accreditation is granted to specific programs at specific levels. Architecture, nursing, law, medicine, and engineering programs are often evaluated through specialized accreditation.

Act utilitarianism Form of utilitarianism wherein the test of utility maximization is applied directly to single acts. In act utilitarianism, an action is evaluated on a case-by-case basis for practical usefulness. Compare to *rule utilitarianism.*

Acute exposure A single exposure to a substance that results in severe biological harm or death. Acute exposures are usually characterized as lasting no longer than a day, as compared to longer, continuing *exposure* over a period of time.

Acute toxicity Any adverse effect that occurs within a short period of time following *exposure*, usually up to 24–96 hours, resulting in biological harm and often death.

Ad hominem Fallacious argument attacking the holder of a view rather than the position itself.

Adaptive management (AM) Also known as "adaptive resource management" (ARM), a structured, iterative process of optimal decision-making in the face of uncertainty, with an aim to reducing uncertainty over time via system monitoring. In this way, decision-making simultaneously maximizes one or more resource objectives and, either passively or actively, accrues information needed to improve future management. AM is often characterized as "learning by doing."

Additive effect Response to exposure to multiple substances that equals the sum of responses of all the individual substances added together (compare with *antagonism* and *synergism*).

Additive manufacturing Three-dimensional (3D) printing; process that builds physical objects layer-by-layer. Contrast with *subtractive manufacturing*.

ADME Adsorption, distribution, metabolism, and elimination (or excretion) of a substance after it is taken up by an organism.

ADMET Adsorption, distribution, metabolism, elimination, and the resulting toxicity of a substance after it is taken up by an organism.

Advanced directive A document informing the health providers and physicians ofthe type and extent of care the patient requests in the event that the patient becomes unable to make medical decisions (e.g., in a coma).

Adult stem cell Undifferentiated cell found in a differentiated tissue that can renew itself and (with certain limitations) differentiate to yield all the specialized cell types of the tissue from which it originated.

Advection Transport of a substance along with the flow of a fluid, for example, transport of a solute by the bulk motion of flowing groundwater or particles in the flow streams of an air mass.

Adverse effect A biochemical change, functional impairment, or pathologic lesion that affects the performance of the whole organism or reduces an organism's ability to respond to an additional environmental challenge.

Adverse outcome pathway Chain of biological events, beginning with a *molecular initiating event*, followed by *key events*, that results in an adverse outcome, for example, neurological dysfunction, endocrine disruption, etc.

Aerobe (1) Microorganism that can survive in the presence of molecular oxygen (O_2). (2) Microorganism that requires sufficient concentrations of O_2 to survive.

Agent-based model (ABM) Artificial intelligence and computational method that simulates actions and interactions among autonomous "agents" to predict how these activities affect an entire system. For example, an agent can be a person who uses a product, and likes the smell, so uses the product in ways and amounts than would be expected per the product label. This higher and broader use may change the person's *exposures*. The term "agent," however, is often used to describe *stressors* (e.g., chemical, physical, and biological agents), so a preferable term for biotechnology and exposure science is "individual-based model" (IBM). Another connotation is the exposed-individual-based model (EIBM), when applied to exposure and risk assessment.

Aggregate exposure Combined exposures to a single *stressor* across multiple *exposure routes* and multiple *exposure pathways*. Compare to *cumulative exposure*.

ALARP As low as reasonably possible; a fundamental principle of risk management designed to determine whether risks are minimized while considering costs and other feasibility factors.

Amoral Lacking any moral characteristics. An amoral act is neither morally good nor morally bad—it simply exists. Contrast with *moral* or *immoral*.

Analogy Comparison of similarities between two things to a conclusion about an additional attribute common to both things. This is a type of inductive reasoning see).

Animal welfare ethics Moral consideration of animals in research, food supply, and other human endeavors.

Amino acid Any of 20 basic building blocks of proteins with a free amino ($-NH_2$) and a free carboxyl ($-COOH$) group and having the basic formula $NH_2 - C_R - COOH$. According to the side group R, they are subdivided into polar or hydrophilic (serine, threonine, tyrosine, asparagine, and glutamine); nonpolar or hydrophobic (glycine, alanine, valine, leucine, isoleucine, proline, phenylalanine, tryptophan, and cysteine); acidic (aspartic acid and glutamic acid); and basic (lysine, arginine, and hystidine). The sequence of amino acids determines the shape, properties, and the biological role of a protein.

Amino group NH_2 attached to carbon structures (e.g., in amines and amino acids).

Ammonification Mineralization process whereby organic nitrogen is converted into ammonium (NH_4^+). The conversion by fungi and bacteria is usually enzymatic, for example, Glu 2- oxoglutarate aminotransferase, Glu dehydrogenase, or Gin synthetase.

Amphiphilic Describing a chemical compound that has both hydrophilic and lipophilic properties.

Amphoteric Able to react as either a weak acid or a weak base.

Amplification (1) Treatment (e.g., use of chloramphenicol) designed to increase the proportion of plasmid DNA relative to that of bacterial (host) DNA. (2) Replicating a gene library in bulk. (3) Duplication of genes within a chromosomal segment. (4) Creation of numerous copies of a segment of DNA by the polymerase chain reaction.

Anabolism Synthesis of complex molecules from simpler molecules with the input of energy. Compare to catabolism.

Analogy Comparison of similarities between two things to a conclusion about an additional attribute common to both things. This is a type of inductive reasoning (see *inductive reasoning*). One of *Hill's criteria*.

Analyte (1) Substance measured in a scientific study. (2) Chemical for which a sample (e.g., water, air, or blood) is tested in a laboratory, for example, to determine the amount of cadmium in soil, the specified amount of soil to be collected and analyzed in the laboratory.

Analytic epidemiology Evaluation of associations between exposure to physical, chemical, and biological agents and disease by testing scientific hypotheses.

Antagonism Effect from a combination of two agents that is less than the sum of the individual effects from each agent ($1 + 1 < 2$). Compare to *synergy*.

Anthropocentrism Philosophy or decision framework based on human beings. View that all and only humans have moral value. Nonhuman species and abiotic resources have value only in respect to that associated with human values. Contrast with *biocentrism* and *ecocentrism*.

Anthropogenic Made, caused, or influenced by human activities. Contrast with *biogenic*.

Anthropometrics Comparison of the measurements (Greek: *metron*) with the capabilities of the human body (Greek: *anthropos*).

Applied mathematics Mathematical techniques typically used in the application of mathematical knowledge to domains beyond mathematics itself.

Archetype Symbolic imagery. Psychologist Carl Jung applied the term to the generalized patterns of images that form the world of human representations in recurrent motifs, passing through the history of all culture.

Artificial general intelligence (AGI) Theory that artificial intelligence can learn and reason across any intellectual task identically to that of a human.

Artificial intelligence (AI) Computing technology that simulates human learning, problem solving, creativity, and comprehension.

Association Relationship, not necessarily causal, between two variables. The antecedent variable comes before and is associated with an outcome; however, it may or may not be the cause of the outcome. For example, mean birth weight of minority babies is less than that of babies of the general population. Ethnicity is an antecedent of low birth weight, but not the cause. Other factors, for example, nutrition, smoking status, and alcohol consumption, may be the causal agents.

Attenuation rate The rate at which a contaminant is removed. This is not a rate constant but a rate with typical units of $\mu g\ L^{-1}\ year^{-1}$.

Attenuation The process by which a chemical compound's concentration decreases with time, through sorption, degradation, *dilution*, and/or *transformation*. The term applies to both destructive and nondestructive contaminant removal.

Attributable risk The rate of a disease in exposed individuals that can be attributed to the *exposure*. This measure is derived by difference between the rate (usually *incidence* or *mortality*) of the disease among persons not exposed to the suspected agent and the corresponding rate among exposed individuals.

Attribute (1) In data management, characteristic property of an *entity*. (2) *In geographic information systems*, nonspatial information about a geographic feature stored in a table and linked to the feature by a unique identifier. For example, attributes of a landfill include its name, volume, and loading rate at the gate. In GIS, attribute is also used to denote information associated with a unique value of a *raster* cell.

Autonomy Individual's capability of deliberation about personal goals and of acting under the direction of such deliberation. To respect autonomy is to give weight to autonomous persons' considered opinions and choices while refraining from obstructing their actions unless they are clearly detrimental to others.

Autosome Any chromosome other than the sex chromosomes or the mitochondrial chromosome.

Average daily dose (ADD) Dose rate averaged over a pathway-specific period of exposure expressed as a daily dose on a per-unit-body-weight basis. The ADD is usually expressed as $mg\ kg^{-1}\ day^{-1}$ or other mass-time units.

Baby (1) Very young (e.g., birth to one year of age) child. (2) A child that has not yet been born (see *fetus*). For example, expectant mothers often refer to their baby's kicking and other movements.

Bacteria Unicellular microorganisms that exist either as free living organisms or as parasites, ranging from beneficial to harmful to humans.

Bayes theorem Rule that relates a current probability to a prior probability. $P(A|B) = \frac{P(B|A)P(A)}{P(B)}$, where A and B are events; P(A) and P(B) are probabilities; P(B| A), and P(A| B), and P(B| A) are conditional probabilities, that is, the probability of A given B is true and the probability of B given A is true, respectively.

Bayesian Statistical approach that addresses probability inference, named after Thomas Bayes (*An Essay Towards Solving a Problem in the Doctrine of Chances*, 1763) to decision-making and inferential statistics that deals with probability inference (i.e., using knowledge of prior events to predict future events). In a Bayesian network, priors are updated by additional data that yield posterior probabilities that are often more robust than classical probabilities.

Bayesian belief network (BBN) Cause-and-effect tool represented by a probabilistic graphic model of a set of random variables and their conditional independencies.

Beatitudes Declarations of blessedness. Often associated with Christ's Sermon on the Mount (Matthew 5:1–7:29).

Behaviorism Viewpoint founded by psychologist John Watson (*Psychology as the Behaviorist Views It*, 1913) holding that behaviors can be conditioned and measured. Behaviorism is based solely on observable behaviors, considering cognition and mood to be overly subjective.

Belmont Report "Ethical Principles and Guidelines for the Protection of Human Subjects of Research" by the National Commission for the Protection of Human Subjects of Biomedical and Behavioral Research. Released on April 18, 1979. Articulated the basic ethical principles and guidelines that should assist in resolving the ethical problems that surround the conduct of research with human subjects. Expounded three basic ethical principles: respect for persons, beneficence, and justice.

Benchmarking Systematic comparison of an organization's risk management practices, capabilities, and performance metrics against established standards and best practices.

Beneficence Promotion of good for others and contribution to their welfare. Obligation to protect subjects from harm. The principle of beneficence requires that probable risks to research subjects be minimized and the potential for benefit to the subjects and/or to society as a whole be maximized. One of three principles stemming from the Belmont Report (see).

Benefit-cost analysis (or cost-benefit analysis) Method designed to determine the feasibility or utility of a proposed or existing project. Yields a benefit-cost ratio (see).

Benefit-cost ratio (BCR) Weighted benefits divided by weighted costs; used to compare and differentiate among project alternatives. Gross BCR < 1 is undesirable. The greater the BCR, the more acceptable the alternative.

Benevolence (1) Disposed to do good, to act charitably, to possess virtue. (2) Goodness and moral excellence. Compare to beneficence.

Best practice (1) Optimal service to the client. (2) Treatment is appropriate, accepted, and widely used according to expert consensus; it embodies an integrated, comprehensive, and continuously improving approach to care (medicine). It is morally obligatory that healthcare practitioners provide patients with the best practice (also known as standard therapy or standard of care).

Bias (1) Unfair influence (sociology). (2) Systematic error in one direction, such as the positive bias of a scale that reads 1 mg too high (instrument error) or the negative bias in interpretations of lesions reported by a physician performing the procedure ("operator bias") that consistently miss some lesions. Bias makes the reported values less accurate.

Big data Datasets so large and complex that traditional software is unable to store, process, and analyze them, so artificial intelligence tools are often needed.

Bio Prefix indicating "life" (Greek).

Bioaccumulation The process whereby certain substances build up in living tissues.

Bioaccumulation factor (BAF) Ratio of a tissue concentration of substance in an organism to its concentration in the environment (usually water) where the organism lives. BAF indicates a compound's potential to accumulate in tissue through exposure to both food and water. Compare to *biomagnification* and *bioconcentration factor*.

Bioactivation Process by which an organism's metabolic or other endogenous processes increase the toxicity of a substance after uptake. In context, may simply be "activation."

Bioaerosol Airborne *particulate matter* comprised of living or dead organisms (e.g., microbes) or their products (e.g., dander, spores, cysts, and pollen).

Bioaugmentation Addition of beneficial microorganisms into groundwater to increase the rate and extent of *biodegradation*. Part of a *bioremediation* strategy.

Bioavailability The degree to which a substance becomes available to the *target* tissue after administration or exposure.

Biocentrism View that all life has moral value. Contrast with *anthropocentrism*.

Biochemical oxygen demand (BOD) A standard test to assess wastewater pollution due to organic substances. BOD_5 is based on measurement of the oxygen used under controlled conditions of temperature (20°C) and time (5 days). Compare to *chemical oxygen demand*.

Biochemodynamics The physical, chemical, and biological processes that transport and transform substances.

Bioconcentration factor (BCF) Ratio of the concentration of substance in an organism's tissue versus its concentration in the environment (usually water) in situations where the organism is exposed exclusively to water. BCF measures the compound's potential to accumulate in tissue through direct uptake from water (excludes uptake from food). Compare to *bioaccumulation factor*.

Biodiversity Number and variety of organisms in given system (e.g., wetland or forest). Usually, lower biodiversity indicates that the system is stressed and in poor condition.

Biodegradation Breakdown of a contaminant, usually catalyzed by enzymes produced by organisms. Term usually applies only to microorganisms. Compare to *phytodegradation*.

Bioenergetics Energy flow and transformation through living systems. It can be within an organism, for example, cellular energetics, or within and between levels of *biological organization*, for example, *trophic states* (energy transfer from producers to first-order consumers, to second-order consumers, to and from decomposers, etc.).

Bioengineering See *biological engineering*.

Biocentrism View that all life has moral value. Contrast with *anthropomorphism*.

Bioethics (1) Inquiry into ethical implications of biological research and applications. (2) Ethical inquiry into matters of life, especially biomedical and environmental ethics (preferred definition in this text). (3) Biomedical ethics.

Biomechanics Structure, function, and motion of the mechanical aspects of biological systems based on classical physics and engineering mechanics to understand how organisms interact with their environments.

Biogenic Made, caused, or influenced by natural processes. Contrast with *anthropogenic*.

Bioinformatics Management and analysis of data using advanced computing techniques applied to biological research and inquiry.

Biosystem Living organism or a system of living organisms that are able to interact with other organisms directly or indirectly.

Biosystem engineering (bioengineering) (1) Application of biological sciences to achieve practical ends. (2) Integration of physical, chemical, or mathematical sciences and engineering principles for the study of biology, medicine, behavior, or health to advance fundamental concepts; to create knowledge for the molecular to the organ systems levels; and to develop innovative biologics, materials, processes, implants, devices, and informatics approaches for the prevention, diagnosis, and treatment of disease; for patient rehabilitation; and for improving health (from the National Institutes of Health).

Biological engineering Combination of biomedical and biosystem engineering (see both) to develop useful biology-based technologies that can be applied across a wide spectrum of societal needs, including diagnosis, treatment, and prevention of disease; design and fabrication of materials, devices, and processes; and enhancement and sustainability of environmental quality.

Biological response Manner and type of effect in an organism (e.g., disease, change in metabolism, and homeostasis).

Biologically based dose response (BBDR) model Predictive model that describes biological processes at the cellular and molecular level linking the target organ dose to the adverse effect.

Biomarker A chemical, physical, or biological characteristic that indicates biological condition. The biomarker may be a chemical to which an organism is exposed (e.g., lead in blood), a metabolite of the chemical (e.g., cotinine in blood as an indication of exposure to nicotine), or a biological response (e.g., an increase in body temperature as a result of exposure to a pathogen).

Biomedical engineering Application of engineering principles to medicine, including drug delivery systems, therapeutic systems, and medical devices.

Biomedical ethics (1) Application of moral reasoning to address questions that emerge in healthcare and clinical situations. (2) Study of morality in biomedical situations.

Biomedical testing Investigations to determine whether a change in a body function might have occurred because of exposure to a hazardous substance.

Biostatistics Application of statistical tools to interpret biological and medical data.

Biota Any living creature: plant (flora), animal (fauna), or microbial.

Biotechnology Use of living creatures to produce things of value to humans (e.g., drugs and food).

Bioterrorism Use of living agents to cause intentional harm to people (e.g., anthrax spores or pathogenic viruses) and society (e.g., agricultural pests).

Blastocyst Preimplantation embryo of about 150 cells that make up a sphere of an outer layer of cells (i.e., trophoblast) that will form the placenta, a fluid-filled cavity (blastocoel), and the inner cell mass, that is, a cluster of cells on the interior from which the stem cells are derived. Human embryonic stem cells are frequently taken from 4 to 5-day-old blastocysts.

Blockchain Secure, shared database wherein the data are saved in blocks that are permanently chained jointly so that past records cannot be altered.

Body burden Total amount of a substance in the body.

Bottom-up View where fundamental components are first considered, working upward to larger perspectives. Contrast with *top-down*.

Butterfly effect Sensitive dependence on initial conditions. Metaphor for the extreme sensitivity of chaotic systems (see *chaos*), in which small changes or perturbations lead to drastically different outcomes. The phrase is derived from a butterfly flapping its wings in California and thereby initiating a change in weather patterns that results in the formation of a thunderstorm in Nebraska (from Edward Lorenz, in his 1963 paper, "Deterministic Nonperiodic Flow". *Journal of Atmospheric Sciences*. **20**: 130–141; although in his presentation to the New York Academy, it was not a butterfly but a seagull's flapping of the wing that posited as the initial condition. Later in 1972, Lorenz used the butterfly in the example).

Cancer Disease of heritable, somatic mutations affecting cell growth and differentiation, characterized by an abnormal, uncontrolled growth of cells. Malignant growth or tumor caused by abnormal and uncontrolled cell division.

Carcinogen Physical or chemical agent that induces cancer.

Cardinal virtues Four paramount virtues in classical philosophy: justice, prudence, fortitude, or temperance.

CAS registration number An organization that indexes information published in "Chemical Abstracts" by the American Chemical Society and provides index guides by which information about particular substances may be located in the abstracts when needed. CAS numbers identify specific chemicals, so they can be important in queries about chemical toxicity and environmental characteristics.

Cation exchange capacity (CEC) Ability of soil, sediment, or other solid *matrix* to exchange *cations* with a fluid. Very important measure of soil productivity and root behavior.

Case-control study An epidemiologic study contrasting those with the disease of interest (cases) to those without the disease (controls). The groups are then compared with respect to exposure history to ascertain whether they differ in the proportion exposed to the chemical(s) under investigation.

Case study Evaluation of an actual occurrence of events to describe specific environmental and health conditions and past exposures.

Catabolism Metabolism in which larger, more complex molecules are broken down into smaller, simpler molecules with the release of energy. Compare to *anabolism*.

Categorical imperative Central theme of Immanuel Kant's deontological ethics (see) that sets one principle from which all specific moral imperatives are derived: "Act only according to that maxim by which you can at the same time will that it should become a universal law" (*Groundwork of the Metaphysic of Morals* [*Grundlegung zur Metaphysik der Sitten*], 1785).

Catalyst Substance that is not affected by a reaction and helps to initiate or accelerate it.

Central nervous system Portion of the nervous system that consists of the brain and the spinal cord.

Chaos theory Exposition of the apparent lack of order in a system that nonetheless obeys specific rules. Condition discovered by the physicist Henri Poincaré around the year 1900 that refers to an inherent lack of predictability in some physical systems (i.e., Poincaré's concept of dynamical instability).

Character Description of a person's attributes, abilities, and traits, especially those related to moral strength. Contrast with *personality*.

Charity Virtue manifested by impartial love. In Christianity, coexists with faith and hope.

Chromosome Structure within a cell's nucleus consisting of strands of deoxyribonucleic acid (DNA) coated with specialized cell proteins and duplicated at each mitotic cell division. Chromosomes transmit the genes of the organism from one generation to the next.

Chi square (X^2) test Statistical test to determine the probability that an observed deviation from the expected event or outcome occurs solely by chance.

Child

(1) Human being less than 18 years of age (United Nations Convention on the Rights of the Child). Subcategories include the following:
- Infant (baby) (ages 01.5)
- Neonate (in first month after birth)
- Toddler (ages 1.5 – 4)
- Middle childhood (ages 4 – 11)
  - Prepubescence (near age 10 – 11)
  - Preadolescence (near age 11 – 13)
- Adolescence and puberty (age 14 – 20)

2. Human being after conception until 18 years of age (same categories as above, but additional category of "unborn child" with subcategories of zygote, embryo, and fetus [see]).

Childhood Time of rapid development of human personality and morality.

Chronic Having a persistent, recurring, or long-term nature. Contrast with *acute*.

Chronic effect An adverse effect on a human or animal in which symptoms recur frequently or develop slowly over a long period of time.

Chronic exposure Multiple exposures occurring over an extended period of time, or a significant fraction of the animal's or the individual's lifetime.

Chronic toxicity Capacity of a substance to cause long-term adverse effects (usually applied to humans and human populations).

-cide Suffix indicating act of killing (Latin: *caedere* –to cut).

Client A person or entity (e.g., company) that seeks service from a professional. The engineer, for example, is required to be a faith agent to the client, meaning that the engineer is responsible for representing the best interests of the client. Contrast with *customer*.

Clone Line of cells genetically identical to the originating stem cell.

Code of ethics Established set of moral expectations of a group, especially of professional societies.

Code of Hammurabi One of the first known collections of 282 moral laws, probably written between 2100 and 1800 BC but credited to Babylonian King Hammurabi (1792–1750 BC).

Coefficient of determination (r^2) Proportion of the variance of one variable predictable from another variable. The ratio of the explained variation to the total variation, which represents the percentage of the data nearest to the line of best fit. For example, if $r = 0.90$, then $r^2 = 0.81$, meaning that 81% of the total variation in one variable (y) can be explained by the linear relationship between the two variables (x and y) as described by the regression equation. Thus, the remaining 19% of the total variation is unexplained.

Coherence Criterion for causality based on the amount and degree of agreement among studies linking cause to effect, especially among various types of studies (e.g., animal testing, human epidemiological investigations, and in vitro studies).

Cohort study Epidemiologic study comparing those with an exposure of interest to those without the exposure. These two cohorts are then followed over time to determine the differences in the rates of disease between the exposure subjects. Also called a prospective study.

Coma Prolonged period of unconsciousness following traumatic brain injury. In this sleep-like state, there is no speech, the eyelids are usually closed, and there is no response to commands. A number of scales (e.g., the Glasgow Coma Scale, Disability Rating Scale, and the Rancho Los Amigos Scale [RLAS]) are used to characterize the severity of coma. The RLAS, for example, uses eight levels in descending order of severity (Source: Miller School of Medicine, University of Miami):

- *Level I*—No response to any stimuli—indicates coma.
- *Level II*—Generalized response, that is, the patient reacts inconsistently and nonpurposefully to stimuli in a nonspecific manner, such as eye blinking, changes in breathing rate, gross body movement, and vocalizatio-n—indicates coma.

- *Level III*—Localized response, that is, the patient reacts specifically but inconsistently to stimuli, such as turning head toward a sound or focusing on an object presented and following simple commands in an inconsistent, delayed manner—not considered a coma, but stimulation techniques appropriate through Level III.
- *Level IV*—Confused-agitated, that is, the patient is in a heightened state of activity with severely decreased ability to process information. The patient is detached from the present and responds primarily to his/her own internal confusion. Behavior is often bizarre.
- *Level V*—Confused, inappropriate, non-agitated, that is, the patient appears alert and is able to respond to simple commands fairly consistently, but responds to more complex commands in a nonpurposeful, random manner and is agitated by external stimuli.
- *Level VI*—Confused-appropriate, that is, the patient shows goal-directed behavior but is dependent on external input for direction. He/she follows simple directions and shows carryover for tasks that have been relearned, such as self-care activities. Responses may be incorrect due to memory problems, but they are appropriate to the situation.
- *Level VII*—Automatic-appropriate, that is, the patient appears appropriate and oriented but goes through daily routines automatically and has shallow recall of what he/she has been doing. The patient shows increased, but superficial awareness of self and other people, demonstrates decreased judgment and problem-solving abilities, lacks realistic planning for the future, and requires at least minimal supervision for learning and safety purposes. Judgment and other higher-level cognitive abilities remain compromised.
- *Level VIII*—Purposeful and appropriate, that is, the patient is alert and oriented, able to recall and integrate past and recent events, is aware of and responsive to the environment, and needs no supervision once learning has occurred. He/she may continue to show decreased reasoning, tolerance for stress, judgment in emergencies or unusual circumstances, and decreased social, emotional, and intellectual capacities.

Common sense Innate, sound judgment, not specially resulting from formal training. However, engineers nurture common sense by blending formal, technical training with experience and mentorship.

Community Assemblage of two or more biotic populations of different species that reside in the same spatial area.

Compartment (1) In *pharmacokinetics* and *toxicokinetics*, the specific location within an organism where a substance may reside, and from which it may move, for example, blood, fat, or bone tissue. (2) *Environmental medium.*

Compartmental model (1) Model that predicts or characterizes the transport and fate of a compound within an organism (e.g., moving from blood to tissues, transformed by metabolism, and detoxified or *bioactivated* during the path to elimination). (2) Similar model accounts for transport and fate, usually at a larger scale in the environment, for example, the quantity and change of a chemical as it moves from the air to the water to the sediment to aquatic biota and within the *food chain.*

Compartmentalization Viewing a system by its individual components. This can be problematic when an engineer does not consider the system as a whole (e.g., when the structural engineer and soil engineer do not collaborate on selecting the best and safest combination of materials and structures suited to a soil type, or when a biomedical engineer does not work closely with various specialized healthcare professionals in a clinical setting to adapt a realistic device to the comprehensive needs of the patient). Compartmentalization can be good when it allows the engineer to focus adequate attention on the components (see *bottom-up*), so long as the design is properly built into a system.

Competence (1) Skill in practice. For professionals, competence is requisite to ethical practice. (2) Sufficient velocity for a fluid to carry a load (especially a stream's ability to carry solids).

Complexity Relative measure of uncertainty in achieving functional requirements or objectives. Designers are frequently expected to reduce the complexity of engineered systems.

Computational Use of numerical methods and algorithms to integrate data, solve problems, and analyze information.

Computational paradigm Discovery based on simulations of physical and natural phenomena and mechanisms.

Concentration (1) Quantity of substance per unit volume (fluid) or per unit weight (solid *matrix*, e.g., soil, sediment, tissue). (2) Method of increasing the dissolved solids per unit volume of solution, for example, via evaporation of the liquid. (3) Increasing suspended solids per unit volume of *sludge* via sedimentation or dewatering. (4) Method in analytical chemistry of increasing the proportion of an *analyte* in a solvent to improve detection.

Confidence, confidence level (1) Client's trust in a professional. (2) Amount of certainty that a statistical prediction is accurate. Physical sciences may differ from social sciences in what is considered acceptable confidence; for example, the former may require 99% while social scientific research may consider 95% to be acceptable. Depending on the application, engineering research ranges in acceptable confidence level (e.g., structural fatigue research may require higher confidence levels than environmental research).

Conflict of interest Situation wherein professional or personal obligations or personal or financial interests detract from a professional's ability to carry out one's responsibilities and duties fairly.

Confounder Factor that distorts or masks the true effect of risk factors in an epidemiologic study. A condition or variable that is both a risk factor for disease and associated with an exposure of interest. This association between the exposure of interest and the confounder (a true risk factor for disease) may make it falsely appear that the exposure of interest is associated with disease. For example, a study of low-birth-weight children in low-income families must first address the confounding effects of tobacco smoking before ascribing the actual risk associated with income.

Conjugation In *genetic engineering*, transferring genetic material between bacteria through direct cell-to-cell contact or through a bridge between the two cells.

Conscience (1) Set of ethical principles held by a person that govern one's thoughts, motivations, and actions. (2) Awareness of these principles. (3) In psychoanalytic theory, the part of the superego (see) that informs the ego (see) on what should be done.

Consequentialism Ethical theory with the perspective that the value of an action derives solely from the value of its consequences. Consequentialists hold that the consequences of a particular action form the basis for any valid moral judgment about that action, so that a morally right action is an action that produces good consequences. One of three major theories of normative ethics (see), along with virtue ethics and deontological ethics (see).

Contingent probability Probability that an event will occur as a result of one or more previous events. Also known as conditional probability.

Contractarianism Ethical theory based on adherence to the social contract. Includes libertarianism and Rawlsism (see both).

Control group Group used as the baseline for comparison in epidemiologic studies or laboratory studies. This group is selected because it either lacks the disease of interest (case-control group) or lacks the exposure of concern (cohort study). Also known as a reference group.

Cooking Irresponsible data manipulation of retaining only those results that fit the study objectives, hypothesis, or theory and discarding others. This is done in an attempt to improve statistical significance and confidence, for example, a better Chi square or correlation coefficient (r), or coefficient of determination (i.e., r^2).

Copyright Legal documentation of written intellectual property.

Correlation coefficient (r) Statistical measurement of the strength and the direction of a linear relationship (association) between two variables.

Creativity Ability to create knowledge, to see patterns, and to make connections so as to approach problem-solving and opportunities in an inventive and novel way.

Critical path Systems engineering of activities, decisions, and actions that must be completed on schedule and at a sufficient level of quality for the entire project to be successful.

Cross-sectional study Epidemiological study of observations representing a particular point in time. Contrast with *longitudinal study*.

Cumulative distribution function In statistics, decision theory, and probability theory, the probability that a random variable X with a given probability distribution will be found to have a value less than or equal to x.

Cumulative exposure Combined exposure to multiple *stressors* via multiple *exposure pathways* reaching a single biological *target*. Compare to *aggregate exposure*.

Curatorial artificial intelligence Use of artificial intelligence (AI) to collect, analyze, and filter data and information. The result is information that has been curated for a specific user group. Compare to *generational AI*.

Customer Buyer of goods and services.

Cytochrome P450 (CYP) Enzymes that use iron to oxidize substances, often as part of the body's strategy to dispose of potentially harmful substances by making them more water-soluble. Varying versions of CYP are used to identify different enzymatic activities, so they are important in *toxicodynamics* and *toxicokinetics modeling*.

Cytokine Non-antibody proteins released by a cell in response to inducing stimuli, which are mediators that influence other cells. Produced by lymphocytes, monocytes, macrophages, and other cells.

Data Plural of datum. Gathered facts from which conclusions can be drawn. Collection of raw, unorganized numbers, observations, facts, or symbols.

Data center Physical facility used by organizations to centralize and maintain their critical applications, data, and computing infrastructure consisting of rows of high-capacity and powerful servers, processors, and memory modules to run applications and process data.

Data quality objective (DQO) Qualitative and quantitative statements that clarify and define the type of data needed in a study and specify tolerable levels of potential decision errors and uncertainty to support the study objectives.

Database Structured set of *data* held in a computer, with rules of accessibility, ranging from confidential to *open*.

Dataset (1) *Database*. (2) *Data* selected and arranged in rows and columns for processing by statistical software.

Data structure Systematic method of organizing *data*.

Death (1) Cessation of life (see); legal definition: "the cessation of life; permanent cessations of all vital functions and signs" (Black's Law Dictionary). (2) Whole body death occurs with complete cessation of breathing and heartbeat, leading to anoxia (no oxygen), then ischemia (no blood flow), with death occurring with ischemia in the brain. (3) Brain death, which has generally displaced whole body death as technology has temporarily prevented brain ischemia even with cessation of cardiovascular function (e.g., during a transplant operation), occurs when the brain ceases to function as indicated by lack of response to external stimuli, absence of body movements, cessation of spontaneous breathing, and two isoelectric (flat) electroencephalogram (EEG) readings separated by 24 hours [Ad Hoc Committee of the Harvard Medical School to Examine the Definition of Brain Death, published in 1968, in the *Journal of the American Medical Association*, 205 (337)].

Decision space The set of all options from which the decision-maker can choose. A restrictive decision, for example, a binary decision of yes or no, has a smaller decision space than one with numerous approaches available to the decision-maker.

Decision tree Diagram indicating various steps to different outcomes. Supports the optimal course of action in situations where several possible alternatives have uncertain outcomes.

Deductive reasoning A conclusion is necessitated by previously known facts. If the premises are true, the conclusion must be true. Starting from general knowledge and moving to specifics (e.g., from cause to effects). Contrast with *inductive reasoning*.

Deep ecology Environmental movement initiated in 1972 by Norwegian philosopher, Arnie Naess, that advocates radical measures to protect the natural environment irrespective of their effect on the welfare of humans (opposite of *anthropocentrism*).

Deep learning Subset of machine learning (see) using multilayered neural networks analogous to the human brain.

Demand Quantity of a good or service that society chooses to buy at a given price.

Deontology or deontological ethics (Greek: *deon* meaning obligation) Ethical theory basing right and wrong on duty.

Depuration Cleansing of a previously dosed organism by ending the dosing completely.

Dermal exposure Contact between a chemical and the skin.

Dermal toxicity The ability of a pesticide or toxic chemical to poison people or animals by contact with the skin.

Descriptive epidemiology Study of the amount and distribution of a disease in a specified population by person, place, and time. Compare to *analytic epidemiology*.

Descriptive ethics Study of what a group actually holds to be right and wrong and maintains as ideals. Contrast with *normative ethics*.

Deterministic model Model based on expectations in the absence of random or measurement error, for example, stochasticity (compare to *probabilistic model*); intended to represent processes to lead to an outcome. The deterministic model gives the same results for a specific set of inputs. The model is parameterized to the extent that its components are defined and the extent to which they can be combined to determine the outcome. For example, a predator–prey model can be based on the number of prey, the number of predators, and numerous parameters to explain the interactions between the prey and predator. For example, the change in prey and change and predators with respect to time can simply be expressed as differential equations. The predictive power and uncertainty are influenced by the assumptions, for example, food is never scarce, prey are the exclusive food supply for the predator, rates of change are proportional to population size, environmental conditions do not change, predators are never satiated, etc. Since the model is based on differential equations, it must be deterministic. The model can be improved as better and more specific information is made available, linking the parameters to outcomes.

Developing nation Nation at the lower end of socioeconomic and technological advancement; formerly known as third-world nation.

Developmental toxicity Adverse effects on developing organism that may result from exposure prior to conception (either parent), during prenatal development, or postnatally until the time of sexual maturation. The major manifestations of developmental toxicity include death of the developing organism, structural abnormality, altered growth, and functional deficiency.

Device (medical) Instrument, apparatus, implement, machine, contrivance, implant, in vitro reagent, or other similar or related article, including any component, part, or accessory, which is intended for use in the diagnosis of disease or other conditions or in the cure, mitigation, treatment, or prevention of disease in man or other animals or intended to affect the structure of any function of the body and which does not achieve its primary intended purpose through chemical action and which is not depended upon being metabolized for the achievement of its primary intended purposes (Federal Food, Drug and Cosmetics Act, 1938). The last sentence of this definition helps to distinguish a device from a drug. Both are regulated, but differently, by the U.S. Food and Drug Administration.

Dignity A person's worthiness of respect, privacy, autonomy, and self-worth.

Disability adjusted life year (DALY) Metric used in *life cycle analysis* to represent the years of life lost (shortened) plus years limited due to disability that result from exposure to a substance or other stress. One lost year of "healthy" life. The sum of DALYs across the population (i.e., burden of disease) is a measurement of the gap between current health status and an ideal health situation where the entire population lives to an advanced age, free of disease and disability.

Disaster (Latin: *dis* and *astrum* meaning "bad star"). A relative term meaning a catastrophic event that wreaks great destruction. However, the term is not exclusive to large-scale events, such as hurricanes or earthquakes, but can also include small scale events with highly negative consequences, such as an engineering or medical failure where one or a few people are impacted but which has other implications (malpractice, bad publicity, blame, etc.).

Disease Abnormal and adverse condition in an organism (although sometimes applied metaphorically, such as the "disease" of poverty).

Disparate effect Health outcome, usually negative, that is disproportionately high in certain members of a population, such as an increased incidence of certain cancers in minority groups.

Disparate exposure Exposure (see) to a physical, chemical, or biological agent that is disproportionately high in certain members in a population, such as the higher-than-average exposure of minority children to lead.

Disparate susceptibility Elevated risk of certain members of a population (e.g., genetically predisposed) to the effects of a physical, chemical, or biological agent; can lead to disparate effects (see).

Distribution (1) In pharmacokinetics and toxicokinetics, the transfer of a substance within an organism. The "D" in *ADME*. (2) In statistics and probability theory, see *probability distribution*.

Distributive justice Just allocation of goods, services, and utility within society. Focuses more on outcome than on procedures (i.e., procedural justice).

Doctrine of the mean Confucian, 23-chapter book on usefulness of a golden way (Tao) to gain virtue.

Dose Amount of a substance available for interactions with metabolic processes or biologically significant receptors after crossing the outer boundary of an organism. Potential dose is the amount ingested, inhaled, or applied to the skin. Applied dose is the amount presented to an absorption barrier and available for absorption (although not necessarily having yet crossed the outer boundary of the organism). Absorbed dose is the amount crossing a specific absorption barrier (e.g., the exchange boundaries of the skin, lung, and digestive tract) through uptake processes. Internal dose is a more general term denoting the amount absorbed without respect to specific absorption barriers or exchange boundaries. The amount of the chemical available for interaction by any particular organ or cell is termed the delivered or biologically effective dose for that organ or cell.

Dose-response Relationship between a quantified exposure (dose) and the proportion of subjects demonstrating specific biologically significant changes in incidence and/or in degree of change (response).

Drone Aircraft controlled remotely or autonomously without a human pilot on board.

Drug Substance intended for use in the diagnosis, cure, mitigation, treatment, or prevention of disease, which is regulated by the U.S. Food and Drug Administration. Contrast with *device* and *nutritional supplement* (see both).

Dry lab (1) In silico research (contrast with wet lab). (2) Walkthrough prior to actual laboratory work (step preceding wet lab). (3) Unethical practice of forging (making up) data.

Dual use (1) Science, engineering, and technology designed to provide both military and civilian benefits. (2) Research and technology that simultaneously benefit and place society at risk (e.g., biotechnological advances that advance vaccines that also increase the risks of bioterrorism).

Dynamical instability See *chaos theory*.

Ecocentrism Philosophy or decision framework based on the whole ecosystem rather than a single species. Contrast with *anthropocentrism*.

Ecofeminism View that oppression of women and environmental degradation are linked. French feminist Françoise D'Eaubonne coined the term in 1974 in an attempt to characterize how male-dominated societies have inflected violence on both women and nature.

Educated men and women Term used to distinguish college-educated persons, included in certain graduation ceremonies.

Effectiveness Measure of the extent and degree to which a design achieves a goal (i.e., effectiveness of design is measured with respect to design specifications and objectives).

Efficiency Ratio of total energy or mass output to total energy or mass input, expressed as a percentage. Treatment or removal efficiency is the product of the contaminant mass prior to treatment (I) times the contaminant mass after treatment (E) divided by I. To express efficiency as a percentage, these values are multiplied times 100: $\frac{I \times E}{I} \times 100$

Efficacy Ability or capacity to generate a desired effect.

Ego The self. In psychoanalytic theory, the ego represents the id (see) to the outer world, mediating between the id and superego (see).

Egoism Theoretical framework holding that self-interest dictates ethical decision-making (also known as psychological egoism or ethical egoism).

Elasticity (1) Ability of a body to return to its original shape after being stressed (engineering). (2) Degree to which an item's price change results from a unit change in supply ("supply elasticity") or a unit change in demand ("demand elasticity") (economics).

Elitism Sense of entitlement based on class and socioeconomic status.

Embryo (1) Human being from the time of fertilization until the end of the eighth week of gestation, thereafter, considered to be a fetus. (2) Developing organism from the time of fertilization until significant differentiation has occurred, when the organism becomes known as a fetus. An organism in the early stages of development.

Embryonic stem cell Undifferentiated cells from the embryo with the potential to become all cell types found in the body. Embryonic stem cells are found in the inner cell mass of blastocysts (see).

Embryonic stem cell line Embryonic stem cells that have been cultured in vitro allow proliferation without differentiation for months to years.

Emergence (1) In systems sciences, the complex systems exhibit properties that the individual components do have. For example, vehicular traffic exhibits patterns that the vehicles themselves do not possess. (2) In artificial intelligence, the unexpected, complex abilities that arise in models as they grow in scale and complexity, without having been explicitly programmed.

Emerging technology Rapidly evolving or new innovations characterized by widespread potential to disrupt and transform existing systems.

Emotive theory of values Theory holding that judgments about values are simply emotional expressions based on the inclinations and attitudes of the individual doing the judging.

Empathy Ability to place someone in another person's (or other species) mental or spiritual position.

Empiricism Position of modern science that all knowledge about the physical world is based on observation and sensory experiences. See *a posteriori knowledge*.

Endocrine system Chemical messaging system in organisms used for regulation by secretion of hormones by glands that are sent through the circulatory system to cells where the hormones bind to receptors.

Ends justifying means Rationale for excusing unethical approaches to achieve worthy ends (at least in the view of those undertaking the actions).

Engineering (1) Application of scientific and mathematical principles to practical ends, especially the design, manufacture, and operation of structures, machines, processes, and systems. (2) The profession that implements definition #1.

Environmental equity Fair distribution of environmental risks across population groups.

Environmental impact statement (EIS) Document prepared by a government agency detailing the potential effects resulting from a major action being considered by that agency. In the United States, the EIS is required under the National Environmental Policy Act and is usually preceded by an *environmental assessment*.

Environmental justice (EJ) Fair treatment of all people regardless of race, color, national origin, or income with respect to the development, implementation, and enforcement of environmental laws, regulations, and policies. An environmental justice community is a neighborhood, zip code, or census tract in which residents face disproportionate pollution and health hazards, compounded by socioeconomic vulnerabilities.

Engineering optimization Use of optimization (see) techniques to meet specific engineering design goals.

Environmental racism Disproportionately negative treatment of members of ethnic and racial minority groups in decisions regarding polluting agents in their neighborhoods and communities. The term was coined in 1982 by then-director of the United Church of Christ's Commission for Racial Justice, Rev. Ben Chavis.

Environmental engineering Subdiscipline of engineering (usually civil engineering) concerned with applications of scientific principles and mathematics to improve the condition of the environment.

Environmental science Systematic study of the environment and its components and processes (e.g., nutrient cycling, pollutant transport, and adverse effects).

Environmentalism (1) Advocacy in the protection of the environment. (2) Philosophy underpinning this advocacy. Such advocacy may or may not be scientifically based (i.e., differs from environmental science and environmental engineering).

Epidemiology Study of the causes, distribution, and control of disease in populations.

Epigenetics Concern with mechanisms that regulate gene activity.

Epistemology Philosophical inquiry into the nature of knowledge.

Error (1) Mistake. (2) In statistics, the difference between a reported value and the actual value (see *bias*).

Ethical optimization Process of selecting an action by balancing technical and moral drivers and constraints to achieve a result that meets user needs and that is the most ethical choice among options.

Ethics (1) Set of moral principles. (2) Study of morality and moral decision-making.

Eugenics Study of hereditary (genetic) improvement of the human race by controlled selective breeding. It has been used as a rationale for genocide and other immoral acts (e.g., in Nazi Germany).

Euthanasia Deliberate use of medical means to cause death.

Event Set of outcomes that are preceded and linked to an earlier set of outcomes (probability theory).

Evil An immoral act (noun) or morally bad (adjective). Opposite of good.

Ex situ Moved off-site (e.g., contaminated soil transported to an incinerator for treatment).

Ex vivo Outside the body, frequently the equivalent of in vitro (see).

Experiment Investigation to support or reject a hypothesis or to increase knowledge about a phenomenon.

Exposome Totality of human environmental, nongenetic *exposures*, beginning at conception and complementing the *genome*. The exposome includes the biology, activities, and locations that account for the person's lifetime exposure.

Exposure Contact with a substance through various pathways, especially oral, dermal, and inhalation.

Exposure assessment Identification and evaluation of the human population exposed to a toxic agent, describing its composition and size, as well as the type, magnitude, frequency, route, and duration of exposure.

Exposure pathway Physical course (e.g., air, food, soil, and drinking water) that a substance follows from its source to the exposed human or nonhuman organism.

Exposure route The physiological means by which a substance enters an organism, for example, inhalation, ingestion, or dermal contact.

Exposure scenario Set of facts, assumptions, and inferences regarding how exposure occurs to support risk assessors in evaluating, estimating, or quantifying exposures.

Exposed-individual-based model (EIBM) *Agent-based model* applied to exposure and risk assessments. Coined by author to avoid confusion regarding the term "agent."

Expression (genetic expression) Effect on cell resulting from the gene's instructions in transcription.

Extraordinary means Treatments and medical care that are unduly burdensome or sorrowful, such as amputation, or beyond the economic means of the person, or that only prolong the suffering of a dying person. If such measures are considered to be morally *extraordinary*, they are not obligatory. Recent advances in medicine, however, complicate the decision whether to undergo or forego medical treatment, since medicine can now save many people who would simply have been allowed to die in the past. Thus, the moral permissibility and obligation are not always straightforward, wherein some may consider ordinary means to be extraordinary and *vice versa.*

Fact That which can be shown to be true, to exist, or to have occurred.

Failure Lack of success as indicated by design specifications and measures of success.

Faith (1) Belief beyond that is based on logical proof or material evidence. (2) Christian virtue of secure belief in God and trusting acceptance of God's will.

Faithful agent A trustee or representative authorized by a person to be entrusted with that person's best interests. Thus, in engineering codes of ethics, a faithful agent will avoid conflicts of interest (see), will maintain appropriate confidentiality, and in other ways guard the trust given by the client.

False positive Positive finding of a test when, in fact, the true result was negative. For example, a drug screen shows that a person has used opiates, even though the person has not. Type I error (see).

False negative Finding of the absence of a condition (e.g., disease) in a test when, in fact, the disease is present. For example, a lung cancer screen shows that the patient has no cancer, but at a level of detection below the screen, the cancer has cancer cells in the lung. Type II error (see).

Fetus (fetus) Developing mammal (including human) subsequent to embryonic stage and preceding birth.

Fishbone diagram See *Ishikawa diagram.*

Forging Inventing or concocting research results and reporting methods and findings of fictitious or exaggerated experiments. Synonymous with dry lab (definition 3).

Food chain A biological system in which individuals in a trophic level feed on organisms in the trophic level below theirs. As such, energy is transferred from level to level.

Food web Complex of interrelated *food chains* in an ecological community.

Free energy (ΔG) Intrinsic energy in a substance available to do work, especially to drive chemical transformations. Also known as Gibbs free energy.

Fuzzy logic System dealing with the partial truths, assigning values ranging from completely true to completely false.

Game theory Decision-making under conditions of uncertainty and interdependence; considering the characteristics of players, strategies, actions, payoffs, and outcomes at equilibrium.

Gamete Reproductive cell (i.e., an egg or a sperm).

GANTT chart Graphical display of a project or program showing each task as a horizontal bar with its length being proportional to time needed for completion.

Gene Ordered sequence of nucleotides located in a particular position on a particular chromosome, representing the fundamental unit of heredity.

Generative artificial intelligence (gen AI) Type of AI that is able to create original content in response to user prompts. Compare to *curatorial AI.*

Genetic engineering (GE) Modification of the structure of genetic material in a living organism, involving the production and use of *recombinant DNA.*

Genetically modified organism (GMO) Organism whose genetic material has been changed in a way that does not occur under natural conditions through cross-breeding or natural recombination.

Genetics Scientific investigation of heredity.

Genome Entire genetic complement, that is, all of the hereditary material possessed by an organism.

Genome editing Genetic engineering by inserting, replacing, or removing DNA from a genome using artificially engineered nucleases.

Genomics Study of genes, including their functions.

Geoengineering Large-scale, up to planetary scale, actions to address climate change and other global problems.

Germ theory Paradigm that diseases are caused by singular, proximate, pathogenic microbes. Displaced miasma theory (see) in the late 19th century.

***Gestalt* theory** (German: form) View that perception and other psychological phenomena must be understood for their overall patterns and forms, as opposed to the individual components.

Gestation Time from fertilization of the ovum to birth. Also known as uterogestation.

Golden mean Optimal position between extremes (deficiency and excess) expressed by Aristotle and other ancient Greeks.

Golden rule Empathetic view; ethics of reciprocity—Do unto others as you would have them do unto you (Matthew 7:12).

Good Moral excellence; an action or behavior representing such excellence. Opposite of evil (see).

Good Samaritan Helpful and sympathetic person modeled after the person in Jesus Christ's parable (Luke 10:25–37).

Grand rounds Training sessions for physicians and other health care providers covering health topics likely to be encountered in practice.

Green engineering Design, commercialization, and use of processes and products, which are feasible and economical while minimizing the generation of pollution at the source and the risk to human health and the environment (U.S. Environmental Protection Agency).

Green hospital Healthcare facility that incorporates sustainable practices into its operations (e.g., recycling, green purchasing, and minimization of wastes).

Gross domestic product (GDP) Total monetary value of all final goods and services produced within a nation's borders during a given period of time.

GUI Graphical user interface.

Half-life ($T_{1/2}$) Time needed for half the quantity of a substance taken up by a living organism to be metabolized and eliminated by normal biological processes. Also called biological half-life.

Harm Damage to another person or creature.

Harm principle John Stuart Mill's recommendation that utilitarianism's (see) premise of greatest good is restricted (e.g., by law) when others are harmed.

Hazard Potential source of harm.

Hazardous waste Solid waste with a hazardous characteristic or is otherwise listed as a hazardous waste in regulations.

Healthy worker effect Situation where there are fewer deaths for workers in an industry compared with the U.S. population, usually because severely ill and chronically disabled are excluded from employment. The general population may not be the best comparison group for workers.

Hedonism Belief that pleasure is the greatest good and highest aspiration of humankind. In early utilitarian thinking, hedonism was the basis for utility or good.

Hierarchy of needs Psychological construct proposed by Abraham Maslow ("A Theory of Human Motivation," 1943) holding that as humans meet their basic needs, they are successively able to meet higher (growth) needs, until finally reaching self-actualization (which includes morality and creativity).

Hill's criteria Minimal conditions necessary to establish a causal relationship between two items, presented by British medical statistician Sir Austin Bradford Hill (1897–1991) as a means of finding causal links between a specific factor (e.g., exposure to air pollution) and specific adverse effects (e.g., asthma). These criteria, originally recommended for occupation settings but now applied to numerous health and environmental problems, are meant to be guidelines rather than inviolable rules of epidemiology (see).

Hallucinations (AI) Occurrence in which a large language model (LLM) or generative AI system generates output that is factually incorrect, nonsensical, or completely untruthful, despite presenting it with total confidence.

Hyperscale Extreme-scale architecture of a data center that is optimized for cloud computing and massive machine learning training sets.

Hypothetico-deductive method Method of logical deduction, attributed to Karl Popper (*The Logic of Scientific Discovery*, 1934), limiting scientific discovery to that which is testable, requiring an approach that formulates hypotheses, a priori, with the intent of rejecting these hypotheses. The method assumes that a hypothesis can never truly be proved but at best can be corroborated.

Hysteresis (1) Changes that occur depending on the direction taken in a pathway, for example, a material may behave differently in the same temperature range when cooled than when heated in that same range. In mechanics, the changes of a body as it returns to its original shape after being stressed. (2) Failure to return to previous condition, such as due to an energy loss that always occurs under cyclical loading and unloading of a spring, proportional to the area between the loading and unloading load-deflection curves within the elastic range of a spring (engineering), or the failure of a variable to return to its initial equilibrium after a temporary shock (economics).

Holism (1) Theory that living matter is made up of organic or unified wholes that are greater than the simple sum of their parts. In this view, wholes have some priority over the elements, members, individuals, or parts composing them. Holistic views propose that the meaning and truth of claims cannot be assessed separately but must be assessed as part of theories, bodies of theory, or worldview. (2) Holistic investigation or system of treatment.

Homeostasis (1) Ability of an organism to self-regulate functions, inherent trend toward stability. (2) In engineering, computer science, and the social sciences, any system that self-regulates to maintain stability.

Honesty Virtue of acting with truthfulness, openness, and fairness.

Hope (1) To anticipate a positive outcome. (2) In Christianity (1 Corinthians 13), the desire and search for future goodness that is difficult but possible to attain with God's help. Coexists with faith and love (charity).

Hormone Chemical released by glands of the endocrine system (see).

Humors Ancient Greek's explanation of personality, especially how the four fluids become unbalanced. Galen (c.190) extended these four fluids to characterize temperament: blood (sanguine), black bile (melancholic), yellow bile (choleric), and phlegm (phlegmatic). These characterizations remain today, although the connection to the fluids is no longer supported (e.g., a happy person is considered to be sanguine and a sad person is melancholic, and an unpleasant person is considered to be bilious).

Hypethico-deductive method Method of logical deduction, attributed to Karl Popper (*The Logic of Scientific Discovery*, 1934), limiting scientific discovery to that which is testable, requiring an approach that formulates hypotheses, a priori, with the intent of rejecting these hypotheses. The method assumes that a hypothesis can never truly be proved but at best can be corroborated.

Iatrogenic Caused by an act of the treating physician (e.g., prescribed drug's side effects lead to a stroke).

Id In psychoanalytic theory, the most primitive of the three divisions of the psyche, having to be mediated by the ego and superego (see both).

Ideal utilitarianism Form of utilitarianism holding the view that the value of an action derives solely and completely from the value of the action's consequences.

Immoral Contrary to established moral principles.

In silico Based on information, usually using computational methods, rather than using actual materials being studied. To some extent, in silico research is an alternative to in vivo and in vitro research (see both), which is desirable in the case of limiting animal research and in reducing risks in humans who undergo in vivo procedures.

In situ Taking place where it is found (e.g., cleaning up hazardous waste sites where they exist, rather than moving the materials off-site).

In utero In the womb (e.g., fetal alcohol syndrome results from the unborn child's exposure to alcohol and its metabolites during gestation [see]).

In vitro Outside of the organism (literally, "in glass, " i.e., in a test tube).

In vitro fertilization Union of an egg and sperm outside the body and in an artificial environment.

In vivo Inside the organism (e.g., experiments within a rat to observe biochemical responses to a chemical dose).

Incidence or incidence rate Number of new cases in a defined population within a period of time.

Index of biological integrity (IBI) Method of indicating the quality of aquatic systems. Usually, the total number of organisms and the number of different species present are inventoried, followed by the application of an index, or scale, that lists organisms according to their sensitivity to pollution.

Induced termination of pregnancy Synonym of abortion (see).

Inductive reasoning Starting from a specific experience and drawing inferences from the specific set of facts or instances to a larger number of instances (generalization). Conclusions are drawn from the perspective that all individuals of a kind have a certain character on the basis that some individuals of the kind have that character. Contrast with *deductive reasoning*.

Inference Reasoning that one statement (the conclusion) is derived from one or more other statements (the premises). See *syllogism*.

Infinite scrolling Web design used by social media (see) that continuously loads new content automatically as user scrolls down.

Informatics Application of computational and other technologies to access and to enhance information; one means of turning data into information (see both).

Information Processed and organized data (see). Value-added data as a step toward knowledge.

Information processing theory Cognitive processes consisting of stages of encoding, storage, and retrieval of data.

Institutional review board Group of scientists and nonscientists who review proposed studies to ensure that the rights of study participants are protected.

Instrumental value Worth based on usefulness. In biomedical ethics, the perspective of whether a human life has value that depends on usefulness (Will the baby be loved?–Will the elderly person continue to enjoy life?) is an instrumental viewpoint. In environmental ethics, use of the term environmental resource implies that ecological value is based on the utility of the ecosystem (e.g., wetlands as breeding areas for game fish, as retention areas to prevent floods, and as sinks for carbon to prevent global warming). Contrast with *intrinsic value*.

Integrity Moral soundness based on honesty and upright character (see).

Intellectual property Proprietary knowledge protected by law, for example, patents and copyright.

Interoperability Ability of computer systems, software, and databases to exchange and make use of information without substantially recoding and rework. Standard that allows completely different products to communicate with each other and use the shared data without requiring special effort from the end-user. For example, continuous emissions monitoring systems (CEMS) can be integrated with open data acquisition and handling systems (DAHS) inside industrial smokestacks to ensure that pollution control technologies are operating efficiently and to prevent tampering with monitoring data.

Intervention Direct involvement of corrective action to change existing conditions for the better (i.e., both an engineering and medical concept).

Intrinsic value Worth based on existence, not usefulness. All humans have intrinsic value in contemporary morality. In biomedical ethics, however, there is no unanimity of thought about the intrinsic value of an embryo or a fetus, or a person nearing end of life. Those subscribing to sanctity of life viewpoints see intrinsic value of any human being (beginning with the human zygote and ending in natural death). In environmental ethics, there is not unanimity of thought about nonhuman species. For example, the loss of a species is morally wrong based on the value of the existence of the species, not its actual or potential value (e.g., as a cure for cancer or as a food source for a food species). Contrast with *instrumental value*.

Intuition Direct perception of meaning without conscious reasoning. Compare to *deductive* and *inductive reasoning*.

Ishikawa diagram Graphical technique for identifying cause-and-effect relationships among factors in a given situation or problem. Also known as a fishbone diagram.

Junk science Term applied to questionable data or information used to support advocacy positions, or factually correct data and information misapplied to support advocacy positions.

Justice The fair and equal treatment of people. Providing what is deserved. Distribution of goods and services based on fairness and what is deserved.

Key event (KE) Events in an *adverse outcome pathway*, following a *molecular initiating event*, that are required for the outcome and that are linked to other key events by *key event relationships*.

Knowledge Familiarity, awareness, or understanding gained through experience or study. A necessary step toward wisdom.

Land ethic Ethical view proposed by Aldo Leopold (*A Sand County Almanac*, 1949) that "reflects the existence of an ecological conscience, and this in turn reflects a conviction of individual responsibility for the health of land." This is a precursor to ecocentrism (see).

Large language model (LLM) Type of artificial intelligence that applies neural networks (see) to forecast word sequences to understand, process, and generate human language by analyzing large datasets.

Latency Period of time between disease occurrence and detection, sometimes used interchangeably with induction.

Law of diminishing returns Economic principle espoused by Thomas Malthus (*Essay on the Principle of Population*, 1798) stating that when a fixed input is combined in production with a variable input using a given technology, increases in the quantity of the variable input will eventually depress the productivity of the variable input. Malthus proposed this as a law from his pessimistic idea that population growth would force incomes down to the subsistence level.

Law of supply and demand Economic principle stating that, in equilibrium, prices are determined so that demand equals supply; thus, changes in prices reflect shifts in the demand or supply curves.

Left brain/right brain thinking Differentiation between where incoming stimuli are processed in the brain. Verbal, sorting, and detail-oriented processing occurs in the left, and the spatial, intuitive, and nonverbal processing occurs in the right.

Lethal concentration (LC_x) Concentration of a substance in air in which X% of test animals die, for example, the median lethal concentration is LC_{50}: a common measure of *acute toxicity*.

Lethal dose (LD_x) Amount of a substance delivered to a test animal in a single dose that kills X%, for example, the median lethal dose is LC_{50}: a common measure of *acute toxicity*.

Leviathan Treatise (full title: *Leviathan or the matter, forme, and power of a common-wealth ecclesiasticall and civill*) written by Thomas Hobbes in 1651, expounding the theory that absolute government was the exclusive vehicle needed to balance personal interests and desires with their rights of life and property. Thus, members of society enter into a social contract (see) with the sovereign.

Libertarianism Form of contractarianism (see) holding that individuals ought to have complete freedom of action, providing these actions do not infringe on the same freedom of others. Thus, government is circumscribed.

Liberty Freedom of choice, autonomy, and immunity, especially from government's arbitrary exercise of authority.

Life Period from onset (i.e., conception) to end (i.e., death) of a unique organism. Antonym can be either death (see) or nonliving.

Life cycle analysis Life cycle assessment (LCA): Consideration and quantification of total environmental and energy impact of a product or process, beginning at or before raw material extraction through use, disposal, recycling, and post-use.

Life cycle impact assessment (LCIA) Part of a *life cycle assessment* that interprets releases of substances and use of resources in terms of harm to the environment, human health, and resources.

Life cycle inventory (LCI) Phase of life cycle assessment in which inputs and outputs are compiled and quantified for a product throughout its life cycle. LCI analysis includes collection of the data needed.

Lifetime average daily dose (LADD) Estimated dose to an individual averaged over a lifetime of 70 years, used in assessments of *carcinogenic* risk.

Linearity Following the mathematical equation for a line ($y = mx + b$, where m is the slope and b is the y-intercept). Also, used to describe the degree to which data points approximate the line of best fit (linear regression).

Logic Branch of philosophy addressing inference (e.g., using a syllogism [see] to determine the validity of an ethical argument).

Logical positivism Twentieth-century philosophical view that all meaningful statements are either analytic or can be verified conclusively using observation and experimentation, and that non-verifiable (e.g., metaphysical) theories have no meaning (also known as logical empiricism). Even for the physical sciences (let alone, theology), logical positive comes undone, since much of scientific theory is not analytic or even empirically verifiable (e.g., Newton's general law of gravitation).

Longitudinal study Epidemiological study using data gathered at more than one points in time, for example, after an exposure or a medical intervention. Contrast with *cross-sectional study*.

Love (1) Complex mix of emotions, behaviors, and beliefs associated with strong feelings of affection, protectiveness, warmth, and respect for another person, animal, or entity. (2) In theological thought, the divinely infused habit to cherish the Creator for his own sake above all things and fellow humans for the sake of the Creator. The greatest of the three Christian virtues (also known as charity [see]).

Low birth weight Weight at birth of human baby below what is considered the ideal range, often less than 5 pounds and 8 ounces (2500 g). Compare to *very low birth weight*.

Lowest-observed-adverse-effect level (LOAEL) Lowest exposure level at which there are biologically significant increases in frequency or severity of adverse effects between the exposed population and its appropriate control group.

Machine learning (ML) Computational process to learn new things by inference; ML is used to feed data to AI and train it over time.

Macroethics Expectations of an entire profession, for example, the engineering profession's positions regarding emerging technologies, social justice, or sustainability.

Malfeasance Wrongdoing, misconduct, or misbehavior.

Maleficence Causing evil or harm to another.

Malevolence Evil intent.

Malpractice Unprofessional activity, especially which harms a patient (medicine) or client (law or design).

Malthusian population theory View first articulated by Rev. Thomas Robert Malthus (1766–1834) that the human population would increase at a geometric rate and the food supply at an arithmetic rate. The view was overly pessimistic and did not account for technological advances (e.g., the Green Revolution).

Mass balance Assessment includes a quantitative estimation of the mass loading to the dissolved plume from various sources, as well as the mass attenuation capacity for the dissolved plume.

Mechanism of action (MOA) Specific biochemical interaction through which a substance produces its intended effect, in the case of a pesticide or drug, or toxic reaction, in the case of a toxic substance. The mechanism is usually characterized by specific molecular targets to which the substance binds, such as to an enzyme or to a receptor. For example, numerous pesticides' mechanism of action is by inhibition of neurotransmitters, for example, acetylcholine. Organochlorine pesticides often alter movement of ions across the nerve cell membranes, changing the ability of the nerve to fire. Organophosphate and carbamate pesticides act primarily at the synapses, altering the regulation of the transmission of signals between neurons.

Mendelism Heredity theory underlying classical genetics, proposed by Roman Catholic monk and scientist Gregor Mendel in 1866.

Mental model Template in one's mind, based on past experience and current knowledge, to decipher and use sensory information.

Mentor Guide or advisory guide to a more junior colleague.

Metabolism Act of a living organism converting and degrading a substance from one form to another (known as a metabolite).

Metabonomics Study of total metabolite pool (especially using computational methods).

Metadata The quality and other characteristics of the data: "data about the data."

Metaphor Figurative comparison of known things to other, more difficult to understand concepts.

Miasma Dominant disease theory from the 17th century through much of the 19th century that posited the cause of disease to be from foul emanations from air, soil, and water. Displaced by germ theory (see) as standard disease paradigm in the 19th century.

Microethics Expectations of the individual professional practitioner or researcher. Compare to *macroethics*.

Minimax theorem Key convention of game theory holding that lowest maximum expected loss in a two-person zero-sum game equals the highest minimum expected gain. It is a useful technique to address uncertainties in decision-making.

Model Mathematical function with parameters that can be adjusted so the function closely describes a set of empirical data. A mechanistic model usually reflects observed or hypothesized biological or physical mechanisms and has model parameters with real-world interpretation. In contrast, statistical or empirical models selected for particular numerical properties are fitted to data; model parameters may or may not have real-world interpretation. When data quality is otherwise equivalent, extrapolation from mechanistic models (e.g., biologically based dose-response models) often carries higher confidence than extrapolation using empirical models.

Modifying factor (MF) A factor used in the derivation of a reference dose or reference concentration. The magnitude of the MF reflects the scientific uncertainties of the study and database not explicitly treated with standard uncertainty factors (e.g., the completeness of the overall database). An MF is greater than zero and less than or equal to 10, and the default value for the MF is 1. Use of a modifying factor was generally discontinued in 2004. Compare to *uncertainty factor (UF)*..

Modularity Ability of a dataset, model, or model subset to be used in numerous models and applications, as opposed to stand-alone systems.

Molecular descriptor Information encoded in the molecular structure into values that can be used to characterize a compound, such as those used in a quantitative structure-activity relationship.

Molecular docking Description of the "best-fit" orientation of a ligand that binds to a particular protein of interest.

Molecular initiating event (MIE) The first step in an *adverse outcome pathway*.

Molecular pharming Use of *genetically modified organisms* to produce pharmaceuticals. Application of genetic engineering that introduces genes, primarily of human and animal origin, into plants or farm animals to produce medicinal substances. Premise is using plants as efficient chemical factories for producing antibodies, vaccines, blood proteins, and other therapeutically valuable proteins.

Moral (1) Pertaining to the judgment of goodness or evil of human action and character. (2) Often, an adjective for goodness or ethically acceptable actions (opposite of immoral).

Moral development Progression of morality and ethics in an individual, growing from one based on rewards and punishment to principled conscience.

Moral hazard Actuarial risk resulting from uncertainty about the honesty of the insured, especially the risk that one party to a contract will change behavior to the detriment of the other party once the contract has been concluded (important to the insurance industry and legal profession).

Moral sense Motivation deriving logically from ethical or moral principles that rule a person's judgments and actions.

Morality Distinction between what is right and wrong.

Morbidity State of disease.

Morbidity rate Statistical measure of the frequency or prevalence of a specific disease, injury, or health condition within a defined population over a specific period of time.

Mores Morally significant norms of a society.

Mortality rate Proportion of a population that dies during a specified time period. Also called death rate.

Multipotent stem cells Stem cells that can give rise to several other cell types, but those types are limited in number. Hematopoietic cells (i.e., blood stem cells that can develop into several types of blood cells) are examples. Scientists have long held the opinion that differentiated cells cannot be altered or caused to behave in any way other than the way in which they have been naturally committed, but recent findings have called this assumption into question (see Nanog).

Nanog Protein central to the unique properties of embryonic stem cells. The protein is named after the mythological Celtic land of the ever young, *Tir nan Og*, and is required for the special ability of stem cells to multiply without limit while remaining able to make many different types of cells. Researchers are investigating whether transcription can produce nanog in adult stem cells, making them pluripotent. That is, they will behave just like embryonic stem cells.

Nanoscale or nano-sized Having at least one dimension <100 nm.

Nanomaterial Substances with *nanoscale* dimensions, for example, nanocarbon, nanosilver, and nanocerrium.

Naïve realism Describe the belief that physical objects continue to exist when they are no longer perceived. Contrast with *solipsism*.

Nanotechnology Science and engineering addressing the design and production of extremely small (< 100 nanometers in diameter) devices and systems fabricated from individual atoms and molecules.

Nash equilibrium Solution concept of game theory wherein equilibrium is reached so that each player's action is optimal given the actions of the other players; that is, all players are executing dominant strategies.

Natural law View that natural order exists in the human world, which is good; so, people must not violate this order.

Negative paradigm Most unethical action or case possible. In line drawing, the negative paradigm is the polar opposite of the positive paradigm (see).

Negative utilitarianism Permutation of utilitarianism (see) where the metric of morality is the prevention of the greatest evil or the promotion of the least harm for the greatest number.

Nerve Enclosed, cable-like bundle of nerve fibers or axons.

Neural network ML (see) model that maps inputs to outputs by using interconnected nodes, so-called artificial neurons, in layers to recognize patterns in weights and to predict outcomes by continuous adjustments of internal weights via training instead of reliance on explicit programmed rules.

Neurotechnology Techniques and apparatus used to analyze and treat the nervous system, especially the brain (e.g., implantation of devices in the brain to allow paralyzed patients to move a robotic arm).

New Testament Second half of the Christian Bible that includes the teachings of Jesus Christ and that of his followers.

Nicomachean ethics Aristotle's exposition of virtue ethics: 10 books based on his lectures at the Lyceum (either edited by or dedicated to his son, Nicomachus).

No-observed-adverse-effect level (NOAEL) Highest exposure level where there are no biologically significant increases in the frequency or severity of adverse effects between the exposed population and its appropriate control.

Nonmaleficence Affirmation of doing no harm or evil.

Nongovernmental organization (NGO) Entity that advocates or represents positions, including scientific, legal, and medical perspectives, without a governmental mandate. Examples include Doctors without Borders, Resources for the Future, and Engineers without Borders.

Normative ethics Study of ethics concerned with classifying actions as right and wrong. Contrast with *descriptive ethics*.

Norms Expectations of behavior of individuals within society.

Nosocomial Contracted in the hospital (e.g., infection from an airborne microbial exposure while being treated for a broken arm).

Nutritional supplement A dietary supplement intended to be ingested in pill, capsule, tablet, or liquid form; not represented for use as a conventional food or as the sole item of a meal or diet; and labeled accordingly. It is regulated by the U.S. Food and Drug Administration as a food and not as a drug.

Odds ratio Ratio of the odds of disease among the exposed compared with the odds of disease among the unexposed. For rare diseases, such of cancer, the odds ratio can provide an estimate of relative risk.

Ockham's razor Principle espoused by medieval nominalist William of Ockham that entities are not to be multiplied beyond necessity. The principle encourages asking whether any proposed kind of entity is necessary.

Offsetting behavior Inadvertent attenuation or reversal of a risk management action due to a reduction of care by those targeted for risk reduction. For example, if actions are taken to reduce bacterial growth in food supplies, and consumers consider the foods to be safer to the extent that they are less careful in food preparation, this behavior offsets at least some of the safety margin of the risk reduction efforts.

Omics Shorthand term for computational, biological subfields for describing very large-scale data collection and analysis, all with the suffix "omics" (e.g., genomics, proteomics, and metabonomics).

Onus Burden of responsibility.

Opportunity risk Likelihood that a better opportunity will present itself after an irreversible decision has been made (e.g., prohibiting research in an emerging technology may prevent exposure to a toxic substance to a few, but in the process, the cure for a disease may be lost).

Optimal range Range of success, below and above which are unacceptable (e.g., trivalent chromium must be taken within the optimal range because intake at too low of a dosage leads to a nutritional deficiency and too high a dosage leads to toxicity).

Optimization Selecting the best design for the conditions. The "best" is determined by the designer based on one or more variables (e.g., a heart valve may have three key variables: flow rate, reliability, and durability—the engineer would design the valve by optimizing these three variables to achieve the best performance).

Organ Completely differentiated unit of an organism that provides a certain, specialized function.

Organism A living entity, consisting of one or more cells (unicellular and multicellular organisms, respectively).

Outlier Value that is markedly smaller or larger than other values in a dataset. Can be problematic for researchers since it decreases the coefficient of determination (i.e., r^2).

Pandora's box Metaphor for a prolific source of problems (Greek: All gifted; from mythology of a box given to Pandora by Zeus, who ordered that she not open; Pandora succumbed to her curiosity and opened it; all the miseries and evils flew out to afflict humankind).

Paradox Argument appearing to justify a self-contradictory conclusion by using valid deductions from acceptable premises.

Parametrics Descriptors of an entire population, without the need of inference. Compare to *statistics*.

Pareto efficient Resource allocation wherein there is no rearrangement that can make anyone better off without making someone else worse off.

Partial birth abortion Late-term abortion, clinically known as intact dilation and extraction (IDX or Intact D&X).

Pathogen Microbe capable of producing disease.

Patient Individual receiving medical attention, such as treatment.

Peace engineering Application of technical skills to promote peace.

Pedagogy Instruction techniques used to promote learning.

Pentateuch, The From Greek ("five tools"), the first five books of Judeo-Christian Scripture, also known as the Law or the *Torah*. Moses is credited as the author.

Perception Information and knowledge gained through the senses.

Persistent vegetative state (PVS) Condition wherein a person loses higher cerebral powers but maintains sleep–wake cycles with at least partial hypothalamic and brain stem autonomic functions. Compare to *coma*.

Person Human being.

Personhood Status of having moral rights. Some restrict the status of personhood only to individuals who are self-aware and capable of rational thought and moral agency. However, this can be opposed in that some individuals lose these capacities temporarily and because persons at various stages have yet some or all of these capacities or have lost them.

Personality Set of traits possessed by a person that uniquely influences cognitions, motivations, and behaviors. Contrast with *character*.

Persistence Resistance to degradation in the environment.

Persistent organic pollutant (POP) *Recalcitrant* organic compounds, especially those listed under international treaties and agreement, for example, Stockholm Convention on Persistent Organic Pollutants, in an effort to eliminate or restrict their production and use.

PERT chart "Program Evaluation Review Technique" diagram depicting project tasks and their interrelationships.

Pesticide Substance used to control pests by using its toxic properties.

Pharmacodynamics Manner in which a substance exerts its effects on living organisms (also known as toxicodynamics).

Pharmacokinetics Behavior of substances within an organism, especially by absorption, distribution, biotransformation, storage, and excretion (also known as toxicokinetics).

Philosophy Pursuit of wisdom (from Greek: *philos* for love and *sophia* meaning wisdom).

Photon Fundamental, indivisible unit, that is, quantum (see), of electromagnetic radiation.

Physician One who practices medicine, especially a licensed medical practitioner.

Physiologically based pharmacokinetic (PBPK) model Model used to characterize pharmacokinetic behavior of a chemical. Available data on blood flow rates, and metabolic and other processes that the chemical undergoes within each compartment are used to construct a mass-balance framework for the PBPK model.

Pica Ingestion of nonfood substances (e.g., paint chips, soil, etc.) which can account for disparate exposures (see) in certain subpopulations (e.g., minority children).

Point of departure The dose-response (see) point that marks the beginning of a low-dose extrapolation. This point is most often the upper-bound on an observed incidence or on an estimated incidence from a dose-response model.

Plagiarism Unethical practice of presenting someone else's written works as one's own.

Pluripotent stem cells Cells that can produce all the cell types of the developing body, such as the cells in the inner cell mass of the blastocyst (see).

Polychlorinated biphenyl (PCB) Highly toxic molecule of two benzene rings bonded to each other with chlorine substituents (209 structural variations, known as congeners); presently banned but manufactured by Monsanto for much of the 20th century.

Polycyclic aromatic hydrocarbon Class of products of incomplete combustion comprised of fused aromatic rings. A number are suspected carcinogens (e.g., benzo(a)pyrene).

Positive paradigm Most ethical action or case possible. In line drawing, the positive paradigm is the polar opposite of the negative paradigm (see).

Precautionary principle Risk management (see) approach taken when scientific knowledge is incomplete and the possible consequences are devastating (e.g., global climate change). The principle holds that scientific uncertainty must not be accepted as an excuse to postpone cost-effective measures to prevent a significant problem.

Prescriptive statement Statement that says how things should be, as opposed to a descriptive statement, which says how things are. Contrast with *proscriptive statement*. Codes of ethics contain both prescriptive and proscriptive statements.

Prevalence Proportion of cases that exist within a population at a specific point in time, relative to the number of individuals within that population at the same point in time.

Price elasticity of demand Percentage change in quantity demanded of a good as the result of a given percent change in price (the percentage change in quantity demanded divided by the percentage change in price).

Price elasticity of supply Percentage change in quantity supplied of a good as the result of a given percent change in price (the percentage change in quantity supplied divided by the percentage change in price).

Prisoner's dilemma Game theory scenario wherein two parties each seek self-interests noncooperatively, yielding a result that leaves them both worse off. Tragedy of the Commons (see) is an example.

Probabilistic model Generation of distributions of possible outcomes to estimate or predict an outcome. Compare to *deterministic model*.

Probability Measurement of the likelihood that an event will occur, ranging from 0 (no likelihood whatsoever) to 1 (absolute certainty).

Profession Group (e.g., physicians or engineers) with a common mission, requiring substantial education and training, self-determination of professional requirements to enter, organized into an identifiable professional body, and requiring the adherence to standards of conduct.

Proscriptive statement Prohibition against certain acts. Codes of ethics contain both proscriptive and prescriptive statements (see).

Proteomics Study of proteins in the body, especially the protein complement of the genome (see).

Psychiatry Medical specialty that addresses the origin, diagnosis, prevention, and treatment of mental disorders.

Psychology Study of human behavior.

Public, The Whole collection of people comprising a society.

Punishment Penalty inflicted on a person for committing an act unacceptable to society; something meant to deter an unacceptable act.

Pursuit of happiness One of the unalienable rights articulated in the U.S. Declaration of Independence, along with life and liberty, this pursuit sets the stage of limited government and self-determination of citizens.

Quantitative microbial risk analysis (QMRA) Formal process of estimating human health risks due to exposures to microbes. Like chemical risk assessment, QMRA includes steps for identifying *hazards*, estimating *exposures*, and evaluating *dose-response* but also includes other factors important to microbial adverse outcomes, including person-to-person transmission and immunity.

Quantitative polymerase chain reaction (qPCR) Determination of a polynucleotide by including a known amount of readily distinguished template as an internal standard to compensate for variation in efficiency of amplification.

Quantitative structure-activity relationship (QSAR) Understanding and application of chemical structure to estimate the behavior of a compound, for example, *toxicity* or *persistence*.

Quantum Smallest possible indivisible unit of energy or matter. For example, the photon is a quantum of light, which is a packet of light energy.

Quantum computing Technology that applies quantum physics to solve problems beyond the capacity of supercomputers.

Quantum yield Number of photons required per molecule of carbon dioxide converted to sugar (or per oxygen molecule produced).

qubit (quantum bit) Fundamental unit of information in quantum computing; quantum counterpart to the classical bit.

Racism Belief that one group of human beings is inferior to another as a result of their race.

Random Selected from a statistical population such that each sample has an equal probability of being selected.

Random Forest Machine learning ensemble approach for classification and regression that joins different *decision trees* based on differing attributes to arrive at a result. Each decision tree depends on the values of an independently sampled random vector and applies the same distribution for all trees.

Rationality Philosophical theories concerned with truth, reason, and knowledge.

Rawlsism Form of contractarianism (see) based on the view that "a practice is just if it is in accordance with the principles all who participate in it might reasonably be expected to propose or to acknowledge before one another when they are similarly circumstanced and required to make a firm commitment in advance" (John Rawls, 1957, *Justice as Fairness*, 54 (22), pp. 659–60). A key means of achieving this is through the veil of ignorance (see).

Reasonable person standard Position (legal, engineering, etc.) expected to be held by a hypothetical person in society who exercises average care, skill, and judgment in conduct.

Reasoning Derivation of a conclusion from premises.

Recalcitrance (1) Resistance of a chemical compound to degradation. (2) Inverse of degradability.

Receptor (1) Molecule in a cell or on its surface that binds to a specific substance and causes a specific physiologic effect in the cell. (2) The potentially or actually affected entity exposed to a physical, chemical, or biological agent.

Recessive Quality of a gene that causes it to be expressed only if there are two identical copies or, for a male, if one copy is present on the X chromosome.

Recombinant Describing material produced by *genetic engineering*.

Recombinant DNA (rDNA) Genetically engineered DNA prepared by transplanting or splicing genes from one individual to another, including from the cells of one species into the cells of a host organism of a different species. Thereafter, the rDNA becomes part of the host's genetic makeup and is replicated.

Redictable Ridiculously predictable (coined by D.J. Vallero and A.C.V. Randall). For example, if a microbe has been genetically modified to degrade crude oil spilled from a tanker, the likelihood that the same microbe will degrade asphalt in adjacent roads is redictable.

Redundancy (1) In engineering, the intentional duplication of necessary components, systems, or functions to increase system reliability and create a fail-safe condition to ensure that operations continue seamlessly if a primary part fails. (2) In information technology, system configurations designed to provide backup components to maintain operations if a primary system fails. Two foundational models are N+1 and 2N, where N represents the base capacity required to run a system.

Reference concentration (RfC) An estimate (with uncertainty spanning perhaps an order of magnitude) of a continuous inhalation exposure to the human population (including sensitive subgroups) that is likely to be without an appreciable risk of deleterious effects during a lifetime. It can be derived from a *NOAEL*, *LOAEL*, or benchmark concentration, with uncertainty factors generally applied to reflect limitations of the data used. Generally used in noncancer health assessments. Durations include *acute*, *short-term*, *subchronic*, and *chronic*.

Reference dose (RfD) An estimate (with uncertainty spanning perhaps an order of magnitude) of a daily oral exposure to the human population (including sensitive subgroups) that is likely to be without an appreciable risk of deleterious effects during a lifetime. It can be derived from a *NOAEL*, *LOAEL*, or benchmark dose, with uncertainty factors generally applied to reflect limitations of the data used. Generally used in noncancer health assessments. Durations include *acute*, *short-term*, *subchronic*, and *chronic*.

Reinforcement Linking a stimulus and response with a reward or punishment, leading to an increased or decreased likelihood that the response will be repeated subsequently when the stimulus is reintroduced (key concept of behaviorism [see]).

Relative risk Ratio of the risk of disease or death among the exposed segment of the population to the risk among the unexposed.

Reliability Probability that a device or system will perform its specified function, without failure under stated environmental conditions, over a required lifetime.

Reliability engineering Designing a system to perform a required function under stated conditions for a stated period of time.

Religion Structured belief system, centered around the supernatural and sacred, and including expectations (tenets) for morality consistent with beliefs.

Renaissance Time from about 1450 to 1600 AD, characterized by advances in the humanities and learning and the birth of modern science in Western civilization.

Respect for nature View expounded by Paul Taylor (*Respect for Nature*, 1986) defending biocentrism (see), based on a systematic and comprehensive account of moral relationships between humans and other living things. This requires an acceptance that all living things have inherent moral value so that respect for nature is the "ultimate moral attitude."

Respect for persons Principle expounded in the Belmont Report (see) that incorporates at least two ethical convictions: first, that individuals should be treated as autonomous agents, and second, that persons with diminished autonomy are entitled to protection. The principle of respect for persons thus divides into two separate moral requirements: the requirement to acknowledge autonomy and the requirement to protect those with diminished autonomy.

Responsible conduct of research (RCR) (1) Ethical approach to research, especially performing research honestly and objectively, with integrity, and with full consideration of the possible effects (e.g., protecting human subjects, animal welfare, etc.). (2) Requirement by public research agencies that grantees and contractors (e.g., National Institutes of Health) properly train their researchers (e.g., PhD students) in research ethics.

Resilience (1) Attribute of ecosystem stability, expressing the system's ability to recover after disturbance. (2) Capacity of an engineered system, material, or infrastructure to anticipate, absorb, adapt to, and recover from disturbances, shocks, or unexpected conditions while maintaining critical functionality.

Reverence for life View introduced by Albert Schweitzer (1875– 1965) based on "world and life affirmation," based on *ehrfurcht vor dem leben* (German: attitude of awe and wonder tempered with fear, summed up by Schweitzer as "I am life which wills to live, in the midst of life which wills to live."

Right of conscience Moral prerogative to exercise professional judgment.

Right of conscientious refusal Moral prerogative of a professional to refuse to engage in activity that he or she considers to be unethical.

Right to die Political and philosophical position that it is the individual's right to decide to live, especially when confronted with a terminal illness or extremely serious, progressive disease; often opposed on the basis of sanctity of life (see).

Right to life Viewpoint holding that human life as a paramount value; central feature of debates concerning abortion, capital punishment, embryonic stem cell research, euthanasia, and assisted suicide. As such, it is a central focus of pro-life advocates.

Rights ethics View that an act is morally right when it is the preferred way to respect the rights of others. John Locke (*Essays on the Law of Nature*, 1664) advocated rights ethics.

Risk Likelihood of an adverse outcome.

Risk characterization The integration of information on hazard, exposure, and dose-response to provide an estimate of the likelihood that any of the identified adverse effects will occur in exposed people.

Risk homeostasis Defeat of built-in factors of safety by asserting a new way to use the products.

Risk management Decision-making process that accounts for political, social, economic, and engineering implications together with risk-related information in order to develop, analyze, and compare management options and select the appropriate managerial response to a hazard.

Risk ratio (RR) See *relative risk.*

Risk shifting Taking an action that changes or reduces the risk of one population but increases the risk in a different population (e.g., banning DDT reduces the risk of cancer in developed nations but increases the risk of malaria in tropical and subtropical developing countries).

Risk trade-off Eliminating or reducing one risk but introducing or increasing a countervailing risk (e.g., reducing the pain of a headache by taking aspirin but increasing the risk of Reye's syndrome or removing mold in buildings may increase worker exposure to asbestos).

Risk-benefit analysis Comparison of risks of various options to their benefits (e.g., health and wildlife risks of applying a pesticide compared to the benefits of crop protection).

Robotics Interdisciplinary technology that involves the design, construction, operation, and application of robots.

Root cause analysis Analytical method for deciding causes to an existing problem based on a comprehensive, systematic reconstruction of events that contributed to the outcome; retrospectively retracing events, that is, a "reverse event tree."

Rule utilitarianism Form of utilitarianism wherein the test of utility maximization is applied to acts only indirectly through some other suitable object of moral assessment, that is, to rules of conduct. Rule utilitarianism bases the morality of an action on whether or not that action follows a universal rule that would have good consequences if everyone followed it (compare to the duty-based concept, *categorical imperative*).

Sanctity of life View that human life is precious from conception to natural death since it is created in the image of the Creator.

Scale Spatial extent from very small (molecular) to very large (planetary).

Schrödinger equation The partial differential equation that governs the wave function of a non-relativistic quantum-mechanical system. Named after the Austrian physicist who published it in 1926, forming the basis for the work that resulted in his Nobel Prize in Physics in 1933.

Science Systematic investigation, through experiment, observation, and deduction, in an attempt to produce reliable explanations of the physical world and its processes.

Scientific method Progression of inquiry: (1) to identify a problem you would like to solve, (2) to formulate a hypothesis, (3) to test the hypothesis, (4) to collect and to analyze the data, and (5) to draw valid conclusions.

Senescence Organism's aging process.

Sensation biology Branch of life sciences related to how organisms detect, process, and respond to stimuli from their environments.

Sensitivity Ability of a test to detect a condition when it is truly present. Compare to *specificity.*

Sequestration (1) Physicochemical process that removes and stores substances, for example, mercury that has been removed from the troposphere and stored in ocean or lake sediments. (2) *Geoengineering* to capture and store *global greenhouse gases*, especially carbon dioxide.

Seven virtues The four cardinal virtues (see) plus the three theological virtues: faith, hope, and love (charity) given in 1 Corinthians 13.

Short-term Describing an *exposure* to an agent that occurs multiple times or continuously for one week (or other designated short time period).

Similarity algorithm Computational tool to characterize the distance between factors, such as physicochemical properties, to estimate similar biological function or activity, that is, nearby chemicals would be expected to have similar activity. Can be part of *quantitative structural activity relationship* models.

Singularity Theoretical point in the future when technological growth becomes uncontrollable and irreversible, such as when artificial intelligence becomes autonomous and surpasses human intelligence by redesigning and upgrading itself exponentially, that is, the so-called self-improvement loop.

Slope factor, cancer Dose-response curve for a substance indicating cancer potency (units = mass per body mass per time, e.g., mg kg^{-1} day^{-1}).

Social contract Agreement among the members of a society and its government delineating and limiting the rights and duties of each. Thomas Hobbes argued that the power of a sovereign is justified by an implied contract wherein the people agree to obey the government in return for a guarantee of peace and security, which is not possible in the "state of nature" (see) conceived to exist without such contract.

Social ecology View that environmental problems are firmly rooted in human social interactions. Social ecologists believe that an ecologically sustainable society can still be socially exploitative.

Social media Internet-based platforms that allow human users to interact, communicate, and share content, especially user-generated content.

Society Collection of human beings that is distinguished from other groups by shared institutions and a common culture.

Solidarity Catholic social teaching principle that all humans are interconnected and responsible for one another as one global human family.

Solipsism Idea that the only one's existence is a certainty, and that true knowledge of anything else is impossible. Contrast with *naïve realism*.

Somatic cell nuclear transfer (SCNT) Method of cloning by transferring the nucleus from a donor somatic cell into an enucleated egg to produce a cloned embryo.

Spatial scale Geographic extent of a resource or problem. Global-scale examples include pandemics, changes in climate, or nuclear threats. Continental-scale examples are shifting biomes and border control between nations. Regional-scale examples include the contamination of rivers or polluting the air. Local-scale examples include crime, job loss, hazardous waste sites, and landfills.

Specialization Degree to which an individual professional concentrates his or her practice into a narrow range of expertise and activities.

Specificity Ability of a test to exclude the presence of a condition when it is truly not present. Compare to *sensitivity*.

Standard deviation (σ) Measure of the spread in a dataset; wider spread means larger standard deviation.

Standardized mortality ratio (SMR) Ratio of the number of deaths observed in the study group to the number of deaths expected based on rates in a comparison population, multiplied by 100.

State of nature According to Thomas Hobbes (*Leviathan*, 1651, human beings would behave "badly" toward one another in the absence of a social contract (see) with government. Hobbes believed that such a state would lead to a "war of every man against every man" and make life "solitary, poor, nasty, brutish, and short."

Statistical significance The probability that a result is not likely to be due to chance alone. By convention, a difference between two groups is usually considered statistically significant if chance could explain it only 5% of the time or less. Study design considerations may influence the a priori choice of a different level of statistical significance.

Statistics Mathematics concerned with collecting and interpreting quantitative data and applying probability theory to estimate conditions of a universe or population from a sample.

Stimulus-response theory Basic concept of behaviorism (see) that based on the stimulus-response theory that a stimulus will cause a response either by pairing a response with a reflective trigger (e.g., Pavlov's famous experiment of inducing salivation in dogs by ringing a bell that was paired with food) or rewarding a response in the presence of a stimulus (e.g., a rat pressing a bar that releases a food pellet).

Stochastic model *Probability distributions* of potential outcomes estimated by allowing for random variation in model inputs over time. For example, a stochastic model of how a population may be exposed to a pollutant can be based on each person in a population being represented as a random sample, so that the rules for determining the characteristics and the exposure for each individual person are simply repeated to model the population. The difference between one person and another results only because there is stochastic variation when sampling from distributions. Thus, the population distribution by simulating the same person many times, that is, population estimates can be obtained by repeating this process for the individual person many times using *Monte Carlo simulation*.

Stochasticity Randomness; state of lacking predictable order but adhering to the laws of probability.

Strain Geometrical expression of deformation caused by the action of stress (see) on a physical body.

Stratified random sample Separation of a sample into several groups and randomly assigning subjects to those groups.

Stress (1) Applied force or system of forces that are apt to strain (see) or deform a physical body. (2) The internal resistance of a physical body to such applied force or system of forces.

Subjectivism Philosophical view that knowledge and value are dependent on and limited by one's subjective experience.

Subtractive manufacturing Process that shapes a physical object by cutting, drilling, or carving away material from a larger solid block (workpiece) until the final part is manufactured. Contrast with *additive manufacturing*.

Structure-activity relationships (SAR) Applying the chemical structure of a molecule to its expected behavior and effects in the environment or in a biological system.

Subchronic exposure Repeated exposure by the oral, dermal, or inhalation route for more than 30 days, up to approximately 10% of the life span in humans (more than 30 days up to approximately 90 days in typically used laboratory animal species). (See also *chronic exposure*.)

Subsidiarity Catholic social teaching principle that social matters are best managed at the smallest, lowest, or least centralized competent level of authority, for example, the family or neighborhood rather than a national or corporate authority.

Superego In psychoanalytic theory, component of human psyche that has incorporated the values and standards of society in conflict with the primitive selfishness of the id (see). This conflict is mediated by the ego (see).

Supererogatory Morally good beyond what is obligatory (duty-based).

Supply Quantity of a good or service that a seller would like to sell at a particular price.

Supply curve Relationship between a good's quantity supplied and the good's price.

Susceptibility Extent to which an individual is prone to the effects of an agent.

Sustainability Processes and activities that are currently useful and that do not diminish these same functions for future generations.

Sustainable design Application of principles of sustainability (see) to structures, products, and system; an aspect of green engineering (see).

Sword of Damocles A constant and impending hazard or peril (from the Greek moral anecdote wherein King Damocles required a guest to sit precariously under a large sword held only by a single horsehair).

Syllogism Argument according to Aristotle's logical theory that includes a major premise, a minor premise, and a conclusion. Ethical syllogisms have factual premise, a connecting premise, an evaluative premise, and a moral conclusion.

Sympathy Compassion: sharing the feelings of others. Compare to *empathy*.

Synderesis (or synteresis) Immediate, intuitive grasp of the fundamental principles of morality.

Synergy Effect from a combination of two agents is greater than the sum of the additive effects from both agents ($1 + 1 > 2$).

System (1) Combination of organized elements comprising a unified whole. (2) In thermodynamics, a defined physical entity containing boundaries in space, which can be open (i.e., energy *and* matter can be exchanged with the environment) or closed (no energy or matter exchange).

Systems biology (1) Discipline that seeks to study the relationships and interactions between various parts of a biological system (e.g., metabolic pathways, organelles, cells, and organisms) and to integrate this information to understand how biological systems function. (2) Treatment of biological entities as systems composed of defined elements interacting in distinct ways to enable the observed function and behavior of that system. Properties of such systems are embedded in a quantitative model that guides further tests of systems behavior.

Synthetic biology Design and synthesis of complex, biologically based systems with structures and functions that do not exist in nature.

Tao "Method" or "way" described in writings of Tao-te ching, Chuang-tzu, and others. Continuous activity between the *yin* and *yang* (see).

Tautology Proposition that is already true by definition, in no need of logical deduction (e.g., the statement "An octagon has eight sides" is a tautology).

Taxonomy Classification system.

Technology (1) Application of scientific knowledge. (2) The apparatus that results from such applications.

Teleology See *consequentialism.*

Temporal scale Range of complexity associated with time. Extremely short temporal scale may be measured in nanoseconds, for example, nuclear reactions, whereas long temporal scales may be measured in millions of years, for example, fossilization of plants to coal.

Temporality Criterion for causality requiring that the cause (e.g., exposure to an infectious agent) precede the effect (e.g., disease).

Terrorism Unlawful use of, or threatened use of, force or violence against individuals or property to coerce or intimidate governments or societies, often to achieve political, religious, or ideological objectives (U.S. Department of Defense).

Thermodynamics, Laws of Physical principles addressing the physical relationships between energy and matter, especially those concerned with the conversion of different forms of energy.

Thought experiment Conception of the consequences of an intervention in the world without actually intervening. In some instances, an actual experiment is not possible or feasible in practice, or perhaps not even in theory because it violates the laws of physics.

Tissue Collection of interconnected cells that carry out a similar function in an organism.

Top-down Starting with the upper levels of organization and working downward to the details. Contrast with *bottom-up*.

Torah Hebrew law as recorded in the Pentateuch, that is, the first five books of the Old Testament.

Total quality management (TQM) Structured management approach, introduced by engineer W. Edwards Deming, based on continued refinements to improve the quality of products and services, informed by ongoing feedback from customers and clients.

Toxicity Extent and degree of biological harmfulness of a chemical, physical, or biological agent, ranging from acute (short-term harm) to chronic (long-term harm) toxicity.

Toxicity pathway Segment of the *adverse outcome pathway* and *mode of action* that begins with the *molecular initiating event* and ends with the cellular effect in a single organism.

Toxicodynamics (TD) The determination and quantification of the sequence of events at the cellular and molecular levels leading to a toxic response to an environmental agent. Often used synonymously with the broader term *pharmacodynamics*, but toxicodynamics is exclusively applied to adverse agents rather than the efficacy of substances.

Toxicogenomics (1) Investigation of genetic influences on how the organisms respond to toxic substances. (2) Study of how genomes respond to environmental stressors or toxicants that integrates genome-wide mRNA expression profiling with protein expression patterns based on bioinformatics to understand the role of gene-environment interactions in disease and dysfunction.

Toxicokinetics (TK) The determination and quantification of the time course of absorption, distribution, biotransformation, and excretion of chemicals. Often used synonymously with the broader term *pharmacokinetics*, but toxicokinetics is exclusively applied to adverse agents rather than the efficacy of substance. The mechanisms are predicted using physiologically based toxicokinetic/toxicodynamic (PBTK/TD) models.

Toxicology Study of harmful interactions between chemical, physical, or biological agents and biological systems.

Tragedy of the Commons Term coined by Garrett Hardin (*Science*, volume 162 [1968]) characterizing the degradation of commonly held resources as a result of selfish self-interest in maximizing utility of each individual using the resource.

Transcription Transfer of information in DNA sequences to produce complementary *messenger RNA (mRNA)* sequences. It is the beginning of the process by which the genetic information is translated to functional peptides and proteins.

Trimming Irresponsible data manipulation of smoothing of irregularities to make observational data appear more accurate and precise than is the case. This is done in an attempt to improve the coefficient of determination (i.e., r^2).

Truth Conformity to fact. Compare to *honesty*.

Type I error Error of rejecting a true null hypothesis. Contrast with *type II error*.

Type II error Error of accepting (not rejecting) a false null hypothesis. Contrast with *type I error*.

Uncertainty Difference between what is known and what is actually the truth. Scientific uncertainty includes error and unknowns, such as those resulting for selecting variables, undocumented variability, and limitations in measurements and models. In science, there is almost always uncertainty. The goal is to decrease uncertainty and to document known uncertainties.

Uncertainty factor (UF) One of several, generally 10-fold, default factors used in operationally deriving the *RfD* and *RfC* from experimental data. The factors are intended to account for (1) variation in *susceptibility* among the members of the human population (i.e., interindividual or intraspecies *variability*); (2) uncertainty in extrapolating animal data to humans (i.e., interspecies uncertainty); (3) uncertainty in extrapolating from data obtained in a study with less-than-lifetime exposure (i.e., extrapolating from subchronic to chronic exposure); (4) uncertainty in extrapolating from a *lowest-observed-adverse-effect level* rather than from a *no-observed-adverse-effect level*; and (5) uncertainty associated with extrapolation when the database is incomplete.

Utilitarianism Theory proposed by Jeremy Bentham (*Principles of Morals and Legislation*, 1789) and James Mill (*Utilitarianism*, 1863) that action should be directed toward achieving the greatest happiness for the greatest number of people (if applied to nonhuman species, this is referred to merely as "greatest number").

Utility (1) Condition of being useful. (2) Useful outcome. (3) Level of enjoyment an individual attains from choosing a certain combination of goods.

Valuation Quantifying or otherwise placing value on goods and services. Monetized valuation uses monetary currency (e.g., gross domestic product), whereas many environmental and quality of life resources are not readily conducive to monetized valuation; for example, old growth forests have nonmonetized value (e.g., habitat, ecological diversity) but little monetized value since they are not used for timber.

Value (1) Principle, standard, or quality that is good for a person to hold. (2) Worth.

Value engineering (VE) Systematic application of recognized techniques by a multi-disciplined team to identify the function of a product or service, establish a worth for that function, generate alternatives through the use of creative thinking, and provide the needed functions to accomplish the original purpose of the project reliably and at the lowest life cycle cost without sacrificing safety, necessary quality, and environmental attributes of the project (U.S. Department of Transportation).

Value of life (1) Economic or moral worth of a human being. (2) Marginal cost of preventing death (social sciences).

Variability True heterogeneity or diversity (e.g., among a population that is exposed to airborne pollution from the same source and with the same contaminant concentration, the risks to each person as a result of breathing the polluted air will vary).

Veil of ignorance Notion introduced by John Rawls (*A Theory of Justice*, 1971) to determine morality of an action, supposing that societal roles can be completely reshaped and redistributed and that the ethical decision- maker does not know his or her reassigned role. This is an attempt to reach distributive justice (see).

Very low birth weight Weight of human baby at birth less than 3 pounds and 5 ounces (1500 g). Compare to *low birth weight*.

Vice Immorality; negative moral trait. Opposite of *virtue*.

Virtue Moral excellence or goodness. Opposite of *vice*.

Virtue ethics Ethical theory that emphasizes the virtues, or moral character, in ethical decision-making. Focuses on what makes a good person, rather than what makes a good action. One of three major theories of normative ethics (see), along with consequentialism and deontology (see).

Vulnerability Condition of an individual or population determined by physical, social, economic, and environmental variables, wherein susceptibility to a hazard increases (e.g., asthmatics are more vulnerable to the effects of some air pollutants than is the average person).

Way of knowing Method of learning. Discovery and reason hold primacy as scientific ways of knowing. Intuition (see), to scientists, is an advanced, integrated means of knowing. All are considered a posteriori knowledge (see). However, theological ways of knowing also include a priori knowledge (see) given by the Creator). Thus, intuition can be a combination of a posteriori and a priori knowledge.

Weight of evidence Strength of data and information supporting a conclusion. When little reliable data are available the weight of evidence is lacking, whereas when numerous studies provide reliable information to support a particular position (e.g., exposure to a chemical associated with a health effect), the weight of evidence is strong.

Whistleblower Informant who exposes wrongdoing within an organization or to an outside authority (e.g., government agency) that has or appears to have the ability to stop the misconduct.

Wilderness A large, natural (or nearly natural) region, not controlled by humans.

Willingness to pay Economic concept meaning the most money that people will give for a good or service, depicted by the total area under a demand curve.

Wisdom Insight, erudition, and enlightenment resulting from the accumulation of knowledge and the ability to discern what is meaningful from what is not.

Yin* and *Yang Concept in Chinese philosophy describing two opposing and equal forces in the universe needed for harmony. Yang includes such principles as strength, liveliness, and brightness, whereas Yin includes weakness, passivity, and darkness.

Zygote Diploid cell that results from the fertilization of an egg cell by a sperm cell.

In addition to the author's definitions, some were drawn from:

1. The Online Ethics Center Glossary: http://onlineethics.org/glossary.html
2. University of Nebraska Medical Center Glossary: http://www.unmc.edu/ethics/words.html
3. L.M. Hinman's "Ethics Update – Glossary": http://ethics.sandiego.edu/Glossary.html
4. R. N. Johnson's "A Glossary of Standard Meanings of Common Terms in Ethical Theory": http://web.missouri.edu/~philrnj/eterms.html
5. U.S. Department of Health and Human Services, Office of Research Integrity: http://ori.dhhs.gov/education/products/rcradmin/glossary.shtml
6. U.S. Department of Health and Human Services, Agency for Toxic Substances and Disease Registry: http://www.atsdr.cdc.gov/glossary.html
7. U.S. Environmental Protection Agency, "Terms of Environment: Glossary, Abbreviations and Acronyms": http://www.epa.gov/OCEPAterms/
8. U.S. Environmental Protection Agency, "Glossary of IRIS Terms": http://www.epa.gov/iris/gloss8.htm
9. Fox Chase Cancer Center, "Glossary of Ethics Terms": http://www.fccc.edu/ethics/Glossary_of_Ethics_Terms.html

In addition, the sources cited in the reference sections following each chapter of this book contain useful definitions of terms.

APPENDIX

D

Example of a technical-ethical optimization for two emerging technology challenges: data centers and geoengineering with attendant Python scripts

Core Directive: Technology optimization is an engineering translation of moral philosophy. Technical performance metrics (speed, latency, output, and margin) shall be evaluated in concert with systemic societal and biosphere consequences.

1 Constraints and drivers

1.1 The post hoc regulation prohibition

The paradigm of "move fast and break things" is rejected. Applying ethical boundaries after a system is built (post hoc regulation) is mechanically inefficient and morally unacceptable for large-scale problems like generative artificial intelligence and geoengineering. Ethical parameters must be injected during the initial "discovery through design" phase as active mathematical constraints inside unified multiobjective design equations.

1.2 The macroethical imperative

Engineering teams shall evaluate technology through the lens of macroethics—the collective, generational impact on public health, demographic equity, and global ecosystems—rather than microethical compliance checklists [1–3].

1.3 Pareto front optimization

When technical goals (e.g., computational speed) collide with ethical obligations (e.g., carbon limits to address global climate change), the system must map a "Pareto front," that is, the set of optimal solutions in a multiobjective problem where no single metric can be improved without sacrificing another. It represents the ultimate limits of the best possible trade-offs between competing goals [4,5]. Optimization is defined here as reaching the absolute highest technical performance *without crossing* predefined ethical boundary lines.

2 Engineering audit

Phase 1: Upstream Sourcing and Lifecycle Assessment (LCA)

- **Hardware Sourcing Audits:** Enforce a cryptographic hardware bill of materials, verifying that zero conflict or child labor was utilized in infrastructure components.
- **Thermodynamic Footprint Accounting:** Factor upstream mining, refinement, and transport energy into the system's lifetime environmental risk equation.

Phase 2: Siting and Local Resource Footprint

- **Resource Displacement Simulation:** Model facility consumption (water, electrical load) against local public grids to ensure operations do not induce utility price spikes or brownouts for neighboring communities.
- **Maximin Infrastructure Integration:** Require that facilities built in infrastructure-vulnerable regions directly co-fund localized grid or utility updates to benefit the least-advantaged demographic.

Phase 3: Real-Time Load Balancing and Orchestration

- **Carbon-Aware Telemetry Hook:** Connect the live system orchestrator to regional grid carbon intensity tracking; automatically defer nonurgent bulk operations during high-emission grid peaks.
- **Proportionality Override Guardrails:** Restrict automatic overrides on dirty grids or tight resource periods *only* to workloads validated as life-critical or public-safety oriented.

Phase 4: Data Governance and Algorithmic Guardrails

- **Absolute Deontic Privacy Barriers:** Implement strict differential privacy pipelines that cannot be disabled to boost model accuracy.
- **Human Agency Veto:** Retain an engineering mechanism blocking any automated performance-optimizing routine from overriding human-in-the-loop oversight.

3 Code

The following unified Python orchestrator represents the production-level software implementation of this optimization, combining multi-objective boundaries, Rawlsian equity allocation, and Aquinas' defensive harm gates into a single automation routine in python:

```
import random
import time
class MacroethicalOptimizationEngine:
 def __init__(self, max_tolerated_pollution=200.0, rawlsian_equity_pool_pct=0.25):
  """
  Executes real-time infrastructure allocation bounded by ethical constraints.
  max_tolerated_pollution: Carbon intensity boundary (gCO2/kWh) - Vallero Biosphere
  Input.
  rawlsian_equity_pool_pct: Guaranteed resource reserve for vulnerable regions - Rawls
  Input.
  """
  self.carbon_boundary = max_tolerated_pollution
  self.equity_fraction = rawlsian_equity_pool_pct
  self.total_system_capacity = 100 # Maximum available operational load capacity
  self.active_queue = []
 def fetch_biosphere_telemetry(self):
  """Simulates real-time tracking of surrounding environmental health."""
  return random.randint(50, 450) # Returns live grid carbon footprint
 def register_system_load(self, client_id, load_weight, is_life_critical=False,
 is_vulnerable_zone=False):
  self.active_queue.append({
   "client": client_id, "load": load_weight,
   "is_life_critical": is_life_critical, "is_vulnerable_zone": is_vulnerable_zone
  })
  print(f"[MANIFEST-REGISTER] '{client_id}' Enqueued. Load: {load_weight} | Critical:
  {is_life_critical}")
 def enforce_manifest(self):
  print("\n================== BEGINNING MACROETHICAL AUDIT LOOP ==================")
  current_pollution = self.fetch_biosphere_telemetry()

  # Calculate Rawlsian protected capacity
  protected_equity_reserve = int(self.total_system_capacity * self.equity_fraction)
  standard_commercial_pool = self.total_system_capacity - protected_equity_reserve

  print(f"[AUDIT] Live Biosphere Pollution: {current_pollution} gCO2/kWh (Limit:
  {self.carbon_boundary})")
  print(f"[AUDIT] Rawlsian Protected Base: {protected_equity_reserve} Nodes |
  Commercial Base: {standard_commercial_pool} Nodes")
```

```
    postponed_workloads = []
    while self.active_queue:
      workload = self.active_queue.pop(0)

      # RULE 1: RAWLS DISTRIBUTIVE EQUITY ROUTER
      if workload['is_vulnerable_zone'] and protected_equity_reserve >= work-
      load['load']:
        protected_equity_reserve -= workload['load']
        print(f" [STATUS: EXECUTE] '{workload['client']}' deployed within Protected
        Rawlsian Reserve.")
        continue
      # RULE 2: AQUINAS HARM GATE (PROPORTIONALITY UNDER ENVIRONMENTAL STRESS)
      if current_pollution > self.carbon_boundary:
        if workload['is_life_critical']:
          print(f" [STATUS: EXECUTE] '{workload['client']}' PROPORTIONAL OVERRIDE.
          Executed under stress to preserve life.")
        else:
          print(f" [STATUS: DEFER] '{workload['client']}' halted. Reason: Violates
          Vallero Biosphere Boundary ({current_pollution} > {self.carbon_boundary}).")
          postponed_workloads.append(workload)

      # RULE 3: STANDARD COMMERCIAL RUN BOUNDED BY CAPACITY
      else:
        if standard_commercial_pool >= workload['load']:
          standard_commercial_pool -= workload['load']
          print(f" [STATUS: EXECUTE] '{workload['client']}' deployed normally under safe
          conditions.")
        else:
          print(f" [STATUS: DEFER] '{workload['client']}' halted due to typical network
          capacity barriers.")
          postponed_workloads.append(workload)
    self.active_queue = postponed_workloads
    print(f"================== AUDIT COMPLETE. DEFERRED QUEUE SIZE: {len(self.active_-
    queue)} ==================\n")
# – EXECUTION ENGINE VERIFICATION –
if __name__ == "__main__":
  # Initialize the MOM execution layer
  mom_engine = MacroethicalOptimizationEngine(max_tolerated_pollution=200.0, raw-
  lsian_equity_pool_pct=0.30)

  # Register diverse institutional workloads
  mom_engine.register_system_load("Commercial Ad Engine", load_weight=20, is_life_
  critical=False)
```

```
mom_engine.register_system_load("Indigenous Language Server", load_weight=15,
is_vulnerable_zone=True)
mom_engine.register_system_load("Metropolitan Trauma Unit Feed", load_weight=25,
is_life_critical=True)
mom_engine.register_system_load("Crypto Arbitrage Script", load_weight=40,
is_life_critical=False)
# Process workloads through the multi-objective governance pipeline
mom_engine.enforce_manifest()
```

Use code with caution.

Macroethical Optimization of Two Emerging Challenges

4 Executive foundations and recent literature

Modern technology optimization faces a structural challenge: optimizing solely for microefficiency (speed, throughput, profit margins) creates severe macro-scale externalities. This compendium codifies recent insights into an operational blueprint for engineers, architects, and corporate boards.

4.1 Key technical-ethical literature (2025–2026)

1. *This text.*
2. ***The AI Optimization Playbook:*** Details enterprise MLOps/LLMOps pipelines, providing practical corporate guardrails to measure return on investment against compliance and transparency requirements [6].
3. ***Magnifica Humanitas*** *(Pope Leo XIV/The Holy See)*: The seminal May 15, 2026, papal encyclical on advanced digital technology. It explicitly warns against reducing human beings to tracking profiles "to be optimized," objects to autonomous warfare, and condemns the hidden human and ecological costs of technical supply chains [7].
4. ***The AI Con***: A sociological and linguistic critique of Big Tech's hype cycle, detailing methods to dismantle algorithmic bias and maintain human agency over prediction models [8].

5 The philosophical-to-code matrices

To engineer ethical applications of emerging technologies, abstract philosophical doctrines must be compiled into binary code logic, dynamic loss functions, and architectural constraints.

5.1 The conceptual philosophy matrix

Philosophical framework	Foundational premise	Optimization mechanism in code	Core system target
Thomas Aquinas	**Doctrine of Double Effect (DDE):** Harmful side effects are tolerated *only* if they are unintended, minimized, and proportional to a grave good.	**Defensive Harm Gate:** Evaluates if technical exceptions or externalities are justified by immediate human necessity.	**External Risk Minimization**
John Rawls	**The Difference Principle:** Systemic inequalities are permissible *only* if they work to the maximum benefit of the least advantaged.	**Maximin Objective Function:** Prioritizes and locks resource allocations to guarantee a baseline of safety for the most vulnerable demographic.	**Distributive Equity**
Immanuel Kant	**The Categorical Imperative:** Act according to universal rules. Treat humans always as ends in themselves, never as a mere means.	**Hard Logical Constraint:** Strict boundaries that cannot be bypassed or compromised for any increase in profit or processing speed.	**User Autonomy and Dignity**
The Commons (Ostrom/Game Theory)	Shared, unregulated resources will inevitably face exhaustion by individual rational actors.	**Incentive Mechanism Design:** Changes environmental rules (quotas, tokens) so self-interest naturally aligns with resource preservation.	**Systemic Sustainability**
Catholic Social Teaching	Technology is an act of cocreation that must serve human dignity, subsidiarity, and ecological stewardship.	**Human-in-the-Loop Architecture:** Mandates decentralized edge processing, transparent data rules, and carbon tracking.	**Authentic Human Flourishing**

5.2 Integration layer

This text rejects post hoc regulation, substituting a methodology where engineers map out the previously mentioned Pareto front—a mathematical boundary representing the absolute best possible trade-offs where a system maximizes technical output *without breaching* its ethical safety thresholds.

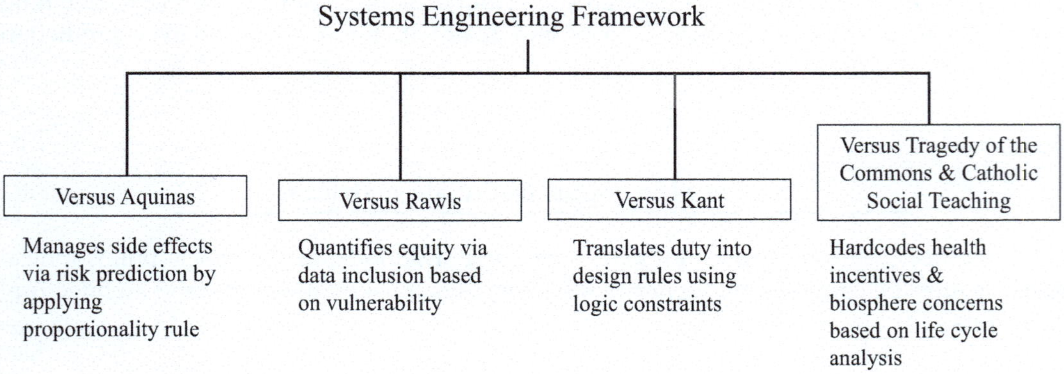

6 Domain-specific operational blueprints: The macroethical optimization of two emerging challenges

6.1 Domain A: enterprise data center optimization

The operational checklist

- **Upstream Lifecycle Assessment (LCA):** Integrate upstream hardware manufacturing energy, chemical processing, and shipping footprints directly into long-term risk equations.
- **Predictive Siting:** Monitor local demographic data to ensure data center water consumption or heat exhaust does not disproportionately degrade the living standards of neighboring, lower-income communities.
- **Carbon Workload Orchestration:** Connect live system orchestrators to regional grid carbon intensity tracking; automatically defer nonurgent bulk operations during high-emission grid peaks.
- **Absolute Deontic Privacy Barriers:** Implement strict differential privacy pipelines that cannot be disabled to boost model accuracy.

Hypothetical production Python implementation: data center orchestrator

Code:
python

```
import random
class DataCenterEthicalOrchestrator:
  def __init__(self, max_carbon_g_kwh=200.0, rawlsian_equity_fraction=0.25):
    self.carbon_threshold = max_carbon_g_kwh # Vallero's Environmental Boundary
    self.equity_fraction = rawlsian_equity_fraction # Rawlsian Difference Parameter
    self.total_compute_nodes = 100
    self.job_queue = []
  def get_live_grid_carbon(self):
    return random.randint(50, 450) # Simulated live gCO2/kWh telemetry
  def queue_workload(self, client_name, compute_load, is_life_critical=False,
  is_vulnerable_zone=False):
    self.job_queue.append({
      "client": client_name, "load": compute_load,
      "is_life_critical": is_life_critical, "is_vulnerable_zone": is_vulnerable_zone
    })
  def optimize_infrastructure(self):
    print("=== Optimizing Data Center Infrastructure (Vallero Systems Framework) ===")
    current_carbon = self.get_live_grid_carbon()
    equity_reserve = int(self.total_compute_nodes * self.equity_fraction)
    commercial_nodes = self.total_compute_nodes - equity_reserve

    print(f"[TELEMETRY] Grid Carbon: {current_carbon} gCO2/kWh | Rawlsian Reserve Set:
    {equity_reserve} Nodes")
```

```
    deferred_queue = []
    while self.job_queue:
      job = self.job_queue.pop(0)

      # 1. RAWLS JUSTICE RULE: Vulnerable zones pull safely from the protected equity
      reserve
      if job['is_vulnerable_zone'] and equity_reserve >= job['load']:
        equity_reserve -= job['load']
        print(f" [EXECUTE] '{job['client']}' run via Protected Rawlsian Reserve.
        Remaining Reserve: {equity_reserve}")
        continue
      # 2. AQUINAS PROPORTIONALITY RULE: If grid is dirty, check if job is critical to
      save life
      if current_carbon > self.carbon_threshold:
        if job['is_life_critical']:
          print(f" [EXECUTE] '{job['client']}' PROPORTIONAL OVERRIDE. Executed on dirty
          grid to preserve human safety.")
        else:
          print(f" [DEFER] '{job['client']}' postponed. Reason: Carbon Intensity
          Violation ({current_carbon} > {self.carbon_threshold}).")
          deferred_queue.append(job)
      else:
        # Grid is clean, process via standard commercial capacity
        if commercial_nodes >= job['load']:
          commercial_nodes -= job['load']
          print(f" [EXECUTE] '{job['client']}' processed normally on clean grid.")
        else:
          print(f" [DEFER] '{job['client']}' postponed due to standard node capacity
          limits.")
          deferred_queue.append(job)
    self.job_queue = deferred_queue
    print(f"=== Loop Complete. Deferred Queue Size: {len(self.job_queue)} ===\n")
if __name__ == "__main__":
  center = DataCenterEthicalOrchestrator()
  center.queue_workload("AdTech Tracker", 20, is_life_critical=False)
  center.queue_workload("Rural Clinic Telehealth Hub", 10, is_vulnerable_zone=True)
  center.queue_workload("Children's Hospital ICU Pipeline", 15, is_life_critical=True)
  center.optimize_infrastructure()
```

Warning: Code is hypothetical, so actual site-specific data and algorithms must be used instead of these illustrative commands.

A techno-ethical site matrix using a multiattribute decision analysis framework. For example, the script can require a 50/50 technical and ethical split by assigning specific weight distributions to performance metrics (e.g., uptime, latency, cooling) and ethical metrics (grid equity, water conservation, land use). It evaluates candidate profiles modeled closely after current multigigawatt US megaprojects, like the data center currently proposed in the Texas Panhandle:

python

```
import pandas as pd
# 1. Define Optimization Weights (50% Technical / 50% Ethical)
optimization_weights = {
  # Technical Performance Metrics (Sum = 0.50)
  "grid_reliability": 0.20,   # Access to redundant high-voltage baseload
  "fiber_latency": 0.15,     # Proximity to dark fiber backbones
  "climate_cooling_tech": 0.15, # Ambient temperature profile to lower HVAC load

  # Ethical & Sustainability Metrics (Sum = 0.50)
  "clean_energy_equity": 0.20, # Renewable mix without triggering consumer rate hikes
  "water_scarcity_impact": 0.15, # Avoidance of vulnerable watersheds/aquifers
  "land_use_justice": 0.15   # Utilization of brownfields over pristine/residen-
  tial land
}
# 2. Input Candidate Profiles (Scored 1 to 10 based on planned multi-GW sites)
candidate_sites = {
  "Texas_Panhandle_Megacampus": {
    "grid_reliability": 8,    # High potential for behind-the-meter gas/wind
    "fiber_latency": 7,     # Requires backhaul infrastructure buildout
    "climate_cooling_tech": 6, # High summer heat peaks
    "clean_energy_equity": 9,  # Abundant wind/solar availability
    "water_scarcity_impact": 4, # High Risk: Sited over the depleted Ogallala Aquifer
    "land_use_justice": 8    # Low density flat farmland, minor community noise friction
  },
  "Ohio_DOE_Nuclear_Brownfield": {
    "grid_reliability": 9,    # Massively robust legacy industrial grid ties
    "fiber_latency": 9,     # Proximity to major trans-American fiber lines
    "climate_cooling_tech": 8, # Temperate climate allows frequent free-air cooling
    "clean_energy_equity": 7,  # Nuclear baseload, but local grid needs massive greening
    "water_scarcity_impact": 9, # Low Risk: Water-abundant watershed
    "land_use_justice": 10   # Ideal: Redeveloping a heavily polluted federal site
  },
  "Southwest_Desert_AI_Hub": {
    "grid_reliability": 7,    # Highly solar-reliant; requires heavy battery storage
    "fiber_latency": 8,     # Excellent transit access lines
    "climate_cooling_tech": 3, # Severe desert heat forcing immense chilling loads
    "clean_energy_equity": 9,  # Co-located solar arrays
    "water_scarcity_impact": 2, # Ethical Failure: Extreme drought zone
    "land_use_justice": 6    # Open space, but risks tribal or ecological disruption
  }
}
```

```
# 3. Process Vector Matrices and Calculate Normalized Scores
site_evaluation_results = {}
for site, metrics in candidate_sites.items():
  # Calculate Sub-Scores (Normalized to a clean 10-point scale)
  technical_subscore = sum(
    metrics[m] * optimization_weights[m] for m in ["grid_reliability", "fiber_latency",
    "climate_cooling_tech"]
  ) / 0.50

  ethical_subscore = sum(
    metrics[m] * optimization_weights[m] for m in ["clean_energy_equity", "water_scar-
    city_impact", "land_use_justice"]
  ) / 0.50

  # Calculate Total Balanced Matrix Score
  total_composite_score = sum(metrics[metric] * optimization_weights[metric] for
  metric in optimization_weights)

  site_evaluation_results[site] = {
    "Technical Optimization Score (Out of 10)": round(technical_subscore, 2),
    "Ethical Optimization Score (Out of 10)": round(ethical_subscore, 2),
    "Total Composite Techno-Ethical Score": round(total_composite_score, 2)
  }
# 4. Generate Dataframe and Rank Locations
matrix_dataframe = pd.DataFrame(site_evaluation_results).T
matrix_dataframe = matrix_dataframe.sort_values(by="Total Composite Techno-Ethical
Score", ascending=False)
# Display Results
print("\n— TECHNO-ETHICAL DATA CENTER SITE EVALUATION MATRIX —")
print(matrix_dataframe.to_string())
```

Warning: Code is hypothetical, so actual site-specific data and algorithms must be used instead of these illustrative commands.

6.2 Domain B: geoengineering and climate intervention optimization

The operational checklist

- **Trans-Regional Impact Accounting:** Enforce multiobjective algorithms where localized cooling benefits must mathematically outweigh the localized severe weather side effects imposed on downstream nations (Aquinas' DDE Checked via Predictive Risk Modeling).
- **Systemic Resiliency Safeguards:** Implement system-level fail-safes ensuring aerosol deployment cannot be abruptly stopped. A sudden termination would violate Kantian duty by triggering a catastrophic, compressed rebound of global temperatures (Termination Shock).

- **Global South Democratic Quorum:** Lock deployment controls under a multiparty cryptographic signature. No climate intervention can fire unless a proportional quorum of representatives from historically climate-vulnerable nations authorizes it (Rawlsian Justice).
- **Mitigation Deterrence Penalties:** Program feedback loops that penalize or halt geoengineering operations if participating entities use the technology as an excuse to stop cutting raw carbon emissions (Tragedy of the Commons).

Hypothetical production python implementation: climate intervention controller

Code:
python

```
class GeoengineeringController:
  def __init__(self, mitigation_reduction_rate_pct=0.05):
    """
    mitigation_reduction_rate_pct: Measures if countries are slacking on cutting emis-
    sions
    because they rely on the geoengineering safety net (Moral Hazard / Commons).
    """
    self.mitigation_rate = mitigation_reduction_rate_pct
    self.authorized_keys = {"Global_North_Quorum": False, "Global_South_Quorum":
    False, "Indigenous_Oversight": False}
  def input_governance_signatures(self, north=False, south=False, indigenous=False):
    """Secures Free, Prior, and Informed Consent (AGU/Rawlsian Principles)."""
    self.authorized_keys["Global_North_Quorum"] = north
    self.authorized_keys["Global_South_Quorum"] = south
    self.authorized_keys["Indigenous_Oversight"] = indigenous
  def evaluate_regional_impacts(self):
    """
    Simulates Vallero's multi-objective predictive risk analysis.
    Calculates if cooling Zone A destroys agriculture in Zone B (Aquinas DDE check).
    """
    return {
      "Zone_A_Cooling_Utility": 85.0,    # High positive effect (intended)
      "Zone_B_Monsoon_Disruption_Risk": 40.0 # High negative effect (unintended side-
      effect)
    }
  def evaluate_deployment(self):
    print("=== Evaluating Planetary Climate Intervention System (Vallero
    Macroethics) ===")
```

```
        # 1. KANTIAN & RAWLSIAN GOVERNANCE GATE (Informed Global Participation)
        if not all(self.authorized_keys.values()):
          print(" [HALT] Deployment Blocked. Reason: Missing required ethical consensus or
          indigenous authorization keys.")
          return
        # 2. TRAGEDY OF THE COMMONS / MORAL HAZARD CHECK
        if self.mitigation_rate < 0.08: # Requires minimum 8% active annual carbon cuts
          print(f" [HALT] Deployment Blocked. Reason: Moral Hazard Detected. Emissions cuts
          too low ({self.mitigation_rate*100}%).")
          print("     Geoengineering cannot substitute active decarbonization.")
          return
        # 3. VALLERO / AQUINAS DOUBLE EFFECT PROPORTIONALITY EVALUATION
        impacts = self.evaluate_regional_impacts()
        if impacts["Zone_B_Monsoon_Disruption_Risk"] > 35.0:
          print(f" [HALT] Deployment Blocked. Reason: Disproportional Harm.")
          print(f"     Intended Cooling Utility ({impacts['Zone_A_Cooling_Utility']}) fails
          to justify ")
          print(f"     the critical down-stream monsoon risk forced onto Zone B ({impacts
          ['Zone_B_Monsoon_Disruption_Risk']}).")
        else:
          print(" [DEPLOY] Aerosol system authorized. Global balance achieved under multi-
          objective consensus.")
if __name__ == "__main__":
  system = GeoengineeringController(mitigation_reduction_rate_pct=0.04) # Insufficient
  carbon cuts

  print("— Test 1: Unilateral Execution Attempt —")
  system.input_governance_signatures(north=True, south=False, indigenous=False)
  system.evaluate_deployment()

  print("— Test 2: Multi-Party Consensus with Moral Hazard Check —")
  system.input_governance_signatures(north=True, south=True, indigenous=True)
  system.evaluate_deployment()

  print("— Test 3: Ethical System Passing All Guardrails —")
  compliant_system = GeoengineeringController(mitigation_reduction_rate_pct=0.10) #
  Strong 10% active carbon cuts
  compliant_system.input_governance_signatures(north=True, south=True, indigenous=True)
  compliant_system.evaluate_deployment()
```

Warning: Code is hypothetical, so actual site-specific data and algorithms must be used instead of these illustrative commands.

7 Global matrix summary

Philosophical framework	Integration layer	Data center variable	Geoengineering variable
Aquinas (Double Effect)	**Predictive Harm Modeling:** Uses high-throughput exposure simulations to mathematically quantify side effects before deployment.	Permitting peak-load backup generators *only* for hospitals/ ICU pipelines.	Stopping atmospheric injections if the resulting cooling causes downstream crop failure.
John Rawls (Justice)	**Technological Justice:** Implements a Maximin design rule using demographic and regional public health data to protect the marginalized.	Locking a minimum server capacity reserve for rural or underfunded clinics during shortages.	Granting global deployment veto keys to developing nations facing severe climate impacts.
Immanuel Kant (Deontology)	**Early-Stage Logic Constraints:** Turns absolute moral duties into unbypasable physical or code barriers during the initial design phase.	Absolute automated data deletion; strict, unmanipulated user consent parameters.	Enforcing unbreakable system loops that physically prevent sudden termination shock.
The Commons/ CST Ecosystems	**Lifecycle and Thermodynamic Accounting:** Treats the entire planet's biosphere as a shared, finite system with long-term metabolic costs.	Restricting water consumption to closed-loop evaporative cooling in high-stress aquifers.	Imposing absolute locks that deny geoengineering access if carbon emission cuts freeze.

8 Governance sign-off

By appending engineering signatures below, the technical team certifies that system optimization loops have been explicitly integrated with predictive harm modeling, technological justice metrics, and biosphere limits.

- **VP of Engineering Infrastructure:** ______________________
- **Principal Systems Architect:** ______________________
- **Lead Data Governance Officer:** ______________________

References

[1] D.A. Vallero, "Macroethics and engineering leadership, Leadership and Management in Engineering 8 (4) (2008) 287–296.

[2] J.R. Herkert, "Ways of thinking about and teaching ethical problem solving: Microethics and macroethics in engineering, Science and Engineering Ethics 11 (3) (2005) 373–385.

[3] B. Allenby, "Macroethical systems and sustainability science, Sustainability Science 1 (1) (2006) 7–13.

[4] D.A. Vallero, "Chapter 11: Economics and feasibility assessments, in: D.A. Vallero (Ed.), Environmental Systems Science, Elsevier, Amsterdam, (2021), pp. 509–542.

[5] A. Panichella, "An improved Pareto front modeling algorithm for large-scale many-objective optimization," in *Proceedings of the Genetic and Evolutionary Computation Conference*, 2022, pp. 565–573.

[6] C. Schiros, S. Kaur, R. Arora, U. Jagannathan, The AI Optimization Playbook: Drive business success with proven AI strategies, best practices, and responsible innovation, Packt Publishing Ltd, London, (2025).
[7] P.L. XIV, Magnifica humanitas: On safeguarding the human person in the time of artificial intelligence, New City Press, New York, (2026).
[8] M.S. Mahnke, Emily M. Bender and Alex Hanna. The AI Con. How to fight big tech's hype and create the future we want. New York: Harper-Collins Publishers. 2025, MedieKultur: Journal of Media and Communication Research 42 (80) (2026) .

Index

Note: Page numbers followed by "*f*," "*t*," and "*b*" refer to figures, tables, and boxes, respectively.

F

P

Printed and bound by CPI Group (UK) Ltd, Croydon, CR0 4YY
25/09/2026
02223045-0003